WATER SCARCITY AND SUSTAINABLE AGRICULTURE IN SEMIARID ENVIRONMENT

WATER SCARCITY AND SUSTAINABLE AGRICULTURE IN SEMIARID ENVIRONMENT

Tools, Strategies, and Challenges for Woody Crops

Edited by

IVÁN FRANCISCO GARCÍA TEJERO

VÍCTOR HUGO DURÁN ZUAZO

ELSEVIER

ACADEMIC PRESS
An imprint of Elsevier

Academic Press is an imprint of Elsevier
125 London Wall, London EC2Y 5AS, United Kingdom
525 B Street, Suite 1800, San Diego, CA 92101-4495, United States
50 Hampshire Street, 5th Floor, Cambridge, MA 02139, United States
The Boulevard, Langford Lane, Kidlington, Oxford OX5 1GB, United Kingdom

Library of Congress Cataloging-in-Publication Data
A catalog record for this book is available from the Library of Congress

British Library Cataloguing-in-Publication Data
A catalogue record for this book is available from the British Library

ISBN: 978-0-12-813164-0

For information on all Academic Press publications visit our website at https://www.elsevier.com/books-and-journals

Publisher: Andre Gerhard Wolff
Acquisition Editor: Nancy Maragioglio
Editorial Project Manager: Billie Jean Fernandez
Production Project Manager: Vijay Bharath .R
Cover Designer: Matthew Limbert

Typeset by TNQ Books and Journals

Whom tauch us, whom learnt with us

CONTENTS

LIST OF CONTRIBUTORS

Isabel Abrisqueta
Centro de Edafología y Biología Aplicada del Segura (CEBAS-CSIC), Murcia, Spain

Juan José Alarcón Cabañero
Centro de Edafología y Biología Aplicada del Segura (CEBAS-CSIC), Murcia, Spain

Pedro Balda
Universitat de les Illes Balears, Palma de Mallorca, Spain

Pilar M. Barrios
University of Cordoba, Cordoba, Spain

Jose María Bayona Gambín
Centro de Edafología y Biología Aplicada del Segura (CEBAS-CSIC), Murcia, Spain

Cristina Besada
Instituto Valenciano de Investigaciones Agrarias, Valencia, Spain

Luis Bonet
Instituto Valenciano de Investigaciones Agrarias, Valencia, Spain

Josefina Bota
Universitat de les Illes Balears, Palma de Mallorca, Spain

Eliemar Campostrini
Universidade Estadual do Norte Fluminense, Rio de Janeiro, Brazil

Angel Carbonell-Barrachina
Universidad Miguel Hernández de Elche, Alicante, Spain

Belén Cárceles Rodríguez
IFAPA Centro "Camino de Purchil", Granada, Spain

Ana Centeno
Technical University of Madrid, Madrid, Spain

Maria M. Chaves
Universidade NOVA de Lisboa, Oeiras, Portugal

Jacinta Collado-González
Universidad Miguel Hernández de Elche, Alicante, Spain

Wenceslao Conejero
Centro de Edafología y Biología Aplicada del Segura (CEBAS-CSIC), Murcia, Spain

Mireia Corell
Universidad de Sevilla, Seville, Spain

Joaquim M. Costa
Universidade de Lisboa, Lisboa, Portugal; Universidade NOVA de Lisboa, Oeiras, Portugal

José Francisco Couceiro
IRIAF, Ciudad Real, Spain

María Cruz-Blanco
IFAPA – Centro "Alameda del Obispo", Córdoba, Spain

Zulma N. Cruz
Instituto Nacional de Ciencias Agrícolas (INCA), San José de Las Lajas, Cuba

María V. Cuevas
Instituto de Recursos Naturales y Agrobiología de Sevilla (IRNAS, CSIC), Seville, Spain

Jefferson R. da Silva
Universidade Estadual do Norte Fluminense, Rio de Janeiro, Brazil

José M. de Paz
Instituto Valenciano de Investigaciones Agrarias, Valencia, Spain

Antonio Diaz-Espejo
Instituto de Recursos Naturales y Agrobiología de Sevilla (IRNAS, CSIC), Seville, Spain

Víctor H. Durán Zuazo
IFAPA Centro "Las Torres-Tomejil", Seville, Spain

Gregorio Egea
Universidad de Sevilla, Seville, Spain

Ricardo Egipto
Universidade de Lisboa, Lisboa, Portugal; INIAV, I.P., Pólo de Dois Portos, Dois Portos, Portugal

José M. Escalona
Universitat de les Illes Balears, Palma de Mallorca, Spain

José E. Fernández
Instituto de Recursos Naturales y Agrobiología de Sevilla (IRNAS, CSIC), Seville, Spain

Victoria Fernández
Technical University of Madrid, Madrid, Spain

Irene Fernández García
University of Cordoba, Cordoba, Spain

Jaume Flexas
Universitat de les Illes Balears, Palma de Mallorca, Spain

José R. Francia Martínez
IFAPA Centro "Camino de Purchil", Granada, Spain

Dionisio Franco Tarifa
Finca Experimental "El Zahorí", Granada, Spain

Clara Gabaldón-Leal
IFAPA – Centro "Alameda del Obispo", Córdoba, Spain

Alejandro Galindo
University of Twente, Enschede, The Netherlands

Baltasar Gálvez Ruíz
IFAPA Centro "Camino de Purchil", Granada, Spain

José M. García
Instituto de la Grasa (IG, CSIC) Campus Universidad Pablo de Olavide, Seville, Spain

J. García Morillo
University of Cordoba, Cordoba, Spain

Iván F. García Tejero
IFAPA Centro "Las Torres-Tomejil", Seville, Spain

Luis Garrote
Universidad Politecnica de Madrid, Madrid, Spain

María del Carmen Gijón-López
IRIAF, Ciudad Real, Spain

Ignacio F. Girón
Instituto de Recursos Naturales y Agrobiología (CSIC), Seville, Spain

Julián C. González
Universidad de Almeria, Almeria, Spain

R. González Perea
University of Cordoba, Cordoba, Spain

Virginia Hernandez-Santana
Instituto de Recursos Naturales y Agrobiología de Sevilla (IRNAS, CSIC), Seville, Spain

Francisca Hernández
Universidad Miguel Hernández de Elche, Alicante, Spain

Ana Iglesias
Universidad Politecnica de Madrid, Madrid, Spain

Diego S. Intrigliolo
Centro de Edafología y Biología Aplicada del Segura (CEBAS-CSIC), Murcia, Spain

Roberta S.N. Lima
Universidade Federal de Alagoas, Arapiraca, Brazil

Carlos M. Lopes
Universidade de Lisboa, Lisboa, Portugal

Ignacio J. Lorite
IFAPA – Centro "Alameda del Obispo", Córdoba, Spain

Jose Francisco Maestre Valero
Universidad Politécnica de Cartagena, Cartagena, Spain

María J. Martín-Palomo
Universidad de Sevilla, Seville, Spain

Sonia Medina
Corporación Universitaria Lasallista, Caldas, Colombia

Hipólito Medrano
Universitat de les Illes Balears, Palma de Mallorca, Spain

Houssem Memmi
Technical University of Madrid, Madrid, Spain; IRIAF, Ciudad Real, Spain

Aida Mérida García
University of Cordoba, Cordoba, Spain

Chandra S. Misra
Universidade NOVA de Lisboa, Oeiras, Portugal

Esther Montes
Universitat de les Illes Balears, Palma de Mallorca, Spain

Marta Maria Moreno
Universidad de Castilla la Mancha, Ciudad Real, Spain

Alfonso Moriana
Universidad de Sevilla, Seville, Spain

Miquel Nadal
Universitat de les Illes Balears, Palma de Mallorca, Spain

Emilio Nicolás
Centro de Edafología y Biología Aplicada del Segura (CEBAS-CSIC), Murcia, Spain

Luis Noguera-Artiaga
Universidad Miguel Hernández de Elche (UMH), Alicante, Spain

Pedro Antonio Nortes Tortosa
Centro de Edafología y Biología Aplicada del Segura (CEBAS-CSIC), Murcia, Spain

Carmen M. Padilla-Díaz
Instituto de Recursos Naturales y Agrobiología de Sevilla (IRNAS, CSIC), Seville, Spain

Margarita Parra
Centro de Edafología y Biología Aplicada del Segura (CEBAS-CSIC), Murcia, Spain

Francisco Pedrero Salcedo
Centro de Edafología y Biología Aplicada del Segura (CEBAS-CSIC), Murcia, Spain

David Pérez-López
Technical University of Madrid, Madrid, Spain

Alfonso Perez-Martin
Instituto de Recursos Naturales y Agrobiología de Sevilla (IRNAS, CSIC), Seville, Spain

Carla Pinheiro
Universidade NOVA de Lisboa, Oeiras, Portugal; Universidade NOVA de Lisboa, Caparica, Portugal

Rafael Porras
IFAPA – Centro "Alameda del Obispo", Córdoba, Spain

Alicia Pou
Universitat de les Illes Balears, Palma de Mallorca, Spain

Emilio C. Poyato
University of Cordoba, Cordoba, Spain

José D.C. Ramalho
Universidade de Lisboa, Oeiras, Portugal; Universidade NOVA de Lisboa, Caparica, Portugal

Weverton P. Rodrigues
Universidade Estadual do Norte Fluminense, Rio de Janeiro, Brazil

Celia M. Rodriguez-Dominguez
Instituto de Recursos Naturales y Agrobiología de Sevilla (IRNAS, CSIC), Seville, Spain; University of Tasmania, Hobart, TAS, Australia

Pedro Rodríguez
Instituto Nacional de Ciencias Agrícolas (INCA), San José de Las Lajas, Cuba

Juan A. Rodríguez Díaz
University of Cordoba, Cordoba, Spain

Carmen R. Rodríguez Pleguezuelo
IFAPA Centro "Camino de Purchil", Granada, Spain

Rafael Romero
Instituto de Recursos Naturales y Agrobiología de Sevilla (IRNAS, CSIC), Seville, Spain

Cristina Romero Trigueros
Centro de Edafología y Biología Aplicada del Segura (CEBAS-CSIC), Murcia, Spain

José S. Rubio
Centro de Edafología y Biología Aplicada del Segura (CEBAS-CSIC), Murcia, Spain

Margarita Ruiz-Ramos
UPM – CEIGRAM, Madrid, Spain

María C. Ruiz-Sánchez
Centro de Edafología y Biología Aplicada del Segura (CEBAS-CSIC), Murcia, Spain

Cristina Santos
IFAPA – Centro "Alameda del Obispo", Córdoba, Spain

Bruce Schaffer
University of Florida, Homestead, FL, United States

José Silvestre
Universidade de Lisboa, Lisboa, Portugal; INIAV, I.P., Pólo de Dois Portos, Dois Portos, Portugal

Arturo Torrecillas
Centro de Edafología y Biología Aplicada del Segura (CEBAS-CSIC), Murcia, Spain

Jose M. Torres-Ruiz
BIOGECO, INRA, University Bordeaux, Pessac, France

Ignacio Tortosa
Universitat de les Illes Balears, Palma de Mallorca, Spain

Juan Vera
Centro de Edafología y Biología Aplicada del Segura (CEBAS-CSIC), Murcia, Spain

Fernando Visconti
Instituto Valenciano de Investigaciones Agrarias, Valencia, Spain

Olfa Zarrouk
Universidade NOVA de Lisboa, Oeiras, Portugal

PREFACE

This publication includes a set of studies that provide a high level of innovation, specifically concerning the sustainability of production systems. Innovation and sustainability are currently the determining factors for the scientific quality of research processes. Thus innovation constitutes the essence of science and culminates in sustainability, which enables it to endure through time.

To write a preface to a scientific publication implies an exercise in reflection that anticipates and at the same time frames what is to be transmitted. While it need not presuppose a detailed explanation of the content that follows, it should lay out the keys to the **why**, to propose, by analysis, the **how**, **when**, and **where**.

In this sense, I would like to emphasize the process of innovation, so significant as well as relevant, which has emerged in recent years and which I have had the opportunity to share with so many professional researchers and planners of public policies who are interested in agricultural production, many of them participants in this publication. Without failing to underscore the importance of the high level of knowledge amassed and so brilliantly displayed here by the authors, I have chosen to place them in the framework of advances made in the sense of a story, in such a way that **data and story** constitute a complementary binomial of this publication.

How then can we tell the story of advances, more than achieved, lived, in these last decades? Only by beginning with the real developments that have affected primarily the **method** if we take into account the procedures, the **devices** if we regard technology, and the **teamwork** if we consider the relation between the different disciplines that enable the diverse perspectives in the interpretation and discussion of the same process. In this particular case, in reference to the effects of the "scarcity of water and sustainable Mediterranean agriculture," we witness the way that methods, technology, and teams have been weaving a network of knowledge that has propelled the advances made, sometimes spectacularly.

Next, I would like to highlight as the essential substrate of progress the necessary interrelations that form over a long water cycle in the continuous soil–plant–atmosphere system as nodes in a dynamic and ever-changing network composed of factors and variables that influence plant production. In this way, we have been able to establish how the soil conditions, in relation to water, and evaporative demands established by the atmospheric characteristics determine the relations between the three elements that prove crucial to determining the production conditions. This way of expressing the vital relations in the plant kingdom lead to a new dimension when they are in turn related to the production processes. This gives rise to the concept of "water footprint," which quantifies the water supply necessary to reach and sustain certain production levels, the optimal condition of which is measured in terms of efficiency—that is, profitable yield with minimum resources in sustainable situations.

We can then understand and above all characterize the prominence that each element has held in the advances made. Furthermore, we can see how the measurement of the relations of the soil–plant–atmosphere continuum has progressed, from static in situ gauges of the classical weather stations, to remote sensors such as satellites or dynamic, prospective devices such as drones, together with more accurate weather forecasts over the short and medium terms. If advances have been achieved in the way of evaluating the conditions, progress has been made simultaneously also in treatment methods to ensure highly efficient production, evident, for example, in the way atmospheric conditions are quantified and even modulated, for example, by innovations in intensive greenhouse cultivation.

If the knowledge of the interrelations established in the aforementioned continuum is increasing thanks to more novel contributions such as greater access to large data sources (big data) and the inclusion of such frontiers as robotization of processes and the "Internet of things," the result is an aggregation of factors that continue to form steadily more complex relations and finally facilitate the most efficient decisions in what are called "production models." This process can be recognized in each of the works documented in this publication.

Over my professional career I have learned, and I wish to transmit this message with special emphasis, that the interpretation of the data should be accompanied by reflection capable of encouraging and giving rise to a comprehensive method respectful of the vital relations between the elements involved and at the same time ensuring their sustainability.

Each of the works presented here embodies in one way or another this approach to research. Thus this publication encompasses the most significant advances that have been made in the realm of water management in Mediterranean agriculture from three different but interdependent perspectives: that of management and the knowledge of future scenarios; that referring to crop yield as the element generating wealth and value and hence fundamental for the sustainability of the rural environment; and that relating to the knowledge of the processes and mechanisms regulating plant responses to water scarcity. Therefore this work seeks to provide a compendium that will serve as a consultation tool for researchers, students, technicians, and academic staff, offering rapid responses to key questions in water management. It can furthermore serve as a bibliographic reference for future research.

In this way the results presented here represent new nodes to develop the network that foments wise decision making in the direction of progress. Therefore I congratulate the authors and also encourage the analysis and discussion of all potential readers, especially public managers and planners in the sphere of agriculture. These are the ones, primarily at the technical and policy levels, who are responsible for ensuring at the public level the underpinnings of sustainability and progress, which are completely compatible with the processes of efficient production, both public and private.

Dr. José Luis Muriel Fernández

PART 1

Water Scarcity - Climate Change and Environmental Challenges

CHAPTER 1

Water Management and Climate Change in Semiarid Environments

Ignacio J. Lorite[1], Margarita Ruiz-Ramos[2], Clara Gabaldón-Leal[1], María Cruz-Blanco[1], Rafael Porras[1], Cristina Santos[1]
[1]IFAPA — Centro "Alameda del Obispo", Córdoba, Spain; [2]UPM — CEIGRAM, Madrid, Spain

Contents

Water Scarcity and Sustainable Agriculture in Semiarid Environment
ISBN 978-0-12-813164-0
https://doi.org/10.1016/B978-0-12-813164-0.00001-6

1. WEATHER AND CLIMATE CHARACTERIZATION OF MEDITERRANEAN ENVIRONMENTS

1.1 Current Weather Characterization

Weather data are essential inputs for the determination of crop water requirements (Ashraf et al., 1997) and therefore for correct water management in Mediterranean environments. Automated weather stations are designed to accurately measure and record standard meteorological variables over extended periods (Hubbard and Sivakumar, 2000), providing continuous observations without human intervention. Observations from automated weather stations are performed for different purposes in an agricultural context, from activities related to meteorological research (Miller and Barth, 2003) and predictions (Lorite et al., 2015), irrigation scheduling (Cruz-Blanco et al., 2014a), or validation of interpolation methods for reference evapotranspiration (ET_O) assessment (Ramírez-Cuesta et al., 2017).

Although weather station networks have been promoted (Miller and Barth, 2003; Estévez et al., 2011), the current number of weather stations is clearly limited (Collins, 2011; Essou et al., 2016). This is a clear barrier to the proper management of water resources, making difficult the risk management of water scarcity and water conflicts (Maeda et al., 2011). In addition, vast areas are located far from well-managed weather stations, avoiding correct weather characterization, and in many areas weather stations are sparsely distributed (Essou et al., 2016). However, for large areas it is necessary to have an elevated number of observation sites due to the heterogeneity of landscapes (Srivastava et al., 2013). To solve these limitations, consideration of simplified approaches for ET_0 assessment not requiring complete weather stations (such as the Hargreaves equation; Gavilán et al., 2006), or interpolation approaches based on data collected from meteorological stations, have provided accurate weather components in areas where high-quality observations from weather stations are not available (Blum et al., 2013; Bojanowski et al., 2014; Ramírez-Cuesta et al., 2017).

Alternative approaches to obtain critical weather variables in those areas where accurate weather-observed data are not available have been developed for temporal or spatial gap filling, obtaining global or regional datasets (Hijmans et al., 2005). Thus methods considering weather forecast and reanalysis data integrated with remote sensing tools (Bojanowski et al., 2013; Cruz-Blanco et al., 2015) have been considered as alternatives to the use of weather stations.

Cammalleri and Ciraolo (2013) and Cruz-Blanco et al. (2015) considered remote-sensing climate data to obtain variables that are used in ET_0 estimation, such as using meteorological data from the MODIS/Terra sensor (Maeda et al., 2011) or using directly the solar radiation product from the LSA SAF platform on board the Meteosat Second Generation satellite (Cristóbal and Anderson, 2013; Cruz-Blanco et al., 2014b). In parallel, weather forecast data are used in different disciplines (Guijarro et al., 2014; Essou et al., 2016), and therefore numerical weather prediction systems (Perera et al., 2014) and European Centre for Medium-Range Weather Forecasts datasets (Wagner et al., 2014), among others, have been exploited to compute ET_0. Thus, for example, the air temperature at 2 m can be obtained (Persson and Grazzini, 2007) considering the ERA Interim database (Dee et al., 2011). Finally, several studies have shown the capability of integrated remote sensing and forecast data to compute ET_0 in heterogeneous areas (Sun et al., 2011; Cristóbal and Anderson, 2012; Cruz-Blanco et al., 2014b). From these results, irrigation scheduling and yield estimations improved significantly compared with the use of limited weather data from weather station networks (Lorite et al., 2015).

1.2 Future Weather Characterization

Climate model outputs are used as inputs for crop models to simulate different processes such as crop development or water balance under future climate conditions. The climate models combine mathematical equations that integrate different ocean–atmospheric dynamics to describe the climate behavior, generating climate projections for future periods. These projections are given globally in a grid scale with the global climate models (GCMs), with a coarse spatial resolution (from 150 km × 150 km to 300 km × 300 km). However, for regional or local studies, outputs from regional climate models (RCMs) can be considered, with a spatial resolution from 50 km × 50 km up to 1 km × 1 km, obtained from GCM through dynamical downscaling techniques (Fig. 1.1). An alternative approach is the use of statistical downscaling, which captures the observed relationships between large-scale atmospheric variables with those observed locally using transfer functions (Wilby et al., 1998), assuming that these relationships will be held in future. Dynamic downscaling reduces the scale of the GCM outputs through the RCM without assuming this hypothesis, taking the boundary conditions from the driving GCM (Castro et al., 2005). However, even with the fine scale that the RCM has, these models present biases regarding the observed values (Dosio and Paruolo, 2011; Ceglar and Kajfež-Bogataj, 2012) that may reduce the projection quality. These biases could be inherited by the driving GCM, or be part of the intrinsic differences in the RCM physical parameterizations (Kjellström et al., 2010; Nikulin et al., 2011). Thus, for example, some authors have shown that the simulated summer temperature in Southern Europe is overestimated (Christensen et al., 2008; Dosio et al., 2012). These kinds of biases can be reduced by several postprocessing techniques. Thus the delta change method that applies the climate change signal from climate models on observations (Hawkins et al., 2013), or computing transfer functions regarding an observational dataset that can be applied to the simulated climate data to generate the corrected projection (Piani et al., 2010a,b; Dosio and Paruolo, 2011; Dosio et al., 2012), are some approaches considered.

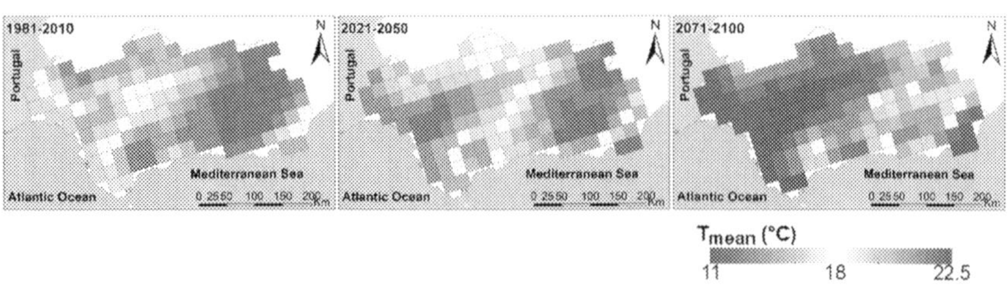

Figure 1.1 Mean temperature (T_{mean}) projected for the baseline (1981–2010), near future (2021–2050), and far future (2071–2100) periods for Andalusia region based on regional climate models from ENS-Spain02 ensemble mean.

Therefore the RCM contains multiple sources of uncertainty, which include the driving forces of future emissions or the differences in parameterization and resolution between different RCMs for the same area (Olesen et al., 2007; Ruiz-Ramos and Mínguez, 2010; Osborne et al., 2013). To manage and quantify the uncertainty of climate projections, ensembles of climate models are used. An ensemble provides a range of climatic changes from its members, instead of a unique value. For example, for a representative location of southern Spain such as Baeza, the consideration of 11 climate models generated a high range of variability in weather components: rainfall reduction from baseline to far future conditions varied from 9.9% to 29.8%, temperature increase varied from 2.4 to 4.2°C, and ET_o varied from 3.7% to 17.8% (Lorite et al., 2017), generating a high variability in the outputs from crop model simulations or impact projections obtained from these values.

Finally, atmospheric concentration of CO_2 is an important component for crop physiology that also impacts on water management. Usually, the mean annual atmospheric CO_2 concentration is based on measures at Mauna Loa Observatory (collected since the 1950s), on data from close to the coast of Antarctica (Etheridge et al., 1996), and future predictions from the National Oceanic and Atmospheric Administration (NOAA) (Raes et al., 2013). Atmospheric CO_2 measurements indicate that the CO_2 ratio has increased by about 25% during the past few hundred years, probably due to industrialization and large-scale land use modification (IPCC, 2013). Studies indicated that CO_2 concentration during 800 years before industrialization was in the range of 275–284 ppm and a major growth of CO_2 concentration (315–329 ppm) was observed in the ice core record with the beginning of the industrial period. Projections for future CO_2 concentration show high uncertainty and depend on the scenario of emissions considered. Conservative projections estimate atmospheric concentrations of CO_2 higher than 550 ppm at the end of the 21st century (Raes et al., 2013).

2. WATER MANAGEMENT IN RAIN-FED AND IRRIGATED AGRICULTURAL SYSTEMS

2.1 Components of Correct Water Management

Water has a critical importance in the semiarid Mediterranean environments due to weather conditions (limited rainfall and high temperatures generating a high ET_o) that usually impose a permanent deficit balance, frequently mitigated using irrigation. Thus water balance is very relevant for correct water management in both rain-fed and irrigated agricultural systems.

Water management evaluation is based on the analysis of the inputs and outputs of the system. Thus inputs are rainfall and irrigation, and outputs are evaporation, transpiration, runoff, and deep percolation. Therefore a correct water management will reduce water losses caused by evaporation, runoff, deep percolation, and inefficient irrigation management (excessive volume or rate of irrigation or low irrigation uniformity). For this evaluation, consideration of components such as soil, irrigation scheduling, irrigation method, or phenology is essential and must be included in any analysis related to water management.

2.2 Soil Management

Soil management has a critical importance in correct water management under Mediterranean conditions. This importance is even more relevant for orchards where bare soil represents an important percentage of the area. Thus consideration of cover crops and conservation practices, such as minimum tillage, provides a significant reduction in water losses by runoff and evaporation, and their use has been increased significantly in the last few years (Jägermeyr et al., 2016). However, these practices require correct management as a wrong implementation could generate negative impacts on water availability and efficiency. Thus cover crops with deficient management could generate severe reductions in yield (Abazi et al., 2013), because an important part of available water is used by cover crops instead of orchards. Also, according to results from field experiments and simulation models, cover crops could generate a negative impact on water resources for trees if they are incorrectly managed, generating reductions of around 25% in olive transpiration (Abazi et al., 2013). However, this impact is diminished if correct cover crop management is carried out, for example, considering early date killing of the cover crop (around March 15th for simulated conditions in Lanjarón, southern Spain; Abazi et al., 2013). In line with this, field experiments have confirmed the importance of soil management for correct water management (Gómez et al., 2009a). For instance, no-tillage systems and removal of weeds with herbicides generated high average annual runoff (average annual runoff coefficient equal to 11.9%); in the opposite case a cover crop of barley reduced runoff significantly (average annual runoff coefficient equal to 1.2%). Conventional tillage provided intermediate values (runoff coefficient of 3.1%; Gómez et al., 2009a). These results concluded that the use of cover crops and minimum tillage can be simple and feasible soil and water conservation practices in olive groves in southern Spain (Gómez et al., 2009b). However, it should always be considered that the cover crop must be established early enough to protect the soil in critical initial years of the grove.

As another example of a sustainable system, no-tillage based, spontaneous vegetation cover and annual recycling of pruning material were able to increase soil water storage from rainfall, compared to a traditional tillage system that produced soil crusting and compacted layers hindering infiltration and percolation of rainfall (Palese et al., 2014). For other woody crops such as apricot the conclusions were similar: limited tillage treatments compared with no-tillage decreased the runoff providing mean irrigation water savings of around 10% (Abrisqueta et al., 2007).

Climate change will enhance the occurrence of severe droughts (Trenberth, 2005) and heavy rain events (Alpert et al., 2002) and then correct management of cover crops can emerge as a key adaptation strategy to improve soil water management. Traditional water balance models have been an excellent tool for evaluating adaptation measures related to soil practices. These models have considered a cascade approach (Lorite et al., 2004) requiring spatial homogeneity in the field. This assumption could be considered correct for extensive crops such as cereals, but not for orchards as water balance differs clearly depending on the location within the plot. Specifically, to solve this knowledge gap, models such as WABOL (Abazi et al., 2013) consider three different zones in the orchard (canopy, cover crop area, and bare soil), each one with specific characteristics in the water balance components.

2.3 Crop Water Requirements

Accurate crop water requirement assessment is a key component for correct water management. Such assessment requires specific instrumentation, resulting in a difficult task, especially when woody crops are analyzed. Thus methodologies for assessing crop water requirements vary from simple water balance approaches and crop coefficients (Allen et al., 1998) to complex energy balance approaches using remote sensing techniques (Allen et al., 2007a,b). However, these methodologies reduce accuracy significantly when applied to orchards (Santos et al., 2012), requiring specific tools for coping with reductions in accuracy. The main difficulty is related to the heterogeneity of fields; canopy and bare soil generate significant differences in evaporation and transpiration, and then require the separation of both terms. Tools such as double-source models could solve this limitation (Cammalleri et al., 2012), although data requirements are excessively elevated for large-scale and practical analysis.

When future climate conditions are considered for the estimation of crop water requirements, simulation models have a special relevance, as they are the unique approaches able to simulate future crop behavior, but always after a calibration and validation procedure based on experimental data.

The core of any simulation model for yield and irrigation requirements assessment must be based on crop transpiration estimation. There are several approaches for assessing crop requirements varying significantly in complexity. Thus Morales et al. (2016) developed a model of potential olive oil production based on a three-dimensional model of canopy photosynthesis and respiration and dynamic distribution of assimilates among organs. In the opposite case of complexity, the FAO-56 procedure provides a simple and useful guide for programming irrigation scheduling for any type of crop (Allen et al., 1998). Specifically for orchards, and based on FAO-56, Allen and Pereira (2009) defined crop coefficients for each woody crop. These crop coefficients must be adjusted to the fraction of ground cover and height using a density coefficient, obtaining very satisfactory results. FAO-66 (Steduto et al., 2012) proposes an alternative approach uniquely considering the percentage of ground cover (Fereres et al., 2012). Similarly, Orgaz et al. (2006) developed functional relationships for calculating the crop coefficient for a given month of the year for any type of olive orchard and then assessed water requirements based on tree transpiration (depending on tree density and canopy volume), evaporation from the soil, and evaporation from the areas wetted by the emitters. Equally, for almond, and as result of a 4-year study on an experimental almond orchard using a large weighing lysimeter, a clear relationship between crop transpiration and intercepted radiation by the canopy has been developed (Espadafor et al., 2015; Espadafor, 2015). This relationship determines site-specific irrigation scheduling adapted to local conditions of the orchard (cultivar, age, and tree spacing, irrigation system, or pruning). Thus daily fraction of photosynthetically active radiation intercepted by the canopy (fIRd) is considered as a predictor of the maximum transpiration for an almond orchard. Thus the ratio between fIRd and K_T (T/ET_o) stayed constant around a value of 1.2. Moreover, considering this value for an almond orchard with 85% intercepted radiation gives $K_T = 1.0$, compatible with the maximum K_c values of 1.1−1.15 published by Girona et al. (2006), Stevens et al. (2012), Goldhamer and Girona (2012), and García-Tejero et al. (2015).

The described approaches are the base for integral models, which have proven to be useful for improving water management for woody crops. Thus the AdaptaOlive model (Lorite et al., 2017) for olive transpiration (T) assessment utilizes the experimental procedure developed by Orgaz et al. (2006) for determining T under optimal irrigation and current weather conditions, based on the fraction of intercepted diffuse radiation. Equally, the WABOL model (Abazi et al., 2013) uses a process-based methodology to describe key mechanisms controlling water balance in rain-fed olive orchards such as runoff, deep percolation, cover crop growth, soil evaporation, and olive and crop transpiration. Similarly, a new version for vineyards, called WABYN, has been developed (Gómez et al., 2016). In both versions, crop requirements are also calculated by using the procedure developed by Orgaz et al. (2006) previously described.

Remote sensing techniques integrated with simulation models arise as a promising tool to overcome problems related to uncertainties in crop coefficient calculation in crops having sparse woody canopies. Compared with traditional K_c-based evapotranspiration estimations, remote sensing methods have a strong advantage in spatial accuracy because of the estimation of evapotranspiration for each pixel of a satellite image, using observed reflected radiation and temperature. Therefore characterization of individual fields, including impacts of canopy density and tree spacing, as well as relative water supply, are better addressed.

Surface energy balance (SEB) approaches based on remote sensing techniques for direct estimation of actual evapotranspiration have been developed over the past 20 years (Norman et al., 1995; Bastiaanssen et al., 1998; Kustas and Norman, 1999; Allen et al., 2007a; Santos et al., 2008; Elhag et al., 2011). These approaches have been applied to estimate evapotranspiration from a variety of sparse woody canopies (Autovino et al., 2016). Authors such as Minacapilli et al. (2009) compared two SEB approaches to estimate the actual evapotranspiration in typical spatially sparse Mediterranean vegetation (i.e., olive, citrus, and vineyards): the one-source SEBAL model (Bastiaanssen et al., 1998), where soil and vegetation are considered as a combined sole source for surface energy exchange, and the TSEB two-source model (Norman et al., 1995), where soil and vegetation components of the surface energy balance are treated separately.

Due to possible biases in model estimates caused by the soil and shadow effects on pixel temperature estimation and uncertainties regarding roughness and aerodynamic profiles, modifications to the initial methodologies have been required when using remote sensing techniques in sparse woody canopies, such as olives. Thus Santos et al. (2012) considered the impact of the procedure estimation for crop height and momentum roughness length on the aerodynamic simulation of the METRIC model (Allen et al., 2007a) for sparse woody canopies. This supports the value of continuing the study on the use of surface energy models to monitor evapotranspiration behavior of olive orchards for the design of customized irrigation schedules considering satellite-based products.

3. IMPACTS OF CLIMATE CHANGE ON WATER MANAGEMENT

3.1 Impact of Climate Change on Transpiration

Climate change for Mediterranean areas is supposed to decrease precipitations and increase temperature and ET_o, as a consequence of rises in atmospheric CO_2 concentration (Fig. 1.2). This rise in [CO_2] could alter the growth of plants, increasing their CO_2 assimilation rates, biomass accumulation, seed yield, light and nutrient efficiency, plant water use efficiency and reducing stomatal reluctance (Saxe et al., 1998), although

Figure 1.2 Olive plants under future weather conditions referred to atmospheric [CO_2] and temperature.

some authors report some degree of plant and tree acclimation (Tuba and Lichtenthaler, 2007; Reich et al., 2016). In any case, woody plants are crucial in the global carbon cycle, being sinks for long-term storage of carbon (Atkinson, 2000). To evaluate responses from various types of trees in a future climate environment, diverse studies under free-air CO_2 enrichment (FACE) systems were made. One of these studies from Tognetti et al. (2002) disposed 5-year-olive trees grown in a FACE system (560 $\mu mol\ mol^{-1}$) performing gas exchange measurements from leaves. Net photosynthesis was bigger in plants with high [CO_2] by 31%−44% and stomatal conductance was lower compared with other plants in elevated [CO_2] (Morison, 1998). As a result, plants in higher [CO_2] had increased water use efficiency by 73%−94% and a reduced stomatal density by 9%−11%. These morphological and physiological responses to high [CO_2] might vary not only between species, but between varieties (Tognetti et al., 1999). These results are similar to other studies referred to grapevine trees, where the effects of applying 550 and 700 $\mu mol\ mol^{-1}$ of [CO_2] (Bindi et al., 2001) were analyzed. In that study, measurements in shoots, leaves, branches, and grapes were made and results showed that high [CO_2] affected crop growth (increasing total biomass by 45%−50%) and fruit quality (acid and sugar content in the middle of the ripening season). However, it is important to note that these effects of atmospheric CO_2 enrichment could be almost cancelled by effects of warmer temperatures expected under future climate conditions (less active photosynthesis or higher respiration). Finally, Tognetti et al. (2005) reported a reduction in stomatal resistance under elevated [CO_2] and a marked reduction in shoot hydraulic conductance in adult grapevine trees, obtaining a variation in the porosity of the xylem and a decreasing xylem efficiency in FACE plants with more time for exposure, assuming an effect on grapevine growth efficiency due to less capacity to supply leaves with water.

As it was previously indicated, transpiration is a key component for crop requirement assessment. Transpiration with optimal irrigation under future conditions is reduced due to the reduction in canopy conductance when [CO_2] increases, according to Field et al. (1995) and later validated by Drake and Gonzàlez-Meler (1997) based on FACE experiments, and by Moriana et al. (2002), Villalobos and Fereres (2004), and Viola et al. (2014) specifically for olive crop. Transpiration efficiency (TE) is defined as the ratio between net assimilation and transpiration and depends on the ambient and inner leaf partial pressure of CO_2 ratio (C_i/C_a) and on vapor pressure deficit, as defined by Beer et al. (2009). C_a is determined by current measurements and future projections of [CO_2] from the NOAA (Raes et al., 2013), and C_i/C_a was measured by Tognetti et al. (2002), defining a value around 0.69 for olive. Equally, Villalobos et al. (2012) found an overall improvement of water use efficiency in olive trees under water deficit and several authors have reported a reduction in C_i/C_a (and then TE increase) when water stress increased (Brodribb, 1996; Bacelar et al., 2007; Gómez del Campo, 2013; López-Bernal et al., 2015).

3.2 Impact of Climate Change on Irrigation Requirements, Yield, and Water Productivity

Initial studies have identified significant increases in irrigation requirements for woody crops such as olive of around 20% on average, comparing baseline with future periods (Rodríguez-Díaz et al., 2007; Tanasijevic et al., 2014; Valverde et al., 2015). However, considering more recent methodologies such as AdaptaOlive it was detected that future irrigation requirements are projected to decrease slightly when deficit irrigation and full irrigation strategies were considered (6.2% and 0.5%, respectively; Lorite et al., 2017). These divergences were caused by consideration of critical processes that reduce the irrigation requirements for olive such as the reduction of crop transpiration under future conditions due to the reduction of canopy conductance, in turn caused by an increase in atmospheric [CO_2].

However, due to the elevated uncertainties associated with physiological components, the quantification of changes in irrigation requirements varies depending on the crop, simulation model, location, and irrigation strategy, requiring a significant increase in experimental and modeling efforts to be able to provide accurate projections of crop water requirements under future climate conditions.

In terms of yield, due to the reduction in rainfall and the increase in temperature and ET_o, the majority of studies describe a serious decrease in yield, even under irrigated conditions (Lobell et al., 2006). However, these changes varied significantly depending on the crops. Thus for wine grapes the yield exhibited very small changes but for crops such as almonds, orange, walnuts, and avocados, projected yield reduction varied from moderate to severe yield declines (Lobell et al., 2006). However, for olive crop under Mediterranean conditions, Lorite et al. (2017) evaluated the impact of climate change on olive yield detecting a null impact under rain-fed conditions and an increase of around 21% for irrigated orchards. In this study the effect of the increase in atmospheric [CO_2] on yield and the effect of water stress during flowering were considered. Morales et al. (2016) confirmed these results with projected yield increases of around 8% for high-density olive orchards when baseline and future periods were compared.

In this agreement, water productivity (WP) and irrigation water productivity (IWP) are associated with yield and irrigation volume applied to crops. Following Lobell et al. (2006), Tanasijevic et al. (2014), and Valverde et al. (2015) a severe reduction in both indicators for crops such as almond and citrus is foreseen. However, with the methodology considered in AdaptaOlive for olive orchards under Mediterranean conditions, significant increases in both indicators were found (Lorite et al., 2017). Thus comparing baseline with far future periods, WP increased around 36% for rain-fed conditions and around 47% for irrigated orchards. Equally, IWP increased around 46% with regulated deficit irrigation strategies but the increase was lower (17%) when full irrigation was considered.

3.3 Modeling of the Impact of Climate Change on Water Management

Consideration of crop simulation models under future climate conditions for evaluating the impact of climate change on water management is affected by limitations in some approaches included in traditional models. For example, when analyzing TE under water stress for olive (Villalobos et al., 2012) or orange (Roccuzzo et al., 2014) at the daily scale, TE is higher for stressed trees than for irrigated ones. As specific functions for olive were not available, models such as AdaptaOlive (Lorite et al., 2017), based on experimental data for different deficit irrigation strategies provided by Bacelar et al. (2007), simulated an increase in TE when stomatal closure caused by water stress happened. However, for almond this model response was not detected (Espadafor, 2015), confirming the conclusions obtained by Romero and Botía (2006).

The future projected yield, in addition to T and TE changes, is influenced by other factors such as the lack of chilling hour requirements, water stress in flowering and/or ripening, heat stress on flowering, or alternate bearing impact. Unfortunately, for almost all these impacts no information is available. Uniquely, the impact of water stress in flowering has been analyzed in detail (Moriana et al., 2003; Rapoport et al., 2012) demonstrating that water deficit during inflorescence development reduces inflorescence and flower number and disturbs ovule development. Using experimental data collected by Rapoport et al. (2012), a linear function was developed for calculating this yield reduction caused by water stress during the flowering stage (Lorite et al., 2017).

To solve the high uncertainty in the results provided by the different crop models, the use of crop model ensembles has been an innovative tool. However, crop model ensembles have only been applied so far for extensive crops, mainly winter cereals, as described Pirttioja et al. (2015) and Ruiz-Ramos et al. (2017), but for orchards the limited number of available crop models avoids consideration of this type of solution.

4. IMPLEMENTATION OF ADAPTATION STRATEGIES UNDER CURRENT AND FUTURE CLIMATES IN SOUTHERN SPAIN

4.1 Current Adaptation Strategies Related to Water Management for Woody Crops

4.1.1 Soil Management Considering Reduced Tillage

Farmers are considering soil management practices related to reduced tillage, mainly in rain-fed agricultural systems to save water and reduce erosion at the field scale. These practices have demonstrated significant reductions in soil evaporation and erosion (Gómez et al., 2003, 2014b; Abazi et al., 2013). However, these new practices also require increasing the use of herbicides (Potter et al., 2015). Farmers are already using these soil practices for crops such as olive, sunflower, wheat, and maize, increasing significantly the area cultivated with these practices over the last few years (MAGRAMA, 2014a).

4.1.2 Consideration of Cover Crops

To avoid erosion processes and save water, the use of cover crops in olive or almond orchards has been increasing in the last few years. Usually, covers of natural weeds, winter wheat, or legumes are common in numerous olive orchards located in southern Spain. In addition, to reduce erosion impacts, an improvement in water management in the orchard is obtained as heavy rainfall events usually generate significant runoff that is reduced when cover crops are present (Abazi et al., 2013). Thus if correct cover crop management is carried out (with mowing/killing dates, for instance, for the south of Spain before April, to avoid water competition with the main crop), rainfall water is stored in the soil, increasing water availability for the tree crop. This increase in stored water at the soil is very important in many areas where rainfall is very limited but heavy rainfall events are frequent.

Finally, an increase in biodiversity has also been reported when cover crops are considered (Gómez et al., 2014a).

4.1.3 Implementation of Regulated Deficit Irrigation Strategies

Mainly for crops such as olive, almond, apple, or peach, consideration of the moment of application of the available irrigation water is critical (Fereres and Soriano, 2007; Pascual et al., 2016). Thus by identifying and adjusting irrigation to those critical periods of the crop when water supply cannot be limited, a very satisfactory performance is achieved, not only saving water, but also reducing the yield in a very limited way. Even for some crops and strategies, quality is increased (for example, for peach, as found by Alcobendas et al., 2013). As result of these strategies, crop profitability and irrigation water productivity are also increased (Egea et al., 2013). Under correct irrigation management, water savings of around 50% can be achieved with reductions in yield of around 20% (Goldhamer et al., 2006; Fereres and Soriano, 2007; Egea et al., 2013). Promotion of these deficit irrigation strategies (such as regulated deficit irrigation, RDI), even with very limited volume (lower than 2000 m^3 ha^{-1}), is considered as an excellent adaptation strategy in a climate change scenario. This recommendation has been previously proposed by several authors such as Fereres and Soriano (2007) and Trentacoste et al. (2015).

Comparison between rain-fed, full irrigation, and RDI strategies for olive showed that the first two options for future periods will not provide satisfactory results, especially considering economic aspects such as net margin (Lorite et al., 2017). Thus net margin for olive was extremely sensitive to decreases in olive oil price and/or increases in water costs and therefore the economic analysis confirms full irrigation as a nonoptimal irrigation strategy, because RDI provided higher net margin. In addition, RDI stabilized the net income year by year compared with rain-fed conditions, providing an additional advantage of the use of this irrigation strategy (Lorite et al., 2017). Previous studies carried out by Fereres and Soriano (2007) and Iniesta et al. (2009) confirmed these results.

The implementation of RDI strategies as a result of the reduction in available water resources will generate an increase in the data requirements for accurate orchard and soil characterization. Thus irrigation scheduling based on RDI requires a full evaluation of the water status of the orchard, especially during critical periods such as flowering, to avoid severe stresses during these crop stages that could impact severely on yield. Consideration of instrumentation to assist in irrigation scheduling is highly recommended, but currently the farmer's skill in using this instrumentation is limited, usually requiring the support of irrigation advisory services (Lorite et al., 2012; Iglesias and Garrote, 2015).

The promotion of RDI strategies is considered as an excellent adaptation strategy under climate change scenarios, but studies related to this issue, such as those developed by Moriana et al. (2003) for olive, Goldhamer et al. (2006) for almond, and García Tejero et al. (2013) for citrus, are still necessary to implement models that predict the productive response of woody crops when they are subjected to different strategies of deficit irrigation. Modeling woody crop behavior in future scenarios of water scarcity will help to develop accurate irrigation schedules to ensure their sustainability and viability.

4.1.4 Extension of the Irrigated Areas With Low Water Allocation

Olive and almond crops are traditionally cultivated in areas under rain-fed conditions in southern Spain (MAGRAMA, 2014a). Many of these areas are clearly limited by the lack of water resources, although temperature and soil conditions are excellent for agriculture. Numerous studies have confirmed the excellent performance and water use efficiency of these crops when limited irrigation supply is provided (Sadras et al., 1991; Moriana et al., 2003; Karam et al., 2007).

In last few decades, around 125,000 ha of olive orchards within the Guadalquivir Valley have been changed from rain-fed to irrigated areas. However, due to water scarcity in the region, irrigation to meet the full crop demand was not a feasible option. Thus a low allocation of around 1500 $m^3\ ha^{-1}$ was designed and then deficit irrigation strategies were implemented in the area. This strategy was promoted by public institutions and has increased the olive yield in the region, improving the net margin for many farmers. If a full demand had been planned, the new irrigated area would have been reduced to less than 50,000 ha without a positive impact on the whole region. Thus the establishment of irrigation with very limited water allocation is currently increasing yield and yield stability, ensuring economic and social sustainability in numerous rural areas. Thus these areas were affected by recurring drought periods with significant reduction in yields and profits, requiring public subventions to settle the rural population.

In future, the clear distinction between rain-fed and irrigated crops will be blurred, probably resulting in a dynamic spatial gradient of irrigation intensity in a region depending on yearly weather and market conditions.

4.1.5 Rearrangement of Irrigation Allocations Within Farms and Irrigation Schemes

Usually, river basin authorities provide a fixed irrigation allocation per surface unit. Thus traditionally under nonlimiting conditions, irrigation allocations of around 5000–6000 $m^3 ha^{-1}$ have been common in southern Spain. A combination of crops with low and high irrigation water demands allow to rearrange the irrigation supply, providing more irrigation water to those crops with higher water productivity (Lorite et al., 2007). Thus crops such as olive, nuts, and horticultural crops are prioritized versus spring cereals such as maize. This redistribution of the available water resources is carried out at the farm (exploitation managed by one farmer where one or several crops are cultivated) and irrigation scheme scale (transferring water resources between exploitations).

4.1.6 Improving Irrigation Water Management at the Field Scale

Suboptimal irrigation scheduling providing incorrect irrigation volume, rates, or timing, or generating low uniformity of the irrigation system, is common in many Mediterranean irrigation districts. One of the main causes of this incorrect irrigation management is farmers' lack of knowledge in relation to crop water demands. In the last few decades, new technologies and services have been developed to help improve irrigation water management (Santos et al., 2008; Lorite et al., 2013b), providing specific irrigation scheduling and early warning systems for identifying water stress at the field scale.

4.1.7 Reductions in Available Irrigation Water Even in Years Without Weather Limitations

Traditionally, in years with nonlimiting water availability, some crops such as maize, sugar beet, cereals, and horticultural crops had generous irrigation allocations, implying null limitations in the available irrigation water. This fact has implied that farmers were not concerned about irrigation management and therefore irrigation efficiency was reduced for those years (García-Vila et al., 2008). Water basin authorities are applying irrigation restrictions even in rainy years to promote the improvement of irrigation efficiency and to save water for drought periods. Thus, depending on the crop, fixed allocations are provided by the water basin authority. These allocations varied from 4500 $m^3 ha^{-1}$ for maize or horticultural crops to around 1500 $m^3 ha^{-1}$ for olive, almonds, winter cereals, or sunflower.

4.1.8 Improvement of the Irrigation Infrastructure

In the last few decades a modernization process in the Andalusian irrigation schemes has generated changes in the irrigation method, changing from the most common irrigation methods, surface or sprinkler, to drip irrigation (MAGRAMA, 2014b). This change has been promoted by the limitation in water resources and has implied a significant increase in irrigation efficiency, reducing losses by runoff and deep percolation (García-Vila et al., 2008).

4.1.9 Incorporation of New Sources of Water for Irrigation

In the last few years the incorporation of new sources of water for existing or new irrigation schemes has been frequent. The most common new alternative has been the use of reclaimed water (RW) for agriculture (Parsons et al., 2010; Petousi et al., 2015). In spite of legal and health restrictions, new irrigation schemes irrigated uniquely with RW have started (such as the "Cuatro Vegas" irrigation scheme in Almeria province, southeastern Spain). Mainly, crops such as olive and horticultural crops have been considered (Pérez-Parra, 2005).

As it was previously indicated, traditional olive orchards are transitioning from traditional rain-fed systems to partially irrigated orchards, and then the identification of new sources of water for irrigation is a recommended adaptation strategy. The utilization of RW has been evaluated compared with freshwater application for irrigating olive orchards. Thus, in an experiment carried out in Greece for olive groves, no differences in trunk diameter, plant height, and photosynthetic activity were found between irrigated treatments with secondary treated wastewater, tertiary treated wastewater, and fresh water (Petousi et al., 2015). Similarly, Ayoub et al. (2016) did not find differences in yield between experiments carried out with fresh water and RW for an olive orchard located at Jordan. Other studies determined no effects on vegetative growth and productivity, but increased salt loads into and beyond the root zone (Segal et al., 2011). Thus when RW is considered, nutrients contained in RW must be computed for fertilization decisions to increase fertilizer application efficiency (Segal et al., 2011). The consideration of RW requires careful monitoring of soil and plants in relation to salts, nutrients, heavy metals, or pathogens (Petousi et al., 2015). Equally, when saline RW is considered, severe impacts for some crops (such as citrus trees) have been reported. Thus long-term irrigation with saline RW for mandarin trees led to a reduction in vegetative growth and prominent reductions in crop fruit load, and therefore yield and water productivity were significantly reduced (Nicolás et al., 2016).

The SuWaNu project (http://www.suwanu.eu) evaluated the strengths, weaknesses, opportunities, and treats for RW in Mediterranean agriculture (Michailidis et al., 2015), determining the use of RW as a very valuable alternative to the limited availability of water resources, mainly for crops such as olive.

4.1.10 Diversification of the Crop Pattern at the Field Scale

The consideration of new (or traditional but not so common nowadays) crops such as legumes (chickpeas and lentils), protein plants (beans and peas), and nuts (almond, pistachios, and walnut trees), and the incorporation of fallow in the annual rotation crop (MAGRAMA, 2014a) are adaptation strategies related to water management. Thus increase in the number of crops generates a higher stability in the farmer's profit and generates water savings to be reallocated to crops with higher water productivity (Lorite et al., 2007). However, this measure also requires additional formation to farmers and an efficient planning of labor, machinery, and inputs in general.

4.1.11 Increase and Promotion of Local and Regional Irrigation Advisory Services

Public authorities are promoting the implantation and use of irrigation advisory services (IAS) to support farmers and technicians for improving irrigation water management at the field scale (Ortega et al., 2005; Montoro et al., 2011; Lorite et al., 2012). Some services provided by the IAS are the generation of irrigation schedules adapted to local field conditions based on the use of remote sensing techniques and modeling (Santos et al., 2008), the measurement and evaluation of irrigation water consumption (Lorite et al., 2013a), irrigation methods training and support for other agriculture practices to be applied in combination with irrigation, such as cultivar selection or support for meeting the requirements for subsidies.

4.1.12 Integrated Crop Management Supported by Decision Systems

As a result of all the previous adaptation strategies (Table 1.1), correct crop management is a very difficult task, demanding high levels of knowledge, information, and experience. Therefore the promotion of decision support systems based on experimentation and big data is a key adaptation strategy (Rinaldi and He, 2014; Kadiyala et al., 2015). Thus these systems must integrate all the potential adaptation options related to crop and cultivar

Table 1.1 Main adaptation strategies considered, positive effect and scale obtained from each one

Adaptation strategy	Positive effect	Scale
Soil management considering reduced tillage	Water saving	Field
	Reduction of runoff and erosion	Field
Consideration of cover crops correctly managed	Reduction of runoff and erosion	Field
Implementation of regulated deficit irrigation strategies	Increase of irrigation water productivity	Field
	Water saving for other crops/uses	Field
Extension of the irrigated area with low water allocation	Increase in yield	Regional
	Increase in stability of yields	Regional
Rearrangement of irrigation allocations within farms and irrigation schemes	Increase in water productivity	Field/irrigation area
Improving irrigation water management at the field scale	Water saving	Field
	Reduction of deep percolation	Field
Reductions in available irrigation water even in years	Water saving	Regional
without weather limitations	Increase in irrigation efficiency	Field
Improvement of the irrigation infrastructure	Water saving	Field/irrigation area
Incorporation of new sources of water for irrigation	Increase in yield	Field/irrigation area
Diversification of the crop pattern at the field scale	Increase in yield stability	Irrigation area/regional
Increase and promotion of local and regional irrigation	Water saving	Irrigation area/regional
advisory services	Increase in yield	Irrigation area/regional
Integrated crop management supported by decision systems	Water saving	Field/irrigation area

selection, soil management, irrigation, fertilization, and pest or disease control and other agronomical practices, avoiding contradictory strategies. In addition, the coordination of adaptation and mitigation strategies to climate change must be promoted (Veroz et al., 2016).

4.2 Adaptation Strategies Under Evaluation and Experimentation

4.2.1 Soil Management Practices for Increasing Albedo to Reduce Air Temperature and Microadvection

Related to soil management, in the last few years innovative research topics have been focused on increasing albedo (the amount of solar radiation reflected from the soil surface) with soil practices such as no-tillage (Davin et al., 2014). This strategy could be very useful for mitigating heat wave impacts on orchard crops with a significant part of bare soil. In southern Spain, crops such as olive or almonds could benefit from this measure.

4.2.2 Cultivar Selection Depending on New Climate Conditions

Studies on phenology under future climate conditions are providing useful information about those crops/cultivars that could show problems related to chilling requirements or the occurrence of frequent extreme events during critical phases such as flowering (De Melo-Abreu et al., 2004). These facts could generate severe crop yield reductions and could be prevented by considering adequate cultivars adapted to local future conditions.

Thus the selection of adequate olive cultivars depending on the location and future climate projections is also a key adaptation strategy, although only a few studies have addressed it (Koubouris et al., 2009; Orlandi et al., 2013; Gabaldón-Leal et al., 2017). For example, for olive, genotypes show different behavior depending on weather conditions, and then potential enhancement of the impact of climate change for some genotype/location combinations (for example, late-flowering cultivars in dry locations) must be considered. Thus analyzing simulated yield under rain-fed conditions for different cultivars, the late flowering olive genotypes showed the lowest yields because the sensitive period (flowering) was late and therefore the probability of heat/water stress damage increased compared with the rest of the genotypes (Gabaldón-Leal et al., 2017). However, under rainy and moderate temperature locations the importance of cultivar selection is reduced (Lorite et al., 2017).

Related to irrigation requirement assessment under RDI strategies, traditional cultivars such as "Arbequina" or "Picual" with early flowering date showed smaller irrigation requirements than late genotypes due to the differences in flowering date (Lorite et al., 2017). This fact, plus the lower probability of occurrence of extreme events during critical periods, confirms that the consideration of cultivars with early flowering such as "Arbequina" is a valuable adaptation strategy for Mediterranean olive orchards.

4.2.3 Identification of Suitable New Cultivation Areas

The combination of maps delimitating areas with enough chilling hours and areas with limited occurrence of heat extreme events during the flowering stage provides an excellent tool to evaluate threats to flowering, identifying suitable new cultivation areas for crops such as olive (Gabaldón-Leal et al., 2017).

In addition, incorporating phenological characteristics of different cultivars to assess crop irrigation requirements and the irrigation water availability could lead to the identification of suitable areas for the cultivation of orchards with RDI currently and under future conditions.

4.2.4 Genetic Improvement

The obtaining of new cultivars adapted to water and heat stress conditions is currently a priority for numerous research groups. Thus new cultivars for olive and almond are under evaluation and hopefully will be available for farmers in the next few years (Vargas et al., 2011; Besnard et al., 2012). Equally, for olive the identification of new cultivars with early flowering date (which would be an interesting behavior under future climate conditions as was previously shown) is an innovative adaptation strategy recently initialized.

4.2.5 Application of Foliar Kaolin Clay Particles to Olive and Other Trees

To reduce leaf temperature, net gas exchange, and water relations, the application of foliar kaolin clay particles has been shown as a useful strategy (Jifon and Syvertsen, 2003; Correia et al., 2014). The application of this technology to Spanish agriculture has been very limited and therefore its application under the regional agronomic and climatic conditions of Andalusia is currently under evaluation, in terms of both performance and cost.

4.2.6 Consideration of Supplementary Irrigation to Reduce the Canopy Temperature

Canopy temperature is related to the water status of the plant (Jackson et al., 1981) and therefore a stressed crop will have a higher canopy temperature compared with the same tree unstressed. In addition, modifications in the energy balance reducing the daily thermal amplitudes occur when irrigation for olive orchards is correctly managed (Andrade et al., 2014). However, studies on this measure are still very limited and additional experimental studies related to this are required.

5. THE ROLE OF WATER MANAGEMENT IN THE MITIGATION OF CLIMATE CHANGE

Some measures for mitigation of climate change are related to water management. Thus these measures are mainly focused on the reduction of greenhouse gas (GHG) emissions and must be promoted in the Mediterranean orchards, but with a previous assessment of their trade-offs with adaptation to avoid contradictory effects.

In Mediterranean climate areas, GHG emissions are expected to rise in the future. In this context it is crucial to find and develop effective mitigation strategies compatible with the adaptation ones to reduce these emissions. To reduce GHG emissions associated with energy, new strategies for irrigation have been promoted to reduce electricity consumption. These strategies are based on the optimization of irrigation water arrangement to the fields, discouraging the traditional pressurized on-demand supply that generates nonefficient and high energy costs (Moreno et al., 2010). This mitigation strategy is carried out at an irrigation district scale. In parallel, promotion of renewable energy sources (mainly solar) for irrigated agriculture is carried out in several areas within Andalusia (González-Perea et al., 2015). Thus some irrigation districts in Andalusia consider uniquely renewable energy sources, reducing water costs and energy use, and allowing extending irrigation to areas without access to traditional energy sources.

Moreover, the selection of an adequate irrigation method contributes to the reduction of GHG emissions. Thus the implantation of drip irrigation systems has been proved to be a useful mitigation strategy (Sanz-Cobeña et al., 2016). Consideration of drip irrigation systems decreases N_2O fluxes by reducing water availability to microorganisms and their capacity to produce N_2O (Vlek et al., 1981; Steenwerth and Belina, 2010; Kennedy et al., 2013). Drip irrigation also decreases the soil water-filled pore space below the optimum range for N_2O production through the denitrification process. In this context, emissions of N_2O in drip-irrigated Mediterranean crop areas have on average 80% lower emissions than sprinkler systems, as detailed by Cayuela et al. (2017). Drip irrigation not only reduces N_2O but could also uptake less CH_4 than sprinkler irrigation (Guardia et al., 2017) due to the increase in methanotrophy with soil water content close to field capacity (Le Mer and Roger, 2001).

Another mitigation measure is related to irrigation and fertilization. Thus irrigated systems receive large amounts of water and nitrogen inputs that produce favorable soil conditions for N_2O production. This is possibly related to denitrification's predominant role in anaerobic soil (Davidson et al., 1991) in irrigated and intensively managed systems (Sanz-Cobena et al., 2012, 2014). For this reason, drip irrigation systems combined with optimized fertilization (fertigation) could reduce emissions by between 30% and 50% (Kallenbach et al., 2010; Schellenberg et al., 2012; Kennedy et al., 2013; Abalos et al., 2014; Vallejo et al., 2014). However, these irrigation techniques could increase NO emissions from leached NO_3, and as a result increase N_2O emissions (Quemada et al., 2013; Sánchez-Martín et al., 2010). Equally, indirect N_2O emissions could be reduced by lower irrigation rates and higher irrigation frequency, and by N supply at the specific moments of the cycle when plants really need it (Quemada et al., 2013).

6. EXTENSION OF ADAPTATION AND MITIGATION STRATEGIES TO CLIMATE CHANGE: THE EXAMPLE OF SOUTHERN SPAIN

6.1 Extension of Adaptation and Mitigation Strategies in Southern Spain

In general, farmers are very reticent to carry out significant changes to their agricultural practices. Many of them carry out traditional practices transferred from previous generations and these cannot be questioned. To solve these limitations, currently, numerous strategies for promotion of adaptation and mitigation strategies are being applied, mainly by public institutions.

Thus promoted by public research institutions, an elevated number of courses, seminars, and technical workshops are taught each year to farmers and technicians to increase knowledge about optimized agronomical practices in Mediterranean environments, adapted to regional requirements and including general, regional, and local adaptation strategies to climate change. For example, the regional government of Andalusia provides a complete program of courses to farmers and technicians that can be found at: http://www.juntadeandalucia.es/agriculturaypesca/ifapa/web/ifapa/servicios/formacion.

Another strategy is related to the promotion of joint research projects associated with climate change and agriculture. Thus funded by public regional and national institutions, some programs have been developed to join different organizations (including farmers' associations, irrigation districts, cooperatives, private companies, and research institutions) in research projects where the main problems of the stakeholders related to agriculture are analyzed. Specific research topics about the impact of climate change on agriculture in Andalusia have been recently considered. In these projects the impact of climate change and the implementation of the adaptation and mitigation strategies described in this chapter specifically for agriculture in Andalusia are being analyzed.

In this context, public regional institutions are promoting the development of irrigation advisory services in numerous areas of the Mediterranean Basin. For example, this strategy has created around new 20 irrigation advisory services in Andalusia. These services were mainly developed for providing irrigation scheduling but they also provide complete farmer support services, such as assessment for obtaining European subsidies and support for fertilization or disease control. This is an excellent tool for the diffusion of adaptation and mitigation strategies among farmers.

In the same vein, public regional institutions are promoting programs for identifying those crops and cultivars with the best performance under different climatic and soil conditions. Within these programs, those cultivars adapted to new climate conditions and drought periods are being recommended to farmers and technicians, providing very high satisfaction to farmers. As an example, in the Andalusia region the results provided by the regional program of experimentation for olive and almond can be found at: http://www.juntadeandalucia.es/agriculturaypesca/ifapa/web/ifapa/productos/transferencia.

Equally, programs of ecological agriculture and integrated production have included specific requirements related to improving irrigation water management and have also considered adaptation and/or mitigation strategies to climate change. These programs provide rules that farmers are required to follow to obtain a certificate of production and thus economical and market profits.

Finally, an additional key factor would be the promotion of the use and development of new tools to improve agricultural water management. These tools include Geographic Information Systems (GIS), remote sensing, decision support systems, and modeling software to develop irrigation scheduling adapted to the specific conditions of each field. Specific tools such as those described by Peragón et al. (2016), who developed a tool for improving irrigation water management to reduce salinization risks in olive orchards, or the tool for improving irrigation and fertilization management in olive (Hidalgo et al., 2015), are some examples of approaches related to the improvement of water management. These services are essential for correct irrigation management, a key component of the adaptation and mitigation strategies under climate change in Mediterranean semiarid regions.

6.2 Impediments for the Implementation of Adaptation and Mitigation Strategies in Southern Spain

The main impediment for the implementation of the proposed adaptation and mitigation strategies is related to farmers' reluctance to carry out changes to their agricultural practices even when evidence of the benefits to adopt the new strategies is shown. However, for other strategies, impediments are caused by national and regional regulations. A summary of the main impediments detected in southern Spain for the implementation of adaptation and mitigation strategies are:

1. Requirements of adapted machinery to carry out direct-seeding practices, increasing cropping costs.
2. Increase in the use of herbicides for no-tillage and minimum tillage practices and with advances in sowing date, generating increases in costs and environmental concerns.
3. Competition for water resources of cover crops with the main crop requires accurate crop management (selection of mowing dates, identification of strategies for cover crop killing, etc.). This fact generates doubts in farmers, discouraging the application of these strategies.
4. Consideration of new crops and agricultural practices could be limited by poor farmer qualification requiring additional training that sometimes is not available.
5. Consideration of deficit irrigation strategies requires careful irrigation water management and farmer knowledge to avoid severe damage to the crops. This fact increases farmers' concerns and reduces the expansion of these strategies.
6. Some improvements related to irrigation infrastructure, reclaimed water use, and energy efficiency practices require elevated initial investments that usually are not possible for individual farmers, requiring public subsidies.

7. CONCLUSIONS

Climate change will severely impact on agriculture, especially in areas under arid and semiarid climate conditions. Thus the forecast reduction in rainfall and the increase in temperatures will generate reductions in yield and profit, as crops will be impacted by water and heat stress. Under this perspective, the correct implementation of adaptation and mitigation strategies will contribute to reduce (or even eliminate) these negative impacts, ensuring the economic and environmental sustainability of the current agricultural systems located in southern Europe. Thus regulated deficit irrigation strategies or correct soil management considering practices of conservation agriculture are some of the measures recommended.

To identify those adaptation strategies more suitable for each crop and location, a correct characterization of the agricultural systems will be critical. Thus crop phenology and sensitivity to water/heat stress, soil characteristics, or availability of water resources are some knowledge requirements that must be considered previous to any implementation of measures of adaptation or mitigation to climate change.

Finally, and although the role of public institutions will have a special relevance in the identification, promotion, and diffusion of the described adaptation and mitigation practices, the implantation of these satisfactory measures against climate change must be the result of joint efforts of all the actors involved in the agricultural systems, from farmers to research institutions, and therefore collaboration among all of them must be promoted.

ACKNOWLEDGMENTS

This study has been financially supported by the project RTA2014-00030-00-00 funded by INIA, FEDER 2014–20 "*Programa Operativo de Crecimiento Inteligente*," FACCE MACSUR—Modelling European Agriculture with Climate Change for Food Security, a FACCE JPI knowledge hub, by AGR-6126 project funded by the *Consejería de Innovación, Ciencia y Empresa (Junta de Andalucía)* and by MULCLIVAR, from the Spanish *Ministerio de Economía y Competitividad* (MINECO) CGL2012-38923-C02-02.

REFERENCES

Abalos, D., Sanchez-Martin, L., Garcia-Torres, L., Van Groenigen, J.W., Vallejo, A., 2014. Management of irrigation frequency and nitrogen fertilization to mitigate GHG and NO emissions from drip-fertigated crops. Science of the Total Environment 490, 880–888.

Abazi, U., Lorite, I.J., Cárceles, B., Martínez Raya, A., Durán, V.H., Francia, J.R., Gómez, J.A., 2013. WABOL: a conceptual water balance model for analyzing rainfall water use in olive orchards under different soil and cover crop management strategies. Computers and Electronics in Agriculture 91, 35–48.

Abrisqueta, J.M., Plana, V., Mounzer, O.H., Mendez, J., Ruiz-Sánchez, M.C., 2007. Effects of soil tillage on runoff generation in a Mediterranean apricot orchard. Agricultural Water Management 93, 11–18.

Alcobendas, R., Mirás-Avalos, J.M., Alarcón, J.J., Nicolás, E., 2013. Effects of irrigation and fruit position on size, colour, firmness and sugar contents of fruits in a mid-late maturing peach cultivar. Scientia Horticulturae 164, 340–347.

Allen, R.G., Pereira, L.S., 2009. Estimating crop coefficients from fraction of ground cover and height. Irrigation Science 28, 17–34.

Allen, R.G., Pereira, L.S., Raes, D., Smith, M., 1998. Crop evapotranspiration: guidelines for computing crop water requirements. In: FAO Irrigation and Drainage, vol. 56. FAO, Rome.

Allen, R.G., Tasumi, M., Trezza, R., 2007a. Satellite-based energy balance for mapping evapotranspiration with internalized calibration (METRIC) – Model. Journal of Irrigation and Drainage Engineering – ASCE 133, 380–394.

Allen, R.G., Tasumi, M., Morse, A., Trezza, R., Wright, J.L., Bastiaanssen, W., Kramber, W., Lorite, I., Robison, C.W., 2007b. Satellite-based energy balance for mapping evapotranspiration with internalized calibration (METRIC) – applications. Journal of Irrigation and Drainage Engineering – ASCE 133, 395–406.

Alpert, P., Ben-Gai, T., Baharad, A., Benjamini, Y., Yekutieli, D., Colacino, M., Diodato, L., Ramis, C., Homar, V., Romero, R., Michaelides, S., Manes, A., 2002. The paradoxical increase of Mediterranean extreme daily rainfall in spite of decrease in total values. Geophysical Research Letters 29, 31.

Andrade, J., Santos, F.L., Correia, M., Afonso, T., 2014. Effects of irrigation and tree spacing on soil and air temperature profiles of olive orchards. Acta Horticulturae 1057, 443–450.

Ashraf, M., Loftis, J.C., Hubbard, K.G., 1997. Application of geostatistics to evaluate partial weather station networks. Agricultural and Forest Meteorology 84, 255–271.

Atkinson, C.J., 2000. Effects of climate change on the behaviour of woody perennials. In: Agrawal, S.B., Agrawal, M. (Eds.), Environmental Pollution and Plant Responses. Lewis Publishers, pp. 19–32.

Autovino, D., Minacapilli, M., Provenzano, G., 2016. Modelling bulk surface resistance by MODIS data and assessment of MOD16A2 evapotranspiration product in an irrigation district of Southern Italy. Agricultural Water Management 167, 86–94.

Ayoub, S., Al-Shdiefat, S., Rawashdeh, H., Bashabsheh, I., 2016. Utilization of reclaimed wastewater for olive irrigation: effect on soil properties, tree growth, yield and oil content. Agricultural Water Management 176, 163–169.

Bacelar, E.A., Moutinho-Pereira, J.M., Gonçalves, B.C., Ferreira, H.F., Correia, C.M., 2007. Changes in growth, gas exchange, xylem hydraulic properties and water use efficiency of three olive cultivars under contrasting water availability regimes. Environmental and Experimental Botany 60 (2), 183–192.

Bastiaanssen, W.G.M., Menenti, M., Feddes, R.A., Holtslag, A.A.M., 1998. A remote sensing surface energy balance algorithm for land (SEBAL): 1. Formulation. Journal of Hydrology 212, 198–212.

Beer, C., Ciais, P., Reichstein, M., Baldocchi, D., Law, B.E., Papale, D., Soussana, J.F., Ammann, C., Buchmann, N., Frank, D., Gianelle, D., Janssens, I.A., Knohl, A., Köstner, B., Moors, E., Roupsard, O., Verbeeck, H., Vesala, T., Williams, C.A., Wohlfahrt, G., 2009. Temporal and among-site variability of inherent water use efficiency at the ecosystem level. Global Biogeochemical Cycles 23 (2), 1–13.

Besnard, G., Anthelme, F., Baali-Cherif, D., 2012. The Laperrine's olive tree (Oleaceae): a wild genetic resource of the cultivated olive and a model-species for studying the biogeography of the Saharan Mountains. Acta Botanica Gallica 159 (3), 319–328.

Bindi, M., Fibbi, L., Miglietta, F., 2001. Free air CO_2 enrichment (FACE) of grapevine (*Vitis vinifera* L.): II growth and quality of grape and wine in response to elevated CO_2 concentrations. European Journal of Agronomy 14, 145–155.

Blum, M., Lensky, I.M., Nestel, D., 2013. Estimation of olive grove canopy temperature from MODIS thermal imagery is more accurate than interpolation from meteorological stations. Agricultural and Forest Meteorology 176, 90–93.

Bojanowski, J.S., Vrieling, A., Skidmore, A.K., 2013. Calibration of solar radiation models for Europe using Meteosat Second Generation and weather station data. Agricultural and Forest Meteorology 176, 1–9.

Bojanowski, J.S., Vrieling, A., Skidmore, A.K., 2014. A comparison of data sources for creating a long-term time series of daily gridded solar radiation for Europe. Solar Energy 99, 152–171.

Brodribb, T., 1996. Dynamics of changing intercellular CO_2 concentration (ci) during drought and determination on minimum functional ci. Plant Physiology 111 (1), 179–185.

Cammalleri, C., Ciraolo, G., 2013. A simple method to directly retrieve reference evapotranspiration from geostationary satellite images. International Journal of Applied Earth Observation and Geoinformation 21, 149–158.

Cammalleri, C., Anderson, M.C., Ciraolo, G., D'Urso, G., Kustas, W.P., La Loggia, G., Minacapilli, G., 2012. Applications of a remote sensing-based two-source energy balance algorithm for mapping surface fluxes without in-situ air temperature observations. Remote Sensing of Environment 124, 502–515.

Castro, C.L., Pielke, R.A., Leoncini, G., 2005. Dynamical downscaling: assessment of value retained and added using the regional atmospheric modeling system (RAMS). Journal of Geophysical Research 110, D05108.

Cayuela, M.L., Aguilera, E., Sanz-Cobena, A., Adams, D.C., Abalos, D., Barton, L., Ryals, R., Silver, W.L., Alfaro, M.A., Pappa, V.A., Smith, P., Garnier, J., Billen, G., Bouwman, L., Bondeau, A., Lassaletta, L., 2017. Direct nitrous oxide emissions in Mediterranean climate cropping systems: emission factors based on a meta-analysis of available measurement data. Agriculture, Ecosystems and Environment 238, 25–35.

Ceglar, A., Kajfež-Bogataj, L., 2012. Simulation of maize yield in current and changed climatic conditions: addressing modelling uncertainties and the importance of bias correction in climate model simulations. European Journal of Agronomy 37, 83–95.

Christensen, J.H., Boberg, F., Christensen, O.B., Lucas-Picher, P., 2008. On the need for bias correction of regional climate change projections of temperature and precipitation. Geophysical Research Letters 35 (20).

Collins, J.M., 2011. Temperature variability over Africa. Journal of Climate 24 (14), 3649–3666.

Correia, C., Dinis, L.T., Pinheiro, R., Fraga, H., Ferreira, H., Gonçalves, I., Costa, J., Oliveira, A., Malheiro, A., Moutinho-Pereira, J., 2014. Climate change and adaptation strategies for viticulture. Journal of International Scientific Publications: Agriculture and Food 2, 424–429.

Cristóbal, J., Anderson, M.C., 2012. Regional scale evaluation of a MSG solar radiation product for evapotranspiration modeling. Hydrology and Earth System Sciences Discussions 9 (7), 8905–8939.

Cristóbal, J., Anderson, M.C., 2013. Validation of a Meteosat Second Generation solar radiation dataset over the northeastern Iberian Peninsula. Hydrology and Earth System Sciences 17 (1), 163–175.

Cruz-Blanco, M., Lorite, I.J., Santos, C., 2014a. An innovative remote sensing based reference evapotranspiration method to support irrigation water management under semi-arid conditions. Agricultural Water Management 131, 135–145.

Cruz-Blanco, M., Gavilán, P., Santos, C., Lorite, I.J., 2014b. Assessment of reference evapotranspiration using remote sensing and forecasting tools under semi-arid conditions. International Journal of Applied Earth Observation and Geoinformation 33, 280–289.

Cruz-Blanco, M., Santos, C., Gavilán, P., Lorite, I.J., 2015. Uncertainty in estimating reference evapotranspiration using remotely sensed and forecasted weather data under the climatic conditions of Southern Spain. International Journal of Climatology 35, 3371–3384.

Davidson, E.A., Hart, S.C., Shanks, C.A., Firestone, M.K., 1991. Measuring gross nitrogen mineralization and nitrification by 15 N isotopic pool dilution in intact soil cores. European Journal of Soil Science 42, 335–349.

Davin, E.L., Seneviratne, S.I., Ciais, P., Olioso, A., Wang, T., 2014. Preferential cooling of hot extremes from cropland albedo management. Proceedings of the National Academy of Sciences of the United States of America 111 (27), 9757–9761.

De Melo-Abreu, J.P., Barranco, D., Cordeiro, M., Tous, J., Rogado, B.M., Villalobos, F., 2004. Modelling olive flowering date using chilling for dormancy release and thermal time. Agricultural and Forest Meteorology 125 (1–2), 117–127.

Dee, D.P., Uppala, S.M., Simmons, A.J., Berrisford, P., Poli, P., Kobayashi, S., Andrae, U., Balmaseda, M.A., Balsamo, G., Bauer, P., Bechtold, P., Beljaars, A.C.M., van de Berg, L., Bidlot, J., Bormann, N., Delsol, C., Dragani, R., Fuentes, M., Geer, A.J., Haimberger, L., Healy, S.B., Hersbach, H., Hólm, E.V., Isaksen, L., Kallbeerg, P., Köhler, M., Matricardi, M., Mcnally, A.P., Monge-Sanz, B.M., Morcrette, J.J., Park, B.K., Peubey, C., de Rosnay, P., Tavolato, C., Thépaut, J.N., Vitart, F., 2011. The ERA-Interim reanalysis: configuration and performance of the data assimilation system. Quarterly Journal of the Royal Meteorological Society 137 (656), 553–597.

Dosio, A., Paruolo, P., 2011. Bias correction of the ENSEMBLES high-resolution climate change projections for use by impact models: evaluation on the present climate. Journal of Geophysical Research 116, D16106.

Dosio, A., Paruolo, P., Rojas, R., 2012. Bias correction of the ENSEMBLES high resolution climate change projections for use by impact models: analysis of the climate change signal. Journal of Geophysical Research 117, D17110.

Drake, B.G., Gonzàlez-Meler, M.A., 1997. More efficient plants: a consequence of rising atmospheric CO_2? Annual Review of Plant Physiology and Plant Molecular Biology 48, 609–639.

Egea, G., Nortes, P.A., Domingo, R., Baille, A., Pérez-Pastor, A., González-Real, M.M., 2013. Almond agronomic response to long-term deficit irrigation applied since orchard establishment. Irrigation Science 31 (3), 445–454.

Elhag, M., Psilovikos, A., Manakos, I., Perakis, K., 2011. Application of the SEBS water balance model in estimating daily evapotranspiration and evaporative fraction from remote sensing data over the Nile Delta. Water Resources Management 25, 2731–2742.

Espadafor, M., 2015. Manejo del agua en plantaciones de almendro: demanda hídrica y respuesta al volumen de suelo mojado en plantaciones jóvenes (Ph.D. thesis). University of Córdoba.

Espadafor, M., Orgaz, F., Testi, L., Lorite, I.J., Villalobos, F.J., 2015. Transpiration of young almond trees in relation to intercepted radiation. Irrigation Science 33 (4), 265–275.

Essou, G.R.C., Brissette, F., Lucas-Picher, P., 2016. Impacts of combining reanalyses and weather station data on the accuracy of discharge modelling. Journal of Hydrology 545, 120–131.

Estévez, J., Gavilán, P., Giráldez, J.V., 2011. Guidelines on validation procedures for meteorological data from automatic weather stations. Journal of Hydrology 402 (1–2), 144–154.

Etheridge, D.M., Steele, L.P., Langenfelds, R.L., Francey, R.J., Barnola, J.M., Morgan, V.I., 1996. Natural and anthropogenic changes in atmospheric CO_2 over the last 1000 years from air in Antarctic ice and firn. Journal of Geophysical Research 101 (D2), 4115–4128.

Fereres, E., Soriano, M.A., 2007. Deficit irrigation for reducing agricultural water use. Journal of Experimental Botany 58 (2), 147–159.

Fereres, E., Goldhamer, D.A., Sadras, V., 2012. Yield response to water of fruit trees and vines: guidelines. In: Steduto, P., Hsiao, T.C., Fereres, E., Raes, D. (Eds.), Crop Yield Response to Water. FAO Irrigation and Drainage Paper, 66. FAO, Rome, pp. 246–295.

Field, C.B., Jackson, R.B., Mooney, H.A., 1995. Stomatal responses to increased CO_2: implications from the plant to the global scale. Plant, Cell and Environment 18, 1214–1225.

Gabaldón-Leal, C., Ruiz-Ramos, M., de la Rosa, R., León, L., Belaj, A., Rodriguez, A., Santos, C., Lorite, I.J., 2017. Impact of changes in mean and extreme temperatures caused by climate change on olive flowering in Southern Spain. International Journal of Climatology 37 (S1), 940–957.

García-Tejero, I.F., Arriaga, J., Durán-Zuazo, V.H., Muriel-Fernández, J.L., 2013. Predicting crop-water production functions for long term effects of deficit irrigation on citrus productivity (SW Spain). Archives of Agronomy and Soil Science 59, 1591–1606.

García-Tejero, I.F., Hernández, A., Rodríguez, V.M., Ponce, J.R., Ramos, V., Muriel, J.L., Durán-Zuazo, V.H., 2015. Estimating almond crop coefficients and physiological response to water stress in semiarid environments (SW Spain). Journal of Agricultural Science and Technology 17 (5), 1255–1266.

García-Vila, M., Lorite, I.J., Soriano, M.A., Fereres, E., 2008. Management trends and responses to water scarcity in an irrigation scheme of Southern Spain. Agricultural Water Management 95, 458–468.

Gavilán, P., Lorite, I.J., Tornero, S., Berengena, J., 2006. Regional calibration of Hargreaves equation for estimating reference ET in a semiarid environment. Agricultural Water Management 81 (3), 257–281.

Girona, J., Mata, M., Del Campo, J., Arbonés, A., Bartra, E., Marsal, J., 2006. The use of midday leaf water potential for scheduling deficit irrigation in vineyards. Irrigation Science 24 (2), 115–127.

Goldhamer, D.A., Girona, J., 2012. Almond. In: Steduto, P., Hsiao, T.C., Fereres, E., Raes, D. (Eds.), Cropyield Response to Water, Irrigation and Drainage. Paper No. 66. FAO, Rome.

Goldhamer, D.A., Viveros, M., Salinas, M., 2006. Regulated deficit irrigation in almonds: effects of variations in applied water and stress timing on yield and yield components. Irrigation Science 24 (2), 101–114.

Gómez del Campo, M., 2013. Summer deficit-irrigation strategies in a hedgerow olive orchard cv. 'Arbequina'. Effect on fruit characteristics and yield. Irrigation Science 31 (3), 259–269.

Gómez, J.A., Battany, M., Renschle, C.S., Fereres, E., 2003. Evaluating the impact of soil management on soil loss in olive orchards. Soil Use and Management 19, 127–134.

Gómez, J.A., Álvarez, S., Soriano, M.A., 2009a. Development of a soil degradation assessment tool for organic olive groves in Southern Spain. Catena 79 (1), 9–17.

Gómez, J.A., Sobrinho, T.A., Giráldez, J.V., Fereres, E., 2009b. Soil management effects on runoff, erosion and soil properties in an olive grove of Southern Spain. Soil and Tillage Research 102 (1), 5–13.

Gómez, J.A., Campos, M., Guzman, G., Castillo-Llanque, F., Giraldez, J.V., 2014a. Use of heterogeneous cover crops in olive orchards to soil erosion control and enhancement of biodiversity. In: ELS-The Earth Living Skin: Soil, Life and Climate Changes, EGU – SSS Conference, ELS2014-34-1, Bari, Italy.

Gómez, J.A., Infante-Amate, J., González de Molina, M., Vanwalleghem, T., Taguas, E.V., Lorite, I.J., 2014b. Olive cultivation, its impact on soil erosion and its progression into yield impacts in Southern Spain in the past as a key to a future of increasing climate uncertainty. Agriculture 4, 170–198.

Gómez, J.A., Guzman, G., Lorite, I., 2016. A conceptual water balance model to explore the impact of different soil management on water availability for vineyards under contrasting environments. In: EGU General Assembly Conference Abstracts, vol. 18, p. 8717.

González-Perea, R., Camacho, E., Montesinos, P., Fernández-García, I., Rodriguez, J.A., 2015. Reducing the energy demand in irrigation water supply systems. Experiences from southern Europe. In: 26th Euro-Mediterranean Regional Conference and Workshops "Innovate to Improve Irrigation Performances".

Guardia, G., Cangani, M.T., Sanz-Cobena, A., Junior, J.L., Vallejo, A., 2017. Management of pig manure to mitigate NO and yield-scaled N_2O emissions in an irrigated Mediterranean crop. Agriculture, Ecosystems and Environment 238, 55–66.

Guijarro, J.A., Conde, J., Campins, J., Picornell, M.A., Orro, M.L., 2014. Tendencias del viento, oleaje y temperatura superficial en el Mediterráneo y Atlántico próximos a partir de datos de reanálisis. In: IX Congreso Asociación Española de Climatología, pp. 315–324.

Hawkins, E., Osborne, T.M., Ho, C.K., Challinor, A.J., 2013. Calibration and bias correction of climate projections for crop modelling: an idealised case study over Europe. Agricultural and Forest Meteorology 170, 19–31.

Hidalgo, A.L., Olivo, R.C., Aguirre, C.J., 2015. Concentration ratio and frequency of application level ozone damage black Sigatoka in bananas. Design of a protocol irrigation with ozonated water. Alternativas 16 (2), 66–75.

Hijmans, R.J., Cameron, S.E., Parra, J.L., Jones, P.G., Jarvis, A., 2005. Very high resolution interpolated climate surfaces for global land areas. International Journal of Climatology 25 (15), 1965–1978.

Hubbard, K.G., Sivakumar, M.V.K., 2000. Automated weather stations for applications in agriculture and water resources management: current use and future perspectives. In: Proceeding of an International Workshop, Lincoln, Nebraska, USA.

Iglesias, A., Garrote, L., 2015. Adaptation strategies for agricultural water management under climate change in Europe. Agricultural Water Management 155, 113–124.

Iniesta, F., Testi, L., Orgaz, F., Villalobos, F.J., 2009. The effects of regulated and continuous deficit irrigation on the water use, growth and yield of olive trees. European Journal of Agronomy 30 (4), 258–265.

IPCC, 2013. Climate change 2013: the physical science basis. In: Stocker, T.F., Qin, D., Plattner, G.-K., Tignor, M., Allen, S.K., Boschung, J., Nauels, A., Xia, Y., Bex, V., Midgley, P.M. (Eds.), Contribution of Working Group I to the Fifth Assessment Report of the Intergovernmental Panel on Climate Change. Cambridge University Press, Cambridge, United Kingdom and New York, NY, USA, p. 1535.

Jackson, R.D., Idso, S.B., Reginato, R.J., Pinter, P.J., 1981. Canopy temperature as a crop water stress indicator. Water Resources Research 17, 1133–1138.

Jägermeyr, J., Gerten, D., Schaphoff, S., Heinke, J., Lucht, W., Rockström, J., 2016. Integrated crop water management might sustainably halve the global food gap. Environmental Research Letters 11 (2), 1–14.

Jifon, J.L., Syvertsen, J.P., 2003. Kaolin particle film applications can increase photosynthesis and water use efficiency of 'Ruby Red' grapefruit leaves. Journal of the American Society for Horticultural Science 128 (1), 107–112.

Kadiyala, M.D.M., Nedumaran, S., Singh, P., Chukka, S., Irshad, M.A., Bantilan, M.C.S., 2015. An integrated crop model and GIS decision support system for assisting agronomic decision making under climate change. The Science of the Total Environment 521, 123–134.

Kallenbach, C.M., Rolston, D.E., Horwath, W.R., 2010. Cover cropping affects soil N_2O and CO_2 emissions differently depending on type of irrigation. Agriculture, Ecosystems and Environment 137, 251–260.

Karam, F., Lahoud, R., Masaad, R., Kabalan, R., Breidi, J., Chalita, C., Rouphael, Y., 2007. Evapotranspiration, seed yield and water use efficiency of drip irrigated sunflower under full and deficit irrigation conditions. Agricultural Water Management 90 (3), 213–223.

Kennedy, T.L., Suddick, E.C., Six, J., 2013. Reduced nitrous oxide emissions and increased yields in California tomato cropping systems under drip irrigation and fertigation. Agriculture, Ecosystems and Environment 170, 16–27.

Kjellström, E., Boberg, F., Castro, M., Christensen, J.H., Nikulin, G., Sánchez, E., 2010. Daily and monthly temperature and precipitation statistics as performance indicators for regional climate models. Climate Research 44, 135–150.

Koubouris, G.C., Metzidakis, I.T., Vasilakakis, M.D., 2009. Impact of temperature on olive (*Olea europaea* L.) pollen performance in relation to relative humidity and genotype. Environmental and Experimental Botany 67 (1), 209–214.

Kustas, W.P., Norman, J.M., 1999. Evaluation of soil and vegetation heat flux predictions using a simple two-source model with radiometric temperature for partial canopy cover. Agricultural and Forest Meteorology 94, 13–29.

Le Mer, J., Roger, P., 2001. Production, oxidation, emission and consumption of methane by soils: a review. European Journal of Soil Biology 37, 25–50.

Lobell, D.B., Field, C.B., Cahill, K.N., Bonfils, C., 2006. Impacts of future climate change on California perennial crop yields: model projections with climate and crop uncertainties. Agricultural and Forest Meteorology 141 (2–4), 208–218.

López-Bernal, A., García-Tejera, O., Vega, V.A., Hidalgo, J.C., Testi, L., Orgaz, F., Villalobos, F.J., 2015. Using sap flow measurements to estimate net assimilation in olive trees under different irrigation regimes. Irrigation Science 33 (5), 357–366.

Lorite, I.J., Mateos, L., Fereres, E., 2004. Evaluating irrigation performance in a Mediterranean environment. I. Model and general assessment of an irrigation scheme. Irrigation Science 23 (2), 77–84.

Lorite, I.J., Mateos, L., Orgaz, F., Fereres, E., 2007. Assessing deficit irrigation strategies at the level of an irrigation district. Agricultural Water Management 91, 51–60.

Lorite, I.J., García-Vila, M., Carmona, M.A., Santos, C., Soriano, M.A., 2012. Assessment of the Irrigation Advisory Services' recommendations and farmers' irrigation management: a case study in Southern Spain. Water Resources Management 26, 2397–2419.

Lorite, I.J., Santos, C., García-Vila, M., Carmona, M.A., Fereres, E., 2013a. Assessing irrigation scheme water use and farmers' performance using wireless telemetry systems. Computers and Electronics in Agriculture 98, 193–204.

Lorite, I.J., García-Vila, M., Santos, C., Ruiz-Ramos, M., Fereres, E., 2013b. AquaData and AquaGIS: two computer utilities for temporal and spatial simulations of water-limited yield with AquaCrop. Computers and Electronics in Agriculture 96, 227–237.

Lorite, I.J., Ramírez-Cuesta, J.M., Cruz-Blanco, M., Santos, C., 2015. Using weather forecast data for irrigation scheduling under semi-arid conditions. Irrigation Science 33 (6), 411–427.

Lorite, I.J., Gabaldón-Leal, C., Belaj, A., de la Rosa, R., León, L., Ruiz-Ramos, M., Santos, C., 2017. Evaluation of the olive response and adaptation strategies to climate change under semi-arid conditions. Agricultural Water Management (Submitted for publication).

Maeda, E.E., Wiberg, D.A., Pellikka, P.K.E., 2011. Estimating reference evapotranspiration using remote sensing and empirical models in a region with limited ground data availability in Kenya. Applied Geography 31 (1), 251–258.

MAGRAMA, 2014a. Encuestas sobre superficies y rendimientos de cultivos (ESYRCE). Encuesta de marco de áreas de España. Ministerio de Agricultura, Alimentación y Medio Ambiente, Madrid.

MAGRAMA, 2014b. Encuesta sobre superficies y rendimientos de cultivos. Informe sobre regadíos en España. Ministerio de Agricultura, Alimentación y Medio Ambiente, Madrid.

Michailidis, A., Papadaki-Klavdianou, A., Apostolidou, I., Lorite, I.J., Pereira, F.A., Mirko, H., Buhagiar, J., Shilev, S., Michaelidis, E., Loizou, E., Chatzitheodoridis, F., Restoy, R.C., Lopez, A.L., 2015. Exploring treated wastewater issues related to agriculture in Europe, employing a quantitative SWOT analysis. Procedia Economics and Finance 33 (15), 367–375.

Miller, P.A., Barth, M.F., 2003. Ingest, integration, quality control, and distribution of observations from state transportation departments using MADIS. In: 19th International Conference on Interactive Information and Processing Systems for Meteorology, Oceanography and Hydrology (IIPS).

Minacapilli, M., Agnese, C., Blanda, F., Cammalleri, C., Ciraolo, G., D'Urso, G., Iovino, M., Pumo, D., Provenzano, G., Rayo, G., 2009. Estimation of actual evapotranspiration of Mediterranean perennial crops by means of remote-sensing based surface energy balance models. Hydrology and Earth System Sciences 13, 1061–1074.

Montoro, A., López-Fuster, P., Fereres, E., 2011. Improving on-farm water management through an irrigation scheduling service. Irrigation Science 29, 311–319.

Morales, A., Leffelaar, P.A., Testi, L., Orgaz, F., Villalobos, F.J., 2016. A dynamic model of potential growth of olive (*Olea europea* L.) orchards. European Journal of Agronomy 74, 93–102.

Moreno, M.A., Córcoles, J.I., Tarjuelo, J.M., Ortega, J.F., 2010. Energy efficiency of pressurized irrigation networks managed on-demand and under a rotation schedule. Biosystems Engineering 107 (4), 349–363.

Moriana, A., Villalobos, F.J., Fereres, E., 2002. Stomatal and photosynthetic responses of olive (*Olea europaea* L.) leaves to water deficits. Plant, Cell and Environment 25 (3), 395–405.

Moriana, A., Orgaz, F., Pastor, M., Fereres, E., 2003. Yield responses of a mature olive orchard to water deficits. Journal of the American Society for Horticultural Science 128 (3), 425–431.

Morison, J.I.L., 1998. Stomatal response to increase CO_2 concentration. Journal of Experimental Botany 49, 443–452.

Nicolás, E., Alarcón, J.J., Mounzer, O., Pedrero, F., Nortes, P.A., Alcobendas, R., Romero-Trigueros, C., Bayona, J.M., Maestre-Valero, J.F., 2016. Long-term physiological and agronomic responses of mandarin trees to irrigation with saline reclaimed water. Agricultural Water Management 166, 1–8.

Nikulin, G., Kjellström, E., Hansson, U., Jones, C., Strandberg, G., Ullerstig, A., 2011. Evaluation and future projections of temperature, precipitation and wind extremes over Europe in an ensemble of regional climate simulations. Tellus A 63, 41–55.

Norman, J.M., Kustas, W.P., Humes, K.S., 1995. Two source approach for estimating soil and vegetation energy fluxes from observations of directional radiometric surface temperature. Agricultural and Forest Meteorology 77, 263–293.

Olesen, J.E., Carter, T.R., Díaz-Ambrona, C.H., Fronzek, S., Heidmann, T., Hickler, T., Holt, T., Mínguez, M.I., Morales, P., Palutikof, J.P., Quemada, M., Ruiz-Ramos, M., Rubaek, G., Sau, F., Smith, B., Sykes, M.T., 2007. Uncertainties in projected impacts of climate change on European agriculture and terrestrial ecosystems based on scenarios from regional climate models. Climatic Change 81, 123–143.

Orgaz, F., Testi, L., Villalobos, F.J., Fereres, E., 2006. Water requirements of olive orchards—II: determination of crop coefficients for irrigation scheduling. Irrigation Science 24 (2), 77—84.

Orlandi, F., Garcia-Mozo, H., Dhiab, A.B., Galán, C., Msallem, M., Romano, B., Abichou, M., Dominguez-Vilches, E., Fornaciari, M., 2013. Climatic indices in the interpretation of the phenological phases of the olive in mediterranean areas during its biological cycle. Climatic Change 116 (2), 263—284.

Ortega, J.F., de Juan, J.A., Tarjuelo, J.M., 2005. Improving water management: the irrigation advisory service of Castilla-La Mancha (Spain). Agricultural Water Management 77, 37—58.

Osborne, T., Rose, G., Wheeler, T., 2013. Variation in the global-scale impacts of climate change on crop productivity due to climate model uncertainty and adaptation. Agricultural and Forest Meteorology 170, 183—194.

Palese, A.M., Vignozzi, N., Celano, G., Agnelli, A.E., Pagliai, M., Xiloyannis, C., 2014. Influence of soil management on soil physical characteristics and water storage in a mature rainfed olive orchard. Soil and Tillage Research 144, 96—109.

Parsons, L.R., Sheikh, B., Holden, R., York, D.W., 2010. Reclaimed water as an alternative water source for crop irrigation. Hortscience 45 (11), 1626—1629.

Pascual, M., Villar, J.M., Rufat, J., 2016. Water use efficiency in peach trees over a four-years experiment on the effects of irrigation and nitrogen application. Agricultural Water Management 164 (2), 253—266.

Peragón, J.M., Delgado, A., Díaz, J.A.R., Pérez-Latorre, F.J., 2016. A GIS-based decision tool for reducing salinization risks in olive orchards. Agricultural Water Management 166, 33—41.

Perera, K.C., Western, A.W., Nawarathna, B., George, B., 2014. Forecasting daily reference evapotranspiration for Australia using numerical weather prediction outputs. Agricultural and Forest Meteorology 194, 50—63.

Pérez-Parra, J., 2005. 'Depuración y reutilización de aguas residuales para riego', Curso superior de especialización, Mejora de la eficiencia en el uso del agua en cultivos hortícolas protegidos. Fundación para la Investigación Agrícola en la Provincia de Almería, pp. 447—469.

Persson, A., Grazzini, F., 2007. User guide to ECMWF forecast products. Meteorological Bulletin 161. M3.2 4 (March).

Petousi, I., Fountoulakis, M.S., Saru, M.L., Nikolaidis, N., Fletcher, L., Stentiford, E.I., Manios, T., 2015. Effects of reclaimed wastewater irrigation on olive (*Olea europea* L. cv 'Koroneiki') trees. Agricultural Water Management 160, 33—40.

Piani, C., Haerter, J.O., Coppola, E., 2010a. Statistical bias correction for daily precipitation in regional climate models over Europe. Theoretical and Applied Climatology 99, 187—192.

Piani, C., Weedon, G.P., Best, M., Gomes, S.M., Viterbo, P., Hagemann, S., Haerter, O., 2010b. Statistical bias correction of global simulated daily precipitation and temperature for the application of hydrological models. Journal of Hydrology 395, 199—215.

Pirttioja, N., Carter, T.R., Rötter, R.P., et al., 2015. FACCE MACSUR Reports. A Crop Model Ensemble Analysis of Temperature and Precipitation Effects on Wheat Yield across a European Transect Using Impact Response Surfaces, vol. 6. D-C4.4.3.

Potter, T.L., Bosch, D.D., Strickland, T.C., 2015. Tillage impact on herbicide loss by surface runoff and lateral subsurface flow. The Science of the Total Environment 530—531, 357—366.

Quemada, M., Baranski, M., Nobel de Lange, M., Vallejo, A., Cooper, J., 2013. Practices to reduce nitrate leaching and increase nitrogen use efficiency in irrigated agriculture. In: General Assembly Conference, Abstracts, Viena, p. 4472.

Raes, D., Steduto, P., Hsiao, T.C., Fereres, E., 2013. Aquacrop (Chapter 2, Users guide. Update).

Ramírez-Cuesta, J.M., Cruz-Blanco, M., Santos, C., Lorite, I.J., 2017. Assessing reference evapotranspiration at regional scale based on remote sensing, weather forecast and GIS tools. International Journal of Applied Earth Observation and Geoinformation 55, 32–42.

Rapoport, H.F., Hammami, S.B.M., Martins, P., Pérez-Priego, O., Orgaz, F., 2012. Influence of water deficits at different times during olive tree inflorescence and flower development. Environmental and Experimental Botany 77, 227–233.

Reich, P.B., Sendall, K.M., Stefanski, A., Wei, X., Rich, R.L., Montgomery, R.A., 2016. Boreal and temperate trees show strong acclimation of respiration to warming. Nature 531 (7596), 633–636.

Rinaldi, M., He, Z.L., 2014. Decision support systems to manage irrigation in agriculture. Advances in Agronomy 123, 229–279.

Roccuzzo, G., Villalobos, F.J., Testi, L., Fereres, E., 2014. Effects of water deficits on whole tree water use efficiency of orange. Agricultural Water Management 140, 61–68.

Rodríguez-Díaz, J.A., Weatherhead, E.K., Knox, J.W., Camacho, E., 2007. Climate change impacts on irrigation water requirements in the Guadalquivir river basin in Spain. Regional Environmental Change 7 (3), 149–159.

Romero, P., Botía, P., 2006. Daily and seasonal patterns of leaf water relations and gas exchange of regulated deficit-irrigated almond trees under semiarid conditions. Environmental and Experimental Botany 56 (2), 158–173.

Ruiz-Ramos, M., Mínguez, M.I., 2010. Evaluating uncertainty in climate change impacts on crop productivity in the Iberian Peninsula. Climate Research 44, 69–82.

Ruiz-Ramos, M., Ferrise, R., Rodríguez, A., Lorite, I.J., Bindi, M., Carter, T.C., Fronzek, S., Palosuo, T., Pirttioja, N., Baranowski, P., Buis, S., Cammarano, D., Chen, Y., Dumont, B., Ewert, F., Gaiser, T., Hlavinka, P., Hoffmann, H., Höhn, J.G., Jurecka, F., Kersebaum, K.C., Krzyszczak, J., Lana, M., Mechiche-Alami, A., Minet, J., Montesino, M., Nendel, C., Porter, J.R., Ruget, F., Semenov, M.A., Steinmetz, Z., Stratonovitch, P., Supit, I., Tao, F., Trnka, M., de Wit, A., Rötter, R.P., 2017. Adaptation response surfaces for managing wheat under perturbed climate and CO_2 in a Mediterranean environment. Agricultural Systems (in press).

Sadras, V.O., Whitfield, D.M., Connor, D.J., 1991. Transpiration efficiency in crops of semi-dwarf and standard-height sunflower. Irrigation Science 12, 87–91.

Sánchez-Martín, L., Meijide, A., García-Torres, L., Vallejo, A., 2010. Combination of drip irrigation and organic fertilizer for mitigating emissions of nitrogen oxides in semiarid climate. Agriculture, Ecosystems and Environment 137, 99–107.

Santos, C., Lorite, I.J., Tasumi, M., Allen, R.G., Fereres, E., 2008. Integrating satellite-based evapotranspiration with simulation models for irrigation management at the scheme level. Irrigation Science 26, 277–288.

Santos, C., Lorite, I.J., Allen, R.G., Tasumi, M., 2012. Aerodynamic parameterization of the satellite-based energy balance (METRIC) model for ET estimation in rainfed olive orchards of Andalusia, Spain. Water Resources Management 26 (11), 3267–3283.

Sanz-Cobena, A., Sánchez-Martín, L., García-Torres, L., Vallejo, A., 2012. Gaseous emissions of N_2O and NO and NO_3 leaching from urea applied with urease and nitrification inhibitors to a maize (*Zea mays*) crop. Agriculture, Ecosystems and Environment 149, 64–73.

Sanz-Cobena, A., Lassaletta, L., Estellés, F., Del Prado, A., Guardia, G., Abalos, D., Aguilera, E., Pardo, G., Vallejo, A., Sutton, M.A., 2014. Yield scaled mitigation of ammonia emission from N fertilization: the Spanish case. Environmental Research Letters 9 (12), 125005.

Sanz-Cobeña, A., Abalos, D., Meijide, A., Sanchez-Martin, L., Vallejo, A., 2016. Soil moisture determines the effectiveness of two urease inhibitors to decrease N_2O emission. Mitigation and Adaptation Strategies for Global Change 21, 1131–1144.

Saxe, H., Ellsworth, D.S., Heath, J., 1998. Tree and forest functioning in an enriched CO_2 atmosphere. In: Transley Review, 98, New Phytol, vol. 139, pp. 395–436.

Schellenberg, D.L., Alsina, M.M., Muhammad, S., Stockert, C.M., Wolff, M.W., Sanden, B.L., Brown, P.H., Smart, D.R., 2012. Yield-scaled global warming potential from N_2O emissions and CH_4 oxidation for almond (*Prunus dulcis*) irrigated with nitrogen fertilizers on arid land. Agriculture, Ecosystems and Environment 155, 7–15.

Segal, E., Dag, A., Ben-Gal, A., Zipori, I., Erel, R., Suryano, S., Yermiyahu, U., 2011. Olive orchard irrigation with reclaimed wastewater: agronomic and environmental considerations. Agriculture, Ecosystems and Environment 140 (3–4), 454–461.

Srivastava, P.K., Han, D., Rico Ramirez, M.A., Islam, T., 2013. Comparative assessment of evapotranspiration derived from NCEP and ECMWF global datasets through weather Research and Forecasting model. Atmospheric Science Letters 14 (2), 118–125.

Steduto, P., Hsiao, T.C., Fereres, E., Raes, D., 2012. Crop yield response to water. FAO Irrigation and Drainage Paper 66. FAO, Rome.

Steenwerth, K.L., Belina, K.M., 2010. Vineyard weed management practices influence nitrate leaching and nitrous oxide emissions. Agriculture, Ecosystems and Environment 138, 127–131.

Stevens, R.M., Ewenz, C.M., Grigson, G., Conner, S.M., 2012. Water use by an irrigated almond orchard. Irrigation Science 30 (3), 189–200.

Sun, Z., Gebremichael, M., Ardö, J., de Bruin, H.A.R., 2011. Mapping daily evapotranspiration and dryness index in the East African highlands using MODIS and SEVIRI data. Hydrology and Earth System Sciences 15 (1), 163–170.

Tanasijevic, L., Todorovic, M., Pereira, L.S., Pizzigalli, C., Lionello, P., 2014. Impacts of climate change on olive crop evapotranspiration and irrigation requirements in the Mediterranean region. Agricultural Water Management 144, 54–68.

Tognetti, R., Longobucco, A., Raschi, A., Miglietta, F., Fumagalli, I., 1999. Responses of two *Populus* clones to elevated atmospheric CO_2 concentration in the field. Annals of Forest Science 56, 493–500.

Tognetti, R., Sebastiani, L., Minnocci, A., Raschi, A., 2002. Foliar responses of olive trees (*Olea europaea* L.) under field exposure to elevated CO_2 concentration. In: Proceedings of the Fourth International Symposium on Olive Growing (Valenzano, Italy), vol. 1 and 2(586), pp. 449–452.

Tognetti, R., Raschi, A., Longobucco, A., Lanini, M., Bindi, M., 2005. Hydraulic properties and water relations of *Vitis vinifera* L. exposed to elevated CO_2 concentrations in a Free Air CO_2 Enrichment (FACE). Phyton-Annales Rei Botanicae 45 (3), 243–256.

Trenberth, K.E., 2005. The impact of climate change and variability on heavy precipitation, floods, and droughts. Part 17, Climate Change. In: Encyclopedia of Hydrological Sciences.

Trentacoste, E.R., Puertas, C.M., Sadras, V.O., 2015. Effect of irrigation and tree density on vegetative growth, oil yield and water use efficiency in young olive orchard under arid conditions in Mendoza, Argentina. Irrigation Science 33 (6), 429–440.

Tuba, Z., Lichtenthaler, H.K., 2007. Long term acclimation of plants to elevated CO_2 and its interaction with stresses. Annals of the New York Academy of Sciences 1113 (1), 135–146.

Vallejo, A., Meijide, A., Boeckx, P., Arce, A., García-Torres, L., Aguado, P.L., Sanchez-Martín, L., 2014. Nitrous oxide and methane emissions from a surface drip-irrigated system combined with fertilizer management. European Journal of Soil Science 65, 386–395.

Valverde, P., de Carvalho, M., Serralheiro, R., Rodrigo, M., Ramos, V., Oliveira, B., 2015. Climate change impacts on rainfed agriculture in the Guadiana river basin (Portugal). Agricultural Water Management 150, 35–45.

Vargas, F.J., Romero, M.A., Clave, J., Batlle, I., Miarnau, X., Alegre, S., 2011. Important traits in IRTA's new almond cultivars. Acta Horticulturae 912, 359–365.

Veroz, O., Gil, J., Ordóñez, R., Carbonell, R., Márquez, F., González, E., Holgado, A., Lorite, I., Santos, C., Robles del Salto, J.F., Gómez, M., Triviño, P., 2016. Proyecto Life+ Climagri, integración de estrategias de mitigación y adaptación de cambio climático. Vida Rural 423, 38–44.

Villalobos, F.J., Fereres, E., 2004. Climate change effects on crop water requirements in Southern Spain. II Contrasting meteorological and agronomic viewpoints. In: Jacobsen, S.E., Jensen, C.R., Porter, J.R. (Eds.), Proceedings of VIII Congress of the European Society of Agronomy KVL, Copenhagen, pp. 349–350.

Villalobos, F.J., Perez-Priego, O., Testi, L., Morales, A., Orgaz, F., 2012. Effects of water supply on carbon and water exchange of olive trees. European Journal of Agronomy 40, 1–7.

Viola, F., Caracciolo, D., Pumo, D., Noto, L.V., Loggia, G.L., 2014. Future climate forcings and olive yield in a mediterranean orchard. Water (Switzerland) 6 (6), 1562–1580.

Vlek, P.L.G., Fillery, I.R.P., Burford, J.R., 1981. Accession, transformation, and loss of nitrogen in soils of the arid region. Plant and Soil 58, 133–175.

Wagner, W., Brocca, L., Naeimi, V., Reichle, R., Draper, C., De Jeu, R., Ryu, D., Su, C.H., Western, A., Calvet, J.C., Kerr, Y.H., Leroux, D.J., Drusch, M., Jackson, T.J., Hahn, S., Dorigo, W., Paulik, C., 2014. Clarifications on the 'comparison between SMOS, VUA, ASCAT, and ECMWF soil moisture products over four watersheds in U.S.'. IEEE Transactions on Geoscience and Remote Sensing 52 (3), 1901–1906.

Wilby, R.L., Wigley, T.M.L., Conway, D., Jones, P.D., Hewitson, B.C., Main, J., Wilks, D.S., 1998. Statistical downscaling of general circulation model output: a comparison of methods. Water Resources Research 34, 2995–3008.

CHAPTER 2

Water—Energy Nexus in Irrigated Areas. Lessons From Real Case Studies

Irene Fernández García, Aida Mérida García, Juan A. Rodríguez Díaz, Pilar M. Barrios, Emilio C. Poyato
University of Cordoba, Cordoba, Spain

Contents

1. INTRODUCTION

The world population has grown considerably during the last few decades and it is expected that this trend will continue in the foreseeable future. Thus the world population in 2050 will rise to 9100 million people. To satisfy the nutritional requirements of this increased population, it will be necessary also to raise food production by 70% (FAO, 2009).

Agriculture and more precisely irrigated agriculture play a major role in ensuring that food demands are met, since yields obtained by irrigated crops are much higher than rain-fed ones. The current world irrigated area

Water Scarcity and Sustainable Agriculture in Semiarid Environment
ISBN 978-0-12-813164-0
https://doi.org/10.1016/B978-0-12-813164-0.00002-8

is around 300 million ha, more than twice the irrigated area in 1960. However, water resources are scarce in some areas and the possibilities of increasing the irrigated area further are limited (Alexandratos and Bruinsma, 2012). Hence the adoption of more efficient water distribution networks and on-farm irrigation systems is required to cope with the higher water demand.

2. WATER–ENERGY NEXUS IN IRRIGATED AREAS

Irrigated agriculture is the main user of fresh water, accounting for 70% of the global freshwater withdrawals (Siebert et al., 2010). In Europe, on average, agriculture uses 33% of total water abstraction, although this figure can reach 80% in Southern Europe (European Environment Agency, 2012), where due to favorable climatic conditions irrigated agriculture plays a major role.

Considering the necessity of improving water use efficiency, measures focused on reducing water use in agriculture have been developed. As an example, in Spain, modernization projects of hydraulic infrastructures moving from open channel systems to pressurized networks and adopting more efficient on-farm irrigation systems have been promoted (MAPA, 2002). Thus around 1 million ha of the 3.5 million ha of irrigated areas have been modernized (MARM, 2010). Thanks to the modernization process, surface irrigation has decreased from 42% to 30%, whereas drip irrigation has increased from 30% to 47% from 2002 to 2011. As a result of this transformation, water use for irrigation per unit of irrigated area has been reduced by 21% from 1950 to 2007 (Corominas, 2010). However, the higher water use efficiency achieved by these projects has been linked to an increment of 657% in energy requirements (Fig. 2.1). Most of the pressurized irrigation networks require pumps to extract, in the case of using groundwater resources, or distribute water on fields and these operations are the main energy consumers in these networks.

In addition to the higher energy demand, currently, farmers have to absorb amortization, operation, and maintenance costs of the new irrigation infrastructures (Rodríguez Díaz et al., 2009) and, along with high energy prices, this has led farmers to question the feasibility of irrigation because electricity tariffs have increased considerably in the last few years. This is not a problem peculiar to Spain; it has also happened in other countries where modernization processes were carried out (Brazilian et al., 2011). In Spain, the price of energy for irrigation has increased because of both the

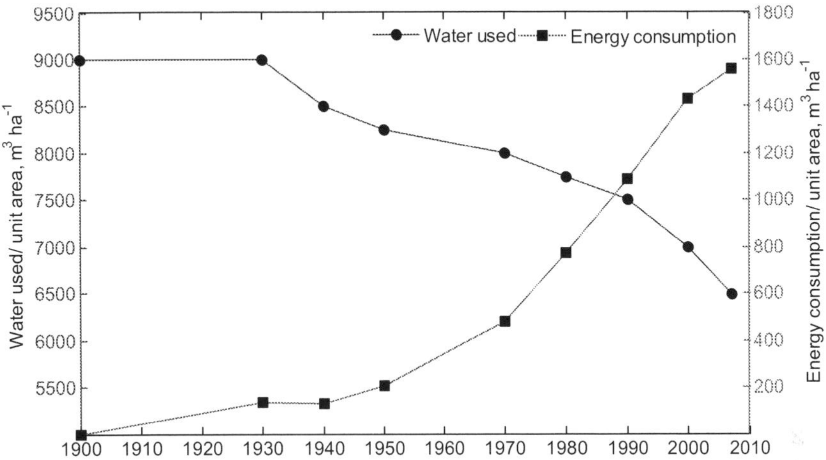

Figure 2.1 Evolution of water used and energy consumption in the irrigation sector. *(Author's elaboration based on Corominas, J., 2010. Agua y Energía en el riego en la época de la sostenibilidad. Ingeniería del Agua 17.)*

liberalization of the electricity market in 2003 and the elimination of special irrigation rates in 2008. Hence costs associated with power and energy have risen significantly.

Traditionally, the impacts of modernization processes were evaluated according to water savings; however, more recent studies have highlighted the necessity of analyzing the water–energy nexus as a whole. The energy dependence of the new pressurized irrigation networks together with the higher fuel and electric energy prices results in an increase of operating costs. In addition, CO_2 emissions by the agricultural sector have also increased, accentuating its contribution to global warming and representing nearly one-fifth of total world CO_2 emissions (Tubiello et al., 2014). In subsequent years, contamination and climate change will have negative impacts on agriculture, which is particularly affected by unpredictable precipitations, extended periods of drought, and unexpected floods (Nelson et al., 2009).

To evaluate the effects of modernization processes, several works focused on the impacts of water and energy use and productivity have been developed. Lecina et al. (2010) evaluated the consequences of modernization processes on water quantity and quality in the Ebro River Basin (northeast Spain), determining that farmers selected more productive crops and hence more intensive agriculture with higher crop water requirements. Likewise, Rodríguez Díaz et al. (2011) evaluated the effects of modernization processes in 10 irrigation districts from southern Spain using indicators related to water, energy, and productivity. They confirmed the high energy requirements of the new pressurized networks. Another analysis concerning the consequences of updating to pressurized irrigation systems was carried out by Fernández García et al. (2014b). They evaluated five irrigation districts from southern Spain considering two periods: one period before modernization and the other after changing to pressurized networks. As a result, they detected an average reduction in irrigation water use of 23%, but water costs were increased by 52%. Similar to the conclusion drawn by Lecina et al. (2010), they also determined that farmers were migrating to more profitable crops, with higher water requirements, to cope with the increased costs.

Therefore the transformation of hydraulic infrastructures to pressurized networks and the adoption of more efficient on-farm irrigation systems have involved a significant reduction in water use. However, energy requirements, another scarce resource, have increased dramatically leading to the necessity to raise strategies focused on optimizing both water and energy resources.

3. ENERGY-SAVING MEASURES FOR IRRIGATION WATER SUPPLY NETWORKS

Concerns about optimizing water and energy use in irrigated areas have led to the development of different strategies aimed at reducing the use of both resources. Usually, these strategies focus on improvements in irrigation scheduling, but also in the optimization of the energy requirements in water distribution networks. These strategies can be sorted into: (1) measures related to energy audits and benchmarking analysis, (2) measures focused on water distribution network management, (3) strategies to optimize pumping station performance, (4) strategies aimed at minimizing the operational cost, (5) actions focused on water distribution design, and (6) smart irrigation systems.

3.1 Energy Audits and Benchmarking Analysis

Energy audits are diagnostic tools that consist in identifying the weak points of water distribution networks by determining a set of indicators, proposing measures to improve the performance of the network, and carrying out an economic evaluation of the proposed actions to prioritize measures to be adopted taking into account the obtained energy savings (IDAE, 2008).

Abadia et al. (2010) developed a methodology for energy audits in irrigated areas. By applying these guidelines to 22 irrigation networks, they determined average energy savings of 14% and economic savings of 21% after adopting measures to improve the performance of these networks. Rodríguez-Díaz et al. (2008) carried out a benchmarking analysis in nine irrigation networks in southern Spain. They applied cluster analysis to classify the irrigation networks into statistically homogeneous groups and ranked the irrigation districts according to a quality index. By this procedure they compared each irrigation network with another with similar characteristics. They highlighted that water use efficiency was higher in those districts where users were charged per unit of irrigation water consumed. A few years later, Fernandez Garcia et al. (2014b) analyzed the same districts using performance indicators, pointing out that energy represents between 40% and 50% of the total water costs. The high energy costs had a direct impact on the irrigation depth, which tends to be significantly smaller than the theoretical irrigation requirements.

3.2 Optimum Management of Pressurized Irrigation Networks

Different strategies focused on the improvement of network management have been proposed. Network sectoring (with one or several supply points) by grouping hydrants into sectors and organizing farmers in irrigation turns according to their power requirements has shown significant energy savings in the irrigation networks studied, with energy savings of around 30% (Jiménez-Bello et al., 2010). In some cases, sectors were already established in the irrigation districts, while in others, methodologies that grouped hydrants according to topological coordinates related to the elevation and the distance of each hydrant with respect to the pumping station were proposed (Carrillo Cobo et al., 2011).

Other measures to reduce the energy consumption in water distribution networks consist in controlling critical points. Critical points are hydrants with high energy requirements due to their elevation or distance with respect to the pumping station. Different methodologies to detect and control critical hydrants were developed (Khadra and Lamaddalena, 2010; Rodríguez-Díaz et al., 2012). The procedure proposed by Rodríguez-Díaz et al. (2012) was based on the calculation of dimensionless coordinates useful to characterize the network's topology and hydraulic design. This method carried out thousands of simulations of the irrigation network considering different water demand patterns or loading conditions. In each

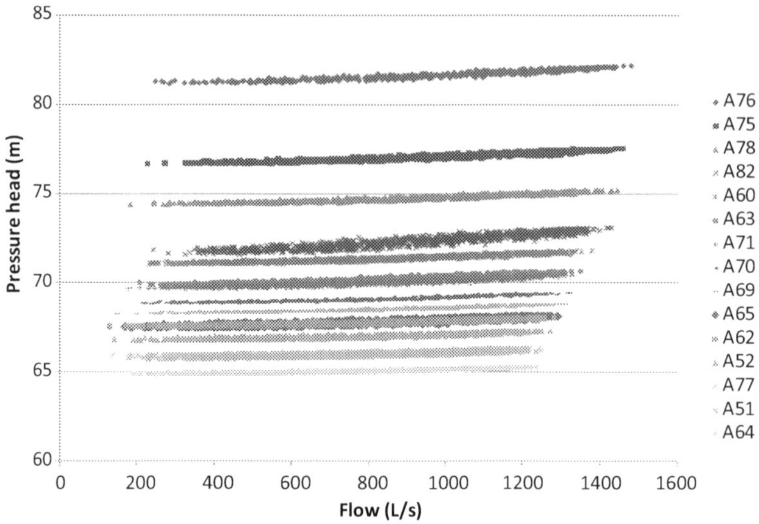

Figure 2.2 Effect of sequentially removing the critical hydrants on the pressure head requirements at the pumping station in an irrigation network. *(From Rodríguez Díaz, J.A., López Luque, R., Carrillo Cobo, M.T., Montesinos, P., Camacho Poyato, E., 2009. Exploring energy saving scenarios for on-demand pressurised irrigation networks. Biosystems Engineering 104, 552–561.)*

simulation, a critical hydrant and its power requirements were determined and its coordinates were estimated. Thanks to this, it is possible to identify which actions should be assumed to improve the performance of the critical hydrants.

Fig. 2.2 shows the effects of sequentially removing critical hydrants on the pressure head in an irrigation network. The pumping station has to provide 10 m additional to satisfy the pressure requirements of only three hydrants. The performance of these hydrants can be improved by installing booster pumps or replacing the undersized pipes with others of greater diameter. Assuming these actions, Rodríguez-Díaz et al. (2012) estimated energy savings of between 10% and 31%.

3.3 Optimum Pumping Station Performance

Pumping for distributing water and extracting groundwater is the main reason for energy consumption in pressurized irrigation networks. The performance of the pumping station in an irrigation network should be adapted to its operation to maximize efficiency according to the flow and pressure requirements. Moreno et al. (2009) developed a methodology to obtain the optimal characteristic and efficiency curves at a pumping station.

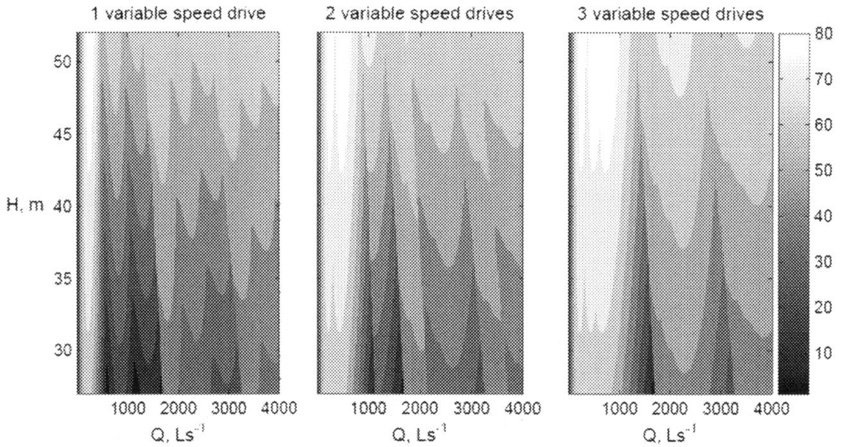

Figure 2.3 Pumping efficiency (%) for 1, 2, and 3 variable speed drives. *(From Fernández García, I., Moreno, M.A., Rodríguez Díaz, J.A., 2014a. Optimum pumping station management for irrigation networks sectoring: case of Bembezar MI (Spain). Agricultural Water Management 144, 150–158.)*

Lamaddalena and Khila (2013) evaluated the energy savings at a pumping station by including variable speed drives. They determined energy savings of 29% and 48% in two Italian irrigation networks. The adaptation of pump performance to network management by sectors was analyzed by Fernández García et al. (2014a). They developed a methodology to optimize the sectoring operation and pressure head, considering the pumping efficiency in Bembézar MI irrigation district (Spain). They also modeled and evaluated the benefits of installing variable speed drives. Fig. 2.3 shows the effects of including from one to three variable speed drives in the pumping station. Pumps without variable speed drives provided high efficiencies for high flow demands. However, for low and medium discharges, efficiencies were frequently low. By including variable speed drives, high efficiencies were also achieved when the flow demand was medium and low.

The methodology developed by Fernández García et al. (2014a), combining sectoring and variable speed drives, led to energy savings of up to 26% in the irrigation network studied.

3.4 Minimization of the Operational Cost

In the foregoing sections, measures focused on the optimization of energy consumption have been raised. However, along with energy consumption, the price of this resource has risen sharply. Hence methodologies to reduce the energy cost at pumping stations have been developed. Córcoles et al. (2015)

proposed a model to optimize the start irrigation time of each hydrant considering scenarios with different daily operation times, trying to avoid the peak periods. They estimated energy cost savings of 36% when the daily operation time was 8 h. Likewise, Fernández García et al. (2016b) focused on the optimization of the operational cost of a pressurized irrigation network considering different management scenarios (sectoring and critical points control) and including the electricity tariff. Similar to Córcoles et al. (2015), they concluded that by avoiding irrigation during peak hours a significant reduction in energy cost could be obtained. They also highlighted that network operation by controlling critical points led to lower operational costs than network management by sectors, despite this latter option being less energy demanding.

3.5 Optimal Design of Pressurized Irrigation Networks

The aforementioned methodologies to reduce the energy consumption in pressurized irrigation networks have been analyzed in networks initially designed for 100% simultaneity in most cases. However, high operational costs lead to the application of reduced water quantities, thus promoting deficit irrigation techniques. Hence the operation of these networks is different from their initial design conditions and they are not optimized under the new operation scenario.

Optimization of the design of irrigation networks has also been considered. Farmani et al. (2007) carried out the optimization of the design of pressurized irrigation systems comparing two scenarios: one scenario in which the on-demand operation is considered, i.e., farmers can irrigate whenever they want, and another scenario in which farmers are organized in irrigation turns. They estimated a reduction of 50% in the design cost when different irrigation turns were considered. González-Cebollada et al. (2011) proposed a new method, recursive design, to determine the minimum redesign cost of pressurized irrigation networks using less simulation time than other procedures.

Nevertheless, these methodologies minimized the design cost of pressurized irrigation networks without considering the reduction of the operational cost. Related to this aspect, Fernández García et al. (2016a) raised a multiobjective approach to minimize both the rehabilitation cost of the network and the annual operational cost considering the on-demand operation. They carried out a cost analysis and selected the best performing solution (with both minimum redesign and operational cost) considering different useful life and recovery factors.

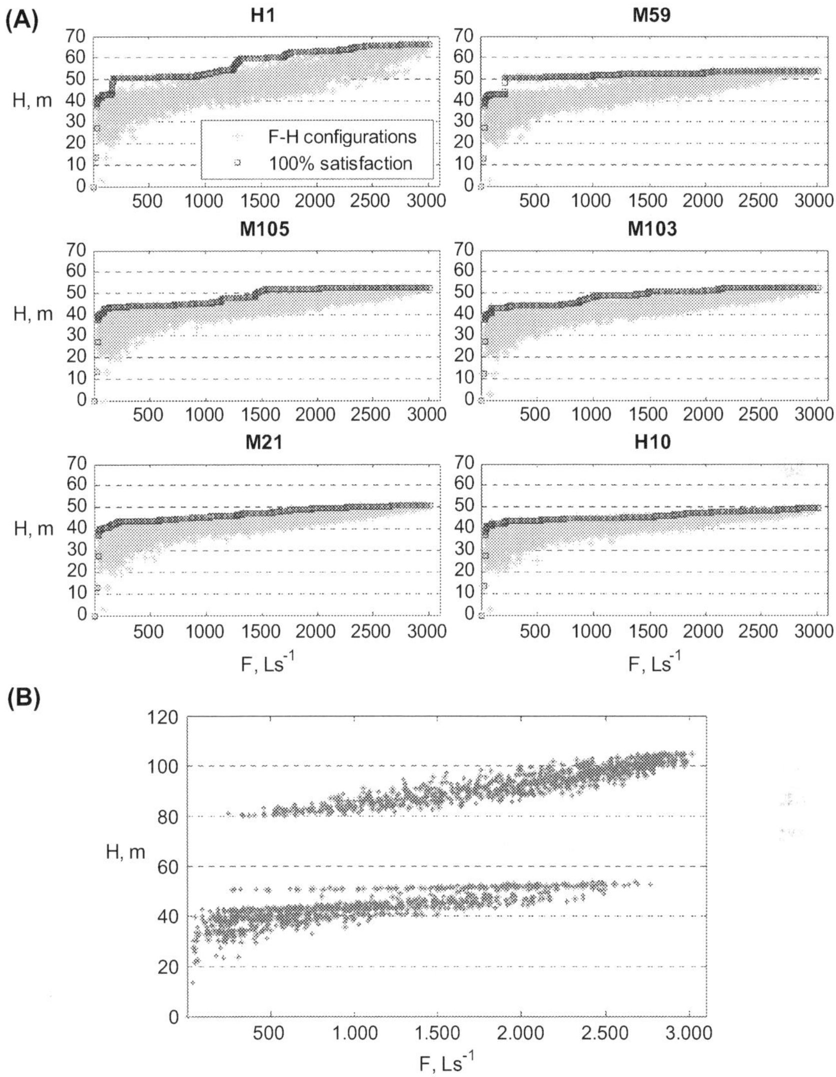

Figure 2.4 Flow (F)–head (H) configurations in an irrigation network after proposing optimal redesigns (H1, m59, m105, m103, m21, and H10) (A) and flow-head configuration in the original design (B). *(From Fernández García, I., Creaco, E., Rodríguez Díaz, J.A., Montesinos, P., Camacho Poyato, E., Savic, D., 2016a. Rehabilitating pressurized irrigation networks for an increased energy efficiency. Agricultural Water Management 164, 212–222.)*

Fig. 2.4 shows the reduction of the required pressure head in a pumping station for different proposed redesigns of the network compared to the required head according to the original design.

3.6 Smart Irrigation Systems

The incorporation of information and communication technologies (ICTs) in agriculture involves a step forward toward a more efficient use of water and energy resources. Progress in the development of different sensors to collect crop, soil, and meteorological information, among others, and big data techniques are encouraging the adoption of precision agriculture. Precision agriculture, and more precisely precision irrigation, consists in applying the right amount of water at the right time and at the right place (Camp et al., 2006). Thus precision irrigation allows the reduction of the irrigation depth and therefore the total water volume that has to be pumped.

Different precision irrigation case studies have been raised. Tsakmakis et al. (2017) applied a methodology focused on precision irrigation, combining meteorological, reference evapotranspiration, and crop growth models to cotton cultivation in northern Greece. They stablished the irrigation scheduling considering short-term meteorological predictions, soil moisture content, and crop water requirements according to its growing stage. By this methodology, water savings of 32% were estimated without losing yield. It is worth noting that reduction in the water applied entails lower energy consumption in most cases. Another example of precision irrigation can be found in García Morillo et al. (2015). They designed a precision irrigation system to be applied to strawberry farms located in southern Spain, combining soil water sensors, smart water meters, programmers, electrovalves, and weather stations along with a more efficient on-farm irrigation system. By their methodology, water use efficiency and energy efficiency were improved. The incorporation of manned and unmanned vehicles (unmanned aerial vehicles and remotely piloted aircraft systems) into agriculture enables information to be obtained about crop water stress, thus determining the right moment of water supply and, along with other sensors, the right amount of water to be applied. In a 70 ha commercial citrus orchard farm in southern Spain, water savings of 25% were estimated by applying this latter methodology, considering an irrigation deficit strategy, whereas savings in water and power costs were higher than 44 € ha^{-1} (Joint Research Centre (JRC) of the European Commission et al., 2014).

4. THE ROLE OF RENEWABLE ENERGY SOURCES IN THE IRRIGATION SECTOR

4.1 Solar Energy for the Irrigation Supply

Renewable energies can provide a sustainable alternative as an energy supply source for the agricultural sector. In particular, in areas with high annual irradiance levels, solar energy has a great potential, as in the Mediterranean region. Different authors have worked on the incorporation of solar energy in the irrigation sector. Hamidat et al. (2003) satisfied the daily crop irrigation needs of small-scale plots (2 ha) with solar energy in the Sahara region. Other researchers have focused on the integration of hybrid energy supply systems (solar and wind energy), concluding a better adaptation with the combination of a winter and a summer crop (Vick, 2010). Carrillo Cobo et al. (2014) proposed a methodology to combine network sectoring and solar energy as a partial energy source, estimating energy cost savings of up to 72%. However, with this methodology, the system continues to be power grid dependent.

In contrast to other renewable energy sources such as windmills, photovoltaic systems provide energy just at the moment the crop needs water (irradiance and evapotranspiration curves are generally parallels). However, irradiation availability is not constant and varies throughout the year and even during the day, which represents a challenge to irrigation management. This intermittent nature makes it necessary to use storage elements to ensure continuous energy availability (Baños et al., 2011). Yahyaoui et al. (2016) developed a model for the management of an off-grid photovoltaic water pumping installation by integrating fuzzy logic into the control system. Considering water needs and energy availability, the system decides the on/off status of the different elements (panels, batteries, and pumps) using the photovoltaic energy and batteries to pump water to a reservoir, finally applying it by gravity to the crop. In the case of small installations, batteries can provide an alternative support. However, for high power requirement facilities, they are not economically viable and photovoltaic systems have to be designed without batteries. This new option, which does not include batteries, is much more complex to manage because it requires synchronization of power production and irrigation in real time. Therefore due to the variability of solar power generation, new and innovative ways of irrigation management are required.

4.2 Example of Solar Irrigation: Case Study for Olive Orchards in the Mediterranean Region

Olives are one of the main crops in the Mediterranean region. Traditionally, the olive tree has been a rain-fed crop. Nevertheless, its intensification and the improvement of its productivity as a result of a complementary water supply (Pastor et al., 1999) have promoted the implementation of irrigation in many agricultural holdings. Furthermore, the rusticity of the crop allows deficit irrigation techniques to be applied, achieving high productivity with irrigation doses that do not entirely satisfy the crop water requirements.

Deficit irrigation consists in exposing the crop to a certain level of water stress without significant yield reduction (Kirda, 2002). Deficit irrigation can be applied in two different ways. The first, consists in the application of a constant reduction of the irrigation volume needed to satisfy the potential crop water requirements. The second, known as regulated deficit irrigation, involves the application of reduced water volumes that vary according to crop sensitivity to water stress in every cycle stage. This strategy is commonly used in permanent crops such as fruit trees, reducing their water consumption during the entire irrigation season. Thus greater reductions are applied in months in which the lack of water has less impact on productivity, while lower reductions are considered in the most critical periods in which the crop is more sensitive to the lack of water (Pérez-Rodríguez and Parras-Cintero, 2014). In the case of olive orchards cultivated in the Guadalquivir River Basin (southern Spain), July and August are considered the least critical months (Melgar et al., 2008; Cabeza et al., 2009), whereas flowering, fructification, and oil accumulation are the most critical phases (Orgaz et al., 2005).

The olive tree irrigation season in the Mediterranean region usually elapses from April to September (months with a significant deficit between crop water demand and precipitation), which coincides with the peak photovoltaic energy production months (Fig. 2.5). This fact facilitates the implementation of solar energy as an alternative energy supply source for off-grid irrigation systems.

A smart solar irrigation system for olive orchards was developed and tested in an experimental field located at the University of Córdoba (southern Spain). There, water is stored in a reservoir and conveyed to the on-farm irrigation network with a 13 kW pump, supplied with a solar energy production system of 15.4 kW. Water is applied to the crop by

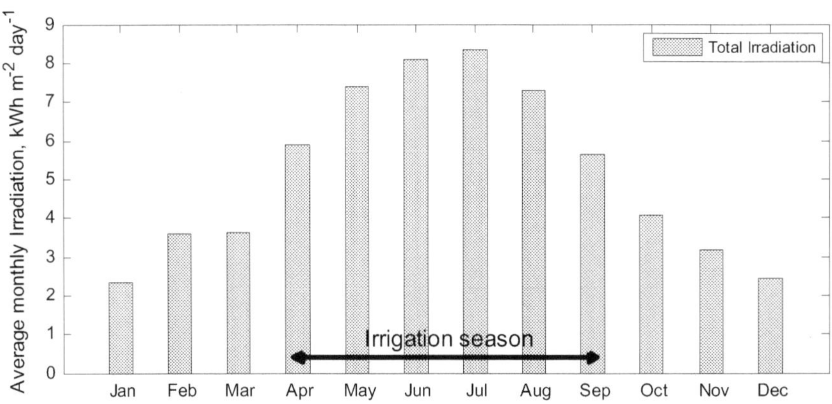

Figure 2.5 Average monthly irradiation in Córdoba, Spain, during 2013 and the irrigation season for olive trees in the Mediterranean region.

pressure-compensating emitters. The on-farm irrigation network is organized into three sectors, each one activated by an electrovalve (Fig. 2.6).

The system carried out a decision-making process in real time for the activation and deactivation of the three electrovalves, taking into account the instant solar power production of the photovoltaic installation. To do this, in an initial phase the irrigation network was modeled on the hydraulic simulator EPANET (Rossman, 2000), and the power requirements that ensured that all emitters received the required pressure according to each sector were estimated. Then, a previous irrigation scheduling was proposed, using a long-term climate dataset.

Priority rules for the activation of the electrovalves were established based on the power thresholds of each sector. According to the irradiance and power availability, the model prioritized irrigation of the most power-demanding sectors, those with higher thresholds. When the irrigation depth of the most demanding sectors was applied, sectors with fewer power requirements continued with their irrigation scheduling. This process was performed until irrigation in all sectors was completed or until the irradiation level was not enough to provide the minimum power required by any sector. At the end of each day, a soil–water balance was stablished considering the real values of effective precipitation, irrigation, and crop evapotranspiration. Then, the irrigation depth for the following day was calculated.

On cloudy days, energy availability is smaller, but also crops require less water. However, it is possible that the system cannot provide all the

Figure 2.6 Smart solar irrigation system located at the University of Córdoba, Spain. Photovoltaic installation (*upper left*); experimental plot (*upper right*); decision support system (*bottom left*); olive orchard plot (*bottom right*).

irrigation depth. In these cases, the system records these water deficits and applies them to the irrigation depths of the following day.

The model was tested in the experimental plot during the 2012/2013 irrigation season (168 days, between April 16 and September 30).

The power thresholds of the three sectors were 5.4, 11.3, and 8.7 kW for S1, S2, and S3, respectively. Thus S1 was the less demanding sector, while S2 was the most restrictive sector. On sunny days, the photovoltaic power production curve fitted to a normal distribution, whereas on partially sunny and cloudy days, the photovoltaic power production experienced large fluctuations. Fig. 2.7 shows the activation sequence of the three sectors making up the irrigation network on June 30 of the 2012/2013 irrigation season. It was a sunny day, so S1 started to irrigate at 09:40 a.m. and worked for 55 min. At 10:40 a.m. the system switched to S3, which irrigated for 40 min, and then it was the turn of S2, which irrigated for 1 h and 5 min. After the most demanding sector finished the application of the full water depth, S3 was activated again and operated for 1 h and 35 min to

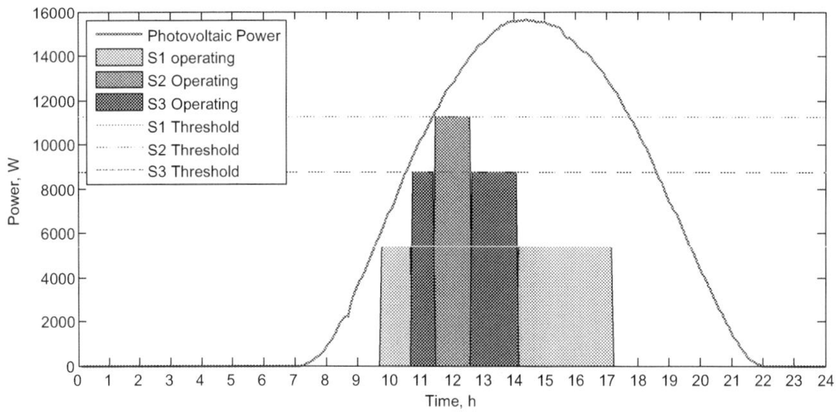

Figure 2.7 Synchronization of the sector irrigation with the photovoltaic power production on June 30 in the experimental plot at the University of Cordoba, Spain.

satisfy the required irrigation time. Finally, S1 was restarted for 3 more hours to fulfill its irrigation requirements.

Fig. 2.8 shows irrigation synchronization on a partly sunny and partly cloudy day (May 21, 2013), in which there were numerous fluctuations in photovoltaic power production. Thus the irrigation sequence adapted to the unsteady energy availability and all sectors fulfilled the irrigation requirements.

The system provided enough energy to satisfy the forecasted irrigation requirements of the three sectors in 92% of the days of the irrigation season.

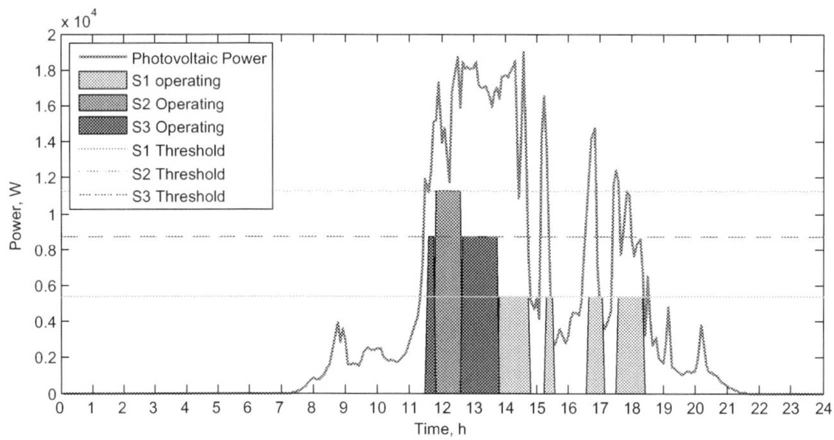

Figure 2.8 Synchronization of the sector irrigation with the photovoltaic power production on May 21 in the experimental plot at the University of Cordoba, Spain.

On the rest of the days the irradiation level was insufficient and some sectors did not have time enough to apply their full water allocation. These deficits of water were then applied in the following days.

5. CONCLUSIONS

Agriculture is facing a major challenge to ensure the food demands of a growing population, making the sustainable use of water and energy resources essential. Irrigated agriculture plays and will play an important role in this challenge and will have to adapt to the new demands.

Measures focused on improving water and energy use efficiency, such as energy audits, benchmarking, network sectoring, critical points control, network redesign, among others, have been widely studied, showing significant savings. However, the incorporation of renewable energy sources in the irrigation sector along with ICTs involves a step forward toward a more sustainable use of these resources.

The use of solar energy requires a change in traditional management because, given that power fluctuates during the day and months, water applications must match the power availability and irrigation scheduling must be adapted in real time. An example of smart solar irrigation in olive orchards was presented, which showed how solar energy can represent a real and viable alternative. The integration of photovoltaic energy as an energy source in irrigation systems for olive orchards is a real, viable, and sustainable alternative in areas with appropriate irradiation levels, providing energy independence and reducing greenhouse gas emissions.

ACKNOWLEDGMENTS

This research is part of the TEMAER project (AGL2014-59747-C2-2-R), funded by the Spanish Ministry of Economy and Competitiveness.

REFERENCES

Abadia, R., Rocamora, M.C., Corcoles, J.I., Ruiz-Canales, A., Martinez-Romero, A., Moreno, M.A., 2010. Comparative analysis of energy efficiency in water users associations. (Estudio comparativo de eficiencia energética en comunidades de regantes). Spanish Journal of Agricultural Research 8 (Suppl. 2), S134–S142.

Alexandratos, N., Bruinsma, J., 2012. World Agriculture Towards 2030/2050 (No. 13–3).

Baños, R., Manzano-Agugliaro, F., Montoya, F.G., Gil, C., Alcayde, A., Gómez, J., 2011. Optimization methods applied to renewable and sustainable energy: a review. Renewable and Sustainable Energy Reviews 15, 1753–1766.

Brazilian, M., Rogner, H., Howells, M., Hermann, S., Arent, D., Gielen, D., Steduto, P., Mueller, A., Komor, P., Tol, R., Yumkella, K., 2011. Considering the energy, water and food nexus: towards an integrated modeling approach. Energy Policy 39, 7896–7906.

Cabeza, C., Jiménez, M.J., Gómez, J.A., Landa, B., Navas, J.A., Orgaz, F., Testi, L., Cornejo, J., Hermosín, M.C., Benítez, E., Campos, M., Cotes, B., Argüeles, A., Cifuentes, V., Beaufoy, G., Duarte, J., Farfán, M.A., Garrido, J.M., Arriaza, M., Bar, A., 2009. Olivar y manejo del agua. In: Olivar Sostenible. Prácticas Para Una Producción Sostenible de Olivar En Andalucía, pp. 2–5.

Camp Jr., C.R., Sadler, E.J., Evans, R.G., 2006. Precision water management: current realities, possibilities and trends. In: Srinivasan, A. (Ed.), Handbook of Precision Agriculture. Haworth Press, Binghamton, New York, pp. 153–183.

Carrillo Cobo, M.T., Rodríguez Díaz, J.A., Montesinos, P., López Luque, R., Camacho Poyato, E., 2011. Low energy consumption seasonal calendar for sectoring operation in pressurized irrigation networks. Irrigation Science 29, 157–169.

Carrillo Cobo, M.T., Camacho Poyato, E., Montesinos, P., Rodríguez-Díaz, J.A., 2014. New model for sustainable management of pressurized irrigation networks. Application to Bembézar MD irrigation district (Spain). Science of the Total Environment 473–474, 1–8.

Córcoles, J.I., Tarjuelo, J.M., Carrión, P.A., Moreno, M.A., 2015. Methodology to minimize energy costs in an on-demand irrigation network based on arranged opening of hydrants. Water Resources Management 29, 3697–3710.

Corominas, J., 2010. Agua y Energía en el riego en la época de la sostenibilidad. Ingeniería del Agua 17.

European Environment Agency, 2012. Towards Efficient Use of Water Resources in Europe. https://doi.org/10.2800/95096.

FAO, 2009. High Level Expert Forum- How to Feed the World in 2050. Rome, Italy.

Farmani, R., Abadia, R., Savic, D., 2007. Optimum design and management of pressurized branched irrigation networks. Journal of Irrigation and Drainage Engineering 133, 528–537.

Fernández García, I., Moreno, M.A., Rodríguez Díaz, J.A., 2014a. Optimum pumping station management for irrigation networks sectoring: case of Bembezar MI (Spain). Agricultural Water Management 144, 150–158.

Fernández García, I., Rodríguez Díaz, J.A., Camacho Poyato, E., Montesinos, P., Berbel, J., 2014b. Effects of modernization and medium term perspectives on water and energy use in irrigation districts. Agricultural Systems 131, 56–63.

Fernández García, I., Creaco, E., Rodríguez Díaz, J.A., Montesinos, P., Camacho Poyato, E., Savic, D., 2016a. Rehabilitating pressurized irrigation networks for an increased energy efficiency. Agricultural Water Management 164, 212–222.

Fernández García, I., Montesinos, P., Camacho Poyato, E., Rodríguez Díaz, J.A., 2016b. Energy cost optimization in pressurized irrigation networks. Irrigation Science 34, 1–13.

García Morillo, J., Martín, M., Camacho, E., Rodríguez Díaz, J.A., Montesinos, P., 2015. Toward precision irrigation for intensive strawberry cultivation. Agricultural Water Management 151, 43–51.

González-Cebollada, C., Macarulla, B., Sallán, D., 2011. Recursive design of pressurized branched irrigation networks. Journal of Irrigation and Drainage Engineering 137, 375–382.

Hamidat, A., Benyoucef, B., Hartani, T., 2003. Small-scale irrigation with photovoltaic water pumping system in Sahara regions. Renewable Energy 28, 1081–1096.

IDAE, Instituto para la Diversificación y Ahorro de la Energía, 2008. Ahorro y Eficiencia Energética en Agricultura de Regadío.

Jiménez-Bello, M.A., Martínez Alzamora, F., Bou Soler, V., Ayala, H.J.B., 2010. Methodology for grouping intakes of pressurised irrigation networks into sectors to minimise energy consumption. Biosystems Engineering 105, 429–438.

Joint Research Centre (JRC) of the European Commission, 2014. Monitoring Agriculture Resources (MARS) Unit H04. In: Zarco-Tejada, P.J., Hubbard, N., Loudjani, P. (Eds.), Precision Agriculture: An Opportunity for EU Farmers – Potential Support with the CAP 2014–20.

Khadra, R., Lamaddalena, N., 2010. Development of a decision support system for irrigation systems analysis. Water Resources Management 24, 3279–3297.

Kirda, C., 2002. Deficit Irrigation scheduling based on plant growth stages showing water stress tolerance. In: Deficit Irrigation Practices. FAO, pp. 3–10.

Lamaddalena, N., Khila, S., 2013. Efficiency-driven pumping station regulation in on-demand irrigation systems. Irrigation Science 31, 395–410.

Lecina, S., Isidoro, D., Playán, E., Aragüés, R., 2010. Irrigation modernization and water conservation in Spain: the case of Riegos del Alto Aragón. Agricultural Water Management 97, 1663–1675.

MAPA, Ministerio de Agricultura Pesca y Alimentación, 2002. REAL DECRETO 329/2002, de 5 de abril, por el que se aprueba el Plan Nacional de Regadíos.

MARM, Ministerio de Medio Ambiente y Medio Rural y Marino, 2010. Estrategia Nacional para la Modernización sostenible de los regadíos H2015. informe de sostenibilidad ambiental.

Melgar, J.C., Mohamed, Y., Navarro, C., Parra, M.A., Benlloch, M., Fernández-Escobar, R., 2008. Long-term growth and yield responses of olive trees to different irrigation regimes. Agricultural Water Management 95, 968–972.

Moreno, M.A., Planells, P., Córcoles, J.I., Tarjuelo, J.M., Carrión, P.A., 2009. Development of a new methodology to obtain the characteristic pump curves that minimize the total cost at pumping stations. Biosystems Engineering 102, 95–105.

Nelson, G.C., Rosegrant, M.W., Koo, J., Robertson, R., Sulser, T., Zhu, T., Ringler, C., Msangi, S., Palazzo, A., Batka, M., Magalhaes, M., Valmonte-Santos, R., Ewing, M., Lee, D., 2009. Cambio Climático: El impacto en la agricultura y los costos de adaptación. Instituto Internacional de Investigación sobre políticas Alimentarias, Washington.

Orgaz, F., Villalobos, F., Testi, L., Pastor, M., Hidalgo, J.C., Fereres, E., 2005. Programación de riegos en plantaciones de olivar. Metodología para el cálculo de las necesidades de agua de riego en el olivar regado por goteo. In: Cultivo Del Olivo Con Riego Localizado. Junta de Andalucía y ediciones Mundi-Prensa, Sevilla, pp. 83–137.

Pastor, M., Castro, J., Mariscal, M.J., Vega, V., Orgaz, F., Fereres, E., Hidalgo, J., 1999. Respuestas del olivar tradicional a diferentes estrategias y dosis de agua de riego. Investigación Agraria Producción y Protección Vegetal 14 (3).

Pérez-Rodríguez, J.M., Parras-Cintero, J., 2014. Manual práctico de riego del olivar de almazara. Centro de Investigaciones Científicas y Tecnológicas de Extremadura. CICYTEX, Badajoz.

Rodríguez Díaz, J.A., López Luque, R., Carrillo Cobo, M.T., Montesinos, P., Camacho Poyato, E., 2009. Exploring energy saving scenarios for on-demand pressurised irrigation networks. Biosystems Engineering 104, 552–561.

Rodríguez Díaz, J.A., Camacho Poyato, E., Blanco Pérez, M., 2011. Evaluation of water and energy use in pressurized irrigation networks in Southern Spain. Journal of Irrigation and Drainage Engineering 137, 644–650.

Rodríguez-Díaz, J.A., Camacho-Poyato, E., López-Luque, R., Pérez-Urrestarazu, L., 2008. Benchmarking and multivariate data analysis techniques for improving the efficiency of irrigation districts: an application in Spain. Agricultural Systems 96, 250–259.

Rodríguez-Díaz, J.A., Montesinos, P., Camacho Poyato, E., 2012. Detecting critical points in on-demand irrigation pressurized networks – a new methodology. Water Resources Management 26, 1693–1713.

Rossman, L., 2000. EPANET 2. Users Manual. US Environmental Protection Agency (EPA), USA.

Siebert, S., Burke, J., Faures, J.M., Frenken, K., Hoogeveen, J., Döll, P., Portmann, F.T., 2010. Groundwater use for irrigation – a global inventory. Hydrology and Earth Systems Sciences 14, 1863–1880.

Tsakmakis, I., Kokkos, N., Pisinaras, V., Papaevangelou, V., Hatzigiannakis, E., Arampatzis, G., Gikas, G.D., Linker, R., Zoras, S., Evagelopoulos, V., Tsihrintzis, V.A., Battilani, A., Sylaios, G., 2017. Operational precise irrigation for cotton cultivation through the coupling of meteorological and crop growth models. Water Resources Management 31, 563–580.

Tubiello, F.N., Salvatore, M., Cóndor-Golec, R.D., Ferraca, A., Rossi, S., Biancalani, R., Federici, S., Jacobs, H., Flammini, A., 2014. Agriculture, Forestry and Other Land Use Emissions by Sources and Removals by Sinks. Climate, Energy and Tenure Division. FAO.

Vick, B.D., 2010. Developing a hybrid solar/wind powered irrigation system for crops in the great plains. In: Paper Presented at the 39th ASES National Solar Conference 2010, SOLAR 2010, vol. 1, pp. 159–183.

Yahyaoui, I., Tadeo, F., Segatto, M.V., 2016. Energy and water management for drip-irrigation of tomatoes in a semi- arid district. Agricultural Water Management 183, 4–15.

CHAPTER 3

Water Footprint Accounting for Improving Irrigation Management in Olive Trees

R. González Perea, J. García Morillo, Juan A. Rodríguez Díaz, Pilar M. Barrios, Emilio C. Poyato
University of Cordoba, Cordoba, Spain

Contents

1. INTRODUCTION

Freshwater scarcity has been recognized as one of the most crucial environmental concerns, assuming that over two-thirds of the world's population will be affected by water scarcity by 2025 (UNESCO, 2009). The growing freshwater shortage is already evident in many parts of the world (Hoekstra et al., 2012).

Water resource availability is a limiting factor for economic development in many water-stressed countries. The Mediterranean region is one of the most water-scarce regions in the world, mainly in southern and eastern countries, such as Spain, where the expansion of irrigated production, coupled with tourism and urbanization, has created significant water supply challenges (Daccache et al., 2014).

Water Scarcity and Sustainable Agriculture in Semiarid Environment
ISBN 978-0-12-813164-0
https://doi.org/10.1016/B978-0-12-813164-0.00003-X

At the global scale, most water use occurs in agricultural production, mainly for irrigation. Thus irrigated agriculture is highly dependent on water resource availability and efficient water use. This entails increasing water productivity ($t\ m^{-3}$) and improvements in irrigation efficiency. Raising water productivity in agriculture ("more crop per drop") can contribute to reducing the pressure on global freshwater resources (Monaghan et al., 2013).

One of the most frequently used tools to implement such an approach is the water footprint (WF), a concept that was first introduced by Hoekstra (2003). As an indicator, the WF measures the appropriation of water resources by human activity by evaluating water consumption and the impact on water quality. This process is also called "water footprint accounting" or "water footprinting" and is part of a wider evaluation process known as "water footprint assessment" (Hoekstra et al., 2011).

WF accounting is the phase in which data are collected and accounts are developed. It is a suitable procedure to appraise the water use and crop yield relationship. As such, it is a measure of crop water efficiency and hence crop production efficiency.

However, studies carried out on WF accounting only provide information about the water used per unit of agricultural production, but not about the irrigation process or whether the water is used efficiently. Therefore the WF indicator should be accompanied by other indicators to evaluate if the irrigation water is being used in the best way possible (García Morillo et al., 2015).

Relative irrigation supply (RIS) is a widely used indicator to assess irrigation management (Levine, 1982). RIS is the ratio of the total annual volume of water diverted or pumped for irrigation and the total theoretical irrigation needs of the crops calculated per irrigation season. These indicators have become a powerful tool for assessing irrigation management at irrigation district and farm levels (Rodríguez Díaz et al., 2008).

Unlike in this work, most WF studies have been developed at the global, national, or regional level using annual or monthly average data provided by available literature or public databases. However, when strategies to reduce WFs and to improve water use efficiency are developed, data must be collected at the farm level and on-site studies are also required.

In this chapter, a methodology to carry out a joint evaluation of water use and management through the analysis of indicators such as crop water footprint (WF_c), RIS, and seasonal water volume applied is presented. At the farm scale, this methodology can be used as a diagnostic tool of the irrigation process, identifying inefficiencies, supporting the implementation of improvement measures, and applying the proposed methodology to a real case study: an olive orchard located in Cordoba (southern Spain).

2. THE PERFORMANCE INDICATORS PROPOSAL

2.1 Water Use Indicators: Water Footprint and Crop Water Applied

According to the virtual water concept introduced by Allan (1998), water is divided into three categories: green, blue, and gray.

Green water is defined as the fraction of rainwater stored in the soil and available for the crop to be evapotranspired during plant growth (ET_{green}), which is equivalent to the effective rainfall (P_{eff}) concept. Green water is particularly meaningful for agricultural products.

Blue water is the summation of all the water withdrawals for irrigation from aquifers and surface water sources that are evapotranspired during the crop season (ET_{blue}). Blue water is estimated as the difference between the theoretical crop evapotranspiration (ET_c) and the effective rainfall. It represents the crop irrigation needs that are supposedly always satisfied. Thus when the effective rainfall is higher than ET_c, ET_{blue} is equal to zero. According to this, blue water is the amount of water that does not return to the environment from which it was initially withdrawn; in other words, it is the water that is "lost in evapotranspiration." This concept is also known as freshwater consumptive uses.

In addition, the gray WF is an indicator of pollution and is defined as the volume of freshwater that is required to assimilate the load of pollutants based on existing ambient water quality standards (Hoekstra, 2003). There is no scientific consensus about the gray WF component and some authors such as García Morillo et al. (2015), González Perea et al. (2016), or Vanham and Bidoglio (2013) do not include it in their WF estimations.

Following Hoekstra's terminology, the WF of any crop (WF_c) is the sum of the green, blue, and gray components, and is usually expressed in $m^3\,t^{-1}$ or $L\,kg^{-1}$ (Eq. 3.1).

$$WF_c = WF_{green} + WF_{blue} + WF_{gray} \tag{3.1}$$

The green component of the WF crop (WF_{green}, $m^3\,t^{-1}$) is the green crop water use (CWU_{green}, $m^3\,ha^{-1}$) divided by the crop yield (Y, $t\,ha^{-1}$) (Eq. 3.2).

$$WF_{green} = \frac{CWU_{green}}{Y} \tag{3.2}$$

CWU_{green} is calculated under the assumption that the crop is produced in rain-fed conditions. Thus CWU_{green} is the effective rainfall, which can be estimated according to the method of the United States Soil Conservation Service (United States Department of Agriculture) (Allen et al., 1998).

The blue component of the water footprint (WF_{blue}, $m^3\ t^{-1}$) is calculated as the total volume of water applied for the whole cropping season (CWA_{blue}, $m^3\ ha^{-1}$) divided by the crop yield (Y, $t\ ha^{-1}$) (Eq. 3.3).

$$WF_{blue} = \frac{CWA_{blue}}{Y} \tag{3.3}$$

Eq. (3.3) represents a coherent procedure to estimate real crop water use based on records of water applied. This approach can be applied to a wide range of irrigation managements such as the full satisfaction of crop irrigation needs or deficit irrigation (González Perea et al., 2016), although other authors use the theoretical irrigation needs to assess WF_{blue} without on-site water use information. CWA_{blue} takes into account all water withdrawals throughout the season. It accounts for the whole water volume applied to achieve the recorded production. It is an indicator of the real water applied by each producer at farm scale. This indicator also counts the indirect water uses needed for crop production such as soil preparation, irrigation to remove salt from the soil (WA_{ss}), frost protection (WA_f), or to apply fertilizers or chemicals (WA_{ch}). Furthermore, the volume of irrigation water applied during the growing period does not often match the theoretical irrigation needs due to irrigation system inefficiencies, inadequate irrigation scheduling, or when deficit irrigation is applied. Thus CWA_{blue} was calculated as follows (Eq. 3.4):

$$CWA_{blue} = WA_g + \sum_{i=1,\ldots,n} WA_i \tag{3.4}$$

where WA_g ($m^3\ ha^{-1}$) is the actual irrigation water applied during the crop-growing period and WA_i is the amount of water used in the different agricultural practices (up to n) cited earlier (soil preparation, salt removal, chemical application, etc.).

The CWA_{blue} indicator informs on the real amount of water used by each farmer instead of the theoretical water evapotranspired by the crop. These values were not estimated from irrigation needs as in other studies but were directly collected from water meters installed on farms. Considering the CWA_{blue}, WF_{blue} becomes WF_{blue} applied ($m^3\ t^{-1}$), which provides information about the irrigation water that is actually used to produce one unit of agricultural production, that is, this indicator links the actual water withdrawn to crop yield.

Crop water consumption or water that is actually evapotranspirated to the atmosphere, CWC_{blue} ($m^3\ ha^{-1}$), was calculated according to Eq. (3.5) and it represents the total irrigation water evapotranspired by the crop.

$$CWC_{blue} = \min[CWA_{blue}, 10 \cdot (ET_c - P_{eff})] \tag{3.5}$$

where P_{eff} and ET_c are expressed in mm $year^{-1}$. ET_c is estimated according to Allen et al. (1998) and calculated as follows:

$$ET_c = \sum_{\substack{i=1...k \\ j=1...l}} Kc_{ij} \cdot ET_{oj} \tag{3.6}$$

where i is the season day index, j is the crop stage index, K_c is the crop coefficient that varies with the crop stage, and ET_o in mm is the estimated reference evapotranspiration.

CWC_{blue} is the difference between ET_c and P_{eff} only when irrigation needs are completely satisfied. In water-scarce regions, deficit irrigation is a common practice, which consists of applying a lower amount of water than crop irrigation needs. In this case, CWC_{blue} is smaller than that difference and equals CWA_{blue} because ET_c is at most the water received by the crop. CWC_{blue} does not consider the irrigation system and management inefficiencies; however, they can be estimated as the difference between CWA_{blue} and CWC_{blue}. Therefore to evaluate WF_c and irrigation management together, CWA_{blue} provides more useful information at the decision-making level, as it is also a reference indicator to compare real crop water use and water rights.

Finally, Eq. (3.7) accounts for the total amount of water (green and blue) used at field level during the whole cropping season (CWU, $m^3\ ha^{-1}$).

$$CWU = CWU_{green} + CWA_{blue} \tag{3.7}$$

Then, the total WF_c was calculated using the recorded information about the water use efficiency of the crop itself, in accordance with the local agro-climatic conditions and the actual irrigation water applied. According to Spanish Water Law (2001) and Water Framework Directive (2000), polluted water should be treated before it returns to the environment to achieve the environmental water quality standards. For this reason, the gray water component has not been included in this study, as the treatments to achieve these water quality standards do not involve an extra water withdrawal. It is unnecessary to put additional pressure on water resources, in particular in water-scarce regions.

2.2 Water Management Indicator: Relative Irrigation Supply

RIS (Levine, 1982) is an indicator that provides information about how irrigation matches theoretical water requirements, as is shown in Eq. 3.8.

$$RIS = \frac{WA_g}{10 \cdot (ET_c - P_{eff})} \tag{3.8}$$

where WA_g ($m^3\ ha^{-1}$) is the actual irrigation water applied during the crop-growing period, ET_c (mm $year^{-1}$) is the crop evapotranspiration, and P_{eff} (mm $year^{-1}$) is the effective rainfall. RIS values lower than 1 indicate deficit irrigation, while values over 1 show overirrigation. In practice, values around 1.2 are desirable to avoid deficit irrigation because of on-farm distribution uniformity and imprecise irrigation scheduling.

3. CASE STUDY

3.1 Study Area and Data Collection

In southern Spain the olive tree is the most important irrigated crop due to its extension. A 10 ha olive orchard (var. *Arbequina*) has been chosen to apply the methodology developed in this chapter. Three irrigation seasons, 2013, 2014, and 2015, were assessed. The experimental field was located in Rabanales Campus, University of Cordoba (southern Spain). The agroclimatic variables for the three irrigation seasons were collected from the weather station located within the study area. The average values of ET_o and precipitation, P, for the three irrigation seasons are shown in Table 3.1. The 2013 season was an extremely wet year, while 2015 was a dry year. The 2014 season can be considered as an average season.

The K_c values of olive trees used in this chapter are shown in Table 3.2 and they were obtained from Pastor and Orgaz (1994). The olive trees in the study area were planted in 2011 with a plant density of 1429 trees ha^{-1}

Table 3.1 Seasonal potential evapotranspiration (ET_o) and precipitation (P) of the study area for 2013, 2014, and 2015

Seasons	2013	2014	2015
ET_o (mm)	1244.6	1249.1	1246.8
P (mm)	803.4	637.1	352.1

Table 3.2 Monthly K_c values for olive trees

Month	K_c
January	0.5
February	0.5
March	0.65
April	0.6
May	0.55
June	0.5
July	0.45
August	0.45
September	0.55
October	0.6
November	0.65
December	0.5

and were grown under two different irrigation allocations (water rights): 1000 $m^3\ ha^{-1}$ (I1) and 2000 $m^3\ ha^{-1}$(I2). Thus Table 3.3 shows crop production for the two irrigation allocations. Crop production data were from Parra Bóccoli et al. (2016). By ANOVA multifactorial analysis, this author determined that irrigation allocation had no significant influence on crop production. The irrigation system was drip irrigation using pressure-compensating emitters, which pressure compensation range varied from

Table 3.3 Crop production (kg ha^{-1}), crop irrigation needs, CWU_{green}, CWA_{blue}, CWC_{blue} indicators, green and blue components of water footprint, and total water footprint for 2013, 2014, and 2015 irrigation seasons

	2013		2014		2015	
	I1[a]	I2[b]	I1	I2	I1	I2
Crop irrigation needs ($m^3\ ha^{-1}$)	5384.6		5673.3		5992.5	
Yield (kg ha^{-1})	3171	2686	4057	6014	10,829	10,986
CWU_{green} ($m^3\ ha^{-1}$)	6427.2	6427.2	5096.8	5096.8	2816.8	2816.8
CWA_{blue} ($m^3\ ha^{-1}$)	1000	2000	1000	2000	1000	2000
CWC_{blue} ($m^3\ ha^{-1}$)	1000	2000	1000	2000	1000	2000
WF_{green} ($m^3\ t^{-1}$)	202.66	239.31	125.63	84.74	26.01	25.64
WF_{blue} ($m^3\ t^{-1}$)	315.32	744.68	246.48	332.54	92.35	182.05
WF ($m^3\ t^{-1}$)	517.97	983.99	372.10	417.29	118.36	207.70

CWA, crop water applied; *CWC*, crop water consumption; *CWU*, crop water use; *WF*, water footprint.

[a]Irrigation allocation 1: 1000 $m^3\ ha^{-1}$.

[b]Irrigation allocation 2: 2000 $m^3\ ha^{-1}$.

10 to 40 m. Deficit irrigation is the irrigation management rule applied in this plantation. It is the most frequent irrigation practice in olive tree production in southern Spain.

3.2 Water Footprint

The applied water recorded during the three irrigation seasons was used to calculate water indicators. Table 3.3 also shows the crop irrigation needs and CWU_{green}, CWA_{blue}, and CWC_{blue} values for the three irrigation seasons. In the wettest season, 2013, crop irrigation needs were the lowest (538.46 mm). On the contrary, the highest crop irrigation needs (599.25 mm) were reached in 2015 (the driest season). As olives trees were cultivated under deficit irrigation conditions, it is considered that the crop used the whole rainfall water stored in the soil. So, CWU_{green} values ranged from 642.72 $m^3\ ha^{-1}$ for the wettest season to 281.68 $m^3\ ha^{-1}$ for the driest season, independently of irrigation allocations. In this study, the CWA_{blue} is just the irrigation allocation (the amount of water that the farmer is authorized to withdraw for irrigation). The sum of CWU_{green} and CWA_{blue} is considerably lower than the crop irrigation needs for the three studied irrigation seasons. Due to deficit irrigation conditions, where the full irrigation requirements were not matched, the water consumed or evapotranspirated by the plant (CWC_{blue}) coincided with the irrigation water applied (CWA_{blue}).

For both irrigation depths, the WF was calculated for each irrigation season. Thus Table 3.3 shows WF_{green} for the three irrigation seasons studied and represents the rainfall water that is evapotranspirated by the crop. WF_{green} values ranged from 25.64 (2015) to 239.31 $m^3\ t^{-1}$ (2013). WF_{green} values are influenced by the total rainfall as well as yield that increased accordingly when more irrigation water was applied.

WF_{blue} relates the volume of water applied by irrigation to crop production. Table 3.3 shows this relationship for the three studied irrigation seasons. As in WF_{green} the lowest value of WF_{blue} was found in the 2015 season (92.35 $m^3\ t^{-1}$) due to the high crop production in this irrigation season. For the other irrigation seasons, WF_{blue} values ranged from 246.48 to 744.68 $m^3\ t^{-1}$.

Finally, Table 3.3 also shows the WF for the three irrigation seasons. The WF values for the first irrigation season ranged from 517.97 $m^3\ t^{-1}$ for I1 to 983.99 $m^3\ t^{-1}$ for I2. In this year, the WF value under I1 conditions was 23% lower than under I2 conditions. In the 2014 season, the WF values for I1 and I2 conditions ranged from 372.10 to 417.29 $m^3\ t^{-1}$, respectively.

It is important to consider that the olive trees were planted in the 2011 season and in the 2015 season they were close to full production. Thus in the 2015 season, crop production was greatly increased (threefold 2013 values). Consequently, the WF values were much lower than in previous seasons (they ranged from 118.36 to 207.70 $m^3\ t^{-1}$ for I1 and I2 irrigation depths, respectively).

3.3 Irrigation Management Indicators

The RIS indicator provides information about how close theoretical irrigation requirements are to irrigation water applied. RIS values lower than 1 indicate deficit irrigation, while values over 1 show overirrigation. In practice, values around 1.2 are desirable to avoid deficit irrigation because of on-farm distribution uniformity and imprecise irrigation scheduling. For irrigation allocation I1, RIS values were 0.186, 0.176, and 0.167 for 2013, 2014, and 2015 irrigation seasons, respectively. However, for irrigation allocation I2 the RIS values rise to 0.371, 0.353, and 0.334 for the three irrigation seasons, respectively. These values show that the olive trees were always cultivated under strong deficit irrigation as irrigation allocations were always below irrigation requirements. However, Moriana et al. (2003) proved that high crop yield can be achieved with small water applications. With I1 and I2 water allocations, crop yields were higher than 10,000 kg ha^{-1} (Table 3.3) in some of the irrigation seasons of this study.

3.4 Comparison With Other Studies at Different Spatiotemporal Levels

WFs can be inventoried at different levels of spatiotemporal detail. At a lower level of detail, the WF is calculated based on global average water use data (blue and green) from available databases and is sufficient for raising awareness or to show rough projections of future global water consumption (Hoekstra et al., 2009). At that level, Mekonnen and Hoekstra (2011) provided WF_{blue} and WF_{green} values for olive orchards at a global scale. These values were 2388 and 11,826 $m^3\ t^{-1}$ for WF_{blue} and WF_{green}, respectively. At national or regional level, water footprinting is based on data from national or regional available literature and databases. This level of accounting is suitable for providing a basis for understanding where hotspots can be expected in local watersheds and for making water allocation decisions. For the different regions of Spain, in the period 1997–2008, Salmoral et al. (2011) performed the estimations of green and blue WFs for olive orchards. The average values of WF_{green} and WF_{blue} of that study in that period were 717.75 and 483.25 $m^3\ t^{-1}$, respectively, which is significantly higher than the estimations carried out in this work based on real data collected at field level.

Thus it is essential that useful WF estimations must be based on local precise data. Accounting relies on the best estimates of actual on-site water use, preferably verified in the field. This high level of spatiotemporal detail is suitable for formulating site-specific strategies to reduce WFs. The WF values presented in this study were obtained at plot level in Cordoba. The WF_{green} values for the 2015 season (olive trees in full production) were 26.01 $m^3\ t^{-1}$ for irrigation allocation I1 and 25.64 $m^3\ t^{-1}$ for irrigation allocation I2, and the WF_{blue} values for the I1 and I2 irrigation allocations were 125.63 and 84.74 $m^3\ t^{-1}$, respectively. Therefore depending on the spatiotemporal level of the study the WF values can be considerably different. Analysis based on national data, which considers neither the real irrigation practices nor yields, usually leads to important overestimations of the WF.

4. CONCLUSIONS

WF_c analysis has important limitations. Although it informs about the water needed to obtain one unit of agricultural production, it does not provide information about irrigation management and water use efficiency. The WF indicator has been analyzed together with RIS, which is an irrigation adequacy indicator. Both indicators have been used to determine the efficiency of water use in the irrigation process of olive trees.

The analysis carried out during three irrigation seasons showed quite similar values of RIS for both irrigation allocations (I1 = 1000 $m^3\ ha^{-1}$ and I2 = 2000 $m^3\ ha^{-1}$). For I1, values ranged from 0.16 to 0.18 and for I2 from 0.33 to 0.37 during the studied seasons. These values show that the olive trees were always cultivated under deficit irrigation conditions and irrigation allocations were always below crop water needs. In 2015, which was a dry year and the plantation was already in full production, there were no significant differences in crop yield for both irrigation allocations (10,829 and 10,986 kg ha^{-1}, respectively). This result shows that, in this particular case, high crop yields can be achieved with small water application depths.

The olive plantation was in full production during the last season (2015). Its production was close to 11,000 kg ha^{-1} for both irrigation allocations. Yields in 2013 and 2014 were much smaller, between 3000 and 6000 kg ha^{-1}. In 2013, the WF value under I1 conditions was 23% lower than under I2 conditions because of the effect of the irrigation over crop yield. During 2014, the WF values for both allocations were quite similar ranging from 372.1 to 417.29 $m^3 t^{-1}$, respectively. In 2015, the yield was greatly increased and consequently the WF values were much lower, from 118.36 to 207.7 $m^3 t^{-1}$. Nevertheless, these WF values are quite different to those calculated by other authors for this crop in Spain (1200 $m^3 t^{-1}$). However, previous works were based on global or national databases, which do not consider the real irrigation practices and actual yields. Thus this work demonstrates that the spatiotemporal scale is essential for the estimation of the WF. The use of local precise data, collected at farm level, is essential for an accurate estimation of the WF and for the development of measures to improve on-farm water use efficiency.

Finally, irrigated agriculture is recognized as a water-intensive sector due to its high dependency on the use and management of water resources. It is therefore essential that the farming sector takes a responsible approach toward the sustainable use and conservation of fresh water.

ACKNOWLEDGMENTS

This research was supported by an FPU grant (Formación de Profesorado Universitario) from the Spanish Ministry of Education, Culture and Sports to Rafael González Perea.

This work is part of the TEMAER project (AGL2014-59747-C2-2-R), funded by the Spanish Ministry of Economy and Competitiveness.

REFERENCES

Allan, J.A., 1998. Virtual water: a strategic resource global solutions to regional deficits. Ground Water 36, 545–546.

Allen, R.G., Pereira, L.S., Raes, D., Smith, M., 1998. Crop Evapotranspiration: Guidelines for Computing Crop Water Requirements. Irrigation and Drainage, FAO Irrigation and Drainage, Rome, Italy.

Daccache, A., Knox, J.W., Weatherhead, E.K., Daneshkhah, A., Hess, T.M., 2014. Implementing precision irrigation in a humid climate – recent experiences and on-going challenges. Agricultural Water Management 147, 135–143.

García Morillo, J., Rodríguez Díaz, J.A., Camacho, E., Montesinos, P., 2015. Linking water footprint accounting with irrigation management in high value crops. Journal of Cleaner Production 87, 594–602.

González Perea, R., Camacho Poyato, E., Montesinos, P., García Morillo, J., Rodríguez Díaz, J.A., 2016. Influence of spatio temporal scales in crop water footprinting and water use management: evidences from sugar beet production in Northern Spain. Journal of Cleaner Production 139, 1485–1495.

Hoekstra, A.Y., 2003. Virtual water trade. In: Proceedings of the International Expert Meeting on Virtual Water Trade, vol. 12, pp. 1–244. Int. Expert Meet. Virtual Water Trade.

Hoekstra, A.Y., Chapagain, A.K., Aldaya, M.M., Mekonnen, M.M., 2009. Water Footprint Manual State of the Art 2009. Water Footpr. Netw, p. 131.

Hoekstra, A.Y., Chapagain, A.K., Aldaya, M.M., Mekonnen, M.M., 2011. 'The Water Footprint Assessment Manual'. Setting the Global Standard.

Hoekstra, A.Y., Mekonnen, M.M., Chapagain, A.K., Mathews, R.E., Richter, B.D., 2012. Global monthly water scarcity: blue water footprints versus blue water availability. 7. https://doi.org/10.1371/journal.pone.0032688.

Levine, G., 1982. Relative Water Supply: An Explanatory Variable for Irrigation Systems. Technical report.

Mekonnen, M.M., Hoekstra, A.Y., 2011. The green, blue and grey water footprint of crops and derived crop products. Hydrology and Earth System Sciences 15, 1577–1600.

Monaghan, J.M., Daccache, A., Vickers, L.H., Hess, T.M., Weatherhead, E.K., Grove, I.G., Knox, J.W., 2013. More "crop per drop": constraints and opportunities for precision irrigation in European agriculture. Journal of the Science of Food and Agriculture 93, 977–980. https://doi.org/10.1002/jsfa.6051.

Moriana, A., Orgaz, F., Pastor, M., Fereres, E., 2003. Yield responses of a mature olive orchard to water deficits. Journal of the American Society for Horticultural Science 128, 425–431.

Parra Bóccoli, A., Muñoz Diez, C., Morello Parra, P., 2016. Ensayos comparativos de densidades, variedades y orientación en olivar en seto.

Pastor, M., Orgaz, F., 1994. Riego deficitario del olivar. Agricultura 746, 768–776.

Rodríguez Díaz, J.A., Camacho Poyato, E., López Luque, R., Pérez Urrestarazu, L., 2008. Benchmarking and multivariate data analysis techniques for improving the efficiency of irrigation districts. An application in Spain. Agricultural Systems 96, 250–259.

Salmoral, G., Aldaya, M.M., Chico, D., Garrido, A., Llamas, R., 2011. The water footprint of olives and olive oil in Spain. Spanish Journal of Agricultural Research 9, 1089–1104.

Spanish Water Law, 2001. Real Decreto Legislativo 1/2001, de 20 de julio, por el que se aprueba el texto refundido de la Ley de Aguas, pp. 26791–26817.

UNESCO, 2009. 'The United Nations World Water Development', Report 3-Water in a Changing World. Paris, France.

Vanham, D., Bidoglio, G., 2013. A review on the indicator water footprint for the EU28. Ecological Indicators 26, 61–75.

Water Framework Directive, 2000. 'Directive 2000/60/EC of the European Parliament and of the Councils of 23 October 2000 Establishing a Framework for Community Action in the Field of Water Policy' 2000/60/EC, pp. 1–73.

CHAPTER 4

Local and Collective Actions for Adaptation to Use Less Water for Agriculture in the Mediterranean Region

Ana Iglesias, Luis Garrote
Universidad Politecnica de Madrid, Madrid, Spain

Contents

1. ADDRESSING THE ADAPTATION CHALLENGE

There is a general consensus about the unsustainability of the present model of water use for agriculture in Mediterranean regions and about the need to reach a balance between rural development, food security, and the environment (Iglesias and Buono, 2009; Iglesias et al., 2011a,b; Vorosmarty et al., 2000). Climate change will likely affect rural societies and the environment creating or reinforcing new challenges to water management. In particular, sustainable management of freshwater resources—especially focusing on the availability and quality of water—is one of the main challenges to our present agricultural model that relies on providing additional water to crops to compensate for the hot and arid climate.

Throughout history, societies have shown a great ability to adapt to changing conditions, with or without a conscious response by citizens and governments. However, it is likely that the changes imposed by climate change in the future will exceed the limits of autonomous–endogenous

Water Scarcity and Sustainable Agriculture in Semiarid Environment
ISBN 978-0-12-813164-0
https://doi.org/10.1016/B978-0-12-813164-0.00004-1

adaptation, and that policies will be required to support and enable different sectors of society to cope with similar changes.

This chapter aims to provide an understanding of the potential local and collective adaptation strategies to reduce the use of water for agriculture in Mediterranean countries in view of climate change. Following this introduction that outlines the main challenges, the chapter assesses the risks of climate change to water resources and then evaluates local and collective actions toward adaptation. Finally, the chapter draws on the results presented to suggest some policy implications related to the adaptation of water resources for agriculture to climate change in the Mediterranean region.

2. CLIMATE CHANGE INCREASES THE PRESSURE TO USE LESS WATER FOR AGRICULTURE

There have been several thousand studies on the potential impacts of climate change on water resources with many different approaches, and the knowledge of the potential impacts is clear and unambiguous (IPCC, 2014). The projected impacts pose challenges for many water-dependent activities and magnify the regional differences in Mediterranean rural communities (Iglesias et al., 2007; Garrote et al., 2015a). Overall, the studies are consistent in the intensification of unsustainability of maintaining the present model of water management for agriculture (Garrote et al., 2015a; Gleeson et al., 2012; Huntjens et al., 2010).

The risk of climate change evolves from changes in precipitation to impacts on socioecological systems. Changes in precipitation are probably the most important factor determining the likely impacts on the Mediterranean region. Despite forecasts of increased total annual precipitation in some regions, evapotranspiration is expected to increase in response to increased temperatures. Significant increased temporal and spatial variability of extreme weather is expected, increasing flooding and drought frequency leading to competition for water (Feyen et al., 2012; Iglesias and Garrote, 2014). Sea-level rise will inundate coastal areas; rising sea levels may also lead to salinization of the water supply and soil. A decrease in water availability is predicted together with an increase in water demand, leading to potential conflict between users. Decreasing water resources in some areas may affect soil structure, while reduced soil drainage may lead to increased salinity. However, an increase in frequency and intensity of floods is predicted in some areas where significant winter rainfall is likely. These changes are expected to reduce the diversity of Mediterranean species.

The risk of climate change, the expected intensity of effects, and the potential consequences for agroecosystems and rural areas are presented in Table 4.1. Most studies agree that climate change will likely have the following common consequences across the Mediterranean region (EEA, 2017): (1) increased demand for agricultural water throughout the region due to expected increases in crop evapotranspiration in response to increased temperatures in all regions; (2) increased water shortages, particularly in the spring and summer months, therefore increasing the water requirement for irrigation in all areas; (3) increased water quality deterioration due to higher water temperatures and lower levels of runoff in some regions, particularly in summer, imposing further stress in agricultural-irrigated areas; and (4) increased risk of flooding in winter due to the expected concentration of precipitation in this period, affecting significant areas in the low-lying coastal areas.

3. LOCAL AND COLLECTIVE ACTIONS TOWARD ADAPTATION

Adaptation is about preparing people and their assets for the impacts of climate change. It is concerned with minimizing adverse effects, or maximizing new opportunities, through taking actions that either anticipate or react to changing climatic conditions. Adaptation actions are fundamental to reducing the environmental, social, and economic costs of climate change.

Several hundred studies have reflected on the management actions and policies that may be appropriate for redressing the impact of climate change on water resources (see Iglesias and Garrote, 2015 for a full review of studies). The studies show that, for the water sector, planned interventions must consider both supply-side and demand-side solutions (Iglesias and Garrote, 2015; Garrote et al., 2015a,b; Gleick and Palaniappan, 2010). Various types of adaptation can be distinguished. Some studies have highlighted the distinction between local and collective actions and the importance of codesigning adaptation plans (García de Jalón et al., 2014; Ivey et al., 2004). Local measures are taken by public or private actors, such as farmers, farmers groups, and local water managers, in response to actual or expected water scarcity, with or without the active intervention of policy (García de Jalón et al., 2013). Collective measures are usually policy driven, that is, the result of a deliberate policy decision. Collective actions are therefore associated water management authorities, or public agencies, and may be linked to public investment.

Table 4.1 Risk of climate change, expected intensity of effects, potential consequences for agroecosystems and rural areas, and confidence level

Risk of climate change	Expected intensity of negative effects	Potential consequences for agroecosystems and rural areas	Confidence level
Challenges to the sustainability of water resources	Changes in hydrological regime Differences in water needs Increased water shortage	Variations in hydrological regime Decreased availability of water Risks of water quality loss Increased risk of soil salinization Conflicts among users Groundwater abstraction, depletion, and decrease in water quality	High
Increased irrigation requirements	High in areas already vulnerable to water scarcity	Increased demand for irrigation Decreased yield of crops	High
Increased water and soil salinity and erosion	High for southern countries	Decrease in water quality from nutrient leaching Decreased crop yields Land abandonment Increased risk of desertification Loss of rural income	High
Increased expenditure in emergency and remediation actions	High for regions with low adaptation capacity	Loss of rural income Economic imbalances	Medium
Biodiversity loss	High for vulnerable regions	Loss of natural adaptation options Modified interaction among species	Medium

Table 4.2 summarizes potential adaptation actions that can be taken at the local or collective level, based on the database reported by Iglesias and Garrote (2015). Local needs and capacities are based on the potential for capacity to develop new irrigation systems or implement improved technology for irrigation, desalinization, water reuse technology, alternatives to groundwater management, water harvesting, capacity to develop insurance, or capacity to develop water markets (Oweis and Hachum, 2005; Glenk and Fischer, 2010; Iglesias et al., 2011c, 2017).

Although local needs determine the scenario for adaptation, cooperation and collective actions are always a priority for adaptation that includes water resources management, as shown, for example, in the case of transboundary water management (Ma et al., 2008). Upscaling local initiatives is often impossible, but knowledge transfer should play a major role in the development of adaptation strategies, especially the strategies that include local resilience as a major component of the adaptation assessment needs (World Bank, 2010).

The integration of water demands in conjunctive systems allows for the joint management of surface and groundwater resources to overcome dry periods and thus builds robustness into water resource systems (Pulido-Velazquez et al., 2011). This is achieved through integration of a large number of demands and diversification of supply sources in combined systems. The sources of water supply from different origins can have very different characteristics. Resources of a different nature (e.g., surface and groundwater) show highly significant differences in terms of variability and reliability (Garrote et al., 2014). Systems that integrate a large number of demands and supply sources can best respond to situations of scarcity through integrated water resources management, using every resource for purposes that are more appropriate depending on its amount, regularity, and reliability (Garrote et al., 2014; Pulido-Velazquez et al., 2011).

Barriers to adaptation are well known and are usually driven by the high cost of implementing the measure and the lack of incentives associated with promoting adaptive capacity. Adaptation is unlikely to be facilitated through the introduction of new and separate policies, but rather by the revision of existing policies that currently undermine adaptation and by the strengthening of policies that promote adaptation (Iglesias et al., 2011c). Finding common ground between competing claims is a serious challenge to policy development. Nevertheless, this challenge needs to be addressed to ensure the coherence and efficiency of policy measures under a changing climate.

Table 4.2 Adaptation measures related to agricultural water management in response to the risk of climate change in Mediterranean countries

Adaptation measure related to agricultural water management in response to the risk of climate change	Effect of implementing the adaptation measure	Local or collective action and consequences of the action
(1) Implement regional adaptation plans	Enhances effectiveness of adaptation measures	Collective Improves adaptive capacity
(2) Improve monitoring and early warning	Mitigates the consequences of adverse events	Collective Improves adaptive capacity
(3) Improve coordination planning	Enhances effectiveness of adaptation measures	Collective Improves adaptive capacity
(4) Innovation and technology	Improves effectiveness of adaptation measures and reduces costs	Collective Improves adaptive capacity
(5) Innovation: water use efficiency	Increases water availability	Local Improves water availability
(6) Improve soil moisture retention capacity	Increases water use efficiency	Local Improves water availability
(7) Small-scale water reservoirs on farmland	Increases water management flexibility at the local level	Local Improves water availability
(8) Improve reservoir capacity	Increases management flexibility and water availability at the regional level	Local Improves water availability
(9) Water reutilization	Increases water availability	Local Improves water availability
(10) Improve water charging and trade	Decreases inefficient use of water	Local Improves water availability
(11) Renegotiation of allocation agreements	Improves water use efficiency	Collective Improves water availability
(12) Set clear water use priorities	Improves water use efficiency	Collective Improves water availability

(13) Integrate demands in conjunctive systems	Increases management flexibility and water availability	Collective Improves water availability
(14) Create/restore wetlands	Reduces flood peaks	Collective Improves flood resilience
(15) Enhance flood plain management	Reduces flood vulnerability	Collective Improves flood resilience
(16) Improve drainage systems	Reduces extent and duration of flooding	Collective Improves flood resilience
(17) Farmers as "custodians" of floodplains	Decreases risk of flood damage	Collective Improves flood resilience
(18) Hard defences	Decreases risk of flood damage	Local Improves flood resilience
(19) Increase rainfall interception capacity	Reduces flood peaks at the local level	Local Improves drought resilience
(20) Introduce drought-resistant crops	Improves agronomic water use efficiency	Local Improves drought resilience
(21) Insurance to floods or drought	Decreases economic losses to the farmer	Local Improves drought and flood resilience
(22) Change in crops and cropping patterns	Decreases economic risk to farmers	Collective Improves irrigation savings
(23) Improve practices to retain soil moisture	Decreases the need for additional water to crops	Local Improves irrigation savings
(24) Develop climate change-resilient crops	Mitigates impacts of climate change	Local Improves irrigation savings

Continued

Table 4.2 Adaptation measures related to agricultural water management in response to the risk of climate change in Mediterranean countries—cont'd

Adaptation measure related to agricultural water management in response to the risk of climate change	Effect of implementing the adaptation measure	Local or collective action and consequences of the action
(25) Introduce new irrigation areas	Develops new agricultural land	Collective Improves food security
(26) Improve nitrogen fertilization efficiency	Reduces agricultural diffuse pollution	Local Improves environmental health
(27) Soil carbon management and reduced tillage	Reduces soil erosion and improves soil water retention capacity	Local Improves environmental health
(28) Protect against soil erosion	Reduces land degradation	Local Improves environmental health
(29) Increase water allocation for ecosystems	Improves ecosystem services, effective at the global level	Collective Improves environmental health
(30) Maintain ecological corridors	Improves biodiversity with positive global consequences	Collective Improves environmental health
(31) Improve crop diversification	Improves biodiversity	Collective Improves environmental health

A number of studies have highlighted that incorporating local and collective interest is essential to reach effective adaptation policy (De Stefano et al., 2014), reflecting on the growing demand for the processes used to allocate resources that are transparent, based on scientific evidence, and can deliver outcomes that are in the public's interest. A better understanding of how stakeholders perceive climate change, adaptation policies, and the factors or predictors influencing their support for adaptation policies can be a helpful tool in the development of these decisions and policies (García de Jalón et al., 2013).

Iglesias and Garrote (2015) show that the more interesting adaptation options in terms of their benefit-to-effort ratio are the following: improving coordination planning, setting clear water use priorities, and increasing water allocation for ecosystems (Palmer et al., 2009). It is important to highlight that these three adaptation options should be implemented at a policy-scale level (Iglesias and Garrote, 2015). On the other hand, the authors also show that the adaptation options most beneficial at a farm scale are the improvement of drainage systems and small-scale water reservoirs on farmland (Iglesias and Garrote, 2015).

4. CONCLUSIONS

The capacity to adapt to environmental change is implicit in the concept of sustainable development. Climate change adds to the many economic and social challenges already being faced by Mediterranean societies. The unavoidable impacts of climate change put current activities, certainly at the level of individual land and water managers, at significant risk, therefore making imperative the development of both local and collective actions to minimize the risk of increased water scarcity. The Intergovernmental Panel on Climate Change (IPCC, 2014) recognizes that some adaptation action is occurring, but on a very limited basis, and affirms the need for extensive adaptation across nations and across sectors to address impacts and reduce vulnerability.

An optimistic future of Mediterranean rural areas depends on whether agriculture is able to manage and consume water in a sustainable way (Iglesias and Garrote, 2015). This would require a set of local and collective actions, ranging from education to technology and policy. Even when management actions are well defined, demonstration and training efforts are necessary (Quevauviller et al., 2012). For example, improving the efficiency of irrigation or introducing water metering may only be options for societies that already have an understanding of alternative technologies and that know how to encourage implementation.

Water for agriculture often competes with water for other uses. Therefore successful adaptation of water for agriculture often requires combined efforts from other sectors, including financial, rural development, trade, industry and environment, among others. Water policy in many regions has evolved to integrate stakeholders.

The implementation of adaptation options may be a challenge at the local or collective level by individual farmers, water managers, and policies. In the short term, social barriers may limit the adoption of low-cost and technically feasible measures (Quevauviller et al., 2012). Long-term measures that require infrastructure, technology, or governance changes are often difficult to justify in political terms.

ACKNOWLEDGMENTS

We acknowledge the financial support of the European Commission BASE project (grant agreement no.: ENV-308337) of the 7th Framework Programme (http://base-adaptation.eu).

REFERENCES

De Stefano, L., Hernandez Mora, N., Iglesias, A., Sánchez, B., 2014. Water for rice farming and biodiversity: exploring choices for adaptation to climate change in Doñana, Southern Spain. In: Stucker, D., Lopez-Gun, E. (Eds.), Adaptation to Climate Change through Water Resources Management: Capacity, Equity, and Sustainability. Routledge/Earthscan, Oxford, UK.

EEA, 2017. Climate Change, Impacts and Vulnerability in Europe 2017. An indicator-based report. European Environment Agency. EEA Report.

Feyen, L., Dankers, R., Bodis, K., Salamon, P., Barredo, J.I., 2012. Fluvial flood risk in Europe in present and future climates. Climatic Change 112 (1), 47–62.

García de Jalón, S., Iglesias, A., Cunningham, R., Pérez Díaz, J.I., 2014. Building resilience to water scarcity in Southern Spain: a case study of rice farming in Doñana protected wetlands. Regional Environmental Change 14, 1229–1242. https://doi.org/10.1007/s10113-013-0569-5.

García de Jalón, S., Iglesias, A., Quiroga, S., Bardají, I., 2013. Exploring public support for climate change adaptation policies in the Mediterranean region: a case study in Southern Spain. Environmental Science and Policy 29, 1–11.

Garrote, L., Iglesias, A., Garrido, A., 2014. Agriculture and water: some thoughts on adaptation to climate change from a policy perspective. In: Markandya, A., Galarraga, I., Sainz de Murieta, E. (Eds.), Routledge Handbook on Economics of Adaptation to Climate Change, first ed. Routledge, London, ISBN 978-0-415-63311-6, p. 464.

Garrote, L., Granados, A., Iglesias, A., 2015a. Strategies to reduce water stress in Mediterranean river basins. Science of the Total Environment 543 (2016), 997–1009.

Garrote, L., Iglesias, A., Granados, A., Mediero, L., Martin-Carrasco, F., 2015b. Quantitative assessment of climate change vulnerability of irrigation demands in Mediterranean Europe. Water Resources Management 29 (2), 325−338. https://doi.org/10.1007/s11269-014-0736-6.

Gleeson, T., Wada, Y., Bierkens, M.F.P., van Beek, L.P.H., 2012. Water balance of global aquifers revealed by groundwater footprint. Nature 488, 197−200.

Gleick, P., Palaniappan, M., 2010. Peak water limits to freshwater withdrawal and use. Procceedings of National Academy of Sciences, 1004812107v1−8.

Glenk, K., Fischer, A., 2010. Insurance, prevention or just wait and see? Public preferences for water management strategies in the context of climate change. Ecological Economics 69, 2279−2291.

Huntjens, P., Pahl-Wostl, C., Grin, J., 2010. Climate change adaptation in European river basins. Regional Environmental Change 10, 263−284.

Iglesias, A., Buono, F., 2009. Towards sustainability of water policies in Mediterranean countries: approaches in the SWAP project. Current Opinion in Environmental Sustainability 1 (2), 133−140.

Iglesias, A., Garrote, L., 2014. Drought in the light of climate change in the Mediterranean area. In: Quevauvalier, P. (Ed.), Prevention of Hydrometeorological Extreme Events − Interfacing Sciences and Policies. Willey.

Iglesias, A., Garrote, L., 2015. Adaptation strategies for agricultural water management under climate change in Europe. Agricultural Water Management 155, 113−124.

Iglesias, A., Garrote, L., Diz, A., Schlickenrieder, J., Martin-Carrasco, F., 2011a. Rethinking water policy priorities in the Mediterranean region in view of climate change. Environmental Science and Policy 14, 744−757.

Iglesias, A., Mougou, R., Moneo, M., Quiroga, S., 2011b. Towards adaptation of agriculture to climate change in the Mediterranean. Regional Environmental Change 11 (S1), 1−8.

Iglesias, A., Quiroga, S., Diz, A., 2011c. Looking into the future of agriculture in a changing climate. European Review of Agricultural Economics 38 (3), 427−447.

Iglesias, A., Sánchez, B., Garrote, L., López, I., 2017. Towards adaptation to climate change: water for rice in the coastal wetlands of Doñana, Southern Spain. Water Resources Management 31, 629−645.

Iglesias, A., Garrote, L., Flores, F., Moneo, M., 2007. Challenges to manage the risk of water scarcity and climate change in the Mediterranean. Water Resources Management 21 (5), 227−288.

IPCC, 2014. In: Climate Change 2014: Fifth Assessment Report of the Intergovernmental Panel on Climate Change. Cambridge, United Kingdom and New York, NY, USA.

Ivey, J., Smithers, J., de Loe, R., Kreutzwiser, R., 2004. Community capacity for adaptation to climate-induced water shortages: linking institutional complexity and local actors. Environmental Management 33 (1), 36−47.

Ma, J., Hipel, K., De, M., Cai, J., 2008. Transboundary water policies: assessment, comparison and enhancement. Water Resources Management 22 (8), 1069−1087.

Oweis, T., Hachum, A., 2005. Water harvesting and supplemental irrigation for improved water productivity of dry farming systems in West Asia and North Africa. Agricultural Water Management 80, 57−73.

Palmer, M., Lettenmaier, D., Poff, N., Postel, S., Richter, B., Warner, R., 2009. Climate change and river ecosystems: protection and adaptation options. Environmental Management 44 (6), 1053−1068.

Pulido-Velazquez, D., Garrote, L., Andreu, J., Martin-Carrasco, F.J., Iglesias, A., 2011. A methodology to diagnose the effect of climate change and to identify adaptive strategies to reduce its impacts in conjunctive-use systems at basin scale. Journal of Hydrology 405, 110−122.

Quevauviller, P., Barceló, D., Beniston, M., Djordjevic, S., Harding, R.J., Iglesias, A., Ludwig, R., Navarra, A., Navarro Ortega, A., Mark, O., Roson, R., Sempere, D., Stoffel, M., van Lanen, H.A.J., Werner, M., December 1, 2012. Integration of research advances in modelling and monitoring in support of WFD river basin management planning in the context of climate change. Science of the Total Environment 440, 167–177.

Vorosmarty, C., Green, P., Salisbury, J., Lammers, R.B., 2000. Global water resources: vulnerability from climate change and population growth. Science 289, 284–288.

World Bank, 2010. Mainstreaming Adaptation to Climate Change in Agriculture and Natural Resources Management Projects. Available at: http://climatechange.worldbank.org/climatechange/content/mainstreaming-adaptation-climate-change-agriculture-and-natural-resources-management-project.

PART 2

Crop Water Stress Management and Water Use Efficiency Under Drought Conditions

CHAPTER 5

Agronomical Effects of Deficit Irrigation in Apricot, Peach, and Plum Trees

Arturo Torrecillas[1], Mireia Corell[2], Alejandro Galindo[3], David Pérez-López[4], Houssem Memmi[4,6], Pedro Rodríguez[5], Zulma N. Cruz[5], Ana Centeno[4], Diego S. Intrigliolo[1], Alfonso Moriana[2]

[1]Centro de Edafología y Biología Aplicada del Segura (CEBAS-CSIC), Murcia, Spain; [2]Universidad de Sevilla, Seville, Spain; [3]University of Twente, Enschede, The Netherlands; [4]Technical University of Madrid, Madrid, Spain; [5]Instituto Nacional de Ciencias Agrícolas (INCA), San José de Las Lajas, Cuba; [6]IRIAF, Ciudad Real, Spain

Contents

1. INTRODUCTION

Stone fruit trees are deciduous, drupe fruit species that are cultivated all around the world. Although there are a great number of cultivated fruit trees that could be included in this chapter, peach, apricot, and plum are probably the most important because of their cultivated area and irrigation needs in the main producer countries. Water needs for these species are high, with crop coefficients at midseason around 1 (Steduto et al., 2012), and could be greater in high-density orchards or when using cover crops (Fereres and Goldhamer, 1990).

The number of countries that cultivate these species is very high; however, according to production, Asia and Europe are the most important zones. Fig. 5.1 shows the distribution of 80% of world production for

Water Scarcity and Sustainable Agriculture in Semiarid Environment
ISBN 978-0-12-813164-0
https://doi.org/10.1016/B978-0-12-813164-0.00005-3

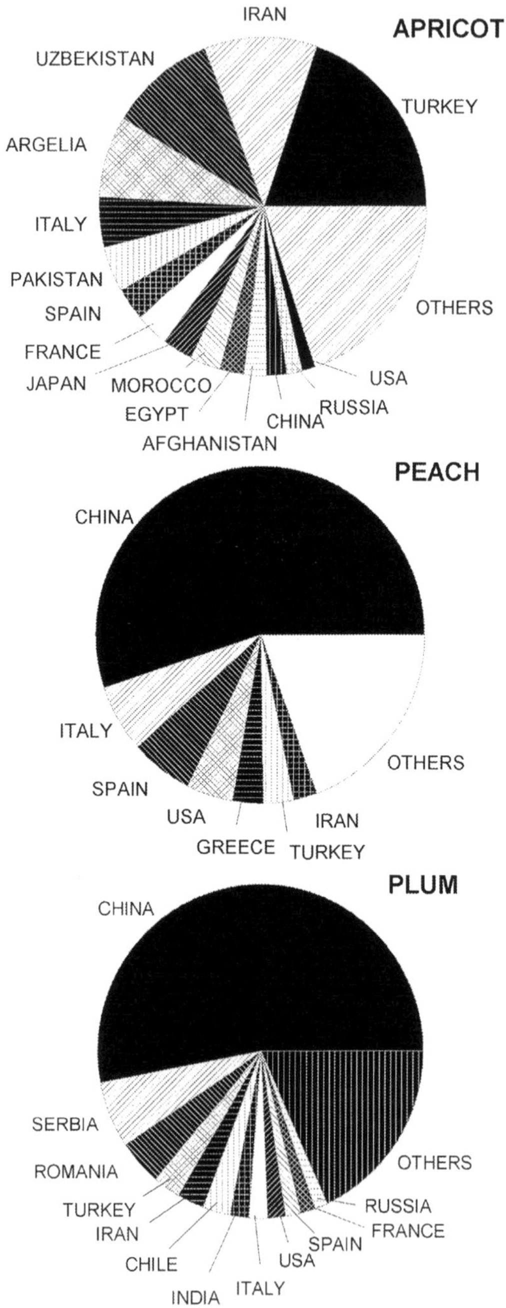

Figure 5.1 Main world producers of apricot, peach, and plum. Sectors represent the percentage of world production of each country. In the three graphs, "Others" represents around 20% of world production. *(Based on FAO (Food and Agriculture Organization of the United Nations), February 2, 2017. Crop Production and Water Use in Agriculture. Available from: http://www.fao.org/faostat/en/#data/.)*

apricot, peach, and plum. For apricot, almost 50% of total world production originates in Turkey, Iran, Uzbekistan, and Argelia; however, a lot of countries have significant roles in the production of this fruit tree and around 12 countries from very different zones generate 30% of total world production (Fig. 5.1).

On the other hand, peach and plum trees present a reduced number of major producers. Eighty percent of world peach production is concentrated in just seven countries, with China, being the most important, at 55% (Fig. 5.1). Mediterranean European countries such as Italy, Greece, Spain, and Turkey are also important zones for peach production. The situation with plum trees is somewhere in between peach and apricot trees. Although China is again the main world producer with around 50%, 11 different countries produce 30% (Fig. 5.1). In this latter group the variability is very high among Mediterranean European countries such as Turkey, Spain, Italy, and France, but also in Continental European countries such as Serbia and Romania, and in American (United States and Chile) or Asian countries (Iran, Russia, and India).

World production and the surface dedicated to these fruit crops have increased in the last 10 years (FAO, 2017). Table 5.1 shows the pattern of world production and dedicated surface together with the average yield and the percentage of water used in agriculture by the main world producers. World production of apricot and peach trees increased by more than 40% from 2003 to 2013, with small changes in the surface involved (Table 5.1).

Table 5.1 Variations in the last 10 years, from 1993 to 2013, in world production (10^6 T) is related to the increase in surface (10^6 ha) and yield (T ha^{-1}) (between brackets the percentage of increase). Average water use in agriculture (%) in 2010 of the main producer countries (Fig. 5.1) is included

	Apricot	Peach	Plum
World production (10^6 T)	From 2.9 to 4.1 (43%)	From 14.9 to 21.6 (47%)	From 9.9 to 11.5 (17%)
World surface (10^6 ha)	From 0.43 to 0.50 (17%)	From 1.42 to 1.54 (8%)	From 2.20 to 2.66 (21%)
World yield (T ha^{-1})	From 6.7 to 8.2 (22%)	From 10.5 to 14.1 (34%)	From 4.5 to 4.3 (−3%)
Water use in agriculture (%)	64.7	68.0	51.2

Based on FAO (Food and Agriculture Organization of the United Nations), February 2, 2017. Crop Production and Water Use in Agriculture. Available from: http://www.fao.org/faostat/en/#data/.

Therefore most of this increase was associated with improving orchard yield, likely related to orchard intensification (Steduto et al., 2012). On the other hand, changes in plum production are small, with an increase in the amount of product likely related to the increase in surface. All these patterns are strongly associated with water management and imply an increase in water needs. The average of water percentage used in agriculture of the world's main producers is around 65% for peaches and apricots and higher than 50% for plums, where continental production is important in comparison to the former fruits. Such an amount of water use in agriculture is even more important when considering the countries with the greatest production, and it reaches values around 80% (i.e., China, Turkey, and Iran).

These data suggest that most of the stone fruit trees probably grow in regions with a strong seasonal water deficit and, according to predictions of climate changes, these needs will increase (Houghton et al., 1995; Collins et al., 2009). In arid and semiarid regions, the occurrence of water deficits in plants is inevitable and the scarce rainfall has to be supplemented by irrigation to ensure regular crop yields and to reduce interannual yield variability (Oweis, 1997). This chapter presents the main effect of water deficit on the yield and quality of the stone fruits apricot, peach, and plum. Strategies regarding irrigation scheduling in conditions of water restriction will also be discussed.

2. DEFICIT IRRIGATION IN STONE FRUITS

Maximum yield and quality, considered only as premium size, were the main objectives of irrigation scheduling in stone fruits until the middle of the 20th century. From the 1970s, the increase in labor costs and water restrictions in some regions raised the possibility of reducing the pruning costs using irrigation restriction as the most efficient tool. This strategy was developed for the first time in peach orchards (Chalmers et al., 1981) and thereafter has spread to the rest of the species (Steduto et al., 2012). Nowadays, moderate water stress conditions are considered for many fruit tree species, not only to save water or to control vegetative growth, but also to increase fruit quality, understood as the improvement of fruit flavor (Behboudian et al., 2011). Deficit irrigation (DI) implies a water stress period. Controlling the moment and level of water stress is critical (Hsiao, 1990) and could produce long-term effects in tree physiology (Fereres and Goldhamer, 1990).

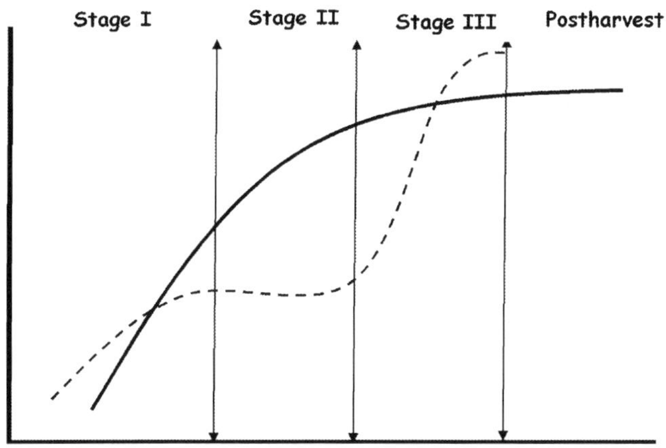

Figure 5.2 Seasonal pattern of fruit development (*dashed line*) and vegetative growth (*solid line*) in stone fruits. Irrigation season is divided into four stages according to fruit development from flowering to rest. The length of these stages is variable between species and even cultivars.

Each phenological stage has a different response and susceptibility to water stress and therefore different effects on the final food product. In irrigation works, the phenological stage is defined according to fruit development. In stone fruits (Fig. 5.2), four different phases are considered: the first rapid fruit growth (stage I), pit hardening (stage II), the second rapid fruit growth (stage III), and postharvest (stage IV). Vegetative growth occurs mainly during stages I and II (Fig. 5.2) and for this reason an effective control of vigor is possible with null or low effects on fruit size (Behboudian and Mills, 1997; Naor, 2006). The main differences between stone fruit species and even cultivars are the length of these periods, which could produce great differences in DI scheduling (Naor, 2006; Steduto et al., 2012).

Control of water stress application, moment, and level is strongly related to yield response. In this sense, irrigation works are grouped in sustained deficit irrigation (SDI) strategies and regulated deficit irrigation (RDI) schedules. SDI scheduling consists of a progressive reduction in tree available water using a percentage of crop evapotranspiration (ET_c) throughout the season. Therefore the level, moment, and duration of water stress changes among orchards and even for different seasons in the same orchard. Conversely, RDI selects the moment and level of water stress (Naor, 2006; Steduto et al., 2012). These scheduling differences between both approaches produce great variations in the relationship yield versus water applied. Girona et al. (2012), for instance, reported two different production functions in peach trees according to DI scheduling. When data from works using sensitive periods or SDI approaches are considered, the relationship is linear, while yield is maximal from an amount of water applied when works of RDI in resistant periods are used (Girona et al., 2012). Therefore the level of water stress, rather than the water applied, is related to the yield response. Partial root drying (PRD) is a good example of attempts to control tree physiology in DI instead of water applied. PRD is based on irrigating only one part of the root zone in each irrigation event, leaving another part to dry to a certain soil water content before rewatering by shifting irrigation to the dry side (Dry and Loveys, 1998; Sepaskhah and Ahmadi, 2010). Abscisic acid of the dry roots reduces tree transpiration, while assimilation is still high and is not affected by the final yield. PRD was very popular in stone fruit irrigation research during the first years of this century, but the real advantage in comparison to RDI is not clear.

The development of continuous plant and soil water status sensors in the last 20 years has opened new lines for DI scheduling. Although the water applied is still the main factor, in many irrigation works there is an increase in the number of experiments that include the use of an indicator derived from these sensors in the schedule. The revision of Fereres and Goldhamer (1990) or Behboudian and Mills (1997), concerning DI in fruit trees, or those of Girona (2002), more specifically concerning DI in peach trees, only discussed the limitation of widely used DI strategies. On the contrary, the last revisions of irrigation management for stone fruits such as Naor (2006), Behboudian et al. (2011), or Steduto et al. (2012) suggest or at least present in some of these stone fruit species threshold values related to plant or soil water status indicators.

3. APRICOT (*PRUNUS ARMENIACA* L.)

Apricots originated on the Russian/Chinese border around 3000 BC and were imported along with peach seeds into Europe via the Silk Road that extended camelback trading to the Middle East. The fruit spread toward the north of India, Punjab, and Tibet and reached Europe from Persia. In Rome, this fruit was not consumed until the 1st century and arrived in Greece during the campaigns of Alexander the Great. The Arabs were responsible for its spreading. Apricot trees are believed to have arrived in America as seeds for growing fruit trees by Spanish explorers and missionaries.

The apricot tree, which is widely cultivated in Mediterranean countries, displays xeromorphic characteristics, such as the ability to endure water stress in the dry season or loss of leaves in winter. For these reasons, despite the crop requiring irrigation for commercial production, it can be grown with marginal water supply.

According to Torrecillas et al. (1999), the mechanisms developed by this crop in response to water deficit are primarily avoidance mechanisms. Stomatal regulation is an early mechanism in place to control water loss via transpiration and avoid reductions in leaf water potential. When a more severe water stress is achieved, other complementary avoidance mechanisms are triggered, such as epinasty, reduced net photosynthesis, and a degree of leaf shedding. Some drought tolerance mechanisms have also been observed in apricot, and they were mainly based on the existence of a highly relative apoplastic water content, low leaf osmotic potentials at the turgor loss point, and active leaf osmotic adjustment, which contribute to the maintenance of leaf turgor during water stress.

Apricot's response to water stress depends more on the physiological processes that take place in the plant than on the water stress level reached and/or its duration (Torrecillas et al., 2000). Nevertheless, from a yield point of view, the apricot tree has two phenological periods that are particularly sensitive or critical to water stress. The first critical period corresponds to the second rapid fruit growth stage and the second to the postharvest period. However, the causes of the reduction in yield are very different. Water stress during the second rapid fruit growth stage induces a reduction in yield due to the smaller fruit size at harvest and earlier ripening. However, water stress immediately after harvest reduces yield the following year, due to an increase in young fruit drop, which leads to significantly lower final fruit set (Torrecillas et al., 2000; Kaya et al., 2010). This latter behavior can be ascribed to the observations of Ruiz-Sánchez et al. (1999), who indicated that the lower final fruit set could be due to the effect of water stress on floral bud initiation and differentiation occurring in this period (Uriu, 1964) and/or to the lower percentage of pollen germination observed in the grains of the following year's bloom. On the other hand, these authors also indicated that apricot trees possess advantageous characteristics that can be used for reduced irrigation practices. These are related to the separation between the main periods of shoot and fruit growth, the apricot's ability to compensate for growth when full irrigation is restored, and the fact that there are several phenological periods in which irrigation withholding does not affect yield and fruit quality.

Water saving of 50% using SDI proved to be detrimental for maintaining the apricot fruit yield over time due to the significant reductions in vegetative growth and the reduction in the number of fruit-bearing points (Pérez-Pastor et al., 2009, 2014). By contrast, RDI treatments provided similar yields to fully irrigated trees in years with moderate water deficits (water saving < 22%) and when the water deficit was applied only before the beginning of the second rapid fruit growth stage (stage III) and 2 months after harvest (Pérez-Pastor et al., 2009, 2014). For this reason, RDI can be considered a practical strategy in areas suffering from permanently limited water resources, since it is possible to obtain similar yields and fruit quality to those obtained in fully irrigated trees (Pérez-Pastor et al., 2009).

Irrigation studies in apricot trees using plant or soil tools are not very common. Bussi and Plenet (2013) showed that, for young apricot trees when irrigation was scheduled using trunk growth rate and maximum daily shrinkage (MDS) measurements, continuous measured derivatives of daily trunk diameter fluctuations, it was possible to achieve considerable reductions in the amount of water applied over the 5 years of the experiment, in comparison with methods in which irrigation was based on ET_c or tensiometric measurements (by −66% and −54%, respectively), without significantly affecting fruit yields or average fruit weights. Also the total soluble solids (TSS) content tended to be higher, and the extent of the red color of the peel was higher with the dendrometric method than under the ET_c or tensiometric methods. The incidence of pitburn was lower with the dendrometric method than with the ET_c or tensiometric methods, and it appeared to be correlated with the percentage of the total water applied before fruit ripening. Water restriction had an impact on the incidence of pitburn by improving calcium concentrations in the apricot fruit. In this sense, calcium delays senescence in fruit by protecting cell membranes against breakdown (McGuire and Kelman, 1984), which allows oxidative enzymes to reach the phenols inside the cell and causes a brown discoloration (Radi et al., 2003). Moreover, water constraints favor the production of some volatile antioxidant compounds in fruit that could stimulate the defense against physiological disorders (Gómez-Rico et al., 2009).

The effect of deficit management in the final quality of fruits has also been poorly studied. Pérez-Pastor et al. (2007) studied the effect of DI on apricot fruit quality at harvest and during storage. In this line, a fully irrigated (100% ET_c) treatment was compared with SDI (50% ET_c) and RDI (100% ET_c during the critical periods and 25% ET_c during the rest of the season) treatments. The main benefits of both DI treatments over the control could be summarized as a slight increase in TSS and firmness at harvest and during cold storage periods. In addition, RDI exhibited lower weight losses and fungal attacks by *Rhizopus* sp. and *Monilinia* sp. during the subsequent 4-day period of retail sale. Consequently, DI practices are a relevant strategy not only to preserve water in apricot trees crops, but also to improve fruit quality and safeguard it during fruit storage.

4. PEACH (*PRUNUS PERSICA* [L.] BATSCH)

Peach is a deciduous tree of the Rosaceae family that grows in the warmer temperate regions of both the Northern and Southern Hemispheres. It bears an edible juicy fruit called peach or nectarine. Probably, it is native of the region of Northwest China, where it was first domesticated and cultivated, possibly between 1000 and 4000 BC. It then spread westward through Asia to the Mediterranean and later to other parts of the world; it was Spanish explorers who took it to the New World (Faust and Timon, 2010). Actually, peach is the fourth most important fruit crop in the world and the second in Europe. Even in areas with relatively high rainfall, irrigation is required to increase peach fruit size and yield (Layne and Tan, 1984).

Similar to most woody crops, peach trees resist dehydration through drought avoidance and tolerance mechanisms. The avoidance of drought in peach trees is related mainly to morphological characteristics (Rieger et al., 2003) as efficient stomatal control, which may minimize water loss by increasing water use efficiency (Rieger and Duemmel, 1992). Also the delay in stomatal opening when water deficit resumes can be considered as another avoidance mechanism to regain full turgor more effectively and to maintain leaf productivity, allowing the crop to compete better in drier areas (Mellisho et al., 2011). Drought tolerance characteristics have also been observed in peach trees. In this sense, the existence of high apoplastic relative water content, a common characteristic in xeromorphic plants (Cutler et al., 1977), and the maintenance of relatively high values under water deficit, could contribute to the maintenance of leaf turgor during stress (Torrecillas et al., 1999). In contrast with other authors (Young et al., 1981; Steinberg et al., 1989; Escobar-Gutierrez et al., 1998), Arndt et al. (2000) and Mellisho et al. (2011) indicated that mature leaves of field-grown peach trees exhibited an active osmotic adjustment. This divergence could be due to the capacity for osmotic adjustment, which varies in function of the species as well as the cultivar (Lakso, 1990; Torrecillas et al., 1996) or because the active osmotic adjustment may be developed only when water deficit is applied gradually over a prolonged period (Turner and Jones, 1980).

Most studies on peach tree response to irrigation have been focused on the importance of increasing water use efficiency under arid or semiarid conditions (Boland et al., 1993a; Boland et al., 1993b). Also RDI was first adopted in peach orchards and its success has been mainly based on the fact that pit hardening (stage II) is quite insensitive to water stress (Chalmers et al., 1981; Li et al., 1989).

The main effect of water restriction on the first fruit growth (stage I) is usually associated with a reduction in fruit size. In contrast with Boland et al. (2000), who indicated that the application of RDI during the early stages of fruit growth (stage I) until the end of shoot growth (stage II) slightly influences fruit size and number, Li et al. (1989), using a very late peach cultivar (Merrill Sundance) in not very arid conditions, reported that irrigation restrictions during stage I increased fruit size and yield, probably due to the favorable climatic conditions (700 mm of rain per year).

Considering that stage II happens frequently together with high rates of vegetative growth, RDI can affect tree crown development (Mitchell and Chalmers, 1982; Boland et al., 2000). This reduction in tree vegetative growth has a very positive impact because (1) it reduces pruning costs, (2) it diminishes the competition between vegetative and reproductive growth, increasing tree efficiency, and (3) it reduces tree-light interception by the tree crown in high-density orchards, improving the economic success of the orchards. Because the earliest studies on the response of fruit trees to RDI were performed in peach orchards, many papers showed that in addition to the effects on tree vigor RDI, stage II provides beneficial effects (Chalmers et al., 1981; Mitchell and Chalmers, 1982; Li et al., 1989; Williamson and Coston, 1990; Boland et al., 1993b, 2000), mainly maintaining the fruit yield and improving the concentration of fruit-soluble solids (Li et al., 1989; Crisosto et al., 1994; Ben Mechlia et al., 2002; Behboudian et al., 2011).

In agreement with Behboudian et al. (2011), Botía et al. (2004) showed that RDI during stage II of peach fruit growth, in midseason Babygold peach cultivar, allowed water to be saved without affecting the yield. Nevertheless, other authors have found different peach responses to irrigation restrictions in stage II. Domingo et al. (2011) showed that irrigation restriction during pit hardening (stage II) in unthinned peach trees (cv. Andross) for the processing industry affected trunk growth, yield, and fruit size. These different behaviors could be due to the effect of other factors, such as fruit load and soil water-holding characteristics. Thus high fruit loads tend to increase fruit growth sensitivity to irrigation restrictions (Berman and DeJong, 1996) and, in soils with low water-holding capacity, irrigation restriction reduces shoot growth quickly (Boland et al., 2000). In this sense, Thakur and Singh (2013) in early (Spring Bright) and midseason (Summer Bright) nectarines, and Valverdu et al. (2012) in "Tardibelle" peach, showed that water deficit during stage II can have a negative effect on fruit size, only if the water deficit persists at the onset of stage III. Moreover, as water deficit increased in both nectarines during stage II, a significant increase in nectarine fruit total sugars, sucrose, and sorbitol and a decrease in total acids and malic acid were noted (Thakur and Singh, 2013), whereas water deficit during peach stage II advanced fruit ripening increased the consumer acceptance of fruit and improved important sensory attributes, such as sweetness and flavor intensity (Valverdu et al., 2012).

The second fruit growth (stage III) is the most sensitive to irrigation restriction from a yield point of view. Domingo et al. (2011) in Andross peach confirmed that fruit size is more sensitive to water deficit during stage III than during stage II of fruit growth. In both cases, a mild irrigation restriction induced reductions in fruit size and yield, which were due to an indirect effect through tree growth when the water deficit occurred during stage II, and to a direct effect of lower water supply, when the water deficit occurred during stage III. In terms of fruit quality, a water deficit during stage III increased TSS juice and flesh firmness, parameters appraised by the peach processing industry (Crisosto et al., 1994; Besset et al., 2001; Domingo et al., 2011). In contrast, Lopez et al. (2010) and Lopez et al. (2012) in an O'Henry peach cultivar (mid—late ripening) showed that a severe water deficit during stage III reduced the marketable value of the fruits, because, in spite of the observed increase in fruit dry matter content, the TSS/TA ratio did not improve, fruit size and fruit juiciness were reduced, and fruit astringency increased with a somewhat salty flavor. Also these authors indicated that yield reduction by the fruit size effect could be partially compensated by applying heavy fruit thinning, even though thinning was not effective in improving fruit chemical characteristics.

Research into the response of peach trees to water deficit from harvest until leaf abscission, the postharvest period, is scarce. Different works suggest two periods with different sensibilities to water deficit: early postharvest, when flower induction and differentiation occurs, and late postharvest (Naor, 2006). Nevertheless, it is known that postharvest irrigation withholding reduces the vegetative growth of early-maturing peach trees (Larson et al., 1988; Johnson et al., 1992; Ghrab et al., 1998), mainly due to the reduction in the growth of waterspouts and lateral shoots, without affecting the growth of fruiting shoots (Dichio et al., 2007). In this sense, Dichio et al. (2007) also showed that an RDI strategy, inducing a moderate water stress during postharvest, improves irrigation efficiency and saves water, while maintaining yield and fruit quality in the following season. On the other hand, severe water stress conditions during early postharvest have been associated with fruit disturbances in the following season (Johnson et al., 1992). However, Conejero et al. (2011) scheduled postharvest RDI, using MDS measurements and inducing a moderate water deficit during the early postharvest period and a more severe water deficit during the noncritical period (late postharvest period), concluding that it was possible to save a considerable amount of water (35%–42%) without affecting the total yield, yield efficiency components, or the distribution of different peach fruit categories. Qassim et al. (2013) showed that substantial water savings can be made in "Tatura 204" peach orchards by the combination of cutting back postharvest irrigation to 20% of crop water requirement and fruit thinning to a defined crop load in the subsequent season.

There has been little work done with PRD on peach trees and the results obtained are somewhat discouraging. In this sense, Tapia et al. (2009) indicated that PRD yielded similarly to SDI, but these yields were lower than those in fully irrigated peach trees. Also Goldhamer et al. (2002) did not record significant physiological or horticultural differences between the RDI and their companion PRD treatments, except for less negative early morning stem water potential with PRD at the end of stage II.

Fruit quality in relation to irrigation management considering cold storage has also been studied. DI during fruit growth and postharvest in Viowhite 5 nectarine plants provides fruits with good quality after chilling storage, both in air and in a controlled atmosphere, because, compared to fully irrigated plants, DI nectarines maintain their initial antioxidant capacity level and total phenolic content, as well as increase the total vitamin C content (Falagán et al., 2013). Pliakoni et al. (2010), in early- to midseason-ripening nectarine trees, showed that RDI increases the fruit phenolic content, in particular proanthocyanidins and mainly in sun-exposed peach fruits. Cropping practices focused on increasing light availability, such as reflective mulch, pruning, etc., can help to increase the fruit phenolic content. Falagán et al. (2015) studied the effect of RDI during stage II and late postharvest on fresh-cut, white flesh, extra-early

VioWhite 5 nectarines and concluded that RDI induced a more stable antioxidant capacity and soluble phenolic content during storage at 5°C.

Irrigation scheduling based on water status parameter is not very common, although threshold values for different approaches have been suggested (Table 5.2). Girona et al. (2012) suggest different midday stem water potential values, varying from the most severe water stress conditions (stage II, in late cultivars, or early postharvest, in early cultivars) with values of −1.8 MPa to fully irrigated conditions in the rest of the season, around −1.0 MPa. Such recommendations are easier than the use of ET_c restriction, which should incorporate the type of soil (Girona et al., 2012) and show the advantage of plant water status approaches.

Trunk diameter fluctuation is a continuous measurement more sensitive to drought than the water potential in peach trees (Goldhamer et al., 1999). MDS has been reported as the best indicator of these daily curves in peach trees.

Li et al. (1989) suggested values between 10 and 30 μm of MDS in fully irrigated conditions (Table 5.2). Several authors have suggested threshold values for peach trees and these have been applied successfully in the irrigation scheduling of stage III for potted experiments (50−100 μm, Mercier et al., 2009, or 200−400 μm, Besset et al., 2001). These great differences

Table 5.2 Different threshold values and the phenological stage when they should be applied during the irrigation scheduling of peach orchards for fully irrigated (FI) and deficit-irrigated (DI) periods. Values of midday stem water potential are included only for late cultivars

Stage	Threshold	References
Midday stem water potential		
Stage I (FI)	−0.9 MPa	Girona et al. (2012)
Stage II (DI)	−1.8 MPa	
Stage III (FI)	−1.0 MPa	
Postharvest (DI)	−1.8 MPa	
Maximum daily shrinkage (MDS)		
Fully irrigated conditions	10−30 μm	Li et al. (1989)
Stage III (DI)	50−100 μm	Besset et al. (2001)
	200−400 μm	Mercier et al. (2009)
MDS signal		
Stage II and postharvest (DI)	1.4	Conejero et al. (2011)
	2	Mirás-Avalos et al. (2016b)

between MDS threshold values during stage III (Table 5.2) in experiments performed in the same location (Middle Rhône Valley, France) but with different cultivars suggest that the absolute value of MDS is not an adequate parameter. Since MDS is strongly related to the evaporative demand, this indicator should be normalized with fully irrigated MDS expected in each condition (Goldhamer and Fereres, 2001). Experiments that have used this normalized indicator (hereinafter, the MDS signal) for peach orchards reported a great reduction of the water applied (around 40%) in stage II and postharvest (Conejero et al., 2011; Mirás-Avalos et al., 2016a) with MDS signals between 1.4 and 2. These latter works estimated the MDS signal with an overirrigated treatment, which is unsuitable for commercial orchards. In this way, baselines that estimated fully irrigated MDS have been published for most fruit trees (Ortuño et al., 2010), including peach trees (Conejero et al., 2007; Mirás-Avalos et al., 2016b), although they have not been validated out of the orchard where they have been obtained.

5. PLUM (*PRUNUS DOMESTICA* L.; *PRUNUS SALICINA* L.)

Plum has many species, but only two of them have worldwide commercial significance: the hexaploidy European plum (*Prunus domestica* L.) or "prunes," originating in the Caucasian mountains adjacent to the Caspian Sea, and the diploid Japanese plum (*Prunus salicina* L. and hybrids), originating in China. Both species are medium-sized deciduous stone fruit trees, which differ notably in terms of their climatic requirements. European plums are cultivated in temperate climates, as they require winter cold conditions to break endodormancy and are relatively late flowering, while Japanese plums grow better in temperate warmer regions, as they require less chilling temperatures and are usually early flowering. Their productive use also differs, as Japanese plums are mainly grown for fresh fruit, while dried fruit is mainly obtained from European plum varieties.

Commercial varieties of plum tree break dormancy and begin flowering in the Northern Hemisphere between late February and mid-April, depending on the environmental conditions and the cultivar. Bud formation occurs from leaf throughout June on mature trees. Reproductive buds are in a lateral position on terminal shoots or on short shoots called spurs. Flower buds are initiated in the growing season prior to anthesis, and development continues throughout the dormant season until the following spring, just before bud break. The proportion of spur and terminal shoots varies largely with varieties and species, and so does the proportion of fruit borne on spurs and shoots, which also varies with the tree age. Most commercial plum cultivars are not self-pollinating and therefore the use of pollinators is required. Plum trees bloom very profusely and thinning is required, performed either manually, chemically, or mechanically, to obtain commercial fruit sizes.

Fruit with a fleshy pericarp is classified as drupe and is usually single seeded. Fruit growth follows the typical double-sigmoidal pattern, with rapid exponential growth during the cell division phase (stage I, 30 days or less in duration), followed by a relatively short period of slow growth during pit hardening and embryo development (lag phase, stage II), and finally a second period of rapid cell enlargement prior to harvest (stage III), when the fruit can increase in size by c.40%—60%, although this is linked to degree-days after flowering, in a similar fashion to other *Prunus* species (DeJong and Goudriaan, 1989). Therefore the duration of each stage varies with varieties and locations. During the postharvest period, some shoot growth and carbohydrate storage for reserves are the primary sinks of photosynthesis products, which continue until leaf fall.

In terms of physiological responses to water or salinity stress, few studies have been specifically conducted on plum trees. Ziska et al. (1984) determined that applying saline stress under field conditions resulted in leaf water deficits only during the last stage of fruit growth. During the rest of the season there was a clear reduction in stomatal conductance and hence in leaf transpiration in response to irrigation with high water salinity levels. Furthermore, Naor (2004) also observed that stomatal conductance regulation played a crucial role for controlling plant water relations in response not only to drought, but also to tree crop level. Indeed, Naor (2004) found stomatal conductance to be highly correlated with leaf water potential when leaves were covered by a plastic bag prior to, or immediately after, their detachment. Subsequently, attempts have been made to modify plant water resistance to drought by transgenic overexpression of a cytosolic ascorbate peroxidase gene (Díaz-Vivancos et al., 2016). Promising results in terms of drought tolerance under pot conditions should be now confirmed in the field where the much slower development of water stress might modify the whole plant response to water stress and the agronomic water use efficiency.

The effects of water restrictions on plum tree performance were first reviewed by Intrigliolo and Castel (2012), who briefly summarized the main findings from experiments reported up to that date. Since then, no further studies have been conducted for French prunes where it is concluded that the general strategy recommended for imposing water deficits is to limit such deficits during the early stages of tree and crop development, imposing a more severe stress in mid- and late season.

In Japanese plum trees, further results have been analyzed, reaching more robust conclusions. In mid- or late season-maturing cultivars, such as Black-Gold and Angeleno, both Intrigliolo and Castel (2012) and Samperio et al. (2015a,b) concluded that the best DI strategy is to combine water restrictions during both phase II of fruit growth and postharvest. To avoid reductions in fruit size at harvest, during fruit growth, plant water stress should be strictly limited to phase II and midday stem water potential should not exceed the threshold of −1.5 MPa (Intrigliolo and Castel, 2012, Table 5.3). During postharvest, it seems that for midseason-maturing cultivars different levels of water stress are suggested by Intrigliolo and Castel (2012) and Samperio et al. (2015b). However, this might not be that simple because for the Angeleno variety, plant water stress more severe than −1.35 MPa was not reached (Samperio et al., 2015b), while in Black-Gold, values as low as −2.0 MPa can be recommended (Intrigliolo and Castel, 2012). A good compromise between the two recommended values can be the one reported by Samperio et al. (2015a), who suggested for the postharvest period of the early season-ripening Red-Beaut a threshold midday stem water potential value of −1.65 MPa.

The midday stem water potential appears to be the most adequate indicator of water stress for irrigation scheduling for plum trees. This was noticed from the first seminal work by McCutchan and Shackel (1992). Many attempts have been made to determine the possibility of using trunk dendrometers and particularly MDS as a surrogate for estimating midday

Table 5.3 Threshold values of midday stem water potential for scheduling irrigation in midseason- or late-ripening plum trees for fully irrigated (FI) and deficit-irrigated (DI) periods

Stage	Threshold	References
Midday stem water potential		
All season (FI)	SWP = −0.41−0.12 VPD	McCutchan and Shackel 1992
Stage I (FI)	−0.7 MPa	Intrigliolo and Castel (2012),
Stage II (DI)	−1.5 MPa	Samperio et al. (2015a)
Stage III (FI)	−1.0 MPa	
Postharvest (DI)	−1.65 MPa	

SWP, midday stem water potential (MPa); *VPD*, vapor pressure deficit (kPa).

stem water potential. However, factors such as tree age and tree crop load also influence the relationship between both indicators (Intrigliolo and Castel, 2006, 2007). Samperio et al. (2014) attempted to predict the midday stem water potential using a crop simulation model, namely, CropSyst, adapting it to the peculiarity of woody crops. The results obtained were very satisfactory for the Angeleno cultivar, but for the more vigorous variety, the Red Beaut, the model needed further adjustments. Nonetheless, these results have probably opened a new line of investigation based on attempting to model the soil—plant—water status continuum in plum orchards for obtaining a reasonable prediction of midday stem water potential, useful for scheduling irrigation.

Further results since the review of Intrigliolo and Castel (2012) were also obtained in terms of the long-term effects of DI applied over seven seasons in a young orchard (Intrigliolo et al., 2013). DI, particularly when accompanied by a high crop level, led to smaller trees. This effect was ultimately the cause of the reported carryover effect on yield, which was reduced by 18% in the eighth season, when the previously deficit-irrigated trees were fully watered. Consequently, it is concluded that DI should be applied with caution in young orchards where the canopy is incomplete. Further research is needed in more mature orchards to evaluate the possible detrimental long-term (i.e., more than 5 years) effects of DI on tree productivity.

Interesting results have also been obtained in terms of responses of fruit quality to water restrictions. The general conclusion is that water stress, particularly when applied during fruit growth, improves fruit composition. In this sense, the most recent findings by Maatallah et al. (2015) obtained in cvs. Black-Diamond, Black-Amber, and Black-Start concluded that, when 50% of ET_c is applied during fruit growth, fruit firmness, soluble solids, total pholics, and flavonoids increased. Similarly, Intrigliolo and Castel (2012) also found that fruit from DI applied during stage II of fruit growth was sweeter than fruit from trees irrigated at full ET_c. However, when water stress is applied during postharvest, no effect on fruit quality on the subsequent season's harvest was recorded (Samperio et al., 2015a).

6. SUMMARY AND CONCLUSIONS

Irrigation scheduling allows an improvement in water resources management, which is critical in arid and semiarid conditions. Regulated DI scheduling, controlling the moment and the level of water stress, is more efficient in conditions of water scarcity than sustained DI. This latter irrigation strategy with the same amount of water could induce a severe water deficit in sensitive periods that reduce the quality or quantity of the yield. The drought sensitivity of the phenological stages of these three species is apparently established. The second rapid fruit growth (stage III) is considered the most sensitive to irrigation withdrawal with a reduction in fruit size, although usually an increase in compounds related to fruit quality has also been reported. On the other hand, the pit hardening (stage II) and postharvest stages could be the most adequate to perform a DI strategy if certain limitations are considered. Drought sensitivity during the phenological stages seems changeable, which introduces an important source of uncertainty in irrigation scheduling. This variability is especially important in early-maturing cultivars when fruit development is very short. Such changes have been reported in the postharvest period of peaches, but likely it would be a similar response during the season and in other stone fruit trees. Therefore further work should be performed to revise the definition of critical periods. In this way, the rehydration of the trees could affect the final yield response to irrigation. Severe water stresses or slow rehydration velocity because of poor irrigation management could produce a period of water stress during a sensitive stage. The importance of this delay in the rehydration process is not clear and it could be related to the moment, the level of water stress, and the duration, which supposes a great challenge in future irrigation experiments. This topic is not usually considered in irrigation works, but it could be very common in commercial orchards because recovery could be limited by the irrigation system design or by water availability. Final approaches for DI scheduling should include this information. All these considerations are aimed at continuous monitoring of the soil—plant system and will provide the most efficient irrigation management. The revision of the irrigation works in these three species suggests that an accurate control of water stress could reduce water applications even in sensitive periods, for instance, in peaches, and explain some of the discrepancies between irrigation works. However, a few experiments have been performed using sensors as the main tool of irrigation scheduling, so further work in relation to indicators but mainly threshold values and approaches is needed.

REFERENCES

Arndt, S.K., Wanek, W., Clifford, S.C., Popp, M., 2000. Contrasting adaptations to drought stress in field grown *Ziziphus mauritiana* and *Prunus persica* trees: water relations, osmotic adjustment and carbon isotope composition. Australian Journal of Plant Physiology 27, 985–996.

Behboudian, M.H., Mills, T.M., 1997. Deficit irrigation in deciduous orchards. Horticultural Reviews 21, 105–131.

Behboudian, M.H., Marsal, J., Girona, J., López, G., 2011. Quality and yield responses of deciduous fruits to reduced irrigation. Horticultural Review 38, 149–189.

Ben Mechlia, N., Ghrab, M., Zitouna, R., Ben Mimoun, M., Masmoudi, M., 2002. Cumulative effect over five years of deficit irrigation on peach yield and quality. Acta Horticulturae 592, 301–307.

Berman, M.E., DeJong, T.M., 1996. Water stress and crop load effects on fruit fresh and dry weights in peach (*Prunus persica*). Tree Physiology 16, 859–864.

Besset, J., Genard, M., Girard, T., Serra, V., Bussi, C., 2001. Effect of water stress applied during the final stage of rapid growth on peach trees (cv. Big-Top). Scientia Horticulturae 91, 289–303.

Boland, A.M., Mitchell, P.D., Jerie, P.H., 1993a. Effect of saline water combined with restricted irrigation on peach tree growth and water use. Australian Journal of Agricultural Research 44, 799–816.

Boland, A.M., Mitchell, P.D., Jerie, P.H., Goodwin, I., 1993b. Effect of regulated deficit irrigation on tree water use and growth of peach. Journal of Horticultural Science 68, 261–274.

Boland, A.M., Jerie, P.H., Mitchell, P.D., Goodwin, I., 2000. Long-term effects of restricted root volume and regulated deficit irrigation on peach I. Growth and mineral nutrition. Journal of the American Society for Horticultural Science, 125, 135–142.

Botía, P., Romero, P., Navarro, J.M., Pérez-Pérez, J.G., García Oller, I., Trintran, A.J., 2004. Efecto de diferentes estrategias de riego deficitario controlado sobre la producción y calidad de variedades de media estación de melocotonero (*Prunus persica* L. Batsch). In: Proceeding of the VII Simposium Hispano-Portugués de Relaciones Hídricas en las Plantas, Faro, pp. 53–56.

Bussi, C., Plenet, D., 2013. Effect of three different irrigation regimes on young apricot (*Prunus armen*iaca L. Batsch) trees. Journal of Horticultural Science and Biotechnology, 88, 519–524.

Chalmers, D.J., Mitchell, P.D., Van Heek, L., 1981. Control of peach tree growth and productivity by regulated water supply, tree density and summer pruning. Journal of the American Society for Horticultural Science 106, 307–312.

Collins, R., Kristensen, P., Thyssen, N., 2009. Water Resources across Europe – Confronting Water Scarcity and Drought. European Environment Agency, Copenhagen.

Conejero, W., Alarcón, J.J., Gracía-Orellana, Y., Abrisqueta, J.M., Torrecillas, A., 2007. Daily sap flow and maximum daily trunk shrinkage measurements for diagnosing water stress in early maturing peach trees during postharvest period. Tree Physiology 27, 81–88.

Conejero, W., Mellisho, C.D., Ortuño, M.F., Moriana, A., Moreno, F., Torrecillas, A., 2011. Using trunk diameter sensors for regulated deficit irrigation scheduling in early maturing peach trees. Environmental and Experimental Botany 71, 409–415.

Crisosto, C.H., Johnson, R.S., Luza, J.G., Crisosto, G.M., 1994. Irrigation regimes affect fruit soluble solids concentration and rate of water loss of 'O'Henry' peaches. HortScience 29, 1169–1171.

Cutler, J.M., Rains, D.W., Loomis, R.S., 1977. The importance of cell size in the water relations of plants. Physiologia Plantarum 40, 255–260.

DeJong, T.M., Goudriaan, J., 1989. Modeling peach fruit growth and carbohydrate requirements: reevaluation of the double-sigmoid pattern. Journal of the American Society for Horticultural Science 114, 800–804.

Diaz-Vivancos, P., Faize, L., Nicolas, E., Clemente-Moreno, M.J., Bru-Martinez, R., Burgos, L., Hernandez, J.A., 2016. Transformation of plum plants with a cytosolic ascorbate peroxidase transgene leads to enhanced water stress tolerance. Annals of Botany 7, 1121–1131.

Dichio, B., Xiloyannis, C., Sofo, A., Montanaro, G., 2007. Effects of post-harvest regulated deficit irrigation on carbohydrate and nitrogen partitioning, yield quality and vegetative growth of peach trees. Plant and Soil 290, 127–137.

Domingo, X., Arbonés, A., Rufat, J., Villar, J.M., Pascual, M., 2011. Four years of RDI during stage-II versus stage-III in peaches for processing: yield and quality. Acta Horticulturae 889, 213–220.

Dry, P.R., Loveys, B.R., 1998. Factors influencing grapevine vigour and the potential for control with partial rootzone drying. Australian Journal of Grape and Wine Research 4, 140–148.

Escobar-Gutierrez, A.J., Zipperlin, B., Carbonne, F., Moing, A., Gaudillère, J.P., 1998. Photosynthesis, carbon partitioning and metabolite content during drought stress in peach seedlings. Australian Journal of Plant Physiology 25, 197–205.

Falagán, N., Aguayo, E., Artés-Hernández, F., Artés, F., Gómez, P.A., Peña-Estévez, M.E., Corbalán, M., 2013. Combining deficit irrigation strategies and controlled atmosphere to maintain bioactive compounds in nectarine. Acta Horticulturae 1012, 97–102.

Falagán, N., Artés, F., Gómez, P.A., Artés-Hernández, F., Pérez-Pastor, A., de la Rosa, J.M., Aguayo, E., 2015. Combined effects of deficit irrigation and fresh-cut processing on quality and bioactive compounds of nectarines. Horticultural Science 42, 125–131.

FAO (Food and Agriculture Organization of the United Nations), 2 February 2017. Crop Production and Water Use in Agriculture. Available from: http://www.fao.org/faostat/en/#data/.

Faust, M., Timon, B.L., 2010. Origin and dissemination of peach. Horticultural Reviews 17, 331–379.

Fereres, E., Goldhamer, D., 1990. Deciduous fruit and nut trees. In: Stewart, B.A., Nielsen, D.R. (Eds.), Irrigation of Agricultural Crops. American Society of Agronomy, Madison, pp. 987–1017.

Ghrab, M., Sahli, A., Ben Mechli, N., 1998. Reduction in vegetative growth and fruit quality improvement in the peach variety "Carnival" through moderate watering restrictions. Acta Horticulturae 465, 601–608.

Girona, J., 2002. Regulated deficit irrigation in peach. A global analysis. Acta Horticulturae 592, 335–342.

Girona, J., Fereres, E., Marsal, J., Goldhamer, D.A., Naor, A., Soriano, M.A., 2012. Peach. In: Steduto, P., Hsiao, T.C., Fereres, E., Raes, D. (Eds.), Crop Yield Response to Water. FAO Irrigation and Drainage Paper No 66. FAO, Roma, pp. 392–409.

Goldhamer, D.A., Fereres, E., 2001. Irrigation scheduling protocols using continuously recorded trunk diameter measurements. Irrigation Science 20, 115–125.

Goldhamer, D.A., Fereres, E., Mata, M., Girona, J., Cohen, M., 1999. Sensitivity of continuous and discrete plant and soil water status monitoring in peach tress subjected to deficit irrigation. Journal of the American Society for Horticultural Science 124, 437–444.

Goldhamer, D.A., Salinas, M., Crisosto, C., Day, K.R., Soler, M., Moriana, A., 2002. Effects of regulated deficit irrigation and partial root zone drying on late harvest peach tree performance. Acta Horticulturae 592, 343–350.

Gómez-Rico, A., Salvador, M.D., Fregapane, G., 2009. Virgin olive oil and olive fruit minor constituents as affected by irrigation management based on SWP and TDF as compared to ETc in medium-density young olive orchards. Food Research International 42, 1067–1076.

Houghton, J.T., Meiro-Filho, L.G., Callander, B.A., Harris, N., Kattenburg, A., Maskell, K., 1995. Climate change. In: The Science of Climate Change in Second Assessment Report of the Intergovernmental Panel on Global Change. Cambridge University Press, Cambridge, pp. 367–397.

Hsiao, T.C., 1990. Measurements of plant water status. In: Stewart, B.A., Nielsen, D.R. (Eds.), Irrigation of Agricultural Crops. American Society of Agronomy, Madison, pp. 243–279.

Intrigliolo, D.S., Castel, J.R., 2006. Usefulness of diurnal trunk shrinkage as a water stress indicator in plum trees. Tree Physiology 26, 303–311.

Intrigliolo, D.S., Castel, J.R., 2007. Crop load affects maximum daily trunk shrinkage of plum trees. Tree Physiology 27, 89–96.

Intrigliolo, D.S., Castel, J.R., 2012. Plum. In: Steduto, P., Hsiao, T.C., Fereres, E., Raes, D. (Eds.), Crop Yield Response to Water. FAO Irrigation and Drainage Paper No 66. FAO, Roma, pp. 348–355.

Intrigliolo, D.S., Ballester, C., Castel, J.R., 2013. Carry-over effects of deficit irrigation applied over seven seasons in a developing Japanese plum orchard. Agricultural Water Management 128, 13–18.

Johnson, R.S., Handley, D.F., DeJong, T.M., 1992. Long-term response of early maturing peach trees to postharvest water deficits. Journal of the American Society for Horticultural Science 117, 881–886.

Kaya, S., Evren, S., Dasci, E., Adiguzel, M.C., Yilmaz, H., 2010. Effects of different irrigation regimes on vegetative growth, fruit yield and quality of drip-irrigated apricot trees. African Journal of Biotechnology 9, 5902–5907.

Lakso, A.N., 1990. Interactions of physiology with multiple environmental stresses in horticultural crops. HortScience 25, 1365–1369.

Larson, K.D., DeJong, T.M., Johnson, R.S., 1988. Physiological and growth responses of mature peach trees to postharvest water stress. Journal of the American Society for Horticultural Science 113, 296–300.

Layne, R.E.C., Tan, C.S., 1984. Long-term influence of irrigation and tree density on growth, survival, and production of peach. Journal of the American Society for Horticultural Science 109, 795–799.

Li, S.H., Huguet, J.G., Schoch, P.G., Orlando, P., 1989. Response of peach tree growth and cropping to soil water deficit at various phonological stages of fruit development. Journal of Horticultural Science 64, 541–552.

Lopez, G., Behboudian, M.H., Vallverdu, X., Mata, M., Girona, J., Marsal, J., 2010. Mitigation of severe water stress by fruit thinning in 'O'Henry' peach: implications for fruit quality. Scientia Horticulturae 125, 294–300.

Lopez, G., Girona, J., Marsal, J., 2012. Fruit quality responses to severe water stress during stage III of peach fruit development. Acta Horticulturae 962, 355–362.

Maatallah, S., Guizani, M., Hjlaoui, H., Boughattas, N.E.H., Lopez-Lauri, F., Ennajeh, M., 2015. Improvement of fruit quality by moderate water deficit in three plum cultivars (*Prunus salicina* L.) cultivated in a semi-arid region. Fruits 70, 325–332.

McCutchan, H., Shackel, K.A., 1992. Stem-water potential as a sensitive indicator of water stress in prune trees (*Prunus domestica* L. cv. French). Journal of the American Society for Horticultural Science 117, 607–611.

McGuire, R.G., Kelman, A., 1984. Reduced severity of Erwinia soft rot in potato tubers with increased calcium content. Phytopathology 74, 1250–1256.

Mellisho, C.D., Cruz, Z.N., Conejero, W., Ortuño, M.F., Rodríguez, P., 2011. Mechanisms for drought resistance in early maturing cvar Flordastar peach trees. Journal of Agricultural Science 149, 609–616.

Mercier, V., Bussi, C., Lescourret, F., Génard, M., 2009. Effect of different irrigation regimes applied during the final stage of rapid growth on an early maturing peach cultivar. Irrigation Science 27, 297–306.

Mirás-Avalos, J.M., Pérez-Sarmiento, F., Alcobendas, R., Alarcón, J.J., Mounzer, O., Nicolás, E., 2016a. Maximum daily trunk shrinkage for estimating wáter needs and scheduling regulated deficit irrigation in peach trees. Irrigation Science. https://doi.org/10.1007/s00271-016-0523-7.

Mirás-Avalos, J.M., Pérez-Sarmiento, F., Alcobendas, R., Alarcón, J.J., Mounzer, O., Nicolás, E., 2016b. Reference values of maximum daily trunk shrinkage for irrigation scheduling in mid-late maturing peach trees. Agricultural Water Management 171, 31–39.

Mitchell, P.D., Chalmers, D.J., 1982. The effect of reduced water-supply on peach-tree growth and yields. Journal of the American Society for Horticultural Science 107, 853–856.

Naor, A., 2004. The interactions of soil- and stem-water potentials with crop level, fruit size and stomatal conductance of field-grown 'Black Amber' Japanese plum. Journal of Horticultural Science and Biotechnology 79, 273–280.

Naor, A., 2006. Irrigation scheduling and evaluation of tree water status in deciduous orchards. Horticultural Reviews 32, 111–166.

Ortuño, M.F., Conejero, W., Moreno, F., Moriana, A., Intrigliolo, D.S., Biel, C., Mellisho, C.D., Pérez-Pastor, A., Domingo, R., Ruiz-Sánchez, M.C., Casadesus, J., Bonany, J., Torrecillas, A., 2010. Could trunk diameter sensors be used in woody crops for irrigation scheduling?. A review of current knowledge and future perspectives. Agricultural Water Management 97, 1–11.

Oweis, T., 1997. Supplemental irrigation. In: A Highly Efficient Water Use Practice. ICARDA Editions, Damasco.

Pérez-Pastor, A., Ruíz-Sánchez, M.C., Martínez, J.A., Nortes, P.A., Artes, F., Domingo, R., 2007. Effect of deficit irrigation on apricot fruit quality at harvest and during storage. Journal of the Science of Food and Agriculture 87, 2409–2415.

Pérez-Pastor, A., Domingo, R., Torrecillas, A., Ruiz-Sánchez, M.C., 2009. Response of apricot trees to deficit irrigation strategies. Irrigation Science 27, 231–242.

Pérez-Pastor, A., Ruiz-Sánchez, M.C., Domingo, R., 2014. Effects of timing and intensity of deficit irrigation on vegetative and fruit growth of apricot trees. Agricultural Water Management 134, 110–118.

Pliakoni, E.D., Nanos, G.D., Gil, M.I., 2010. Two-season study of the influence of regulated deficit irrigation and reflective mulch on individual and total phenolic compounds of nectarines at harvest and during storage. Journal of Agricultural Food Chemistry 58, 11783–11789.

Qassim, A., Goodwin, I., Bruce, R., 2013. Postharvest deficit irrigation in 'Tatura 204' peach: subsequent productivity and water saving. Agricultural Water Management 117, 145–152.

Radi, M., Mahrouz, M., Jaouad, A., Amiot, M.J., 2003. Influence of mineral fertilisation on the quality of apricot fruit. Agronomie 23, 737–745.

Rieger, M., Duemmel, M.J., 1992. Comparison of drought resistance among prunus species from divergent habitats. Tree Physiology 11, 369–380.

Rieger, M., Lo Bianco, R., Okie, W.R., 2003. Response of *Prunus ferganensis*, *Prunus persica* and two interspecific hybrids to moderate drought stress. Tree Physiology 23, 51–58.

Ruiz-Sánchez, M.C., Egea, J., Galego, R., Torrecillas, A., 1999. Floral biology of Búlida apricot trees subjected to postharvest water irrigation withholding. Annals of Applied Biology 135, 523–528.

Samperio, A., Moñino, M.J., Marsal, J., Prieto, M.H., Stöckle, C., 2014. Use of CropSyst as a tool to predict water use and crop coefficient in Japanese plum trees. Agricultural Water Management 146, 57–68.

Samperio, A., Prieto, M.H., Blanco-Cipollone, F., Vivas, A., Moñino, M.J., 2015a. Effect of post-harvest regulated deficit irrigation in 'Red Beaut' Japanese plum: tree water status, vegetative growth, fruit yield, quality and economic return. Agricultural Water Management 150, 92–102.

Samperio, A., Moñino, M.J., Vivas, A., Blanco-Cipollone, F., García-Martín, A., Prieto, M.H., 2015b. Effects of deficit irrigation during stage II and post-harvest on tree water status, vegetative growth, yield and economic assessment in "Angeleno" Japanese plum. Agricultural Water Management 158, 69–81.

Sepaskhah, A.R., Ahmadi, S.H., 2010. A review on partial root-zone drying irrigation. International Journal of Plant Production 4, 241–258.

Steduto, P., Hsiao, T.C., Fereres, E., Raes, D., 2012. Crop Yield Response to Water, FAO Irrigation and Drainage Paper No 66. FAO, Roma.

Steinberg, S.L., Miller Jr., J.C., McFarland, M.J., 1989. Dry matter partitioning and vegetative growth of young peach trees under water stress. Australian Journal of Plant Physiology 17, 23–36.

Tapia, L.M., Larios, A., Abrisqueta, I., Mounzer, O., Vera, J., Abrisqueta, J.M., Ruiz-Sánchez, M.C., 2009. Peach deficit irrigation. Fruit yield and water use efficiency analysis. Reviones Fitotécnicas Mexicanas 33, 1–5.

Thakur, A., Singh, Z., 2013. Deficit irrigation in nectarine: fruit quality, return bloom and incidence of double fruits. European Journal of Horticultural Science 78, 67–75.

Torrecillas, A., Alarcón, J.J., Domingo, R., Planes, J., Sánchez-Blanco, M.J., 1996. Strategies for drought resistance in leaves of two almond cultivars. Plant Science 118, 135–143.

Torrecillas, A., Galego, R., Pérez- Pastor, A., Ruiz-Sánchez, M.C., 1999. Gas exchange and water relations of young apricots plants under drought conditions. Journal of Agricultural Science 132, 445–452.

Torrecillas, A., Domingo, R., Galego, R., Ruiz-Sánchez, M.C., 2000. Apricot tree response to irrigation withholding at different phenological periods. Scientia Horticulturae 85, 201–215.

Turner, N.C., Jones, M.M., 1980. Turgor maintenance by osmotic adjustment: a review and evaluation. In: Turner, N.C., Kramer, P.J. (Eds.), Adaptation of Plants to Water and High Temperature Stress. John Wiley & Sons, London, pp. 87–103.

Uriu, K., 1964. Effect of post-harvest soil moisture depletion on subsequent yield of apricots. Journal of Horticultural Science 84, 93–97.

Valverdu, X., Girona, J., Echeverria, G., Marsal, J., Behboudian, M.H., Lopez, G., 2012. Sensory quality and consumer acceptance of 'Tardibelle' peach are improved by deficit irrigation applied during stage II of fruit development. HortScience 47, 656–659.

Williamson, J.G., Coston, D.C., 1990. Planting method and irrigation rate influence vegetative and reproductive growth of peach planted at high-density. Journal of the American Society for Horticultural Science 115, 207–212.

Young, E., Hand, J.M., Wiest, S.C., 1981. Diurnal variation in water potential components and stomatal resistance of irrigated peach seedlings. Journal of the American Society for Horticultural Science 106, 337–340.

Ziska, L.H., Hutmacher, R.B., Hoffman, G.J., DeJong, T.M., 1984. Changes in leaf water status associated with salinity in mature, field-grown *Prunus-salicina*. Physiologia Plantarum 77, 141–149.

FURTHER READING

Intrigliolo, D.S., Castel, J.R., 2010. Response of plum trees to deficit irrigation under two crop levels: tree growth, yield and fruit quality. Irrigation Science 28, 525–534.

CHAPTER 6

Deficit Irrigation Management in Early-Maturing Peach Crop

María C. Ruiz-Sánchez, Isabel Abrisqueta, Wenceslao Conejero, Juan Vera
Centro de Edafología y Biología Aplicada del Segura (CEBAS-CSIC), Murcia, Spain

Contents

1. PEACH CROP IN MEDITERRANEAN CONDITIONS

Peach (*Prunus persica* [L.] Batsch) is the third most important fruit crop in the world, after apple and pear, and the second most important in Europe. Two European Mediterranean countries (Italy and Spain) contribute more than 20% to world production. Spain with an annual production of 1.57 Mt is the second peach producing country after China. In Spain, peach is the most cultivated sweet fruit (86,118 ha) and the second noncitric fruit, after almond (FAOSTAT, 2016). Its cultivation has grown in recent years, especially in more southern regions, given its better adaptation to the climate, the lower risk of frost, and the lower incidence of diseases. The areas in question can host varieties with few requirements of winter cold; these are the early-maturing cultivars.

In a water scarcity scenario as a result of climatic change, the motto "more crop per drop" applies. Agriculture has to produce more food with

Water Scarcity and Sustainable Agriculture in Semiarid Environment
ISBN 978-0-12-813164-0
https://doi.org/10.1016/B978-0-12-813164-0.00006-5

fewer inputs (water, energy) in a sustainable manner. To meet this, not only water application systems, but also irrigation scheduling by means based on a precise knowledge of the crop water requirements must be improved. Drip irrigation (the most important innovation in agriculture since the 1960s) is used in arid and semiarid areas, because it allows the optimal use of water and fertilizer. Irrigation scheduling by Food and Agriculture Organization (FAO) methodology estimates crop water requirements (ET_c) based on the evapotranspiration of the crop reference (ET_o) and the specific coefficients of each crop (K_c) (Allen et al., 1998). On the other hand, K_c determination for local conditions involves water balance studies using weighing or drainage lysimeters in which all inputs (irrigation, precipitation, drainage, runoff, and soil water content [SWC] variations) are considered (Abrisqueta et al., 2013; Ayars et al., 2003; Girona et al., 2011). Under these conditions, SWC monitoring and soil depth exploration by roots have become fundamental factors for the precision irrigation of crops.

Deficit irrigation is frequently adopted in an attempt to improve water use efficiency in areas with scarce water resources, and is considered an alternative to traditional irrigation scheduling approaches that fully meet plant water requirements (Fereres and Soriano, 2007; Ruiz-Sánchez et al., 2010). Among these water-saving strategies, regulated deficit irrigation (RDI), applied since the 1980s in fruit trees (Chalmers et al., 1981), is based on reducing irrigation water in periods or phenological stages in which water deficit does not affect (or has little effect on) final yield, while 100% of water needs are supplied during the critical periods.

The literature coincides with the view that water deficit intensities during the final stages of fruit growth (stage III for the double-sigmoid fruits) lead to reduction of fruit diameter, which adversely affects overall production. It is therefore considered the most sensitive period for irrigation. Stage II of fruit growth (slow fruit growth rate), in medium- and late-maturing peach cultivars, and the postharvest period are candidate periods for applying reduced irrigation in RDI strategies (Boland et al., 1993; Girona et al., 2004; Ruiz-Sánchez and Girona, 1995). Sustained and partial root-zone drying (PRD) deficit irrigation regimes have also been used to cope with water shortage and efficient water use in fruit trees (Girona et al., 2005; Goldhamer et al., 2002; Johnson et al., 1992; Pérez-Pastor et al., 2016).

General agreement exists concerning the potential water, energy, and labor savings possible with precision irrigation: a real-time decision-making process that takes into consideration soil texture variability, plant root depth, soil water storage, plant phenological stage, atmospheric water demand, and forecast rainfall using sensors. The adoption of precision irrigation will increase food production, crop productivity, water use efficiency, and CO_2 absorption, while lowering energy and labor costs and reducing the environmental footprint. Furthermore, the transfer of this technology to the irrigation community would represent a great and increasingly necessary step forward for agriculture.

In this chapter the responses of early-maturing peach trees to different deficit irrigation strategies are shown in terms of plant–soil water relations, vegetative growth, and yield. The growth pattern and the determination of local crop coefficients under Mediterranean conditions are also described. An automated irrigation system, based on SWC criteria according to sound plant responses for crop yield and quality, is also depicted.

2. EXPERIMENTAL CONDITIONS: PLANT MATERIAL, TREATMENTS, AND MEASUREMENTS

The experiments were performed over a 10-year period (2004–14) in an early-maturing peach tree (*P. persica* [L.] Batsch *cv.* Flordastar, on GF-677 rootstock) orchard, located in an experimental 0.8 ha plot in Santomera-Murcia (southeast Spain): 38°06′N, 1°02′W. The soil is highly calcareous, stony, and 0.9 m deep, with a clay-loam texture. Trees were spaced 5 m × 5 m.

Following normal commercial practices the peach fruits were hand thinned about 1 month after full bloom and harvested in early May. The trees were pruned each year in mid-December, and irrigated by a drip irrigation system consisting of a single lateral line per tree row, with eight emitters per tree, placed 0.5 m from the trunk, each providing 2 L h^{-1}. Irrigation was initiated in mid-February during flowering and suspended in early November.

Several irrigation treatments, distributed in a completely randomized design with four replications (each consisting of one row of 13 trees), were considered:

- Control (T-100) irrigated to satisfy crop evapotranspiration (100% ET_c), estimated from the crop reference evapotranspiration (ET_o), with the Penman–Monteith equation (Allen et al., 1998), and local crop coefficients (Abrisqueta et al., 2013).

- Overirrigated in excess of the estimated crop evapotranspiration (120% ET_c) to ensure nonlimiting soil water availability using the K_c-FAO-56 coefficient.
- Sustained deficit irrigation at 50% of ET_c (T-50) throughout the season.
- Treatment, partial root-zone drying (T-PRD), irrigated at 50% of ET_c on one side of the trees, while the other side was kept dry. For this, two lateral lines per tree row were used, with four 2 L h^{-1} emitters, located on different sides of the tree, and the irrigation was managed to apply water to each side alternately every 2 weeks, or, when rainfall occurred, until the soil water potential reached −50 kPa for 1 week, according to Watermark tensiometers (Abrisqueta et al., 2008).
- Treatment, regulated deficit irrigation (T-RDI), irrigated to fully cover 100% ET_c during the fruit growth period, with the irrigation reduced to 70% and 25% ET_c during the initial and late postharvest period, respectively.
- Automatic control of irrigation based on SWC threshold values (T-soil) measured by capacitance Frequency Domain Reflectometry (FDR-type) probes (Vera et al., 2013).
- Nonirrigated (T-0) throughout the season, except during the fruit growth period (irrigated 100% ET_c).

All trees received the same amount of fertilizer applied through the irrigation system, which was remotely controlled by a head unit programmer acting on electrovalves, and the irrigation water volumes were measured using inline flow meters for each treatment.

Agrometeorological variables were recorded every 15 min by an automatic weather station located close to the peach orchard with real-time access via the web. Volumetric SWC was measured continuously with capacitance probes and discontinuously with neutron probes, from 0.2 to 0.8 m soil depth, in access tubes installed within the wetted area of the second emitter of one tree per replication and treatment.

Midday stem water potential (Ψ_{stem}) was measured at midday in leaves located on the north face of the tree and near the trunk, using a pressure chamber. For that purpose, mature leaves were placed in aluminum foil bags for at least 2 h prior to the measurements (Shackel et al., 1997).

As regards plant growth, the measurements included: shoot length (with a measuring tape in four tagged shoots per tree), trunk diameter (with forest caliper and linear variable displacement transducers 30 cm above the graft union), winter pruning weight, plant cover (area covered by canopy tree with image analysis of photographs), fruit diameter (in tagged fruits, with an electronic digital caliper), and root length (with the minirhizotron method and soil core sampling). Yield and fruit quality was determined annually in mature peach fruits harvested at two to three picking dates. More details are described in Abrisqueta et al. (2010) and Vera et al. (2013).

3. PHENOLOGY AND PEACH TREE GROWTH PATTERN

Knowledge of the plant phenology calendar is essential for successful crop management practices since it permits the scheduling of specific irrigation, fertilization, and hormonal or phytosanitary applications. Application of regulated deficit irrigation also requires an accurate knowledge of the critical phenological periods during which the sensitivity of plants to water stress is maximal. For these reasons, a detailed knowledge of plant phenological processes is essential to profit from new varieties while improving irrigation water use efficiency.

The growth pattern and the phenological stages of early-maturing peach trees, using the traditional nomenclature (Baggiolini, 1980) and the universal scale (to be known as the BBCH scale), which uses a decimal code to describe the growth stages (Lancashire et al., 1991), were studied by Mounzer et al. (2008a). According to this study, the heat requirements (growing degree hours, GDH), calculated considering optimum, base, and critical temperatures of 25, 4, and 36°C, respectively, helped to predict the time needed to reach each phenological stage (e.g., 6244 GDH to reach full bloom and 27,106 GDH before the fruit could be harvested).

The root growth of the early-maturing peach trees continued throughout the season and was asynchronous with that of shoots: the maximum shoot growth rate coincided with the lowest root growth (Fig. 6.1). The root growth rate slowed down during the maximum fruit growth period and increased after harvest, which indicates that the fruits act as strong sinks of photoassimilates (Abrisqueta et al., 2008). Root and trunk growth rates increased immediately after harvest (Fig. 6.1) when the canopy was large enough to direct the photoassimilates to organs that would ensure the following season's yield.

Phenological changes of the aboveground parts of tree crops can be gathered by measuring plant cover. Several studies have demonstrated the relationship between plant water requirements and plant cover, estimated as leaf area (Williams and Ayars, 2005) or as intercepted radiation (Auzmendi et al., 2011; Ayars et al., 2003).

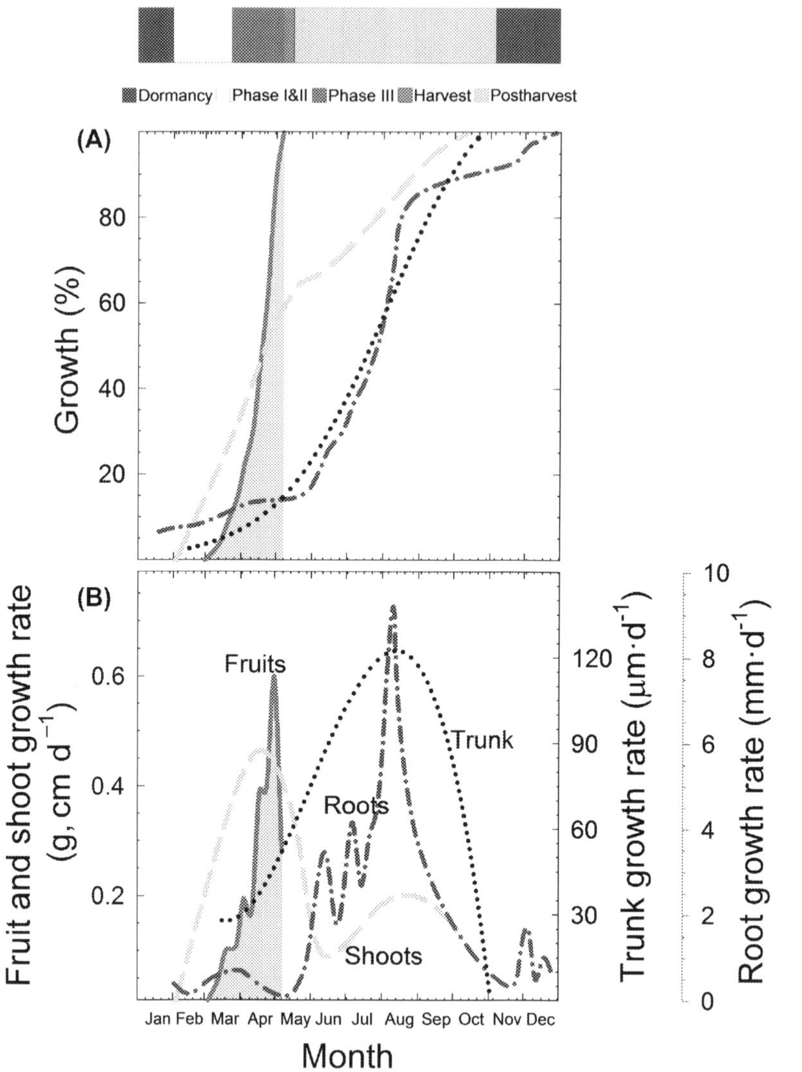

Figure 6.1 Seasonal evolution of fruit weight ( ), shoot length (— —), trunk diameter (········), and root length (—·—·-) of early-maturing peach trees: (A) percentage of the annual growth and (B) daily growth rate.

Seasonal changes of early-maturing peach tree cover were evaluated by zenithal image analysis, a simple but robust method that uses image analysis of photographs taken early morning 7 m above the tree crown (Fig. 6.2A). Peach tree cover increased from February to midsummer, with a short decrease being observed just after harvest, due to an occasional hailstorm,

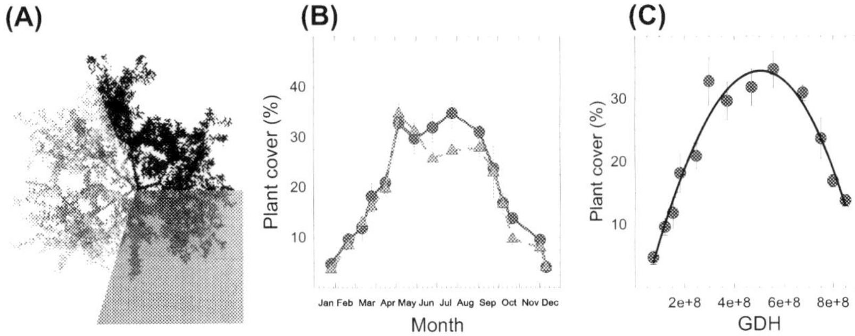

Figure 6.2 (A) Zenithal image analysis of early-maturing peach tree; (B) seasonal pattern of peach tree cover in winter (●) and summer pruning (▲) treatments; (C) peach tree cover as a function of growing degree hours (GDH) accumulation.

after which a continuous decrease in tree cover was noted (Fig. 6.2B). The effect of the summer pruning treatment decreased plant cover (Fig. 6.2B), which allowed alleviating plant water consumption, as indicated by the decrease of 0.2–0.4 MPa in the Ψ_{stem} values (Abrisqueta et al., 2014). Similarly, summer pruning improved late-season peach tree water status (López et al., 2006). The seasonal trend of plant cover related to GDH provides a standard scale that depends on the plant phenology (Fig. 6.2C).

4. PEACH CROP WATER REQUIREMENTS IN MEDITERRANEAN CONDITIONS

4.1 Crop and Basal Coefficients

Plant water requirements, referred to as crop evapotranspiration (ET_c), are commonly estimated from the crop reference evapotranspiration (ET_o), and a specific coefficient for each crop (K_c): ET_c (mm) = ($ET_o \times K_c$) − P, where both ET_o and P (effective precipitation) refer to previous conditions (a week or a month). The diversity of parameters affecting the K_c coefficient usually requires validation through local field experiments.

For this reason, a 4-year study was carried out in an adult early-maturing peach tree orchard. Trees from the overirrigated treatment were grown in drainage lysimeters equipped with irrigation (I) and drainage (D) water meters and access tubes for monitoring the SWC variations (ΔS) by neutron probe. Peach ET_c was estimated using the mass conservation water balance equation: $ET_c = P + I - \Delta S - D$ and $K_c = ET_c/ET_o$.

The K_c values increased in February at the beginning of irrigation, and lasted until early May (day of the year [DOY] 0–120, from 0.07 to 0.58).

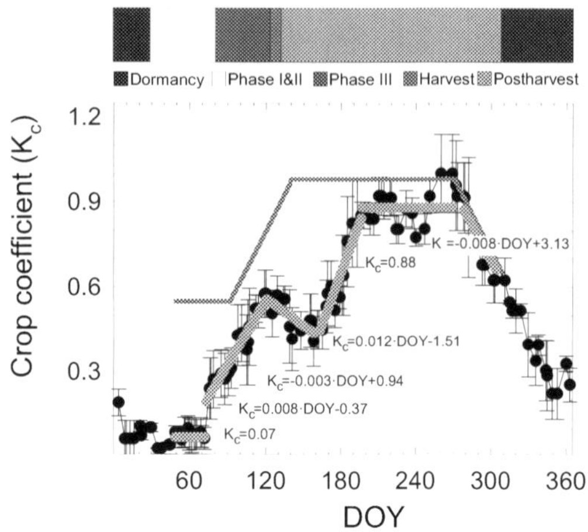

Figure 6.3 Seasonal changes of local crop coefficient (K_c) of drip-irrigated early-maturing peach trees calculated on the basis of a water balance in Santomera (Murcia), Spain (●), compared with Food and Agriculture Organization (FAO)-56 values (▬). K_c values were lineally adjusted equations (~) for each phenological stage. *DOY*, day of the year.

A short decrease was noted at the end of May, reflecting adjustments in tree water use in response to harvest, followed by recovery in early June (DOY 165) until a maximum K_c was reached in August (0.88). This was followed by a rapid decline in K_c in mid-November (DOY 300), when irrigation was ended (Fig. 6.3). The highest water consumption period (up to 90% of annual requirements) occurred after harvest. The seasonal trend of K_c showed important deviations compared with that mentioned in other sources (Ayars et al., 2003; FAO-56), which emphasizes the need for local experiments for accurate irrigation management.

In a subsequent study, K_c values for mature, full yield peach trees were split into basal crop (K_{cb}) and soil evaporation (K_s) coefficients. The curvilinear behavior of K_{cb} reflected the phenological stages: K_{cb} increasing from 0.15 to 0.45 from bloom to harvest, and then a dramatic increase to reach a peak value of 1.0 by mid-July due to the appearance and growth of water sprouts. After that, leaf drop caused a continuous decrease in K_{cb}, reaching 0.15 by the end of the year. Compared with the four-segment linear pattern described in the literature (Allen et al., 1998), peach irrigation water requirements calculated using locally adjusted crop coefficients led to significant water saving. More details can be found in Abrisqueta et al. (2013).

4.2 Irrigation Coefficient

In an attempt to schedule irrigation in an easy, practical, but precise way an experiment to determine the irrigation factor (K) under nonlimited water conditions was performed. This involved the use of capacitance probes to continuously monitor the SWC and measurements of the plant water status (Ψ_{stem}) and canopy cover.

The study was performed in an early-maturing nectarine tree plot (*P. persica* [L.] Batsch, *cv.* Flariba, on GxN-15 rootstock, and 6.5 m × 3.5 m tree spacing), growing in the same soil, climatic and crop management conditions as those of the peach trees described earlier.

Irrigation was automated to fulfill plant water requirements, based on SWC threshold values (measured with capacitance probes located in the main root zone at 0–0.5 m) of 10% of management-allowed depletion (MAD, α) during the fruit growth period and 30% during postharvest to start irrigation and field capacity to end irrigation.

The variation of SWC at 0.70 m depth was not significant, so the drainage was limited to rainfall events, so that drainage can be assumed as negligible. Seasonal Ψ_{stem} values were above those considered as a plant water deficit situation (Abrisqueta et al., 2012), which confirmed the well-adjusted irrigation protocol. For these reasons, the "irrigation/ET_o" ratio, in the absence of rainfall, represented K, which merged crop coefficient, tree cover, and application efficiency.

The monthly K values (Table 6.1) provided annual water requirements for early-maturing nectarine trees of about 660 mm, in the agroclimatic area of the province of Murcia (southeast Spain), with irrigation frequency varying from 1 to 6 days a week, depending on the phenological period. The proposed monthly irrigation factors allow the plant water requirements to be determined and precision irrigation to be managed, thereby increasing the efficiency of water resources and reducing the energy used.

5. PLANT RESPONSES TO DEFICIT IRRIGATION STRATEGIES

At an early stage of plant growth, when early-maturing peach trees were young, deficit irrigation caused a similar statistically significant reduction in

Table 6.1 Monthly irrigation factor (K) for early-maturing *Prunus persica* (L.) trees

March	April	May	June	July	August	September	October
0.31	0.41	0.57	0.72	0.83	0.86	0.77	0.51

Table 6.2 Vegetative growth and yield of 4-year-old early-maturing peach trees in the different deficit irrigation treatments

	Vegetative growth					
Treatment	**Shoot length (m)**	**Trunk diameter (cm)**	**Pruning (kg tree^{-1})**	**Canopy area (m^2 tree^{-1})**	**RLD (cm root cm^{-3} soil)**	**Yield (kg tree^{-1})**
T-100	4.95 a	10.02 a	5.78 a	12.36 a	2.53 a	15.12 a
T-50	3.13 b	8.90 b	3.32 b	9.95 b	0.67 c	17.34 a
T-PRD	3.90 b	8.93 b	3.15 b	8.67 b	1.40 b	13.31 a
T-RDI	3.54 b	8.89 b	2.98 b	9.92 b	—	14.48 a

Mean values followed by different letters in the same column are significantly different at 5% according to LSD test.
RLD, root length density; *T-PRD*, treatment, partial root-zone drying; *T-RDI*, treatment, regulated deficit irrigation.

vegetative growth (shoot length, trunk diameter, and pruning weight) in the sustained (T-50), regulated (RDI), and PRD treatments compared with well-irrigated trees. However, nonsignificant differences were found in total yield between treatments (Table 6.2). In turn, PRD did not offer any additional advantage over a deficit irrigation treatment such as RDI, which allowed 60% water saving without affecting fruit yield (Ruiz-Sánchez et al., 2006). These results revealed the high sensitivity of vegetative growth to soil water deficit, as has been well documented for peach trees (Girona et al., 2005; Johnson et al., 1992).

Water deficit limited root growth in adult early-maturing peach trees, when root length density (RLD) dynamics were evaluated with mini-rhizotrons. Similar seasonal root dynamics, but lower RLD values, were observed in the T-50 compared with T-100 treatments, and an intermediate response was observed in the PRD treatment (Abrisqueta et al., 2008). Also a reduction in the volume of soil explored by the roots in the T-50 treatment compared with the control treatment (T-100) was observed when a soil core trial was performed, taking soil samples from both sides of the tree line (Fig. 6.4); the low SWC values induced a constrained root system, by means of what has been termed a "hydropatterning" mechanism.

Deficit irrigation limited the vegetative growth of adult early-maturing peach trees with an intensity that depended on the part considered: trunk diameter > wood pruning ≥ canopy size > shoot length growth (Abrisqueta et al., 2010). Vegetative growth was clearly limited by the accumulated water deficit as the trees grew old, resulting in smaller trees than in the control treatment, which, in turn, reduced the total peach

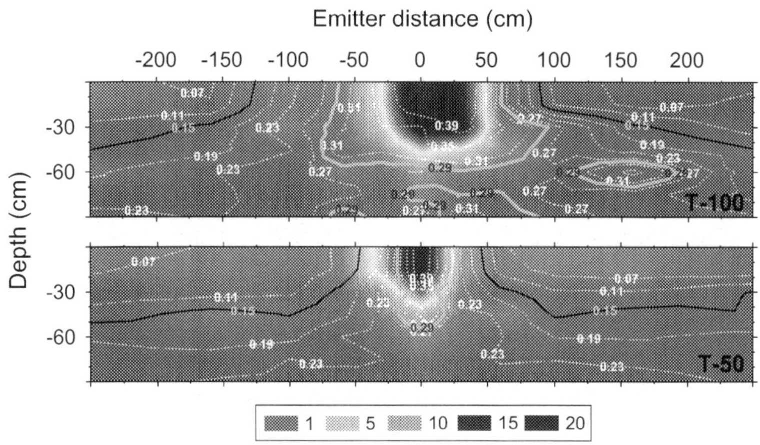

Figure 6.4 Isoline map of soil water content ($m^3\ m^{-3}$) (*dashed black lines*) and root length density (RLD, cm root cm^{-3} soil) (*color gradient*) in control (T-100) and sustained deficit irrigation (T-50) treatments. Light blue (light gray in print version) and *black lines* correspond to soil field capacity and wilting point value, respectively. *DOY*, day of the year.

fruit yield, as Vera et al. (2013) described in early-maturing cultivars, or reduced fruit size, as seen in a long-term study of mid-maturing cultivars (Marsal et al., 2016).

Postharvest is an important period for peach trees since carbohydrates are accumulated and floral organ differentiation occurs (Handley and Johnson, 2000), for which reason water deficits should be managed carefully at this time to avoid reductions in bloom and fruit load. In this sense, deficit irrigation (25% ET_c) applied during postharvest (RDI) decreased early-maturing peach yield in the following year as a result of the fewer fruits per tree, while fruit size remained relatively unchanged (Vera et al., 2013). Similar results were obtained in deficit-irrigated nectarine trees (Naor et al., 2005). However, deficit irrigation (35% ET_c) during stage II allowed water savings of up to 22%–35% without affecting peach yield or final fruit size in medium- and late-maturing cultivars (Girona et al., 2003, 2005).

Water deficits applied throughout the entire irrigation season (T-50) induced a significant reduction in peach yield (Abrisqueta et al., 2010; Girona et al., 2003, 2005; Vera et al., 2013); similar results were obtained using PRD strategies, with no agronomical advantages over other deficit irrigation strategies (Goldhamer et al., 2002; Ruiz-Sánchez et al., 2010). A lower percentage of fruits >61 mm in diameter (commercial grades

A and B) were obtained in T-50 compared with the rest of the treatments (Abrisqueta et al., 2010; Vera et al., 2013).

Deficit irrigation provided firmer fruits with a greater soluble solids content, a more intense reddish color, and lower vitamin C and carotenoid contents, while the phenolic content, sweetness, juiciness, and flavor intensity at harvest increased in comparison with well-irrigated peach trees, both in early- and late-maturing peach cultivars (Alcobendas et al., 2012; López et al., 2010; Tavarini et al., 2011; Vallverdu et al., 2012).

6. IRRIGATION MANAGEMENT BASED ON THE SOIL–PLANT–ATMOSPHERE SYSTEM

When studying the soil–plant–atmosphere system, the climatic water demand should be considered as a "variable and quick" component, the soil as a "steady and slow" component, and the plant as a "reactive and resilient" component. Effective, precise irrigation management must deal with all those issues (Vera et al., 2010).

Standardized agrometeorological sensors allow the accurate evaluation of evaporative demand of the *atmosphere* (air temperature, vapor pressure deficit [VPD], ET_o) (Allen et al., 2011). Some web-based services and apps offer a simplified water balance, taking into account soil texture, root depth, ET_o and precipitation data on a daily or weekly basis (Vera et al., 2017).

The SWC has long been monitored for irrigation management purposes. Some indices derived from the SWC diminution curve can act as indicators of plant water stress initiation (Starr and Paltineanu, 1998; Thompson et al., 2007). For early-maturing peach trees, Abrisqueta et al. (2012) used continuous SWC measurements, obtained with capacitance FDR-type probes, to define the lower limit of the SWC to activate automatic irrigation, which was best defined by the transition between a relatively rapid reduction of SWC in the drying soil to a slower rate, or "breaking point." In the foregoing study it was recommended that the SWC be kept higher than 10% of field capacity during the postharvest period.

The continuous SWC measurements in the T-soil treatment accurately reflected the impact of the different irrigation events on the soil water stored and provided useful information concerning the advance of the wetted front, the depth of the root system activity, and the fate of the applied water (Mounzer et al., 2008b). These data allowed an optimal moisture range for the crop to be set, from which it was possible to precisely adjust the irrigation dose and frequency and to automate the irrigation. When threshold values were precisely defined, slight water deficits only limited vegetative growth but allowed a similar peach yield to that of T-100 trees (Abrisqueta et al., 2010; Vera et al., 2013).

Identification of the level of water stress suffered by *plants* is essential when applying deficit irrigation strategies. Several studies have revealed the value of nontranspiring leaf water potential (Ψ_{stem}) as one of the most accurate plant water status indicators in tree crops (McCutchan and Shackel, 1992; Remorini and Massai, 2003; Shackel et al., 1997).

For peach trees, the minimum Ψ_{stem} values that plants could withstand without penalizing yield were roughly −0.5 to −0.7 MPa during the fruit growth period, and −1.5 MPa for the postharvest period to ensure no impairment of bloom fertility (De la Rosa et al., 2015; Girona and Fereres, 2012), or −2.0 MPa to limit the occurrence of double fruits (Naor et al., 2005). Nevertheless, Ψ_{stem} values should be kept above −0.9 MPa, a value that acts as an indicator of water stress initiation in peach trees (Abrisqueta et al., 2012; Shackel et al., 1997).

However, Ψ_{stem} measurements require destructive sampling and considerable time- and labor-related costs. Alternative methods have been proposed to estimate peach Ψ_{stem} from other plant water status indicators, mainly trunk diameter fluctuations (Conejero et al., 2011; De la Rosa et al., 2014; Do Paço et al., 2012; Goldhamer et al., 1999), which allow data collection to be automated.

In an attempt to estimate the Ψ_{stem} of adult drip-irrigated early-maturing peach trees as a function of the SWC and agrometeorological variables, a study was carried out using data from the long-term deficit irrigation experiment described earlier.

Seasonal trends of Ψ_{stem} showed high and similar values early in the season, coinciding with low evaporative demand conditions, whereas minimum Ψ_{stem} values were observed midsummer, when they ranged from −0.8 MPa in well-irrigated conditions (T-100) to −2.3 MPa in severely water-stressed trees in nonirrigated conditions (T-0 treatment), with moderate water deficit values (−1.2 MPa) observed in trees irrigated at 50% ET_c (T-50) (Fig. 6.5).

A regression analysis highlighted the importance of considering Ψ_{stem} as a function of the water demand of the atmosphere and the SWC resulting from irrigation (supercooled water content being the main contributor). A multiple linear regression equation based on average SWC in the entire soil profile (soil component), GDH (phenology-plant related component),

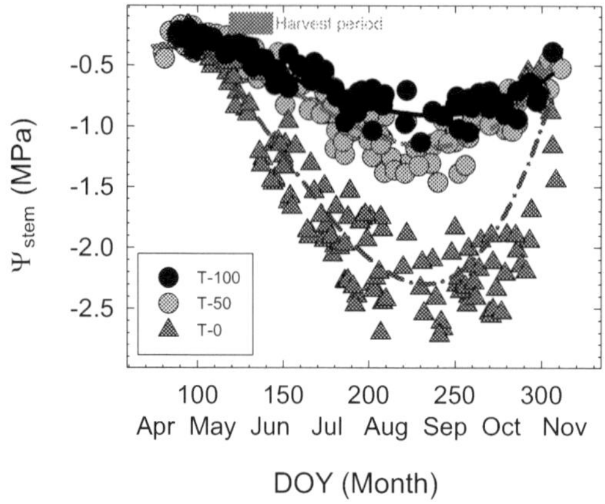

Figure 6.5 Seasonal stem water potential (Ψ_{stem}) in early-maturing peach trees submitted to different irrigation treatments: T-100 (●), T-50 (◉), and T-0 (▲). Lines are the polynomial fit of 5-year data.

and mean daily VPD (atmosphere component) was proposed as an alternative to Ψ_{stem} measurements, explaining 72% of the variance (Abrisqueta et al., 2015).

To guard against the assumptions of linear relationships between the variables involved in the soil–plant–atmosphere system, a hybrid soft computing technique, the Takagi–Sugeno–Kang fuzzy inference system, was applied. Three inputs relevant to Ψ_{stem} were automatically selected: SWC at 0.3 m, the DOY, and mean daily temperature (T_m). Five fuzzy rules were identified based on the fuzzy clusters found in the data, providing an accurate Ψ_{stem} estimation (86% variance explained) (Valdés-Vela et al., 2015).

There was close agreement concerning the seasonal trends of measured Ψ_{stem} and Ψ_{stem} estimated either by multiple linear regression or by fuzzy rules-based models, with greater scattering found from September onward. Both models tended to slightly underestimate Ψ_{stem}, with greater variability being observed at low Ψ_{stem} values (Abrisqueta et al., 2017).

7. CONCLUSIONS

The phenological stages and seasonal growth pattern of early-maturing peach trees grown in Mediterranean conditions revealed an alternating growth pattern between the roots and the canopy. Roots showed continuous growth throughout the year with different growth rates. Root and trunk growth rates increased immediately after harvest when the canopy was large enough to direct the photoassimilates to organs that would ensure the following season's yield. Measurement of the plant cover by image analysis of the tree canopy proved to be effective for estimating the vegetative growth of the peach trees.

K_c, K_{cb}, and K coefficients were calculated from a soil water balance study, under Mediterranean conditions with no water restrictions for an early-maturing peach tree orchard.

Deficit irrigation, regardless of the mode of implementation—sustained (T-50), partial root drying (T-PRD), or regulated (T-RDI)—was seen to reduce peach tree vegetative growth (root, trunk, and canopy) to such an extent that tree size decreased and peach yield was statistically reduced. However, water use efficiency (kg per irrigation water applied) and yield efficiency (kg per trunk sectional area) clearly increased in all deficit irrigation strategies, RDI being the most efficient because of the maximum water saving (irrigation at 35% of the ET_c).

An automatic irrigation system based on SWC status monitoring by capacitance probes (T-soil) with threshold values of MAD = 10% and 30% during the fruit growth stage and postharvest, respectively, allowed feedback irrigation scheduling, which led to a decrease in vegetative growth with no yield penalty.

The statistical models proposed as an alternative to Ψ_{stem} field measurements depended on easily automated inputs and provided useful information on the current plant water status as a guide for irrigation scheduling of early-maturing peach trees under Mediterranean conditions.

It is clear that precision irrigation using sensors in the soil–plant–atmosphere system can increase food production, crop productivity, water use efficiency, and CO_2 absorption, while lowering energy and labor costs, and reducing the environmental footprint.

ACKNOWLEDGMENTS

This study was supported by Spanish National R&D&I Plans (AGL2002-04048-C03-03, AGL2006-12914-C02-01, AGL2009-06981, AGL2010-14861, AGL2013-49047-C2-2-R, and AGL2016-77282-C3-1-R), Séneca Foundation of Murcia (03130/PI/05, 08847/PI/08, and 19903/GERM/15), and Consolider-Ingenio 2010 (CSD2006-0067) grants to the authors. In memoriam of J.M. Abrisqueta García.

REFERENCES

Abrisqueta, I., Abrisqueta, J.M., Tapia, L.M., Munguía, J.P., Conejero, W., Vera, J., Ruiz-Sánchez, M.C., 2013. Basal crop coefficients for early-season peach trees. Agricultural Water Management 121, 158–163.

Abrisqueta, I., Bernal, F.J., Conejero, W., Vera, J., Ruiz-Sánchez, M.C., 2014. Dinámica estacional de la cobertura foliar en *Prunus persica*. In: Actas XXXII Congreso Nacional de Riegos (Madrid), p. 6.

Abrisqueta, I., Conejero, W., Ortuño, M.F., Valdés-Vela, M., Vera, J., Ruiz-Sánchez, M.C., 2017. Estimation of stem water potential in drip-irrigated peach trees. Acta Horticulturae 1150, 97–104.

Abrisqueta, I., Conejero, W., Valdés-Vela, M., Vera, J., Ortuño, M.F., Ruiz-Sánchez, M.C., 2015. Stem water potential estimation of drip-irrigated early-maturing peach trees under Mediterranean conditions. Computers and Electronics in Agriculture 114, 7–13.

Abrisqueta, I., Tapia, L.M., Conejero, W., Sánchez-Toribio, M.I., Abrisqueta, J.M., Vera, J., Ruiz-Sánchez, M.C., 2010. Response of early-maturing peach [*Prunus persica* (L.)] trees to deficit irrigation. Spanish Journal of Agricultural Research 8 (S2), 30–39.

Abrisqueta, I., Vera, J., Tapia, L.M., Abrisqueta, J.M., Ruiz-Sánchez, M.C., 2012. Soil water content criteria for peach trees water stress detection during the postharvest period. Agricultural Water Management 104, 62–67.

Abrisqueta, J.M., Mounzer, O., Álvarez, S., Conejero, W., García-Orellana, Y., Tapia, L.M., Vera, J., Abrisqueta, I., Ruiz-Sánchez, M.C., 2008. Root dynamics of peach trees submitted to partial rootzone drying and continuous deficit irrigation. Agricultural Water Management 95, 959–967.

Alcobendas, R., Mirás-Avalos, J.M., Alarcón, J.J., Pedrero, F., Nicolás, E., 2012. Combined effects of irrigation, crop load and fruit position on size, color and firmness of fruits in extra-early cultivar of peach. Scientia Horticulturae 142, 128–135.

Allen, R.G., Pereira, L.S., Howell, T.A., Jensen, M.E., 2011. Evapotranspiration information reporting: I factors governing measurement accuracy. Agricultural Water Management 98, 899–920.

Allen, R.G., Pereira, L.S., Raes, D., Smith, M., 1998. Crop evapotranspiration. In: FAO Irrigation and Drainage Paper 56. FAO, Rome.

Auzmendi, I., Mata, M., Lopez, G., Girona, J., Marsal, J., 2011. Intercepted radiation by apple canopy can be used as a basis for irrigation scheduling. Agricultural Water Management 98, 886–892.

Ayars, J.E., Johnson, R.R., Phene, C.J., Trout, T.J., Clark, D.D., Mead, R.M., 2003. Water use by drip-irrigated late-season peaches. Irrigation Science 22, 187–194.

Baggiolini, M., 1980. Stades repères du cerisier - Stades repères du prunier. Stades repères de l'abricotier. Stades repères du pêcher. ACTA. Guide Pratique de Défense des Cultures, Paris.

Boland, A.M., Mitchell, P.D., Jerie, P.H., 1993. Effect of regulated deficit irrigation on tree water use and growth of peach. Journal of Horticultural Science 68, 261–274.

Chalmers, D.J., Mitchell, P.D., Van Heek, L., 1981. Control of peach-tree growth and productivity by regulated water-supply, tree density, and summer pruning. Journal of American Society for Horticultural Science 106, 307–312.

Conejero, W., Mellisho, C.D., Ortuño, M.F., Galindo, A., Pérez-Sarmiento, F., Torrecillas, A., 2011. Establishing maximum daily trunk shrinkage and midday stem water potential reference equations for irrigation scheduling of early maturing peach trees. Irrigation Science 29, 299–309.

De la Rosa, J.M., Conesa, M.R., Domingo, R., Pérez-Pastor, A., 2014. A new approach to ascertain the sensitivity to water stress of different plant water indicators in extra-early nectarine trees. Scientia Horticulturae 169, 147–153.

De la Rosa, J.M., Domingo, R., Gómez-Montiel, J., Pérez-Pastor, A., 2015. Implementing deficit irrigation scheduling through plant water stress indicators in early nectarine trees. Agricultural Water Management 152, 207–216.

do Paço, T.A., Ferreira, M.I., Pacheco, C.A., 2012. Scheduling peach orchard irrigation in water stress conditions: use of relative transpiration and predawn leaf water potential. Fruits 68, 147–158.

FAOSTAT, 2016. Food and Agriculture Organisation of the United Nations. Available at: http://www.fao.org/faostat/en/#data/QC.

Fereres, E., Soriano, M.A., 2007. Deficit irrigation for reducing agricultural water use. Journal of Experimental Botany 58, 147–159.

Girona, J., del Campo, J., Mata, M., Lopez, G., Marsal, J., 2011. A comparative study of apple and pear tree water consumption measured with two weighing lysimeters. Irrigation Science 29, 55–63.

Girona, J., Fereres, E., 2012. Peach. In: Steduto, P., Hsiao, T.C., Fereres, E., Raes, D. (Eds.), Crop Yield Response to Water. FAO Irrigation and Drainage Paper No. 66, Rome, pp. 266–280.

Girona, J., Gelly, M., Mata, M., Arbones, A., Rufat, J., Marsal, J., 2005. Peach tree response to single and combined regulated deficit irrigation regimes in deep soils. Agricultural Water Management 72, 97–108.

Girona, J., Marsal, J., Mata, M., Arbonés, A., DeJong, T., 2004. A comparison of the combined effect of water stress and crop load on fruit growth during different phenological stages in young peach trees. Journal of Horticultural Science and Biotechnology 79, 308–315.

Girona, J., Mata, M., Arbonés, A., Alegre, S., Rufat, J., Marsal, J., 2003. Peach tree response to single and combined regulated deficit irrigation regimes under shallow soils. Journal of American Society for Horticultural Science 128, 432–440.

Goldhamer, D.A., Fereres, E., Mata, M., Girona, J., Cohen, M., 1999. Sensitivity of continuous and discrete plant and soil water stress monitoring in peach trees subjected to deficit irrigation. Journal of American Society for Horticultural Science 124, 437–444.

Goldhamer, D.A., Salinas, M., Crisosto, C., Day, K.R., Soler, M., Moriana, A., 2002. Effects of regulated deficit irrigation and partial root zone drying on late harvest peach tree performance. Acta Horticulturae 592, 343–350.

Handley, D.F., Johnson, R.S., 2000. Late summer irrigation of water-stressed peach trees reduces fruit doubles and deep sutures. Hort Science 35, 771.

Johnson, R.S., Handley, D.F., DeJong, T.M., 1992. Long-term response of early maturing peach trees to postharvest water deficits. Journal of American Society for Horticultural Science 117, 881–886.

Lancashire, P.D., Bleiholder, H., van den Boom, T., Langeluddeke, P., Stauss, R., Weber, E., Witzenberger, A., 1991. A uniform decimal code for growth stages of crops and weeds. Annals of Applied Biology 119, 561–601.

López, G., Behboudian, M.H., Vallverdu, X., Mata, M., Girona, J., Marsal, J., 2010. Mitigation of severe water stress by fruit thinning in "O'Henry" peach: implications for fruit quality. Scientia Horticulturae 125, 294–300.

López, G., Mata, M., Arbonés, A., Solans, J.R., Girona, J., Marsal, J., 2006. Mitigation of effects of extreme drought during stage III of peach fruit development by summer pruning and fruit thinning. Tree Physiology 26, 469–477.

Marsal, J., Casadesus, J., López, G., Mata, M., Bellbert, J., Girona, J., 2016. Sustainability of regulated deficit irrigation in a mid-maturing peach cultivar. Irrigation Science 34, 201–208.

McCutchan, H., Shackel, K.A., 1992. Stem-water potential as a sensitive indicator of water stress in prune trees (*Prunus domestica* L. cv. French). Journal of American Society for Horticultural Science 117, 607–611.

Mounzer, O.H., Conejero, W., Nicolás, E., Abrisqueta, I., García-Orellana, Y., Tapia, L.M., Vera, J., Abrisqueta, J.M., Ruiz-Sánchez, M.C., 2008a. Growth pattern and phenological stages of early maturing peach trees under a Mediterranean climate. HortScience 43, 1813–1818.

Mounzer, O.H., Vera, J., Tapia, L.M., García-Orellana, Y., Conejero, W., Abrisqueta, I., Ruiz-Sánchez, M.C., Abrisqueta, J.M., 2008b. Irrigation scheduling of peach trees by continuous measurement of soil water status. Agrociencia 42, 857–868.

Naor, A., Stern, R., Peres, M., Greenblat, Y., Gal, Y., Flaishman, M.A., 2005. Timing and severity of postharvest water stress affect following year productivity and fruit quality of field-grown 'Snow Queen' nectarine. Journal of American Society for Horticultural Science 130, 806–812.

Pérez-Pastor, A., Ruiz-Sánchez, M.C., Conesa, M.R., 2016. Drought stress effect on woody tree yield. In: Ahmad, P. (Ed.), Water Stress and Crop Plants: A Sustainable Approach, first ed., vol. 2. John Wiley & Sons Ltd., UK, pp. 356–374.

Remorini, D., Massai, R., 2003. Comparison of water status indicators for young peach trees. Irrigation Science 22, 39–46.

Ruiz-Sánchez, M.C., Domingo, R., Castel, J.R., 2010. Review. Deficit irrigation in fruit trees and vines in Spain. Spanish Journal of Agricultural Research 8 (S2), 5–20.

Ruiz-Sánchez, M.C., Girona, J., 1995. Investigaciones sobre riego deficitario controlado en melocotoner. In: Zapata, M., Segura, P. (Eds.), Riego Deficitario Controlado. Mundi-Prensa/Unión Europea, pp. 69–95. Colección Cuadernos VALUE I.

Ruiz-Sánchez, M.C., Torrecillas, A., Sánchez-Blanco, M.J., Abrisqueta, J.M., Vera, J., Pedrero, F., Magaña, I., García-Orellana, Y., Ortuño, M.F., Plana, V., Nicolás, E., Conejero, W., Mounzer, O., Alarcón, J.J., 2006. Respuesta fisiológica del melocotonero temprano al riego por desecación parcial de raíces. In: VIII Simposium Hispano-Portugués de Relaciones Hídricas en las Plantas. (Tenerife), pp. 117–121.

Shackel, K.A., Ahmadi, H., Biasi, W., Buchner, R., Goldhamer, D.A., Gurusinghe, S., Hasey, J., Kester, D., Krueger, B., Lampinen, B., McGourty, G., Micke, W., Mitcham, E., Olson, B., Pelletrau, K., Philips, H., Ramos, D., Schwankl, L., Sibbett, S., Snyder, R., Soutwick, S., Stevenson, M., Thorpe, M., Weinbaum, S., Yeager, J., 1997. Plant water status as an index of irrigation needs in deciduous fruit trees. HortTechnology 7, 23–29.

Starr, J.L., Paltineanu, I.C., 1998. Real-time soil water dynamics over large areas using multisensor capacitance probes and monitoring system. Soil and Tillage Research 47, 43–49.

Tavarini, S., Gil, M.I., Tomás-Barberán, F., Buendia, B., Remorini, D., Massai, R., Degl'Innocenti, E., Guidi, L., 2011. Effects of water stress and rootstocks on fruit phenolic composition and physical/chemical quality in Suncrest peach. Annals of Applied Biology 158, 226–233.

Thompson, R.B., Gallardo, M., Valdez, L.C., Fernández, M.D., 2007. Determination of lower limits for irrigation management using in situ assessments of apparent crop water uptake made with volumetric soil water content sensors. Agricultural Water Management 92, 13–28.

Valdés-Vela, M., Abrisqueta, I., Conejero, W., Vera, J., Ruiz-Sánchez, M.C., 2015. Soft computing applied to stem water potential estimation: a fuzzy rule based approach. Computers and Electronics in Agriculture 115, 150–160.

Vallverdu, X., Girona, J., Echeverria, G., Marsal, J., Behboudian, M., López, G., 2012. Sensory quality and consumer acceptance of "Tardibelle" peach are improved by deficit irrigation applied during stage II of fruit development. HortScience 47, 656–659.

Vera, J., Abrisqueta, I., Abrisqueta, J.M., Ruiz-Sánchez, M.C., 2013. Effect of deficit irrigation on early-maturing peach tree performance. Irrigation Science 31, 747–757.

Vera, J., Abrisqueta, I., Conejero, W., Ruiz-Sánchez, M.C., 2017. Precise sustainable irrigation: a review of soil-plant-atmosphere monitoring. Acta Horticulturae 1150, 192–202.

Vera, J., Padilla, A.M., Ruiz-Sánchez, M.C., Abrisqueta, I., Tapia, L.M., Munguía, J.P., Abrisqueta, J.M., 2010. Automatic drip irrigation in peach trees based on soil water content measured by FDR probes. In: Transactions Third Inter. Symposium Soil Water Measurement Using Capacitance, Impedance and Time Domain Transmission, vol. 1, p. 4. Murcia, Spain.

Williams, L.E., Ayars, J.E., 2005. Grapevine water use and the crop coefficient are linear functions of the shaded area measured beneath the canopy. Agricultural and Forest Meteorology 132, 201–211.

CHAPTER 7

Long-Term Physiological and Agronomic Responses of Citrus Irrigated With Saline Reclaimed Water

Emilio Nicolás[1], Cristina Romero Trigueros[1], Pedro Antonio Nortes Tortosa[1], Francisco Pedrero Salcedo[1], Jose María Bayona Gambín[1], Jose Francisco Maestre Valero[2], Juan José Alarcón Cabañero[1]

[1]Centro de Edafología y Biología Aplicada del Segura (CEBAS-CSIC), Murcia, Spain;
[2]Universidad Politécnica de Cartagena, Cartagena, Spain

Contents

1. INTRODUCTION

The future of the sustainability of water resources will depend primarily on trends in water resource renewal (which will be affected by climate change and pollution) and the pressure exerted on water withdrawals (which depends on the evolution of sectoral uses of water, population and urban growth, tourism, and industry), as well as the use of unconventional water resources. Agriculture is the largest consumer of water resources in Europe and is a major economic sector in many of its countries. The use of reclaimed water in agriculture is an important component in water

Water Scarcity and Sustainable Agriculture in Semiarid Environment
ISBN 978-0-12-813164-0
https://doi.org/10.1016/B978-0-12-813164-0.00007-7

conservation strategies, but its effects with different compositions and on different types of soil and crops should be studied for the sustainable irrigation management with this type of water.

Since 1997, the worldwide use of reclaimed water has developed very rapidly. Currently, approximately 4% of all reclaimed water is reused in the world, representing a volume of approximately 14.2 billion m^3 (Global Water Market, 2017). Reuse market studies predict high growth over the next 10 years. Globally, the volume of reclaimed water in some countries such as China is expected to increase significantly. It is estimated that globally the market for reuse was on the verge of expansion and expected to outpace desalination in the future. It is foreseen that, by 2030, water reuse will represent 1.66% (26 billion m^3 $year^{-1}$) of the total water use (Global Water Market, 2017). Globally, agricultural irrigation is the main application for water reuse. For the last 5 years, agriculture has comprised nearly 70% of global water consumption (and over 90% of total water consumption in the least developed economies) (Sonen Capital, 2016).

In the European Union, the total volume of reused treated wastewater in 2006 amounted to 964 million m^3 $year^{-1}$, accounting for 2.4% of the treated urban wastewater (about 1 billion m^3) or less than 0.5% of annual EU freshwater abstraction. No data were available on the current volume of treated wastewater being reused in the European Union; however, the previous study to support the impact assessment estimated the current volume of reused treated wastewater in the European Union to be 1100 million m^3 $year^{-1}$ or 0.4% of annual EU freshwater abstractions (European Commission, 2016a). However, EU potential is much higher, estimated in the order of 6 billion m^3—six times the current volume. Both southern Member States such as Spain, Italy, Greece, Malta, and Cyprus and northern Member States such as Belgium, Germany, and the United Kingdom already have in place numerous initiatives regarding water reuse for irrigation (European Commission, 2016b).

In the case of Spain, the current demand for water is distributed in different sectors (urban, agricultural, industrial, recreation, etc.), again agriculture demand being the most important (around 24,094 Mm^3 $year^{-1}$). In these conditions, and especially in some regions of Spain with a situation of structural water deficit, such as the Mediterranean Basin, the use of reclaimed water in agriculture becomes very important, since it leads to a more efficient management of water resources. In Spain, 347 Mm^3 of water per year was reused in agriculture in 2006, 79.2% of the volume treated (European Commission, 2016b). However, BOI (2015) estimated that the water reuse target for Spain (1200 million m^3 $year^{-1}$) could be achieved by 2018 (European Commission, 2016b).

2. RECLAIMED WATER IN AGRICULTURE

The use of reclaimed water in agriculture has proven to be very important in the, mainly superficial, strategies for the sustainable use of water resources, due to their economic and environmental benefits. Studies on the effects of reclaimed water were initially launched in countries with high technological capacities and a certain economic level such as the United States (California, Florida, Arizona) and Israel.

During the 1970s, studies were started in areas of forests, golf courses, or fodder crops (Bole and Bell, 1978; Burton and Hook, 1979) with the aim of increasing their biomass. In the 1980s, studies began to see the effects on different crops in the area. The use of reclaimed water in fruit trees was successfully managed in crops such as apple (Neilsen et al., 1989a) or peach tree (Basiouny, 1984), in other woody crops such as vine (Neilsen et al., 1989b) or cotton (Feigin et al., 1984), and in some cereals such as maize and alfalfa (Campbell, 1983). In addition, its use was also extended to ornamental species (Hasek, 1986).

In citrus, although several studies have shown advantages and disadvantages of the use of reclaimed water (Zekri and Koo, 1993, 1994; Ramos, 1996; Aucejo et al., 1997; Morgan et al., 2008; Reboll et al., 2000; Pedrero and Alarcón, 2009; Lapeña et al., 2002; Maurer et al., 1995; Meli et al., 2002; Pereira et al., 2011), studies related to salinity nutrition were performed with unrealistic ion ratios of reclaimed water (Maas and Grattan, 1999). Only a few studies on the salinity response to citrus growth have been performed with salts other than NaCl. In the region of Murcia, Spain, long-term studies have been carried out on the effect of the use of reclaimed water in citrus (mandarin and grapefruit), demonstrating that irrigation with this "natural" saline composition tends to accumulate salts within the root zone, and can cause problems of salinity and B accumulation in plant and soil in the long term (Mounzer et al., 2013; Pedrero et al., 2013, 2014; Nicolás et al., 2016). By applying careful monitoring to avoid possible reduction of soil agronomic properties and good management practices, the sustainability of reclaimed water use in agriculture would be guaranteed (Pedrero et al., 2015).

One of the greatest benefits associated with the use of reclaimed water in agriculture is its nutritional value, with consequent savings in mineral fertilizers. Several studies have corroborated this benefit by quantifying such savings. In the same previous studies in mandarin and grapefruit, reclaimed water covered 24% and 15% of N_2 and P_2O_5 requirements, respectively, and all K_2O requirements (Pedrero et al., 2013). Also in the region of Murcia, studies were carried out on horticultural crops under greenhouse conditions (tomato, pepper, and zucchini). According to the application of fertilizers, the highest concentration of nutrients in the reclaimed water compared with transfer water (TW) (superficial resources) represented a saving of N_2, P_2O_5, K_2O, and CaO of 6%, 5%, 8%, and 40%, respectively. Taking into account the market prices of fertilizers, in economic terms it represented a saving of about €160 N ha^{-1}, €150 P_2O_5 ha^{-1}, €350 K_2O ha^{-1}, and €450 CaO ha^{-1}. The use of reclaimed water for irrigation of tomato, pepper, and zucchini crops in hydroponic cultivation is recommended under conventional agricultural practices (hydroponics, drip irrigation, polyethylene greenhouses, etc.) and is microbiologically safe (Pedrero et al., 2014). In this context the most serious problems of reclaimed water in the region of Murcia are salinity and high concentrations of boron. It should be noted that in this experience the reclaimed water did not represent a danger in short-term agricultural use. These fertilizer savings are referred to by other authors in studies of reclaimed water in other horticultural crops. Thus studies on melon (*Cucumis melo* L. cv. Galia) in sandy soil under greenhouse conditions showed a saving of fertilizers, reducing the application of N and K by 40.8% and 17.8%, respectively (Segura et al., 2001; Martínez et al., 2013).

Another concern to take into account is the possible accumulation of phytotoxic elements in the crop, due to its presence in the composition of the reclaimed water. Kalavrouziotis et al. (2005) carried out experiments on onion (*Alium cepa*) and lettuce (*Lactuca sativa*) in greenhouse conditions and observed P accumulation in the soil, while Mn and Zn accumulated mainly in the roots of lettuce and seeds. In another experiment on broccoli (*Brassica oleracea* var. Italica) and Brussels sprouts (*B. oleracea* var. Gemnifera) under the same conditions, the content of P, Zn, and Cd was increased significantly in the soil. Moreover, the concentration of heavy metals in the edible parts of these crops and their microbiological load (fecal coliforms and *Escherichia coli*) in the irrigation reclaimed water was a factor of high sanitary risk (Kalavrouziotis et al., 2008).

Several studies have been carried out on simplified treatments, excluding biological processes for the removal of organic matter and nitrogen, to evaluate the recovery of the nutrients of reclaimed water and avoid the cost related to biological oxidation and sludge production. These prototypes have been tested in the field to evaluate their effectiveness in the production of reclaimed water suitable for reuse in agriculture. In experiments carried out on olive trees, comparing them with olive trees irrigated with surface water, 50% production increases were obtained, improving commercial fruit characteristics (Lopez et al., 2006). In spite of this, there were a number of disadvantages, since this reduction in the biological treatment process can also cause problems in the soil, with a decrease of porosity and, consequently, the retention of water in the soil and its conductivity (Aiello et al., 2007).

3. CASE STUDY. SALINE RECLAIMED WATER EFFECTS ON CITRUS

3.1 Experimental Site and Irrigation Treatments

The experiment was carried out in Campotéjar-Murcia, Spain (38°07′18″N; 1°13′15″W), with a Mediterranean semiarid climate. Annual reference evapotranspiration and rainfall are, on average, 1339 and 278 mm, respectively. Within this area, an orchard of 1 ha was cultivated with two crops. One crop consisted of mandarin trees (*Citrus clementina* cv. "Orogrande") planted in 2000 grafted on Carrizo citrange (*Citrus sinensis* [L.] Osb. × *Poncirus trifoliata* [L.]) rootstock, with a tree spacing of 5 m × 3.5 m. The other crop consisted of "Star Ruby" grapefruit trees (*Citrus paradisi* Macf.) grafted on Macrophylla rootstock (*Citrus macrophylla*) planted in 2005, with a tree spacing of 6 m × 4 m.

Two irrigation treatments based on the source of irrigation water were established. The first treatment was based on irrigation with water pumped from the "Tagus-Segura" water transfer canal, which supplies a large part of the water used in the region of Murcia for both human consumption and irrigation (TW). The TW water had an average electrical conductivity (EC_W) of 1.1 dS m^{-1}. The second treatment was based on irrigation with tertiary RW pumped from a nearby wastewater treatment plant. Treatments TW and RW were applied during the growing season for both citrus species from 2008 onwards.

The irrigation system consisted of a single drip line laid on the soil surface next to each tree row. It provided three pressure-compensating, inline emitters per tree, each discharging 4 L h^{-1}, which were placed 0.85 m from the trunk and spaced 0.9 m apart in the mandarin trees and placed 1 m from the trunk and spaced 1 m apart in the grapefruit trees. The irrigation doses were scheduled on the basis of the daily crop evapotranspiration (ET_c) for the previous week (Pedrero et al., 2015; Nicolás et al., 2016).

The trees were irrigated at 100% ET_c from January to December. The total amount of water applied was measured with inline water flow meters. The irrigation was controlled automatically by a head-unit programmer and electrohydraulic valves. All treatments included application of the same amounts of fertilizer ($N-P_2O_5-K_2O$), applied through the drip irrigation system: 215–100–90 kg ha^{-1} year^{-1} for mandarin trees and 215–110–150 kg ha^{-1} year^{-1} for grapefruit trees. Pest control practices and pruning were those commonly used by growers in the area and were performed equally for the different treatments.

3.2 Reclaimed Water Quality Characterization

Significant differences between TW and RW were observed. RW had the highest salinity and sodicity with average values of EC_w around 3.55 dS m^{-1} and sodium absorption ratio (SAR_w) around 5.98 (meq L^{-1})$^{0.5}$, respectively; whereas TW had lower values, an EC_w of 1.14 dS m^{-1} and SAR_w of 1.40 (meq L^{-1})$^{0.5}$ (Table 7.1). RW also had higher concentrations of NO_3^-, PO_4^{3-}, SO_4^{2-}, Cl^-, K, B, and Na than TW. It is remarkable that Na, Cl^-, and B

Table 7.1 Physical and chemical analyses (electrical conductivity, EC_w; sodium absorption ratio, SAR_w; pH; cations; Na, K, Ca, Mg, and anions; and Cl^-, NO_3^-, PO_4^{---}, and SO_4^{--}) for both transfer water (TW) and reclaimed water (RW). Each value is the average ± SE for the 6 years, with N = 48 for each year's measurements

		2008–13	
	Units	TW	RW
EC_w	dS m^{-1}	1.14 ± 0.15	3.55 ± 0.28
SAR_w	(meq L^{-1})$^{0.5}$	1.4 ± 0.56	5.98 ± 2.03
pH		8.15 ± 1.03	7.93 ± 0.31
Ca	meq L^{-1}	5.76 ± 0.8	8.15 ± 1.3
Mg	meq L^{-1}	4.44 ± 0.7	8.94 ± 1.95
K	mg L^{-1}	7.59 ± 5.85	46.01 ± 9.55
Na	meq L^{-1}	3.98 ± 0.8	20.06 ± 3.7
B	mg L^{-1}	0.12 ± 0.01	0.67 ± 0.12
Cl^-	meq L^{-1}	4.61 ± 0.92	19.80 ± 4.39
NO_3^-	mg L^{-1}	10.32 ± 3.7	25.40 ± 13.7
PO_4^{---}	mg L^{-1}	0.52 ± 0.02	1.57 ± 1.1
SO_4^{--}	meq L^{-1}	9.33 ± 1.6	18.80 ± 6.2

concentrations in RW exceeded the thresholds at which detrimental effects on citrus might be observed: Na > 5.02 meq L^{-1} (Bernstein et al., 1956), Cl^- > 6.71 meq L^{-1} (Ayers and Westcot, 1985), and B > 0.5 mg L^{-1} (Grieve et al., 2012). Bearing in mind the higher concentration of NO_3^-, PO_4^{3-}, and K in RW, this water resource could have provided (after subtracting the average TW nutrient content for the whole experiment) 9.2% of N, 4.5% of P, and more than 100% of the trees' K requirements. Therefore the use of RW could significantly reduce fertilizer application and its associated costs, and prevent soil contamination (US-EPA, 2004), considering that an excess of these could be lost through leaching (Romero-Trigueros et al., 2014).

3.3 Plant Physiological Responses to Irrigation With Saline Reclaimed Water

In our study, both TW and RW treatments had similar stem water potential (Ψ_{stem}) values throughout the experiment and all species maintained Ψ_{stem} values above −1 MPa (except grapefruit in August), which indicated good water status for citrus irrigated with TW or RW (Figs. 7.1C and 7.2C). Therefore the average annual values of Ψ_{stem} were similar in both species and irrigation water quality (Table 7.1).

Irrigation with saline water, in case it exceeds certain toxic threshold levels, has been usually associated with stomatal closure (Gimeno et al., 2012), which in turn can cause (1) low transpiration rates, (2) a reduction of the CO_2 available for photosynthesis (García-Sánchez and Syvertsen, 2006), (3) an increase in leaf ion accumulation (Brumos et al., 2009; Mouhaya et al., 2010), and hence (4) a reduction in fruit yield (Murkute et al., 2005). In this study, in both mandarin and grapefruit, despite the saline stress induced to plants irrigated by RW, differences between A or g_s were hardly observed throughout the experiment (Figs. 7.1 and 7.2). Both A and g_s were also quite stable between years with mean values for the whole experiment of $A = 4.6$ and 4.4 μmol m^{-2} s^{-1} and $g_s = 59.2$ and 56.8 mmol m^{-2} s^{-1} for TW and RW, respectively, in mandarin and $A = 8.9$ and 8.0 μmol m^{-2} s^{-1} and $g_s = 109.7$ and 92.0 mmol m^{-2} s^{-1} for TW and RW, respectively, in grapefruit. Also, for the whole experiment, RW treatment showed similar intrinsic water use efficiency values for TW and RW in both species (Table 7.2).

These results agree with that presented by Pérez-Pérez et al. (2007) who did not observe any negative effect on leaf water relations or net gas exchange after watering seedlings of Carrizo citrange three times per week

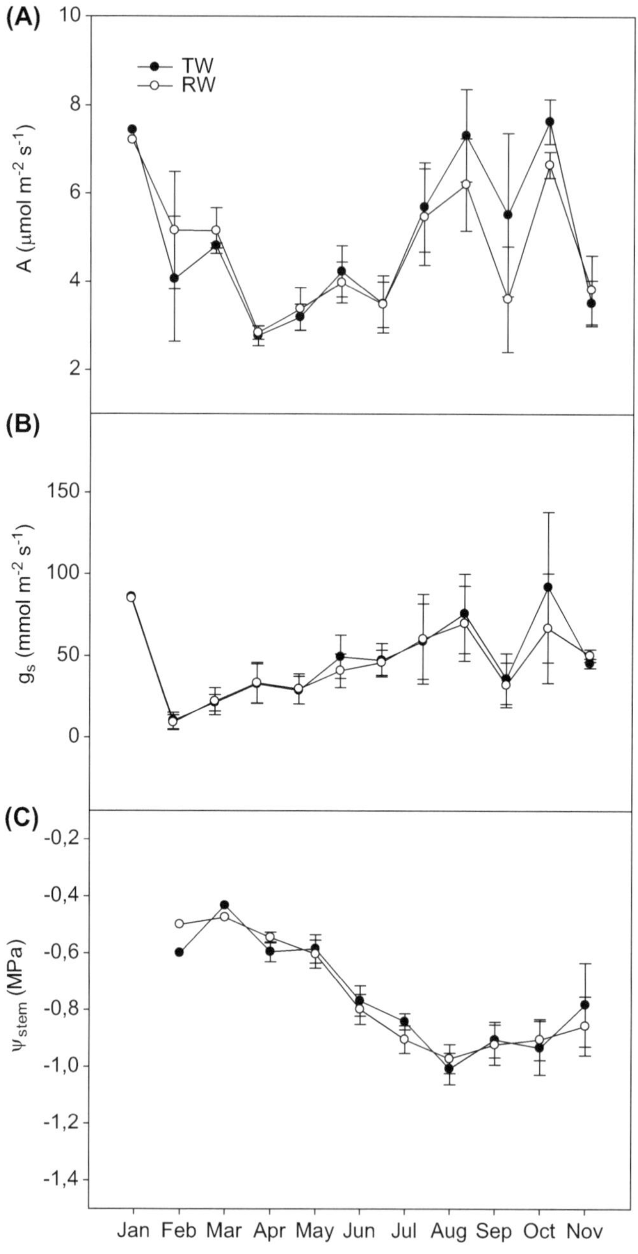

Figure 7.1 Net photosynthesis, *A*, (A) and stomatal conductance, g_s, (B) stem water potential, Ψ_{stem}, (C) in mandarin trees irrigated with transfer (TW) and reclaimed (RW) water. Values are averages ± SE for each month of the 6 individual years with N = 4 for each year.

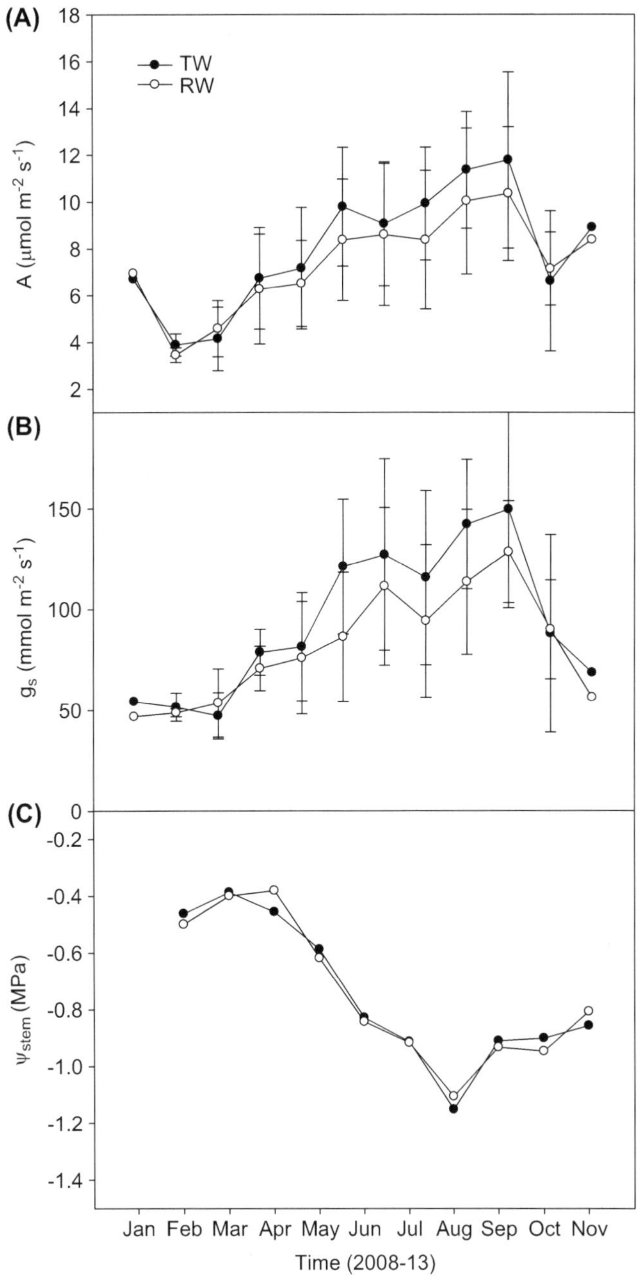

Figure 7.2 Net photosynthesis, A, (A) and stomatal conductance, g_s, (B) stem water potential, Ψ_{stem}, (C) in grapefruit trees irrigated with transfer (TW) and reclaimed (RW) water. Values are averages $\pm$ SE for each month of the 6 individual years with N $=$ 4 for each year.

Table 7.2 Average annual values of reference evapotranspiration (ET_o), rainfall, and irrigation volume applied in 2008–13 for mandarin and grapefruit trees. For each treatment, mean midday stem water potential (Ψ_{stem}), midmorning rate of net photosynthesis (A), stomatal conductance (g_s), and intrinsic water use efficiency (WUE_i) are also shown

Year	ET_o (mm)	Rainfall (mm)	Rainfall October –December (mm)	Irrigation (mm year^{-1})	Ψ_{stem} (MPa)		A (μmol m^{-2} s^{-1})		g_s (mmol m^{-2} s^{-1})		WUE_i (μmol mol^{-1})	
					TW	RW	TW	RW	TW	RW	TW	RW
Mandarin	1339	280.0	86.4	585.5	0.88	0.91	4.6	4.4	59.2	56.8	76.9	76.7
Grapefruit	1339	280.0	86.4	452.2	0.78	0.78	8.9	8.0	109.7	92.0	81.4	87.0

with 100 mL of 100 mM NaCl. As expected, the lowest Ψ_{stem} values were observed in summer, mainly due to the predominant high evaporative demand and the hot and dry winds in the region.

3.4 Impact of Saline Reclaimed Water on Citrus Yield and Water Productivity

Since the beginning of the experiment, the use of RW in adult mandarin led to significant reductions in the tree canopy with respect to the control treatment, which reached an average reduction of 18% during the 6-year experimental period (Table 7.3). Overall, each year the use of RW led to notable reductions in crop fruit load (not significant in 2009; see Pedrero et al., 2014). However, such reductions were more prominent during the seasons 2011−13, when the average reduction in crop fruit load was 43.7% for the RW versus TW treatment, much higher than the 18.7% reduction observed during the first 3 years of the experiment. Each season, these lower crop fruit loads resulted in greater yield reductions, although these differences were only significant since 2010 (Pedrero et al., 2014). For the 2008−10 seasons, the yield reduction for RW treatment was 12.2% as compared to TW, whereas that difference was, on average, 40.1% for the 2011−13 seasons. In the short term, this lesser difference might be explained by a lower reduction in crop fruit load, which was counterbalanced by a significant fruit weight increase (increase of 7.6%) observed in RW treatment as compared to TW treatment (Table 7.3). However, since 2012 the absence of increase in fruit weight combined with the significant reduction in crop fruit load in RW treatment provoked important yield reductions.

In contrast to mandarin, in grapefruit, following the same trend observed in the short-term study (Pedrero et al., 2015), neither the tree canopy, the fruit crop load, the fruit weight, nor the yield were affected by irrigation with saline RW (Table 7.4). As a result, water productivity (WP) between treatments was also similar, as the amount of water used for irrigation in TW and RW treatments was the same. Not finding significant differences in these parameters might be justified by (1) the behavior of plant water status and gas exchange parameters, as significant differences between treatments were not observed, and (2) the rootstock used (*C. macrophylla*), which probably improved the resilience to saline stress.

It is also noteworthy that the two different periods observed coincided with the trees' development throughout the 6-year experiment, i.e., from young (2008−10) to adult (2011−13) trees. In this sense, if the 2008−10 and 2011−13 means for tree canopy, crop load, and yield are compared in

Table 7.3 Tree canopy (m^3), crop load (fruit $tree^{-1}$), yield (kg $tree^{-1}$), fruit weight (g), and irrigation water productivity (WP; kg m^{-3}) for mandarin trees in 2008–13 for the two irrigation treatments (transfer water, TW and reclaimed water, RW). Values are averages ± SE for the 3 individual years with N = 4 for each year. For each row, means with same letter are not significantly different according to Tukey's HS test ($P < .05$)

Year	Treatment	Tree canopy	Crop load	Yield	Fruit weight	WP
2008–10	TW	15.2 ± 0.4b	790.9 ± 111.0b	79.3 ± 9.0a	100.5 ± 1.9a	7.6 ± 1.5a
	RW	11.9 ± 1.0a	643.2 ± 50.1a	69.6 ± 7.1a	108.1 ± 2.9b	6.7 ± 1.2a
2011–13	TW	19.2 ± 1.0b	726.8 ± 87.7b	96.7 ± 34.4a	132.2 ± 12.5a	8.7 ± 3.8a
	RW	16.3 ± 0.5a	408.8 ± 135.9a	57.9 ± 21.7a	140.9 ± 28.4a	5.2 ± 2.4a
2008–13	TW	17.2 ± 0.7b	758.8 ± 99.3b	88.0 ± 21.7a	116.3 ± 7.2a	8.1 ± 2.6a
	RW	14.1 ± 0.7a	526.0 ± 93.0a	69.6 ± 14.4a	124.5 ± 15.6a	5.9 ± 1.8a

Table 7.4 Tree canopy (m^3), crop load (fruit $tree^{-1}$), fruit weight (g), yield (kg $tree^{-1}$), and irrigation water productivity (WP; kg m^{-3}) for grapefruit trees in 2008–13 for the two irrigation treatments (transfer water, TW and reclaimed water, RW). Values are averages ± SE for the 3 individual years with N = 4 for each year. For each row, means with same letter are not significantly different according to Tukey's HS test ($P < .05$)

Year	Treatment	Tree canopy	Crop load	Yield	Fruit weight	WP
2008–10	TW	2.2 ± 1.1a	193.1 ± 43.9a	68.6 ± 10.1a	359.2 ± 16.1a	9.2 ± 1.7a
	RW	2.3 ± 1.0a	159.1 ± 31.0a	61.3 ± 6.1a	380.1 ± 37.2a	8.6 ± 3.4a
2011–13	TW	5.4 ± 1.2a	496.2 ± 111.0a	168.6 ± 24.2a	367.1 ± 33.4a	11.8 ± 1.9a
	RW	5.8 ± 0.9a	555.7 ± 177.2a	168.8 ± 39.3a	373.2 ± 50.3a	12.0 ± 3.6a
2008–13	TW	3.80 ± 1.1a	344.65 ± 77.4a	118.60 ± 17.1a	118.60 ± 24.7a	10.50 ± 1.8a
	RW	4.05 ± 0.9a	357.40 ± 104.1a	115.05 ± 22.7a	115.05 ± 43.7a	10.30 ± 3.5a

the second stage, when the trees were adult, the tree canopy, crop load, and yield increased by 145%, 157%, and 146% for TW treatment and by 152%, 149%, and 175% for RW treatment, respectively. It is remarkable that irrigation with saline RW in citrus has been generally associated with a reduction in vegetative growth, fruit crop load, and fruit yield (Nicolás et al., 2016; Pérez-Pérez et al., 2015).

4. CONCLUSIONS

Increased pressure on water resources due to population growth, water resource pollution, climate change impacts, etc. is raising interest in wastewater reuse, mainly in drought-prone regions where RW can become a major potential source for irrigation as it contains large amounts of organic matter and many nutrients essential for plant growth, and under appropriate management allows reduced fertilizer application rates. As disadvantages, incorrect irrigation management may lead to soil salinization problems or to reductions in the physiological and agronomic response of the plant.

In this 6-year long-term study in grapefruit and mandarin, irrigation with saline RW did not affect significantly either the physiological (net gas exchange and plant water status) or agronomic (fruit crop load, yield, and WP) parameters.

In summary, this study shows the long-term feasibility of using saline RW to irrigate citrus and also their capacity to adapt to the conditions of this study. However, irrigation with saline RW must be performed cautiously as it will only be successful with irrigation management measures and intensive monitoring to avoid damaging the soil and trees.

In the near future, the agricultural sector has to be prepared to depend more and more on alternative water resources such as RW. In this sense, the effects of this water on the crops' sustainability should be further investigated.

ACKNOWLEDGMENTS

This study was supported by three projects granted to the authors, CICYT (AGL2010-17553), CICYT (AGL2013-49047-C2-2-R), and CICYT (AGL2016-77282-C3-1-R).

REFERENCES

Aiello, R., Luigi Cirelli, G., Consoli, S., 2007. Effects of reclaimed wastewater irrigation on soil and tomato fruits: a case study in Sicily (Italy). Agricultural Water Management 93, 65–72.

Aucejo, A., Ferrer, J., Gabaldón, C., Marzal, P., Seco, A., 1997. Toxicity in citrus plantations in Villareal, Spain. Water, Air, Soil and Pollution 94, 349–360.

Ayers, R.S., Westcot, D.W., 1985. Water Quality for Agriculture, Irrigation and Drainage. FAO, Rome.

Basiouny, F.M., 1984. The use of municipal treated effluent for peach tree irrigation. In: Proceedings of the Florida State Horticultural Society, vol. 97, pp. 345–347.

Bernstein, L., Brown, J.W., Hayward, H.E., 1956. The influence of rootstock on growth and salt accumulation in stone-fruit trees and almonds. In: Proceedings of the American Society for Horticultural Science, vol. 68, pp. 86–95.

BIO by Deloitte, 2015. Optimising Water Reuse in the EU, Final Report – Part I, Prepared for the European Commission – DG ENV. In collaboration with ICF and Cranfield University. https://doi.org/10.2779/603205.

Bole, J.B., Bell, R.G., 1978. Land application of municipal sewage wastewater: yield and chemical composition of forage crops. Journal of Environmental Quality 7, 222–226.

Brumos, J., Colmenero-Flores, A., Conesa, P., Izquierdo, P., Sánchez, G., Iglesias, M.F., López-Climent, A., Gómez-Cadenas, A., Talón, M., 2009. Membrane transporters and carbon metabolism implicated in chloride homeostasis differentiate salt stress responses in tolerant and sensitive *Citrus* rootstocks. Functional Integrative Genomics 9, 293–309.

Burton, T.M., Hook, J.E., 1979. A mass balance study of application of municipal wastewater to forest in Michigan. Journal of Environmental Quality 8, 589–596.

Campbell, W.F., 1983. Alfalfa, sweet corn, and wheat responses to long-term application of municipal wastewater to cropland. Journal of Environmental Quality 12, 234–249.

European Commission, 2016a. EU-Level Instruments on Water Reuse Final Report to Support the Commission's Impact Assessment. Prepared by Amec Foster Wheeler Environment & Infrastructure UK Ltd., IEEP, ACTeon, IMDEA and NTUA, ISBN 978-92-79-62616-6. https://doi.org/10.2779/974903.

European Commission, 2016b. Inception Impact Assessment on the Initiative "Minimum Quality Requirements for Reused Water in the EU (New EU Legislation)". http://ec.europa.eu/environment/water/reuse.htm.

Feigin, A., Vaisman, I., Bielorai, H., 1984. Drip irrigation of cotton with treated municipal effluents: II nutrient availability in soil. Journal of Environmental Quality 13, 234–238.

García-Sánchez, F., Syvertsen, J.P., 2006. Salinity reduces growth, gas exchange, chlorophyll and nutrient concentrations in diploid sour orange and related allotetraploid somatic hybrids. Journal of Horticultural Science and Biotechnology 77, 379–386.

Gimeno, V., Simón, I., Nieves, M., Martínez, V., Cámara-Zapata, J.M., García, A.L., García-Sánchez, F., 2012. The physiological and nutritional responses to an excess of boron by Verna lemon trees that were grafted on four contrasting rootstocks. Trees 26, 1513–1526.

Global Water Market, 2017. Meeting the world's water and wastewater needs until 2020. Global Water Intelligence 18 (5).

Grieve, C.M., Grattan, S.R., Maas, E.V., 2012. Plant salt tolerance in agricultural salinity assessment and management. In: Wallender, W.W., Tanji, K.K. (Eds.), ASCE Manuals and Reports on Engineering Practice No. 71, second ed. American Society of Civil Engineers (ASCE), Reston, VA, pp. 405–459.

Hasek, R.F., 1986. Water conservation and recycling in ornamentals production. HortScience 21 (1), 35–38.

Kalavrouziotis, I.K., Kanatas, P.I., Papadopoulos, H., Bladenopoulou, S., Koukoulakis, P.H., Leotsinides, M.N., 2005. Effects of municipal reclaimed wastewater on the macro and microelements status of soil and plants. Fresenius Environmental Bulletin 14 (11).

Kalavrouziotis, I.K., Robolas, P., Koukoulakis, P.H., Papadopoulos, A.H., 2008. Effects of municipal reclaimed wastewater on the macro- and micro elements status of soil and of *Brassica oleracea* var. Italica, and *B. oleracea* var. Gemmifera. Agricultural Water Management 95, 419–426.

Lapeña, L., Escorín, V., Cerezo, M., Flors, V., García, P., 2002. Riego con agua residual urbana depurada en la Plana de Castellon: Un estudio de diez años. Dpto. Ciencias Experimentales ESTCE. Área de Fisiología Vegetal. Universitat Jaume I de Castellón.

Lopez, A., Pollice, A., Lonigro, A., Masi, S., Palese, A.M., Cirelli, G.L., Toscano, A., Passino, R., 2006. Agricultural wastewater reuse in southern Italy. Desalination 187, 323–334.

Maas, E.V., Grattan, S.R., 1999. Crop yields as affected by salinity. In: Skaggs, R.W., van Schilfgaarde, J. (Eds.), Agricultural Drainage. Agron. Monograph 38. ASA, CSSA, SSSA, Madison, WI, pp. 55–108.

Martínez, S., Suay, R., Moreno, J., Segura, M.L., 2013. Reuse of tertiary municipal wastewater effluent for irrigation of *Cucumis melo* L. Irrigation Science 31, 661–672.

Maurer, M.A., Davies, F.S., Graetz, D.A., 1995. Reclaimed wastewater irrigation and fertilization of mature 'Redblush' grapefruit trees on spodosols in Florida. Journal of the American Society for Horticultural Science 120, 394–402.

Meli, S., Porto, M., Belligno, A., Bufo, S., Mazzatura, A., Scopa, A., 2002. Influence of irrigation with lagooned urban wastewater on chemical and microbiological soil parameters in a citrus orchard under Mediterranean condition. Science of the Total Environment 285, 69–77.

Morgan, K.T., Wheaton, A., Parsons, L.R., Castle, W.S., 2008. Effects of reclaimed municipal waste water on horticultural characteristics, fruit quality, and soil and leaf mineral concentration of *Citrus*. HortScience 43, 459–464.

Mouhaya, W., Allario, T., Brumos, J., Andres, F., Froelicher, Y., Luro, L., 2010. Sensitivity to high salinity in tetraploid citrus seedlings increases with water availability and correlates with expression of candidate genes. Functional Plant Biology 37, 674–685.

Mounzer, O., Pedrero, F., Nortes, P.A., Bayona, J.M., Nicolás, E., Alarcón, J.J., 2013. Transient soil salinity under the combined effect of reclaimed water and regulated deficit drip irrigation of Mandarin trees. Agricultural Water Management 120, 23–29.

Murkute, A.A., Sharma, S., Sing, S.K., 2005. Citrus in terms of soil and water salinity: a review. Journal of Scientific and Industrial Research 64, 393–402.

Neilsen, G.H., Stevenson, D.S., Fitzpatrick, J.J., Brownlee, C.H., 1989a. Nutrition and yield of young apple trees irrigated with municipal waste water. Journal of the American Society for Horticultural Science 114, 377–383.

Neilsen, G.H., Stevenson, D.S., Fitzpatrick, J.J., 1989b. The effect of municipal wastewater irrigation and rate of N fertilization on petiole composition, yield, and quality of Okanalgran Riesling grapes. Canadian Journal of Plant Science 69, 1285–1294.

Nicolás, E., Alarcón, J.J., Mounzer, O., Pedrero, F., Nortes, P.A., Alcobendas, R., Romero-Trigueros, C., Bayona, J.M., Maestre-Valero, J.F., 2016. Long-term physiological and agronomic responses of mandarin trees to irrigation with saline reclaimed water. Agricultural Water Management 166, 1–8. Corrigendum 169:193.

Pedrero, F., Alarcón, J.J., 2009. Effects of treated wastewater irrigation on lemon trees. Desalination 246, 631–639.

Pedrero, F., Mounzer, O., Alarcón, J.J., Bayona, J.M., Nicolás, E., 2013. The viability of irrigating mandarin trees with saline reclaimed water in a semi-arid Mediterranean region: a preliminary assessment. Irrigation Science 31, 759–768.

Pedrero, F., Maestre, J.F., Mounzer, O., Alarcón, J.J., Nicolás, E., 2014. Physiological and agronomic mandarin trees performance under saline reclaimed water combined with regulated deficit irrigation. Agricultural Water Management 146, 228–237.

Pedrero, F., Maestre-Valero, J.F., Mounzer, O., Nortes, P.A., Alcobendas, R., Romero-Trigueros, C., Bayona, J.M., Alarcón, J.J., Nicolás, E., 2015. Response of young 'star ruby' grapefruit trees to regulated deficit irrigation with saline reclaimed water. Agricultural Water Management 158, 51–60.

Pereira, B.F.F., He, Z.L., Stoffella, P.J., Melfi, A.J., 2011. Reclaimed wastewater: effects on citrus nutrition. Agricultural Water Management 98, 1828–1833.

Pérez-Pérez, J.G., Syvertsen, J.P., Botía, P., García-Sánchez, F., 2007. Leaf water relations and net gas exchange responses of salinized Carrizo citrange seedlings during drought stress and recovery. Annals of Botany 100 (2), 335–345.

Pérez-Pérez, J.G., García-Sánchez, F., Robles, J.M., Botía, P., 2015. 'Star ruby' grapefruit and 'clemenules' mandarin trees show different physiological and agronomic responses to irrigation with saline water. Irrigation Science 33 (3), 191–204. https://doi.org/10.1007/s00271-014-0459-8.

Ramos, C., 1996. El riego con aguas residuales. Aprovechamiento del agua depurada en la Comunidad Valenciana. Sanejament d'Aigües-Generalitat Valenciana, pp. 49–63.

Reboll, V., Cerezo, M., Roig, A., Flors, V., Lapeña, V., García-Agustín, P., 2000. Influene of wastewater vs groundwater on young *Citrus* trees. Journal of the Science of Food and Agriculture 80, 1441–1446.

Romero-Trigueros, C., Nortes, P.A., Alarcón, J.J., Nicolás, E., 2014. Determination of ^{15}N stable isotope natural abundances for assessing the use of saline reclaimed water in grapefruit. Environmental Engineering and Management Journal 13 (10), 2525–2530.

Segura, M.L., Moreno, R., Martínez, S., Pérez, J., Moreno, J., 2001. Effects of wastewater irrigation on melon growth under greenhouse conditions. Acta Horticulturae 559, 345–352.

Sonen Capital, 2016. Water Investment Primer. Sonen Capital. http://www.sonencapital.com/wp2015/wp-content/uploads/2016/06/16WIP.pdf.

US EPA (U.S. Environmental Protection Agency) & USAID (U.S. Agency for International Development), 2004. Guidelines for Water Reuse. Washington, DC.

Zekri, M., Koo, R.C.J., 1993. A reclaimed water citrus irrigation project. In: Proceedings of the Florida State Horticultural Society, vol. 106, pp. 30–35.

Zekri, M., Koo, R.C.J., 1994. Treated municipal wastewater for citrus irrigation. Journal of Plant Physiology 17, 693–708.

CHAPTER 8

Persimmon (*Diospyros kaki*) Trees Responses to Restrictions in Water Amount and Quality

Diego S. Intrigliolo[1], Fernando Visconti[2], Luis Bonet[2], Margarita Parra[1], Cristina Besada[2], Isabel Abrisqueta[1], José S. Rubio[1], José M. de Paz[2]
[1]Centro de Edafología y Biología Aplicada del Segura (CEBAS-CSIC), Murcia, Spain;
[2]Instituto Valenciano de Investigaciones Agrarias, Valencia, Spain

Contents

1. INTRODUCTION

Persimmon (*Diospyros kaki* L.) came originally from China where its cultivation began several centuries BC, and from there it was extended to Japan and Korea (George et al., 1997; Llácer and Badenes, 2002; Mowat and George, 1996). For decades, Japan has developed an important genetic selection program, which is the origin of many important persimmon cultivars nowadays cropped in the world such as Hachiya (PCA, pollination constant astringent), Fuyu (PCNA, pollination constant nonastringent), and Jiro (PCNA). In Mediterranean and related semiarid areas, numerous varieties have been developed from spontaneous mutations giving rise to many local cultivars spread throughout several countries. Persimmon

Water Scarcity and Sustainable Agriculture in Semiarid Environment
ISBN 978-0-12-813164-0
https://doi.org/10.1016/B978-0-12-813164-0.00008-9

cultivars are classified according to flesh darkening on seed formation into pollination constant and pollination variant, and according to astringency loss on ripening into astringent and nonastringent.

The genus *Diospyros* belongs to the Ebenaceae family and means "food of the gods" (from the Greek). Most of the species are native to tropical and subtropical regions. They are evergreen, but some such as *D. kaki*, *Diospyros lotus* L., and *Diospyros virginiana* L. are deciduous and have adapted to temperate climates (Mowat and George, 1996). The majority of the varieties suitable for fruit consumption belong to *D. kaki*, while *D. kaki* itself, *D. lotus*, and *D. virginiana* are usually used as rootstocks.

According to the Food and Agriculture Organization (FAO) (FAOSTAT, 2017), worldwide production in 2014 was 5.2 M tonnes with 1.0 Mha of area harvested, which means five times more production and six times more cultivated area than 30 years ago (Fig. 8.1).

Nowadays, China clearly accounts for the highest percent of world production (Fig. 8.2), with more than 70% of the total and with 3.7 M tonnes and 0.9 Mha. China's production is based on a wide diversity of varieties, generally with low yield, which are intended for fresh consumption and industrial processing as dehydrated persimmon (Zhang et al., 2009). South Korea and Japan (Fig. 8.2) follow China in the world ranking producing nonastringent varieties such as "Fuyu." The Far East countries are the top producers around the world accounting for nearly 90% of the

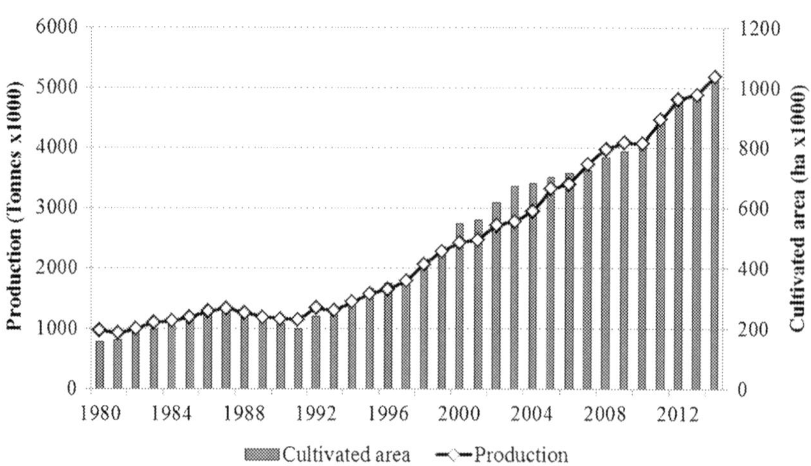

Figure 8.1 Worldwide trend in persimmon cultivation from 1980 until present (FOSTAT, 2017).

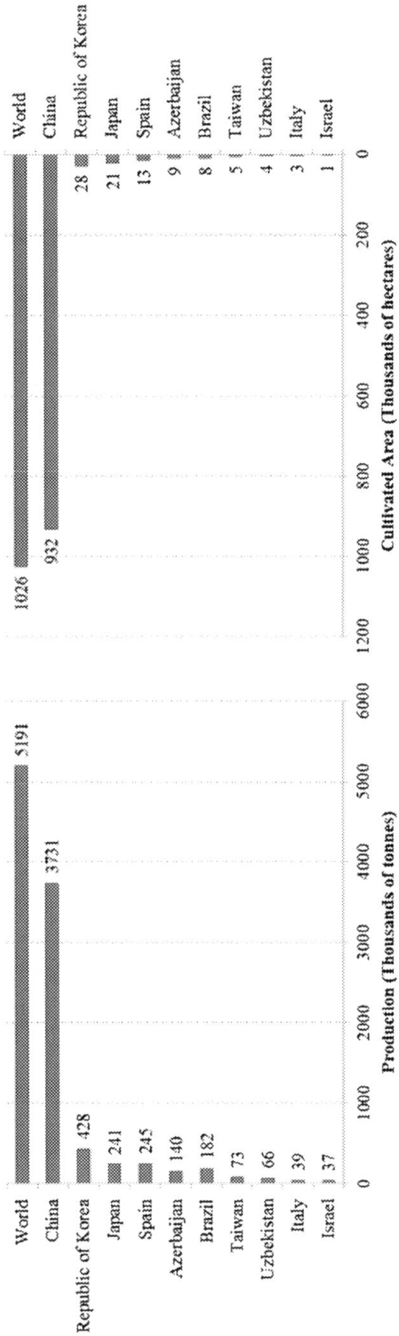

Figure 8.2 World production and cultivated area from the main producing countries in 2014.

total (Llácer and Badenes, 2002). Outside the Far East, Spain, Azerbaijan, and Brazil produced a joint 10% of world persimmon harvest in 2014 (FAOSTAT, 2017).

In the early 1960s only five countries, China, Japan, South Korea, Italy, and Brazil, had more than 100 ha and/or produced over 1000 tonnes per year (Llácer and Badenes, 2002). The evolution of the persimmon cropped area acreage and production since 1980 in the world top producers can be found in Tables 8.1 and 8.2. In this sense, South Korea, Spain, and Brazil have experienced a notable increment in cultivated area and yield throughout the years, whereas Japan and Italy have undergone a decrease in cultivated area and total production.

Regarding the Mediterranean areas, Spain represents the greatest and fastest expansion of the crop, based on the cultivation of two astringent varieties "Rojo Brillante" (Bright Red) and "Triumph." "Rojo Brillante" trees are very productive with just standard cultural practices needed. The fruits are large and appealing because of their color, aroma, and flavor, and it is easy to remove the astringency without compromising firmness (Arnal et al., 2002; Llácer and Badenes, 2009).

2. PHENOLOGY AND GROWTH CYCLE

The *Diospyros* genus includes tree species with round and sympodial branching. It is a vigorous tree that can reach 17 m in height if unpruned. However, i to facilitate cropping, tree height should not surpass 5–6 m. The leaves are alternate, oval, and hard, approximately 13–15 cm long, bright green, and, in some varieties, the leaves turn to a yellow, orange, and red color in the fall before harvesting.

Buds are alone and located in axils of the leaves. During the dormant period, vegetative buds are closed and covered by greenish scales and a gray-white fuzz. During springtime, scales begin to elongate and fall down allowing the emergence of young sprouts with leaves, new buds, nodes, internodes, and apices. The same structures are observed in mixed buds. Both types of buds differ in their form: mixed buds have a conical form with a broad base more or less sharp pointed, whereas the vegetative ones are very sharp pointed. Along the trunk and branches there are also latent buds, but they only grow in the absence of axil buds or when the trees are subjected to biotic and abiotic stresses such as very severe pruning, cold weather, attacks of pathogens, etc. (Giordani et al., 2015).

Table 8.1 Evolution of persimmon harvested area (ha) from the main producer countries since 1980 according to FAO (FAOSTAT, 2017)

	1980	1985	1990	1995	2000	2005	2010	2014
China	110,000	190,000	161,500	268,000	469,000	619,000	729,903	931,907
Japan	29,400	29,800	29,500	28,000	25,000	23,700	22,400	21,300
Republic of Korea[a]	6,590	9,838	13,581	25,009	31,193	26,831	31,808	27,988
Brazil	4,051	3,692	3,960	4,819	6,230	8,309	8,755	8,323
Italy	3,300	3,313	3,055	2,878	2,227	2,720	2,583	2,531
Spain			56	830	2,300	2,885	5,827	13,370
Azerbaijan					3,900	5,932	7,704	8,712
World	154,432	239,055	215,169	332,654	547,283	701,629	820,519	1,025,989

[a]South Korea.

Table 8.2 Evolution of persimmon production (thousands of tonnes) from the main producer countries since 1980 according to FAO (FAOSTAT, 2017)

	1980	1985	1990	1995	2000	2005	2010	2014
China	560	680	625	969	1592	2185	2876	3731
Japan	265	290	286	254	279	286	189	241
Republic of Korea[a]	32	97	96	195	288	364	391	428
Brazil	40	44	47	52	63	165	167	182
Italy	61	56	69	61	42	51	48	39
Spain			1	11	33	52	125	245
Azerbaijan					70	109	142	140
World	968	1186	1158	1573	2428	3315	4070	5191

[a]South Korea.

Persimmon is a low-chill fruit that requires an accumulation of 100–200 h of temperatures below 7°C during the fall and winter months so that vegetative and mixed buds initiate shoot development.

Depending on the combination of different factors such as plant nutritional content, intensity of winter pruning applied, number of buds in the plant, environmental conditions, soil fertility, and irrigation water amount and quality, bud development will produce:

- Water sprout or "wood branch" with high vigor that normally has a vertical growth, the buds of which will produce only sprouts (wood buds).
- Fruiting mixed branch with variable vigor that grow in an oblique, horizontal, or pendulous manner. They have mixed buds in their basal zones.
- Spurs or short branches with abundant mixed buds.

Persimmon trees have a very distinctive reproductive system: dioecious (male and female flowers on different trees), monoecious (male and female flowers on the same tree), and hermaphrodite. However, most commercial cultivars are dioecious with only female flowers, but some are hermaphrodites and are used as pollinators (García-Carbonell et al., 2002). Flowers are generally located in the axils of leaves and are often unnoticed because of their greenish and whitish to yellowish color (Spongberg, 1979). Female flowers are single, large, and pendulous and easily distinguished from male flowers by their large dark green four-lobed calyx; male flowers are smaller and produced in cymose clusters of 3–5 flowers together. Flowering occurs in late spring or early summer. Flower buds are produced along the mixed branches usually in the current year's growth (George et al., 1997).

Table 8.3 Phenological stages of persimmon "Rojo Brillante" cultivar in Spain

	Dec.	Jan.	Feb.	Mar.	Apr.	May	Jun.	Jul.	Aug.	Sep.	Oct.	Nov.
Dormancy												
Bud development												
Bud burst												
Shoot growth												
Flowering and pollination												
Fruit set												
Fruit growth												
Maturity of fruit												
Leaves fall												
Beginning of mixed buds												

The vegetative growth of persimmon trees starts after completion of the cold requirement for dormancy, in the middle or at the end of spring (Mowat and George, 1996) (Table 8.3). The growth period is extended during the summer for over 100–130 days. Persimmon trees grow vigorously with a moderate to high rate, with a tendency to grow to a large size. Depending on the intrinsic vigor of the persimmon species and position in the trunk, branches can grow from 10–20 cm to 90–100 cm. Some species or varieties are particularly prolific in growth, which represent difficulties in terms of orchard management, making pruning strategies critical in persimmon trees (Tao and Sugiura, 1992). Tree growth can also be managed using dwarfing rootstocks and interstocks that may induce dwarfing (Tetsumura et al., 2015), but because of the reduced fruit yield, this technique has no commercial applications yet. Growth in the next season may be highly influenced by climatic events (e.g., frost, typhoon) and the fruit biomass per tree. A severe defoliation after frost or a typhoon results in less reserve accumulation in permanent organs (Choi and Kang, 2007). In these conditions, growth during the following year is less vigorous than expected. Removing fruits for severely defoliated trees has no positive effects on tree reserve accumulation (Choi et al., 2015). On the other hand, when the fruit biomass per tree is low, shoot growth may become too excessive. Proper irrigation strategies in persimmon tree orchards are necessary to stablish an optimal balance between vegetative growth and fruit production (Kanety et al., 2014).

In Spain, the fruit growth cycle starts in July and ends in October (Table 8.3). Periods of 175 and 168 days after full bloom at low and high altitudes, respectively, are appropriate timings for the harvest of "Harbiye"

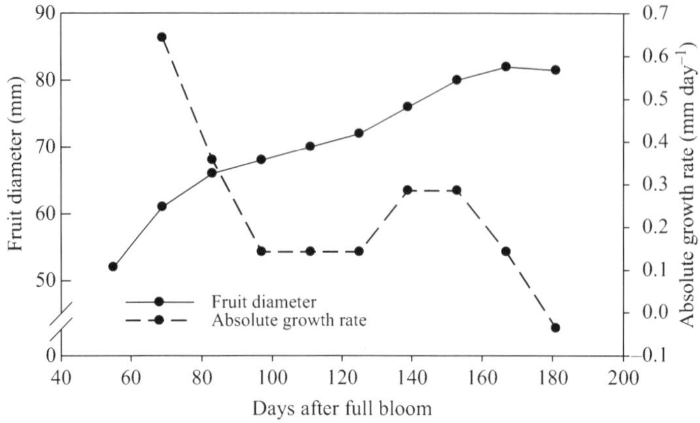

Figure 8.3 Seasonal changes in diameter of fruits and absolute growth rate of "Fuyu" persimmon. Trees bloom from May 22 to 25. *(Adapted from Choi, S.T., Park, D.S., Kim, E.S., Shin, H.Y., Kang, S.M., 2014. Seasonal growth and characteristics of fruits developed from delayed flowers of 'Fuyu' persimmon, Scientia Horticulturae 177, 108–111.)*

persimmon fruits in the east Mediterranean region (Candir et al., 2009). Persimmon fruits have been shown to follow a double-sigmoid growth curve, consisting of two rapid growth stages, stage I and stage III, separated by a period of slow growth (stage II) (Sugiura et al., 1991; Zheng and Sugiura, 1990) (Fig. 8.3). Cell division of persimmon fruits occurs actively at full bloom and the number of cells is significantly related to fruit size at harvest (Hamada et al., 2008). Candir et al. (2009) described the durations of stages I, II, and III of 105–119, 21–35, and 21–42 days, respectively, at low and high altitudes. During these phenological periods, increases in fruit weight averaging approximately 2 g per day are expected (Choi et al., 2013). At the end of stage II, the fruit begins to turn yellow-orange, indicating the occurrence of color break. The greatest increase in total soluble solids content occurs in the final month before harvest during stage III when pronounced changes in skin color and softening of fruits are observed. Climatic differences between persimmon-growing regions cause a wide variation in persimmon fruit growth rate (Mowat and George, 1996). For example, higher average temperatures throughout the fruit-growing season in Kyoto result in full bloom and, consequently, a fruit growth faster than that of trees growing in Tsuruoka. However, harvest times are almost the same between the two sites. This shortening of the period from full bloom to maturity in Tsuruoka is due to the shortening of the second growth stage, which is probably caused by the cooler summer weather (Zheng and Sugiura, 1990). Like pruning, flower or fruit thinning is an important cultural practice that affects fruit and vegetative growth (Choi et al., 2014). This practice improves the fruit size and color and also helps control of fruit yield per tree. Thinning is also used to reduce biennial bearing, particularly if it is carried out at the flowering stage. Fruit growth in particular but also fruit set are sensitive to water stress (Buesa et al., 2013; Naor, 2006), and may be the major cause of low fruit production.

3. WATER REQUIREMENTS

In Mediterranean and related semiarid areas, irrigation is the main external input affecting crop productivity. At the same time, since water is a scarce resource in these areas, efficient irrigation regimes have to be developed.

Irrigation scheduling of crops is often performed based on a soil water balance, in which crop evapotranspiration (ET_c) is estimated using reference evapotranspiration (ET_O) and crop coefficient (K_c), according to the procedure explained in the FAO (Allen et al., 1998). In the case of persimmon trees, however, there are no K_c values (ET_c/ET_O) as reported in Allen et al. (1998).

Studies performed by Kanety et al. (2014) in northern Israel, where *D. virginiana* is more commonly employed as a rootstock, evaluated the irrigation influences on plant growth, yield, and water use of "Triumph" persimmon trees. These authors observed that fruit size at harvest increased linearly until irrigation application reached 80% of the ET_O.

In the Valencia persimmon cropping area (Spain), a study was initiated in 2008 with the aim of determining the water needs of persimmon trees under a Mediterranean climate (unpublished results). This research was carried out using several tools, including sap flow gauges calibrated with whole canopy gas exchange determinations, as first described in Dragoni et al. (2005), and estimations of the soil water balance using soil capacitance probes. The results obtained are still under evaluation and refinement, but based on preliminary data analysis a set of monthly K_c values has been estimated for this area (Table 8.4).

4. TREE PERFORMANCE RESPONSES TO WATER RESTRICTIONS

Under conditions of water shortage, regulated deficit irrigation (RDI) has been shown as a suitable technique that can be used by growers to reduce

Table 8.4 Suggested crop coefficients (K_c) for "Rojo Brillante" persimmon trees grafted on *Diospyros lotus* cropped under standard Mediterranean climate and soil conditions and cultural practices in the Northern Hemisphere. The reported values can be used for ensuring optimum plant water status during the entire growing season

Tree development	Month											
	Jan.	Feb.	Mar.	Apr.	May	Jun.	Jul.	Aug.	Sep.	Oct.	Nov.	Dec.
Mature	0.00	0.00	0.16	0.38	0.47	0.59	0.75	0.80	0.98	0.95	0.50	0.00
Intermediate	0.00	0.00	0.11	0.26	0.33	0.41	0.52	0.55	0.68	0.66	0.35	0.00
Young	0.00	0.00	0.03	0.07	0.09	0.11	0.14	0.15	0.19	0.18	0.10	0.00

the amount of water applied to crops with no or minimal reductions in yield (Ruíz-Sánchez et al., 2010). RDI was developed in the 1980s as a strategy to reduce growth of vigorous trees and to save water (Chalmers et al., 1981; Behboudian and Mills, 1997). Water restrictions must be applied in phenological periods when fruit growth is less sensitive to water deficit (i.e., noncritical) ensuring water availability during the rest of the season to meet full tree water requirements. Therefore the use of RDI techniques requires knowledge of the crop-sensitive periods to deficit irrigation, which are different for each crop depending on its agronomic and physiological characteristics.

In citrus, for instance, RDI has been studied in depth in a large number of species (García-Tejero et al., 2010; Ballester et al., 2011, 2013). The sensitive periods of this crop to deficit irrigation have been clearly identified and the effects of water stress on vegetative and reproductive growth are well documented in the literature (Ruiz-Sánchez et al., 2010). In persimmon, on the other hand, the literature on RDI effects on crop performance is poor and just a few studies have reported the effects of this technique on persimmon yield and its components (Badal et al., 2013; Buesa et al., 2013).

In a study performed on mature persimmon trees, cv. "Rojo Brillante," grafted on *D. lotus* grown in a drip-irrigated commercial orchard located in Valencia, Buesa et al. (2013) showed that persimmon appears to be a sensitive crop to deficit irrigation. Notwithstanding, RDI allowed water savings of up to 20% without any reduction in yield, and fruit weight was always impaired by RDI treatments. Only when RDI trees had a higher number of fruit than control trees (irrigated at full water requirements) were the economic returns not reduced, increasing water use efficiency. As the number of harvested fruit depends on the number of fruit that remains on the tree after the fruit drop period, this period seems to be of particular importance in persimmon trees for the success of an RDI strategy. Results obtained from this study showed that in persimmon trees, RDI is in any case a strategy that can be potentially applied in off-seasons, under low tree crop level, when a moderate water restriction could save water, while increasing the tree fruit-bearing capacity.

A 3-year study conducted by Badal et al. (2013) in the same orchard used by Buesa et al. (2013) showed that moderate plant water stress (stem water potential [Ψ_{stem}] not lower than -1.3 MPa) occurring early in the spring (mid-May to late June, coinciding with the fruit drop waves) decreased fruit drop. Water restrictions applied later in the season, at the end of the fruit drop period, had little effect on reducing fruit drop. This information regarding water stress effects on persimmon trees, as well as the appropriate timing to restrict irrigation in the "Rojo Brillante" cultivar, may be of great utility in Mediterranean and related semiarid areas to increase yield by restricting irrigation applications during late spring. Further research is still needed to elucidate the physiological mechanisms involved in fruit drop, as well as to explore how different water stress levels might modulate this effect in the cv. "Rojo Brillante." More research is also needed to provide growers with tools for predicting the tree-bearing capacity before the second wave of fruit abscission. If tree-bearing capacity could be quantified, growers would have more information to decide whether and when to reduce fruit drop by inducing a certain degree of plant water stress.

Based on the results from the aforementioned studies (Badal et al., 2013; Buesa et al., 2013), a set of monthly K_c values are recommended for irrigation scheduling under RDI (Table 8.5).

Note that RDI irrigation scheduling according to K_c values in Table 8.5 implies applying water restrictions during the June fruit drop, and also during the last phase of fruit growth in case fruit maturation needs to be accelerated, as discussed in depth in the following section. In any case, if water-saving irrigation strategies are implemented, careful water stress monitoring is crucial to avoid harmful impacts on yields. In this sense, Table 8.6 summarizes the threshold values for Ψ_{stem} depending on the phenological period and the degree of water stress.

Previous studies have shown that diurnal trunk shrinkage is a sensitive indicator of plant water stress in persimmon (Badal et al., 2010), but because its variation also depends on the environmental conditions, reference trees need to be used to determine the actual water stress level. Martínez-Gimeno et al. (2017) showed that leaf turgor probes can also be used for determining plant water stress. Different leaf turgor states were associated with different ranges of Ψ_{stem} values according to the intensity of the stress reached by the plants. The obtained results indicate that this technology is a potential decision tool to be used in persimmon orchards for irrigation management in both conventional and RDI orchards.

5. FRUIT QUALITY RESPONSES TO WATER RESTRICTIONS

Different aspects related to fruit quality have been reported to be influenced by irrigation management. Water stress, together with high temperatures

Table 8.5 Suggested crop coefficients (K_c) for mature persimmon trees grown under regulated deficit irrigation under Mediterranean climate

	Month											
	Jan.	**Feb.**	**Mar.**	**Apr.**	**May**	**Jun.**	**Jul.**	**Aug.**	**Sep.**	**Oct.**	**Nov.**	**Dec.**
K_c	0.00	0.00	0.16	0.38	0.47	0.40	0.75	0.80	0.70	0.70	0.50	0.00

Table 8.6 Suggested threshold values for midday stem water potential (Ψ_{stem}, MPa) for "Rojo Brillante" persimmon trees grafted on *Diospyros lotus* cropped under standard Mediterranean climate and soil conditions and cultural practices in the Northern Hemisphere

	Phenological period	
Degree of water stress	**Flowering and fruit set (spring)**	**Phases II and III of fruit growth (summer and early fall)**
Adequate	Above −0.55	Above −0.75
Light	−0.55 to −0.75	−0.75 to −0.90
Mild	−0.75 to 0.95	−0.90 to −1.20
Moderate	−0.95 to −1.15	−1.20 to −1.40
Severe	Below −1.15	Below −1.40

during floral differentiation, can induce the formation of double or malformed pistils within floral primordia (George et al., 1997; Wanatabe 1985). The effect of shading on persimmon fruit malformation was studied by Zilkah et al. (2009) on "Triumph" persimmon grown under three water irrigation regimes applied from summer to winter (50%, 100%, and 150% of the recommended amount). The authors reported that split was the most frequent deformation type under 100% and 150% of irrigation, while under 50% irrigation the fruits displayed mainly asymmetric deformation. This trend was especially significant among mature fruit grown under conditions of exposure (uncovered trees). Moreover, fruit under shading (by covering the trees with black net screens) was less malformed than that on exposed trees, and this effect was more obvious under the regime of deficit irrigation (50% of the recommended amount). Therefore, according to Zilkah et al. (2009), it appears that fruit deformation events may be reduced under sufficient irrigation and shading conditions.

An effect of deficit irrigation on fruit growth and maturation has been reported in different persimmon cultivars. A three-season study conducted on the astringent cultivar "Rojo Brillante" revealed that the response of fruit to short-term RDI (50% of the recommended amount) depends on the fruit growth stage (Intrigliolo et al., 2011). Restricted irrigation carried out during phase II or III of fruit growth induced fruit maturation, which showed up as an increment in skin color and reduced flesh firmness. However, this effect was not observed when water restriction had taken place during phase I of fruit growth. The fact that the same degree of water stress applied during stage II or III of fruit growth might accelerate ripening is particularly important as the first fruit that reaches the market normally has higher economical value. In addition, under Mediterranean and related semiarid climates, persimmon production is based on a few varieties, and it is also important to have the opportunity to modulate the ripening period by using different irrigation regimes. However, it should be considered that water deficit can also reduce fruit weight, but it can be reduced irrespective of the period of water restriction. Intrigliolo et al. (2011) also showed that short periods of deficit irrigation did not affect the effectiveness of post-harvest astringency-removing treatment before commercialization.

The effect of RDI applied for longer periods (50% of the recommended amount from June to October) was applied on the cultivar "Triumph" and the effect of shading was also evaluated by Ben-Arie et al. (2008). Similar to the foregoing exposed results, the fruit growth rate significantly decreased in "Triumph" fruits subjected to water restriction irrespective of the use of cover net (Ben-Arie et al., 2008). Moreover, skin color development was enhanced significantly with reduced irrigation on exposed trees only, when water stress was the greatest. Overall, the fruit quality of "Triumph" was adversely affected by deficit irrigation, but this effect was alleviated by using net shading. In fact, Ben-Arie et al. (2008) reported an increase in fruit sunburn in those trees where deficit irrigation was applied.

Excessive irrigation has also been reported to have an effect on fruit quality. According to Ben-Arie et al. (2008) an excess of watering increased soft blossom-end disorder. This disorder consists in the premature softening of the blossom end of the fruit. It is usually noticeable in the orchard prior to harvest in the form of enhanced reddening of the affected tissue, but sometimes it is not evident until after storage (Fig. 8.4).

According to Ben-Arie et al. (2008) this disorder is related to over-irrigation and poor soil aeration conditions, which would result in the exposure of certain parts of the root system to hypoxic conditions. Under such conditions the formation of 1-aminocyclopropane-1-carboxylic acid and ethanol is induced, and they are in turn transported to the blossom end of the fruit via the xylem. This fact would result in flesh softening in this part of the fruit.

6. RESPONSES TO WATER QUALITY RESTRICTIONS

Salinity refers to the amount of major ions in environmental aqueous systems, including soil pore and irrigation waters. These ions are Na^+, Mg^{2+}, Ca^{2+}, Cl^-, SO_4^{2-}, HCO_3^-, more rarely CO_3^{2-}, and in agricultural soils, importantly, K^+ and NO_3^-. They are readily dissolved by rainfall

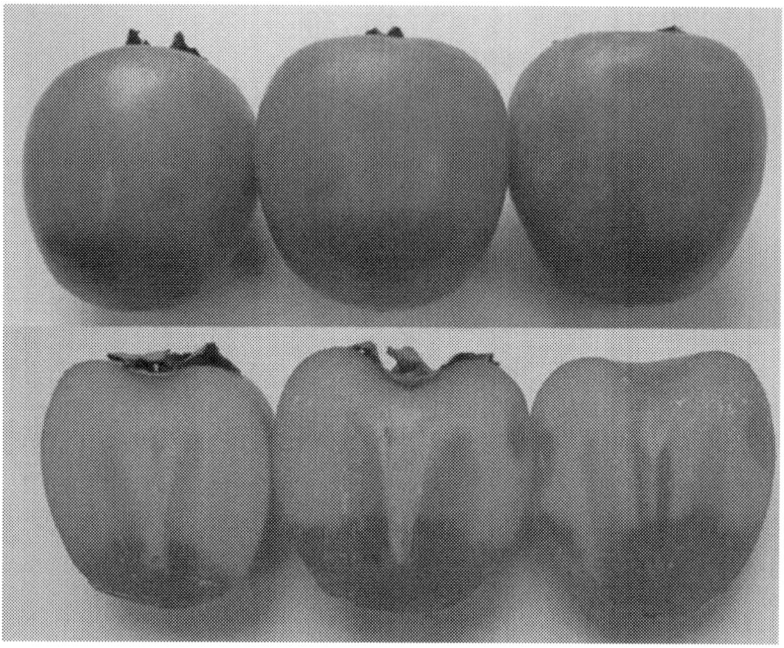

Figure 8.4 Persimmon "Rojo Brillante" fruit affected by the fruit disorder soft blossom end, observed after storage at 15°C for 15 days.

from several earth crust minerals and carried in solution wherever liquid water goes. When irrigation water is applied to soils, salts are concentrated in the soil pore water by plant transpiration and soil evaporation. In case of impaired land drainage, salts hardly leave the soil, and to make matters worse an extra load of underground salts can be brought into it by an irrigation-lifted water table (Grieve et al., 2012). Whatever the case, soil salinity always increases as a result of irrigation, and hence 11% (34 Mha) of world irrigated lands are estimated to be salt affected (FAO, 2011).

Excess salts cause deleterious effects on crop plants, which split into osmotic and ion specific. Osmotic effects are driven by the lowering of the soil water potential that dissolved salts cause, and are almost indistinguishable from drought symptoms (Munns, 2002). Osmotic effects are proportional to the sum of all ion concentrations in the soil water, usually assessed by means of electrical conductivity at 25°C of soil saturation extracts in units of dS m^{-1}. Importantly, osmotic effects do not depend on ion composition. On the contrary, specific ion effects do depend on composition, i.e., on the concentration of certain ions such as chloride [Cl^-], or ion ratios such as $[Na^+]/[K^+]$, $[Cl^-]/[NO_3^-]$, or $[Na^+]/(([Mg^{2+}] + [Ca^{2+}])^{1/2})$ in the soil water. Specific ion effects cause toxicity or nutrient imbalances, and show up as injuries and/or nutrient deficiencies (Grieve et al., 2012). Whatever the case, either osmotic or ion specific, salinity effects eventually bring about growth reduction and yield losses. Salt effects on the morphology and other properties of plant organs, including the quality of edible and/or marketable ones, are also widespread.

Crop plants differ in their sensitivity to salinity and can be classified into four categories, from sensitive to tolerant; however, most woody crops are salt sensitive (Maas and Hoffman, 1977). Persimmon, specifically *D. virginiana*, is not an exception to this general rule since this species has been regarded as such for years (Malcolm and Smith, 1971). However, it is well known that salt tolerance exhibits remarkable variations throughout species within a genus, and even throughout varieties within a species (Grieve et al., 2012). The three most prominent *Diospyros* species used nowadays in agricultural practice as rootstocks for persimmon scions (*D. kaki*) are *D. kaki* itself, *D. virginiana*, and *D. lotus*. They are native to, or have been propagated from, central China, central to southern Japan and South Korea (*D. kaki*) (Yamada et al., 2012), east-central and southeastern United States (*D. virginiana*) (Skallerup, 1953), and various scattered mountainous areas from Japan to Turkey within the belt between 30° and 45° N (*D. lotus*) (Ercisli and Akbulut, 2009; Kiaei and Bakhshi, 2014; Samadi et al., 2009; Giordani, 2002; Maghradze et al., 2012). Common to all these places is the characteristic warm to continental humid climate where neutral to acid well-drained soils often develop (Yamada, 2008). Upon comparison with the world map of Köppen–Geiger climate classification (Rubel and Kottek, 2010), the climate in these places is primarily warm temperate year-round (fully) humid (Cf) and, secondarily, warm temperate winter dry (Cw). Therefore no wonder that the cultivation of persimmon has adapted well to other world areas under Cf climates in southern Brazil (Neuwald et al., 2009), New Zealand (Mowat, 2003), eastern and southeastern Australia (Nissen et al., 2003), and Emilia-Romagna (Italy) (Bellini and Giordani, 2005). The soil moisture regime (SMR) under Cf and Cw climates is commonly udic or perudic (USDA-NCRS 1997), i.e., rainfall plus stored soil water exceeds evapotranspiration, specifically in summer. Therefore since all three agronomically relevant *Diospyros* species have evolved and/or have been bred on rain-leached soils, i.e., with very low salt contents, salinity tolerance has neither been a natural nor an artificial selection trait for any of them, and the salt sensitivity listed for *D. virginiana* can also be easily assumed for both *D. kaki* and *D. lotus*.

The salt sensitivity of the *Diospyros* genus would be no problem provided its cultivation had not been spread to areas under non-Cf and non-Cw climates, mostly where Mediterranean (Cs) and related semiarid conditions (BSk) prevail such as, primarily, Valencia and Huelva (Spain), eastern Uzbekistan, parts of Azerbaijan, central to northern Israel, and Mediterranean regions in Turkey and Campania (Italy) (Table 8.7). According to the Köppen–Geiger classification, Mediterranean climate is described as temperate subtropical summer dry (Cs). Therefore although climate temperature features are similar to Cf and Cw areas and favor persimmon cultivation, the xeric SMR, i.e., summer drought, which features Mediterranean and related semiarid climates (USDA-NCRS, 1997), contrasts with the udic or perudic SMR in the original centers of propagation of persimmon. Summer drought means, on the one hand, that under natural conditions soil leaching is impeded, and even reversed during summer, and on the other hand, that irrigation is essential to sustain the development of most crops, mostly those coming from humid climates.

Because in Mediterranean areas irrigation is required, at least in summer, and irrigation waters always convey dissolved salts, and because soil salts cannot be completely leached by rains or irrigation itself, Mediterranean soils always present some salinity and, in general, the more arid the Mediterranean climate, the more saline the soils are. This is shown by the relationship between soil chloride content in June and the cumulative rainfall since October until May in a representative *D. kaki* cv. "Rojo Brillante" plantation in the Valencia area (Table 8.8). As is shown, subjected to increasingly scarce rainfall during the humid part of the year, soil chloride content, and hence soil salinity, is also higher, and what is more important, leaf chloride build-up is higher too (de Paz et al., 2016).

All three agronomically relevant *Diospyros* species are salt sensitive; however, there are some tolerance differences among them both in quantitative and qualitative terms. Under greenhouse conditions, *D. virginiana* seedlings have been found to better withstand salinity than *D. kaki* ones (Incesu et al., 2014). Specifically, it has been found that *D. virginiana* shows less leaf decline, less leaf chloride and sodium contents, higher leaf chlorophyll concentration, and eventually higher shoot and root dry matter production than *D. kaki* seedlings, and differences became significant as irrigation water salinity reached 100 mM NaCl ($\sim$8.6 dS m^{-1}). Such strong salinity is beyond agronomic interest; however,

Table 8.7 Characteristics of 2014 world persimmon production in Mediterranean and related semiarid climates (Köppen–Geiger classes Cs and BSk)

Site	Area ha^{-1}	Production t^{-1}	Main cultivars[a]	Main rootstocks	References
Azerbaijan[b]	8,700	140,000	Caucasus	*Diospyros lotus?*	FAOSTAT (2016) and Nazarli (2016)
California (USA)	800	10,500	Hachiya, Jiro, Fuyu	*D. lotus*	CCAC (2015) and Marzolo and Lee (2016)
Campania (Italy)	1,400	22,000	Kaki Tipo	*D. lotus*	FAOSTAT (2016) and Bellini and Giordani (2005)
Center, Haifa, North, Tel Aviv (Israel)	1,400	36,600	Triumph	*Diospyros virginiana*	FAOSTAT (2016) and Zilkah et al. (2009)
Isfahan and Tehran regions (Iran)	300	2,500	Shiraz, Karaj, Natanzy	*D. lotus?*	FAOSTAT (2016), Bagheri et al. (2015), Ramin and Tabatabaie (2003) and Shahkoomahally and Ramezanian (2013)
Mediterranean and Black Sea regions (Turkey)[b]	2,200	33,500	Kaki Tipo, Rojo Brillante, Triumph	*Diospyros kaki*	TUIK (2017) and Incesu et al. (2014, 2015)
O'Higgins, Valparaíso, Santiago (Chile)	100	[c]1,300	Jiro, Fuyu, Hachiya	*D. lotus*	FAOSTAT (2016), ODEPA-CIRÉN (2016) and AGVC (2015)
Portugal	230	3,000	Coroa de Rei, Rojo Brillante	*D. lotus*	Regato and Guerreiro (2013) and Bellini and Giordani (2005)
Uzbekistan	4,200	66,000	Chocolate?	*D. lotus?*	FAOSTAT (2016) and Jadrov (1968)
Valencia, Huelva (Spain)	13,400	245,000	Rojo Brillante, Triumph	*D. lotus*	FAOSTAT (2016) and Llácer and Badenes (2009)
Victoria, South, Western (Australia)	130	1,100	Fuyu, Jiro	*D. kaki*	Margetts (2014) and Nissen et al. (2003)
Western Cape (South Africa)	420	6,300	Triumph	*D. virginiana*	Genis (2017), Scheepers (2010), and Rabe (2003)

PCA, pollination constant astringent; *PCNA*, pollination constant nonastringent; *PVA*, pollination variant astringent; *PVNA*, pollination variant nonastringent.
[a]Astringent cultivars: Triumph (PVA), Rojo Brillante (PCA), Hachiya (PCA), Coroa de Rei (PCA?), Shiraz (PCA), Karaj (PCA?), Natanzy (PCA?), Caucasus (PCA?); nonastringent cultivars: Kaki Tipo (PVNA), Chocolate (PVNA), Fuyu (PCNA), Jiro (PCNA).
[b]Includes production in areas under Cf climates.
[c]Estimated value.

Table 8.8 Mean soil and leaf chloride contents in June in a plantation of *Diospyros kaki* cv. "Rojo Brillante" grafted on *Diospyros lotus* in three distinct hydrological years in Valencia (Spain)

Hydrological year	Cumulative rainfall October–May mm^{-1}	Chloride content in June	
		Soil 0–40 cm (mM)	Leaf (%)
2011–12	290	5.1	0.9
2012–13	560	3.8	0.6
2013–14	90	7.3	1.2

these findings along with those shown in Table 8.8 highlight the ability of woody crops to accumulate specific ions or elements such as Na^+, Cl^-, and/or B in various vegetative organs and tissues, especially leaves and calyces, and even when grown in barely saline soils (Bernstein, 1980).

In persimmon leaf, chloride build-up has been observed not only in greenhouse seedlings, but also when grafted grown in farms (Table 8.8). The remarkable ability of *D. lotus* to absorb chloride almost unavoidably leads to over 1% leaf chloride contents at season end in the Valencia cropping area, which shows up as an extensive leaf necrosis (Visconti et al., 2015, 2017). On the contrary, *D. kaki* grafted on *D. virginiana* and exposed to similar soil chloride contents (2–20 mM) experiences some leaf darkening, thickening, and, as salinity increases, wilting, which are all osmotic effects, but seldom leaf necrosis, which is a specific ion toxicity effect. An absence of specific ion toxicity effects is a typical trait of salt-tolerant plants (Munns, 2005) and thus *D. virginiana* can be rated as more tolerant than *D. lotus* just because of this fact. Regarding sodium and boron, leaf contents are, respectively, between 0.01% and 0.05% and between 40 and 140 ppm in commercial plantations of *D. lotus*, and are therefore within the optimum range for persimmon (Bar-Tal et al., 2008; Robinson et al., 1997). In addition, although slightly higher sodium and boron leaf contents are found in persimmons grafted on *D. lotus*, they are, respectively, significantly less correlated and not correlated at all with necrosis symptoms, while leaf chloride contents are highly correlated with them (Visconti et al., 2017).

From slight to moderate irrigation water chloride concentrations (3–6 mM), persimmon yields diminish in commercial plantations, mainly through the decrease in persimmon fruit weight and size (Nissen et al., 2003; Besada et al., 2016). Even more important, chloride phytotoxicity accelerates fruit ripening, which hinders the CO_2 deastringency postharvest treatment in astringent persimmon varieties with eventual harmful consequences in fruit conservation and hence marketability (Besada et al., 2016). At higher chloride concentrations in the irrigation water, fruit numbers also diminish until almost vanishing over 12 mM Cl^- and 2.5 dS m^{-1} for, respectively, *D. lotus* and *D. virginiana* rootstocks (Visconti et al., 2017).

The higher salt tolerance of *D. virginiana* regarding *D. kaki* and especially *D. lotus* can be explained on the basis of chloride exclusion ability (Visconti et al., 2015); in addition, root morphology also seems to play an important role (Visconti et al., 2017). *D. lotus* presents a fibrous root that barely penetrates below the topsoil layer, while both *D. virginiana* and *D. kaki* present a well-developed taproot (Fig. 8.5). Differences between these last rootstocks can be related to different root density and thus soil volume explored: although *D. kaki*'s taproot is regarded as penetrating deeper into the soil than *D. virginiana*'s, its root system seems to be less densely developed (Bellini, 2002). Regarding *D. lotus*, it has been shown that chloride phytotoxicity increases as soil compaction increases in the Valencia cropping area (de Paz et al., 2016), and this fact can be explained on the basis the limited ability of *D. lotus* roots to deepen into the soil. Therefore as soil compaction increases, not only roots but also water and salts remain in the topsoil layer long and concentrated enough to stress the persimmon trees and cause chloride phytotoxicity.

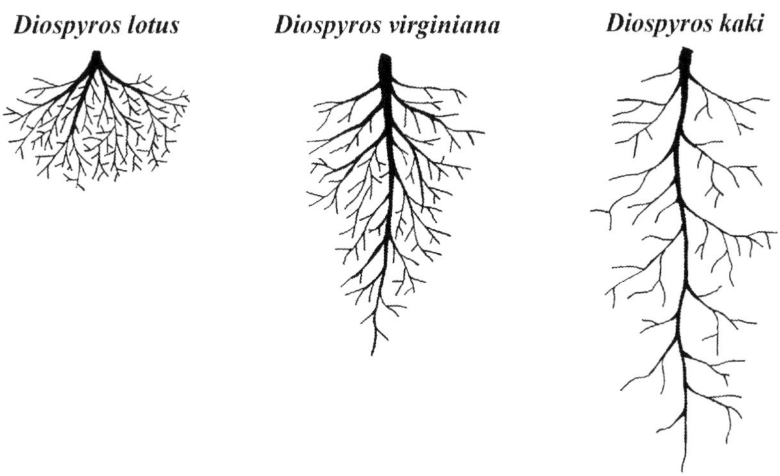

Figure 8.5 Root systems of the three most relevant *Diospyros* species in agriculture.

7. AGRONOMIC PRACTICES TO MITIGATE WATER QUALITY RESTRICTIONS

The most general and traditional method to mitigate salinity problems in agriculture, including chloride phytotoxicity, is to favor salt removal from the root zone by means of leaching (USSL Staff, 1954). This is traditionally accomplished through irrigation with an amount of water in excess of crop requirements, then favoring water percolation through soil, and finally disposing of the drained water. The amount of excess water that must be used to flush enough salts, including specific ions, out of the soil is given by the leaching requirement (LR), which depends mainly on irrigation water salinity and crop salt tolerance (Visconti et al., 2012).

Regarding water quality, to meet full potential yields and to have excellent postharvest processing, persimmons on *D. lotus* should not be irrigated with waters containing over 3 mM Cl^-. In the case of *D. virginiana* rootstock, irrigation waters over 2 dS m^{-1} should be avoided. For persimmons on *D. kaki*, intermediate values for both chloride and salinity could be provisionally recommended as maximum limits; these would be 4 mM Cl^- and 1.8 dS m^{-1}. Note that regarding chloride, these irrigation water quality requirements for persimmon exceed FAOs limits (Ayers and Westcot, 1985) in which until 4 mM Cl^- there is no restriction for use. Regarding salinity, the irrigation water quality requirements for persimmon are well within the FAO's class of slight to moderate restriction for use. Depending on composition, waters with EC_{25} over 1 dS m^{-1} can already contain over 3 mM Cl^- and thus they must be analyzed before use when planning irrigation of persimmons on *D. lotus*. Importantly, chloride input through fertilizers must be avoided. Since KCl is the cheapest potassium fertilizer, farmers usually use it. In persimmon cropping, potassium addition must be made in the form of K_2SO_4 or KNO_3.

For *D. lotus* rootstock and water chloride contents within the range 3–4 mM Cl^- and for *D. virginiana* rootstock and water EC_{25} within the range 2–3 dS m^{-1}, some water excess can be applied and drained to successfully cope with salinity problems. Crop salt tolerance is usually characterized with the threshold–slope model (Maas and Hoffman, 1977). For persimmons grafted on *D. lotus*, threshold and slope parameters have been estimated to be, respectively, 1.2 and 22 dS^{-1} m (Visconti et al., 2017), while on *D. virginiana* they can be estimated as 1.4 and 15 dS^{-1} m, which defines the limit between sensitive and moderately sensitive crops. These parameters can be used in conjunction with various models that range from the traditional Rhoades equation (Rhoades, 1974) to more complex ones to assess the LR (Visconti et al., 2012). In the Valencia area these threshold–slope parameters are included in the online decision support system based on the SALTIRSOIL model for irrigation, crop, and soil management in salt-threatened lands (http://agrosal.ivia.es) (Visconti and de Paz, 2016).

To make leaching an effective method for salt removal, water percolation through soil and eventual drainage should be ensured. In case both requirements are not met, the use of a leaching fraction will unavoidably worsen salinity problems. Enhanced water percolation and drainage are even more necessary for *D. lotus* rootstocks given the inherent shallowness of their root systems (Fig. 8.5). To ease leaching, soil structure must be improved, and general agronomical recommendations are useful for this aim (Gardner et al., 1999). In case of soil compaction due to machinery traffic across the field, one deep plow (chisel, etc.) is recommended to break the compacted soil layers that will, on the one hand, increase the available soil for root exploration, and, on the other hand, increase water percolation and salt leaching through soil. Once compacted soil layers have been removed, soil porosity must be preserved (1) by frequent organic additions to increase the soil organic matter content, (2) by allowing and/or fostering spontaneous winter meadows to grow on lanes between persimmon rows, and (3) by avoiding as much as possible machinery and people traffic in fields when soils are more compaction prone, i.e., when their water contents are over field capacity.

If none of the previous practices is successful to get rid of chloride phytotoxicity in persimmon cropping on *D. lotus*, which is the most spread rootstock under Mediterranean and related semiarid climates (Table 8.7), because of poor water quality and/or restricted soil percolation and drainage conditions (calcareous, clay and/or silt pans, in-depth compacted soil layers, shallow water tables, etc.), there is always the option of changing the rootstock to *D. virginiana*. This replacement can extend the range of water quality and soil conditions under which persimmon cropping is still sustainable. In addition, although persimmon propagation on *D. virginiana* has been found to present various agronomical problems depending on the scion (Hodgson, 1940; Cohen et al., 1991; Bellini, 2002), according to existing data (FAOSTAT, 2016) it would have been successfully carried out in Israel, which is currently the world leader in persimmon yields with 36 t ha^{-1} (Table 8.7).

8. CONCLUSIONS

Persimmon cultivation is gaining importance in many semiarid areas of the world. It is concluded that persimmon water needs can be particularly high during summer, when the crop coefficient can reach values close to or even above 1.0. A table with the suggested crop coefficients to be used has been derived and will be helpful to improve the efficiency in scheduling irrigation. To cope with water scarcity, RDI has been shown as a useful technique for reducing fruit drop in persimmon trees. Late RDI strategies during fruit ripening can accelerate fruit color development resulting in earlier harvesting but with lower fruit weight. On the other hand, water stress during physiological fruit drop can increase fruit set, resulting in higher yield particularly in the off-seasons. The three most prominent *Diospyros* species used as rootstocks for persimmon are sensitive to soil salinity. Specifically, *D. lotus* exhibits a remarkable chloride phytotoxicity. *D. virginiana* avoids leaf chloride build-up and presents salt-tolerance parameters between sensitive and moderately sensitive crops. The use of *D. virginiana* as rootstock for persimmon has not spread due to several agronomical drawbacks. However, adequately managed, its use under Mediterranean and related semiarid climates seems a very promising way to avoid chloride phytotoxicity in persimmons.

ACKNOWLEDGMENTS

Data and results summarized within this chapter have been obtained in the framework of several research projects, including: GVA-IVIA Proyecto Integral Caqui, Contract agreement with the Cooperativa San Bernardo de Carlet, PCIN-2015-263 IRIDA ERA-NET Cofund WaterWorks2014 Call (Water JPI), WEAM4i FP7 GA619061, RTC-2014-2141-2 Hipofrut, RTC-2015-3453-2 RiegoAsesor, and CARM, /GERM/15 SENECA.

REFERENCES

AGVC-Asociación Gremial de Viveros de Chile, 2015. Actualización de la comercialización de plantas frutales, vides yplantines de hortalizas en Chile.

Allen, R.G., Pereira, L.S., Raes, D., Smith, M., 1998. Crop Evapotranspiration. Guidelines for Computing Crop Water Requirements, FAO Irrigation and Drainage Paper 56. Italy, Rome.

Arnal, L., Rio, M.A., Vendrell, M., 2002. Effects of packing line operations on some quality attributes of persimmon fruit cv. 'Rojo Brillante'. In: Bellini, E., Giordani, E. (Eds.), First Mediterranean Symposium on Persimmon. CIHEAM, Zaragoza, Spain, pp. 113–117.

Ayers, R.S., Westcot, D.W., 1985. Water quality for agriculture. In: FAO Irrigation and Drainage Paper 29, Rev. 1, Rome, Italy.

Badal, E., Buesa, I., Guerra, D., Bonet, L., Ferrer, P., Intrigliolo, D.S., 2010. Maximum diurnal trunk shrinkage is a sensitive indicator of plant water stress in *Diospyros kaki* (Persimmon) trees. Agricultural Water Management 98 (1), 143–147.

Badal, E., El-Mageed, T.A., Buesa, I., Guerra, D., Bonet, L., Intrigliolo, D.S., 2013. Moderate plant water stress reduces fruit drop of "Rojo Brillante" persimmon (*Diospyros kaki*) in a Mediterranean climate. Agricultural Water Management 119, 154–160.

Bagheri, M., Esna-Ashari, M., Ershadi, A., 2015. Effect of postharvest calcium chloride treatment on the storage life and quality of persimmon fruits (*Diospyros kaki* Thunb.) cv. "Karaj". International Journal of Horticultural Science and Technology 2, 15–26.

Ballester, C., Castel, J., Intrigliolo, D.S., Castel, J.R., 2011. Response of Clementina de Nules citrus trees to summer deficit irrigation. Yield components and fruit composition. Agricultural Water Management 98, 1027–1032.

Ballester, C., Castel, J., Intrigliolo, D.S., Castel, J.R., 2013. Response of Navel Lane Late citrus trees to regulated deficit irrigation: yield components and fruit composition. Irrigation Science 31, 333–341.

Bar-Tal, A., Yermiyahu, U., Keinan, M., Faingold, I., Ben-Gal, A., 2008. Boron deficiency and toxicity in young persimmon (*Diospyros kaki* L.). Journal of Horticultural Science and Biotechnology 83, 469–473.

Behboudian, M.H., Mills, T.M., 1997. Deficit irrigation in deciduous orchards. Horticultural Reviews 21, 125–131.

Bellini, E., Giordani, E., 2005. Germplasm and breeding of persimmon in Europe. Acta Horticulturae 685, 65–76.

Bellini, E., 2002. Cultural practices for persimmon production. In: Bellini, E., Giordani, E. (Eds.), First Mediterranean Symposium on Persimmon. CIHEAM, Zaragoza, Spain, pp. 39–52.

Ben-Arie, R., Zilkah, S., Klein, I., Gamrasni, D., 2008. Persimmon and environment: soil and water management for high quality fruit production. Advances in Horticultural Science 22 (4), 286–293.

Bernstein, L., 1980. Salt Tolerance of Fruit Crops. USDA, Riverside, California, USA.

Besada, C., Gil, R., Bonet, L., Quiñones, A., Intrigliolo, D., Salvador, A., 2016. Chloride stress triggers maturation and negatively affects the postharvest quality of persimmon fruit. Involvement of calyx ethylene production. Plant Physiology and Biochemistry 100, 105–112.

Buesa, I., Badal, E., Guerra, D., Ballester, C., Bonet, L., Intrigliolo, D.S., 2013. Regulated deficit irrigation in persimmon trees (*Diospyros kaki*) cv.'Rojo Brillante'. Scientia Horticulturae 159, 134–142.

Candir, E.E., Ozdemir, A.E., Kaplankiran, M., Toplu, C., 2009. Physico-chemical changes during growth of persimmon fruits in the East Mediterranean climate region. Scientia Horticulturae 121, 42–48.

CCAC-California County Agricultural Commissioners, 2015. Reports Crop Year 2013–2014.

Chalmers, D.J., Mitchell, P.D., Vanheek, L., 1981. Control of peach tree growth and productivity by regulated water supply, tree density and summer pruning. Journal of the American Society for Horticultural Science 106, 307–312.

Choi, S.T., Kang, S.M., 2007. Effects of defoliation and defruiting in early September on partitioning of nonstructural carbohydrates in 'Fuyu' persimmon at harvest. Horticulture, Environment and Biotechnology 48, 359–364.

Choi, S.T., Park, D.S., Son, X.Y., Park, Y.O., Hong, K.P., Cho, K.S., 2013. Climate-related changes in fruit growth of 'Fuyu' persimmon during the harvest season. Korean Journal of Horticultural Science and Technology 31, 32–37.

Choi, S.T., Park, D.S., Kim, E.S., Shin, H.Y., Kang, S.M., 2014. Seasonal growth and characteristics of fruits developed from delayed flowers of 'Fuyu' persimmon. Scientia Horticulturae 177, 108–111.

Choi, S.T., Ahn, G.H., Kim, E.S., Kang, S.M., Kang, S.K., 2015. Not harvesting 'Fuyu' persimmons after frost damage in autumn has no negative effects on tree storage reserves and growth in the subsequent season. Scientia Horticulturae 190, 94–97.

Cohen, Y., Gur, A., Barkai, Z., Blumenfeld, A., 1991. Decline of persimmon (*Diospyros kaki* L.) trees on *Diospyros virginiana* rootstocks. Scientia Horticulturae 48, 61–70.

de Paz, J.M., Visconti, F., Tudela, L., Quiñones, A., Intrigliolo, D.S., Jordà, M., Bonet, L., 2016. La fitotoxicidad por cloruro en el cultivo del caqui: descripción del problema. Agrícola Vergel: Fruticultura, Horticultura, Floricultura 391, 91–96.

Dragoni, D., Lakso, A.N., Piccioni, R.M., 2005. Transpiration of apple trees in a humid climate using heat pulse sap flow gauges calibrated with whole-canopy gas exchange chambers. Agricultural and Forest Meteorology 130, 85–94.

Ercisli, S., Akbulut, M., 2009. Persimmon cultivation and genetic resources in Turkey. Acta Horticulturae 833, 35–38.

FAO, 2011. The State of the World's Land and Water Resources for Food and Agriculture (SOLAW) – Managing Systems at Risk. Rome and Earthscan, London.

FAOSTAT, 2016. Database. Available at: http://www.fao.org/faostat/en/#data/QC.

FAOSTAT, 2017. FAOSTAT Database. Available from: http://www.fao.org/faostat/en/#home.

García-Carbonell, S., Yagüe, B., Bleiholder, H., Hack, H., Meier, U., Agustí, M., 2002. Phenological growth stages of the persimmon tree (*Dyospiros kaki*). Annals of Applied Biology 141, 73–76.

García-Tejero, I., Romero-Vicente, R., Jiménez-Bocanegra, J.A., Martínez-García, G., Durán-Zuazo, V.H., Muriel-Fernández, J.L., 2010. Response of citrus trees to deficit irrigation during different phenological periods in relation to yield, fruit quality, and water productivity. Agricultural Water Management 97, 689–699.

Gardner, C.M.K., Laryea, K.B., Unger, P.W., 1999. Soil Physical Constraints to Plant Growth and Crop Production. FAO, Land and Water Development Division, Rome, Italy.

Genis, A., 2017. Persimmon-oes kleiner, maar gehalte goed. Landbouweekblad 110.

George, A.P., Mowat, A.D., Collins, R.J., Morley-Bunker, M., 1997. The pattern and control of reproductive development in non-astringent persimmon (*Diospyros kaki* L.): a review. Scientia Horticulturae 70, 93–122.

Giordani, E., Picardi, E., Radice, S., 2015. Morfología y fisiología. In: Badenes, M., Intrigliolo, D., Salvador, A., Vicent, A. (Eds.), El cultivo del caqui, pp. 35–54. ISBN: 978-84-482-6018-7.

Giordani, E., 2002. Varietal assortment of persimmon in the countries of the Mediterranean area and genetic improvement. In: Bellini, E., Giordani, E. (Eds.), First Mediterranean Symposium on Persimmon. CIHEAM, Zaragoza, Spain, pp. 23–37.

Grieve, C.M., Grattan, S.R., Maas, E.W., 2012. Plant salt tolerance. In: Wallender, W.W., Tanji, K.K. (Eds.), Agricultural Salinity Assessment and Management. American Society of Civil Engineers, Reston, Virginia, USA, pp. 405–459.

Hamada, K., Hasegawa, K., Kitajima, A., Ogata, T., 2008. The relationship between fruit size and cell division and enlargement in cultivated and wild persimmons. Journal of Horticultural Science and Biotechnology 83, 218–222.

Hodgson, R.W., 1940. Rootstocks for the oriental persimmon. In: California Avocado Association 1940 Yearbook, vol. 25, pp. 43–44.

Incesu, M., Cimen, B., Yesiloglu, T., Yilmaz, B., 2014. Growth and photosynthetic response of two persimmon rootstocks (*Diospyros kaki* and *D. virginiana*) under different salinity levels. Notulae Botanicae Horti Agrobotanici Cluj-napoca 42, 386–391.

Incesu, M., Yesiloglu, T., Çimen, B., Yilmaz, B., Akpinar, Ç., Ortas, I., 2015. Effects on growth of persimmon (*Diospyros virginiana*) rootstock of arbuscular mycorrhizal fungi species. Turkish Journal of Agriculture and Forestry 39, 117–122.

Intrigliolo, D.S., Bonet, L., Ferrer, P., Besada, C., Salvador, A., 2011. Short-term effects of regulated deficit irrigation of 'Rojo Brilante' persimmon (*Diospyros kaki*) – yield, fruit quality and post-harvest performance. Acta Horticulturae 922, 113–120.

Jadrov, A.A., 1968. On rootstocks for the oriental persimmon in Uzbekistan. Subtropical Kul'tury 6, 123–126.

Kanety, T., Naor, A., Gips, A., Dicken, U., Lemcoff, J.H., Cohen, S., 2014. Irrigation influences on growth, yield, and water use of persimmon trees. Irrigation Science 32, 1–13.

Kiaei, M., Bakhshi, R., 2014. Radial variations of wood different properties in *Diospyros lotus*. Short Communication, Forest Systems 23, 171–177.

Llácer, G., Badenes, M.L., 2002. Persimmon production and market. In: Bellini, E., Giordani, E. (Eds.), First Mediterranean Symposium on Persimmon. CIHEAM, Zaragoza, Spain, pp. 9–21.

Llácer, G., Badenes, M.L., 2009. Production of persimmon in Spain. Acta Horticulturae 833, 39–42.

Maas, E.V., Hoffman, G.J., 1977. Crop salt tolerance – current assessment. Journal of the Irrigation and Drainage Division 103 (2), 115–134.

Maghradze, D., Bobokashvili, Z., Kvaliashvili, V., 2012. Minor and underutilized fruits in Georgia and their wild relatives. Acta Horticulturae 948, 41–48.

Malcolm, C.V., Smith, S.T., 1971. Growing plants with salty water. Journal of the Department of Agriculture, Western Australia 12, 3–6.

Margetts, J., 2014. Australian Persimmon Industry Strategic Investment Plan 2014/15 – 2018/19, PR13004.

Martínez-Gimeno, M.A., Castiella, M., Rüger, S., Intrigliolo, D.S., Ballester, C., 2017. Evaluating the usefulness of continuous leaf turgor pressure measurements for the assessment of Persimmon tree water status. Irrigation Science 35, 159–167.

Marzolo, G., Lee, D., 2016. Persimmon. Agricultural Marketing Resource Center.

Mowat, A.D., George, A.P., 1996. Environmental physiology of persimmons. In: Schaffer, B., Andersen, P.C. (Eds.), Handbook of Environmental Physiology of Fruit Crops. CRC Press Inc., Boca Raton, pp. 195–202. ISBN: 978-08-493-0179-7.

Mowat, A.D., 2003. Characterization of vegetative growth and productivity patterns between New Zealand persimmon orchards. Acta Horticulturae 601, 121–128.

Munns, R., 2002. Comparative physiology of salt and water stress. Plant, Cell and Environment 25, 239–250.

Munns, R., 2005. Genes and salt tolerance: bringing them together. New Phytology 167, 645–663.

Naor, A., 2006. Irrigation scheduling and evaluation of tree water status in deciduous orchards. Horticultural Reviews 32, 111–166.

Nazarli, A., 2016. Welcome Healthiest Fruit of the Yellow Season. Azernews. http://www.azernews.az/lifestyle/105441.html.

Neuwald, D.A., Sestari, I., Saquet, A.A., Sautter, C.K., 2009. Persimmon production and commercialization in Brazil: an overview. Acta Horticulturae 833, 51–56.

Nissen, R.J., George, A.P., Broadley, R.H., Collins, R.J., 2003. A survey of cultivars and management practices in Australian persimmon orchards. Acta Horticulturae 601, 179–185.

ODEPA-CIRÉN, 2016. Catastro Frutícola CIRÉN-ODEPA. http://www.odepa.cl/documentos_informes/catastrofruticola-ciren-odepa/.

Rabe, E., 2003. The New South African Persimmon industry: Rationale and proposed systems. Acta Horticulturae 601, 159–162.

Ramin, A.A., Tabatabie, F., 2003. Effect of various maturity stages at harvest on storability of persimmon fruits (*Diospyros kaki* L.). Journal of Agricultural Science and Technology 5, 113–123.
Regato, M.D., Guerreiro, I.M., 2013. A cultura do diospireiro no Alentejo. Vida Rural – Revista Profissional de Agronegócios 1784, 38–41.
Rhoades, J.D., 1974. Drainage for salinity control. In: van Schilfgaarde, J. (Ed.), Drainage for Agriculture. Soil Science Society of America, Madison, Wisconsin, USA, pp. 433–461.
Robinson, J.B., Treeby, M.T., Stephenson, R.A., 1997. Fruits, vines and nuts. In: Reuter, D.J., Robinson, J.B., Dutkiewicz, C. (Eds.), Plant Analysis: An Interpretation Manual. CSIRO Publishing, Collingwood, Victoria, Australia, pp. 349–382.
Rubel, F., Kottek, M., 2010. Observed and projected climate shifts 1901–2100 depicted by world maps of the Köppen-Geiger climate classification. Meteorologische Zeitschrift 19, 135–141.
Ruíz-Sánchez, M.C., Domingo, R., Castel, J.R., 2010. Deficit irrigation in fruit trees and vines in Spain. Review Spanish Journal of Agricultural Research vol. 8, 5–20.
Samadi, G.R., Miakhel, I.M., Abdiani, S.A., 2009. Status of persimmon cultivation in Afghanistan. Acta Horticulturae 833, 31–34.
Scheepers, S.J., 2010. Advancing Full Production and Increasing Yield in Young "Triumph" Persimmon Orchards. Stellenbosch University, South Africa.
Shahkoomahally, S., Ramezanian, A., 2013. Analytical and interpretation statistical of relationship between total antioxidant activity, ascorbic acid content, total phenolic compounds, soluble tannin and chromatic parameters of persimmon (*Diospyros kaki*) cv. shiraz during cold storage. Agricultural Communications 1, 17–22.
Skallerup, H., 1953. The distribution of *Diospyros virginiana* L. Annals of the Missouri Botanical Garden 40, 211–225.
Spongberg, S.A., 1979. Notes on persimmons, kakis, date plums and chapotes. Journal of the Arnold Arboretum 39, 290–309.
Sugiura, A., Zheng, G., Yonemori, K., 1991. Growth and ripening of persimmon fruit at controlled temperatures during growth stage III. HortScience 26 (5), 574–576.
Tao, R., Sugiura, A., 1992. Micropropagation of Japanese persimmon (*Diospyros kaki* L.). In: Bajaj, Y.P.S. (Ed.), Biotechnology in Agriculture and Forestry. Hightech and Micropropagation, vol. 2. Springer-Verlag, Germany, pp. 424–440.
Tetsumura, T., Ishimura, S., Hidaka, T., Hirano, E., Uchida, H., Kai, Y., Kuroki, S., Uchida, Y., Honsho, C., 2015. Growth and production of adult Japanese persimmon (*Diospyros kaki*) trees grafted onto dwarfing rootstocks. Scientia Horticulturae 187, 87–92.
TUIK, 2017. Turkish Statistical Institute. http://www.turkstat.gov.tr.
USDA-NCRS, 1997. Soil Moisture Regimes Map. US Department of Agriculture, Natural Resources Conservation Service, Soil Science Division, World Soil Resources. Available from: https://www.nrcs.usda.gov/wps/portal/nrcs/detail/soils/use/?cid=nrcs142p2_054017.
USSL Staff, 1954. Diagnosis and improvement of saline and alkali soils. In: Agricultural Handbook No. 60. USDA, Washington DC, USA.
Visconti, F., de Paz, J.M., 2016. DSS-SALTIRSOIL: un sistema de ayuda a la decisión en internet para la obtención de recomendaciones de gestión del riego y el cultivo en tierras amenazadas por salinización. In: Pérez-Urrestarazu, L. (Ed.), Libro de Actas del XXXIV Congreso Nacional de Riegos. AERYD, Sevilla, Spain, pp. 172–181.
Visconti, F., de Paz, J.M., Rubio, J.L., Sánchez, J., 2012. Comparison of four steady-state models of increasing complexity for assessing the leaching requirement in agricultural salt-threatened soils. Spanish Journal of Agricultural Research 10, 222–237.

Visconti, F., de Paz, J.M., Bonet, L., Jordà, M., Quiñones, A., Intrigliolo, D.S., 2015. Effects of a commercial calcium protein hydrolysate on the salt tolerance of *Diospyros kaki* L. cv. 'Rojo Brilante' grafted on *Diospyros lotus* L. Scientia Horticulturae 185, 129–138.

Visconti, F., Intrigliolo, D.S., Quiñones, A., Tudela, L., Bonet, L., de Paz, J.M., 2017. Differences in specific chloride toxicity to *Diospyros kaki* cv. 'Rojo Brilante' grafted on *D. lotus* and *D. virginiana*. Scientia Horticulturae 214, 83–90.

Watanabe, S., 1985. Morphological studies of flower initiation in deciduous fruit trees by scanning electric microscopy. Japanese persimmon, apple, pear and sweet cherry. Bull Yamagata University Agricultural Science, 9, 515–531.

Yamada, M., Giordani, E., Yonemori, K., 2012. Persimmon. In: Badenes, M.L., Byrne, D.H. (Eds.), Fruit Breeding (Handbook of Plant Breeding). Springer, Germany, pp. 663–693.

Yamada, M., 2008. Persimmon propagation, orchard planting, training and pruning in Japan. Advances in Horticultural Sciences 22, 269–273.

Zhang, Q., Guo, D., Luo, Z., Yang, Y., 2009. Current progress on industry and science of persimmon in China. Acta Horticulturae 833, 25–30.

Zheng, G., Sugiura, A., 1990. Changes in sugar composition in relation to invertase activity in the growth and ripening of persimmon (*Diospyros kaki*) fruits. Journal of the Japanese Society for Horticultural Science 59 (2), 281–287.

Zilkah, S., David, I., Lazar, M., Rotbaum, A., Faingersh, E., 2009. The effect of irrigation and shading on fruit deformation of persimmon cv. 'Triumph'. Acta Horticulturae 833, 319–324.

CHAPTER 9

Precision Irrigation in Olive (*Olea europaea* L.) Tree Orchards

José E. Fernández[1], Antonio Diaz-Espejo[1], Rafael Romero[1], Virginia Hernandez-Santana[1], José M. García[2], Carmen M. Padilla-Díaz[1], María V. Cuevas[1]

[1]Instituto de Recursos Naturales y Agrobiología de Sevilla (IRNAS, CSIC), Seville, Spain;
[2]Instituto de la Grasa (IG, CSIC) Campus Universidad Pablo de Olavide, Seville, Spain

Contents

1. TOWARD RATIONAL IRRIGATION

The world's agricultural production has grown between 2.5 and 3 times over the last 50 years to meet rising demands for food and fiber. In that period, however, the cultivated area has grown by 12% only. New, selected cultivars and improvements in the use of agrochemicals, mechanization, and irrigation have contributed to this increase in production per unit area of land. From the 1600 Mha cultivated worldwide in 2006 (11% of the world's land surface), 301 Mha were equipped for irrigation. That 20% of cultivated area under irrigation yielded 45% of food production, but consumed an average of 70% of all water withdrawal (2710 km^3 year^{-1} out

Water Scarcity and Sustainable Agriculture in Semiarid Environment
ISBN 978-0-12-813164-0
https://doi.org/10.1016/B978-0-12-813164-0.00009-0

of a total water withdrawal of 3862 km^3 $year^{-1}$) (FAO, 2011). In many semiarid and arid countries, that percentage is over 80%. The Food and Agricultural Organization (FAO) has projected that in 2050 the area equipped for irrigation will be around 318 Mha, which represents an increase of about 6%. This increase is much slower than in previous decades (between 1961 and 2009 the irrigated area worldwide increased by 1.6% per year), but it may be crucial to face the challenge of feeding the increasing world population (FAO, 2011). This is expected to grow from the current 7.4 billion people to about 9.1 billion in 2050, which will require a 70% increase in overall food production as compared to the years 2005/07 (FAO, 2009). These data impel for a greater efficiency on the agricultural uses of water.

Since the mid-1900s three main advances have been achieved for the optimization of the agricultural use of water. The first was localized irrigation, which became popular after advances made in Israel in the 1950s. Although the first steps on localized irrigation were taken in Germany at the end of the 19th century, it was not until the development of the plastics industry that components for localized irrigation became reliable and cheap enough to be widely used. The second was based on deriving effective deficit irrigation (DI) strategies, designed mainly in arid and semiarid countries from the 1980s. These are based on supplying irrigation amounts (IA) below the irrigation needs (IN) for the maximum potential crop evapotranspiration (ET_c), but supplied in such a way that the impact of reduced irrigation on crop performance is minimized. The third great advance on irrigation is due to new methods and systems for precise assessment of crop water stress and for irrigation scheduling. From the beginning of this century, the development of new sensors and related systems for continuous and automatic recording of main variables associated with water stress, together with the production of inexpensive but powerful and reliable devices for data recording and transmission, and advances in remote imagery, have made possible the development of precision irrigation. This is an effective irrigation approach that is used in olive orchards, as detailed next.

2. IRRIGATION OF OLIVE ORCHARDS

Irrigation management in olive and its benefits to crop development and production have been addressed in several reviews on both biological and physiological aspects of the olive tree (Lavee, 1996; Connor and Fereres, 2005; Fernández, 2014a; Rapoport et al., 2016; Sebastiani et al., 2016) and on water use and irrigation of olive orchards (Xiloyannis et al., 1996; Fernández and Moreno, 1999; Gucci et al., 2012; Carr, 2013; Fernández, 2014b). From the current 10.3 Mha dedicated to olive orchards, c.22% are under irrigation. Some 98% are in the Mediterranean Basin, i.e., in arid or semiarid areas (IOC, 2016). New plantations, both in the Mediterranean Basin and in new olive-producing countries such as Australia and Argentina, are mostly irrigated, and many of them correspond to high density, mechanically harvested orchards (Rallo et al., 2016). Some of these new orchards have plant densities ranging from 450 to 800 trees ha^{-1} (high-density orchards) or over 1500 trees ha^{-1} (super-high-density [SHD] orchards) (Rius and Lacarte, 2010). In both cases, irrigation is needed for acceptable production levels, since rainfall is not enough when plant density is so high. In most olive-growing areas, water for irrigation is scarce and the year-to-year rainfall variability is high (Rallo et al., 2016). This explains why, in most new olive orchards, localized irrigation and DI strategies are common (Fernández, 2014a).

Localized irrigation meant a great step forward in improving irrigation efficiency, i.e., the ratio of the volume of water taken up by the crop to the volume of irrigation water applied. However, the greater water savings in agriculture do not usually come from an optimized irrigation efficiency, but from improving water productivity (WP), which is the ratio of yield (marketable product or net income) to water used by the crop (Molden, 2003; Pereira et al., 2012). In crops for which the production target is biomass production, e.g., alfalfa and sugarcane, the greatest WP values are usually achieved with full irrigation (FI), i.e., when enough water is supplied to replace ET_c. For woody crops, however, DI has proven to be an effective approach for increasing WP, for a number of reasons. First, DI decreases plant transpiration (E_p) through stomatal closure more effectively in fruit trees and vines than in herbaceous field crops, because tall, rough canopies are better coupled to the atmosphere than the short, smooth canopies of field and vegetable crops. Second, for most woody crops, net incomes are not linearly related to biomass accumulation, but to fruit yield and fruit quality (Fereres and Soriano, 2007). DI normally enhances the quality of fruits and derived products (e.g., olive oil and wine), which contributes to increasing the net income of the grower and thus WP. Finally, DI also increases WP through the control of excessive growth. This helps to reduce pruning intensity and the need for phytosanitary treatments. The control of plant vigor is particularly important in SHD orchards (Fernández et al., 2013, 2014a). In fact, DI may increase the productive life of SHD orchards by reducing problems derived from competency among trees for light (Connor et al., 2014). This explains the wide used of both localized irrigation and DI strategies in fruit tree orchards and vineyards.

As for other cropped species, most olive-growing areas show high spatiotemporal variability of plants and soils, which, if not considered when managing irrigation, greatly reduces WP. Thus certain zones within an orchard could receive too much water, with the consequent increases in runoff and drainage, while in others the same water supplies might not be enough to avoid excessive water stress to the crop. Precision irrigation has a great potential for olive orchards because it is a holistic approach designed to manage irrigation effectively under those limiting conditions of high variability and reduced available water (Smith and Baillie, 2009; Fernández, 2017). It is based on the right choice of irrigation system, irrigation strategy, and the method for scheduling irrigation, and implies detailed and continuous recording of the crop and environmental conditions. Both current knowledge of the olive response to water conditions and a variety of sensors and related systems for monitoring tree water status, and for data transmission, combined with remote imagery, allow for precision irrigation in olive orchards.

3. IRRIGATION STRATEGIES

In Mediterranean areas with yearly average values of potential evapotranspiration (ET_o) and precipitation around 1200–1400 mm and 500–600 mm, respectively, about 700 mm, i.e., 7000 $m^3\ ha^{-1}$, are required to replace the ET_c of olive orchards with 100–300 trees ha^{-1}. From those, some 3000–4000 $m^3\ ha^{-1}$ are the IN required to fully irrigate the orchard, the rest being supplied by rainfall (Fernández and Moreno, 1999; Gucci et al., 2012; Fernández, 2014a). For SHD orchards with 1500–2000 trees ha^{-1}, IN is around 5000 $m^3\ ha^{-1}$ (Grattan et al., 2006; Fernández et al., 2013; Padilla-Diaz et al., 2016). In most olive-growing areas, however, the water available for irrigation is below those amounts. Therefore FI can hardly be applied in olive orchards. On the other hand, FI may not lead to the achievement of the greatest net incomes from orchard exploitation. In fact, for most fruit tree species, FI is advisable for the first years after planting only, for the orchard to establish as soon as possible. Later on, irrigation must be managed according to the response of the species to DI. For olive, irrigation can be significantly reduced once the orchard is

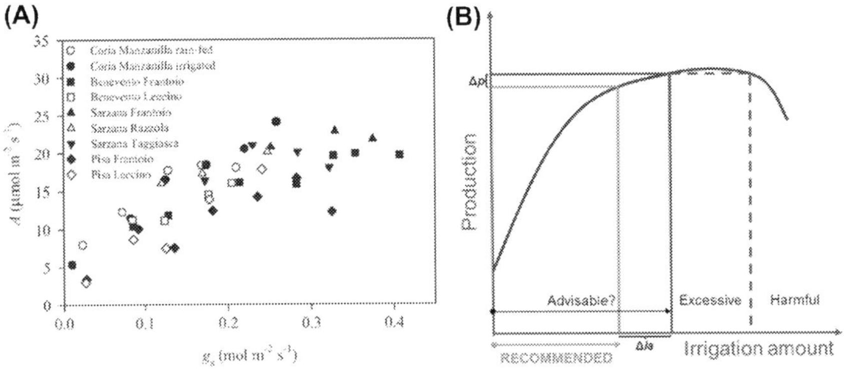

Figure 9.1 (Graph A) Relationship between net CO_2 assimilation (A) and leaf stomatal conductance (g_s) for five different olive cultivars, irrigated and nonirrigated, at four locations in Spain and Italy. Data points represent the average of single measurements averaged for 0.05 mol m^{-2} s^{-1} g_s intervals. (Graph B) Typical response curve of production versus irrigation, applicable to most crops. Δp is the increase in production for an increase in the irrigation amount equal to Δia. Full irrigation could be advisable in crops for which the production target is biomass production (alfalfa, sugar beet, etc.). For fruit trees, however, lower irrigation amounts could be recommended, especially if belonging to orchards with high plant densities. *((A) After Fernández, J.E., Diaz-Espejo, A., d'Andria, R., Sebastiani, L., Tognetti, R. 2008a. Potential and limitations of improving olive orchard design and management through modelling. Plant Biosystems 142 (1), 130–137.)*

established, with a reduced impact on production (Fernández, 2014a; Rallo et al., 2016). This is partly based on the fact that maximum values of net CO_2 assimilation (A) are achieved in olive at stomatal conductance (g_s) values of c.0.2 mol m^{-2} s^{-1}, quite below maximum g_s values occurring in olive trees growing under nonlimiting soil water conditions (Fig. 9.1). Therefore keeping maximum transpiration rates through irrigation is not needed to achieve maximum carbon assimilation, which partly explains the success of DI in olive orchards.

There are a variety of DI strategies suitable to olive orchards. When water for irrigation is very scarce, or in temperate regions with low ET_o or high rainfall, supplementary or complementary irrigation can lead to substantial increases in crop performance, as compared to dry farming (Abd-El-Rahman and El-Sharkawi, 1974; Lavee et al., 1990; Proietti et al., 2012). This irrigation strategy consists in applying a single irrigation event when a fixed threshold for water stress is achieved. No more than two or three irrigation events are applied during the irrigation season. The success of this DI strategy depends to a great extent on the irrigation events being applied at the right moment of the growing cycle. Another DI strategy, called low-frequency DI (Lavee and Wodner, 1991), consists in withholding irrigation until the readily available water is consumed. For olive, the readily available water is often assumed to be 75% of the soil water-holding capacity, i.e., the difference between the soil water content at field capacity and that at wilting point (Orgaz et al., 2017). When the readily available water is consumed, irrigation is applied until field capacity is reached. This is repeated as many times as necessary during the irrigation season. Other strategies widely used in high-density and SHD olive orchards are sustained deficit irrigation (SDI) and regulated deficit irrigation (RDI). SDI consists in supplying a fixed fraction of the IN for FI all throughout the irrigation season. Irrigation is applied daily or 2–4 times per week (Goldhamer et al., 1994; Grattan et al., 2006; Ramos and Santos, 2010). A variation of this strategy is partial root-zone drying, in which water is applied to half of the root zone, switching to the other half every 2–3 weeks (Dry et al., 1996; Fernández et al., 2006a). RDI consists in irrigating daily, or every 2–3 days, at the phenological stages most sensitive to drought, with enough water to replace 80%–100% of ET_c. For the rest of the crop cycle, irrigation is drastically reduced down to one or two irrigations per week only, or even withheld (Chalmers et al., 1981; Goldhamer, 1999; Fernández et al., 2013; Padilla-Díaz et al., 2016). Both SDI and RDI have been tested in different types of olive orchards without clear evidence as to which one is better (Moriana et al., 2003; Iniesta et al., 2009; Ramos and Santos, 2009; García et al., 2013). Our group developed an RDI strategy (Fig. 9.2) for an 'Arbequina' hedgerow olive orchard with 1667 trees ha^{-1} at the Sanabria farm, near Seville (37° 15′ N, −5° 48′ W, 60 m above sea level), which has proven satisfactory (Section 6). We can expect RDI to become more effective than SDI as our knowledge of the physiological and agronomical response of olive to the available water increases. Apart from that, the correct application of RDI requires precise information on the dynamics of crop water stress. This justifies the efforts of the scientific community on the development of new methods for the assessment of water stress and for scheduling irrigation. In the next section we refer to methods allowing for continuous and automatic assessment of tree water stress suitable for precision irrigation.

4. SCHEDULING IRRIGATION

The purpose of scheduling irrigation is to determine both the frequency of irrigation and the amount of water to be supplied in each irrigation event.

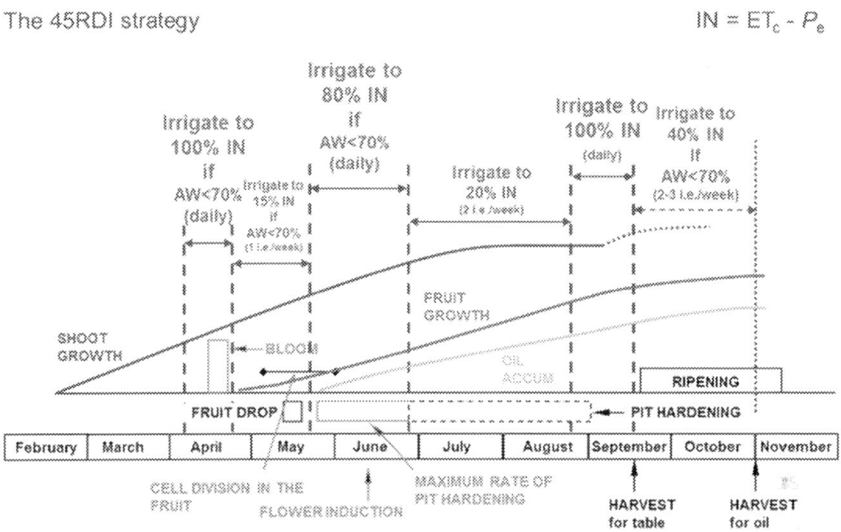

Figure 9.2 Regulated deficit irrigation strategy for the hedgerow olive orchard at the Sanabria farm, designed to supply 45% of the irrigation needs (IN). The orchard, located at 25 km southeast of Seville (south Spain), had 'Arbequina' olive trees planted at 4 m × 1.5 m (1667 trees ha^{-1}). *AW*, Available water in the soil; ET_c, crop evapotranspiration; *IA*, irrigation amount; *i.e.*, irrigation event; P_e, effective precipitation. *(After Fernández, J.E., 2017. Plant-based methods for irrigation scheduling of woody crops. Horticulturae 3, 35. https://doi.org/10.3390/horticulturae3020035.)*

This can be done from either soil water or meteorological measurements, from measurements in the plant, or from a combination of them. The right method will be the one that better allows for the achievement of the production target with the lowest cost in terms of time, effort, and money. Soil water measurements are easy to make and interpret, but the large spatiotemporal variability usually found in cropped soils limits their utility for scheduling irrigation (Jones, 2004, 2007). The crop coefficient approach (Allen et al., 1998), based on meteorological measurements, has been widely used to schedule irrigation in olive orchards (Grattan et al., 2006; Fernández et al., 2006a, 2013). This method requires reliable values of coefficients related to the crop (K_c) and to the fraction of soil covered by the canopy (K_r). Values of both coefficients have been published, but their empirical nature makes risky their use in orchards with different characteristics or at different locations. This issue has been addressed by several authors. Testi et al. (2006) and Orgaz et al. (2006) proposed a model of olive water requirements, which estimates E_p and soil evaporation (E_s) separately, and a new crop coefficient calculated as the sum of three main components, related to tree transpiration (K_p), evaporation from dry areas of the soil surface (K_{s1}), and evaporation from the soil areas wetted by the emitters (K_{s2}). A fourth component can be added, accounting for evaporation of the water intercepted by the canopy (K_{pd}). The model by Testi et al. (2006) takes into account main soil, weather, and plant conditions, such that it is more mechanistic than the crop coefficient approach by Allen et al. (1998). The method by Orgaz et al. (2006) is intended to improve the precision of the calculated IA. In atypical years, average IA values can be corrected by reapplying the method at the end of the month, after knowing the actual rainfall and ET_o values. Also the dual crop coefficient approach (Allen et al., 1998, 2005) is a modified version of the standard crop coefficient approach that allows for a more precise estimation of transpiration in crops with partial cover of the soil. Examples of the use of the dual crop coefficient approach in olive orchards are provided by Cammarelli et al. (2013) and Paço et al. (2014). The crop coefficient approach and its variations are certainly interesting for scheduling irrigation in olive orchards. However, both its empirical nature and the fact that it does not allow for real-time assessment of the crop water needs (meteorological variables must be recorded for certain times before calculating ET_o) curtail its potential for precision irrigation.

The limitations of methods for assessing crop water stress and water needs from soil and meteorological measurements have favored the development of methods relying on plant-based measurements. These have the advantage of integrating the response of the plant to the water status of both the soil and the surrounding atmosphere. Also some plant-based methods allow for continuous and automatic recording and can be easily implemented with data transmission systems. This is the case of methods based on measurements of sap flow (SF), trunk diameter variations (TDV), and leaf turgor pressure (LTP) (Fernández, 2017). The use of SF-related measurements to monitor olive water stress and to schedule irrigation in olive orchards was explored by Fernández et al. (2001, 2008b,c). Fernández et al. (2011a) and Cuevas et al. (2013) combined SF- and TDV-related measurements to further improve irrigation scheduling. Rodriguez-Dominguez (2012) combined measurements of SF and LTP to better understand leaf water relations in the species. Fernández and Cuevas (2010) and Ortuño et al. (2010) reviewed the use of TDV-related measurements in olive and in other species, and Cuevas et al. (2010) and Fernández et al. (2011b) analyzed the performance of different water stress indices derived from TDV-related measurements to assess water stress in olive. Additional work on the use of TDV-related measurements to schedule irrigation in olive was published by Moriana et al. (2010, 2013) and by Girón et al. (2015, 2016). Publications by Ben-Gal et al. (2010), Fernández et al. (2011c), and Padilla-Diaz et al. (2016) report on the usefulness of LTP-related measurements to assess water stress and to schedule irrigation in olive orchards. Egea et al. (2017a) analyzed financial aspects of adopting methods based on SF-, TDV-, or LTP-related measurements for scheduling irrigation in SHD olive orchards. As mentioned earlier, these methods can be combined with remote imagery to identify crop variability within the orchard, which makes them suitable for precision irrigation. This has been addressed for olive by Fernández (2014b, 2017) and Egea et al. (2017b). There is therefore a substantial amount of literature on the fundamentals and potential of these methods to both assess water stress and schedule irrigation in olive orchards of different types. In addition, control theory can be used to further improve irrigation scheduling from any variable from the soil, the plant, or the atmosphere (Romero et al., 2012a). Next we focus on the most recent advances in the use of plant-based measurements.

4.1 Sap Flow

The first measurements of SF in olive were made by Moreno et al. (1996), who explored transpiration and root water uptake in olive trees under changing soil water conditions. Measurements of SF allow for in situ determinations of the dynamics and, if properly calibrated (Fernández et al., 2001, 2006b), absolute values of plant water consumption, which gives them a high potential for scheduling irrigation (Fernández et al., 2008b,c). In the last decade new advances were made concerning the use of SF measurements for assessing the plant water status and water use and for scheduling irrigation. These include the measurement of very low SF velocities and reverse flows. The heat field deformation (HFD) method, developed by Nadezhdina et al. (1998), can be used to measure a wide scale of flow rates, including low, zero, and reverse flows. Burgess et al. (2001) developed the heat ratio method to measure reverse flow in roots and other conductive organs. Later, the calibrated average gradient (CAG) heat-pulse method, developed by Testi and Villalobos (2009), was able to resolve low and reverse flows, thanks to a self-calibrating behavior. The CAG method uses the same probe arrangement as the compensation heat-pulse (CHP) method (Swanson and Whitfield, 1981; Green et al., 2003). Romero et al. (2009, 2012b), Green et al. (2009), and Green and Romero (2012) also used the CHP method to determine low and reverse flows. The HFD method was used by Nadezhdina et al. (2012) and Ferreira et al. (2013) to study, for the first time in olive, SFs related to hydraulic redistribution.

Most approaches for scheduling irrigation from SF-related measurements require detailed training for installing and maintaining the system, and considerable data processing (Fernández, 2014b). This curtails their use in commercial orchards. New approaches, however, point to a wider use of SF-related measurements. This is the case of a new method proposed by Hernandez-Santana et al. (2016) to estimate g_s from SF-related measurements. One of the keys for precision irrigation is the use of a reliable water stress indicator and, among them, g_s reflects finely the level of water stress. However, the use of g_s for irrigation scheduling faces a major limitation because it cannot be automatically and continuously recorded. Nevertheless, Hernandez-Santana et al. (2016) demonstrated that g_s can be estimated continuously from sap flux density (J_s) values normalized by vapor pressure deficit (D). In their approach, estimations of g_s are based on the control that stomata exert on transpiration, and thus on J_s under conditions of high coupling to the atmosphere, which is the case for olive (Moreno et al., 1996; Tognetti et al., 2009). They measured g_s with an infrared gas analyzer, and J_s with the CHP method, in olive trees under different soil and atmospheric water conditions. Robust relationships between g_s and J_s/D were found (Fig. 9.3). Although empirical, those relationships proved that reliable g_s values can be obtained from continuous and automatic SF measurements. This approach proposed by Hernandez-Santana et al. (2016) avoids the uncertainties derived from the required upscaling from measurements at single points in the trunk to the whole tree, which is normally made by using the Penman–Monteith equation (Hernandez-Santana et al., 2015). In addition, it avoids the use of multipoint SF sensors, because J_s measured in the outer part of the xylem, where the new tissues are, was demonstrated to show the best relationship with the new, sun-exposed leaves. If required, a deeper thermocouple can be used to relate J_s to the shaded, old leaves. The approach by Hernandez-Santana et al. (2016) is sensitive enough to allow the evaluation of the g_s response to the two main driving variables of transpiration under water stress: atmospheric and soil water deficit (Fig. 9.4).

This new approach allows for the first time the automated estimation of g_s in olive orchards, which has a great potential to schedule irrigation. In addition, and from the close relation between g_s and CO_2 assimilation (A) and therefore to biomass production and yield (Fernández et al., 2008a; Flexas et al., 2013), the derived g_s can be input in photosynthesis models to

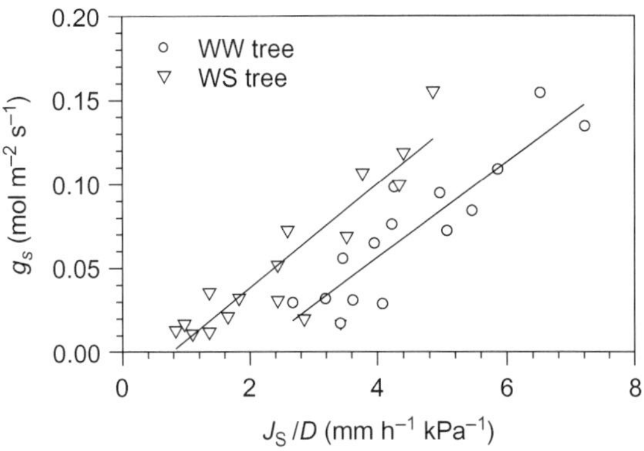

Figure 9.3 Relationship between stomatal conductance (g_s) measured in new, sun-exposed leaves and sap flux density (J_s) at 5 mm below the cambium divided by the air vapor pressure deficit (D) for two olive trees, one with its irrigation needs fully fulfilled (WW tree, *circles*) and another one water stressed (WS tree, *triangles*). Both relationships are highly significant ($P < .0001$) and strong ($R^2 = 0.84$ for WW tree and 0.80 for WS tree). The trees belonged to the hedgerow olive orchard at the Sanabria farm. Each point represents the average of four measurements conducted in two trees, every 1.5 h from dawn to dusk, in two different days. *(After Hernandez-Santana, V., Fernández, J.E., Rodriguez-Dominguez, C.M., Romero, R., Diaz-Espejo, A., 2016. The dynamics of radial sap flux density reflects changes in stomatal conductance in response to soil and air water deficit. Agricultural and Forest Meteorology 218–219, 92–101.)*

estimate the impact of irrigation on yield. Based on the tight coupling of g_s and A, López-Bernal et al. (2015) derived an approach to estimate the canopy A in olive trees from SF-related measurements. The complexity of the approach and the need for accurate absolute values of the actual sap fluxes suggest that the approach can be used for research purposes only. In any case, the approaches of Hernandez-Santana et al. (2016) and López-Bernal et al. (2015) show new ways of using SF measurements for proper management of irrigation in olive orchards.

4.2 Trunk Diameter

The diameter of the trunk changes depending on the hydration level of the plant, such that TDV records are being used in olive orchards, as well as in many other species, both to assess water stress and to schedule irrigation (Cuevas et al., 2010, 2013; Fernández et al., 2011a,b; Girón et al., 2015, 2016; Agüero Alcaras et al., 2016). Olive has a negligible nocturnal transpiration, since the stomata close after sunset and the leaf epidermis is covered by a waxy cuticle that prevents water losses through the leaf surface (Fernández, 2014a). Thus during the night the olive tree rehydrates by

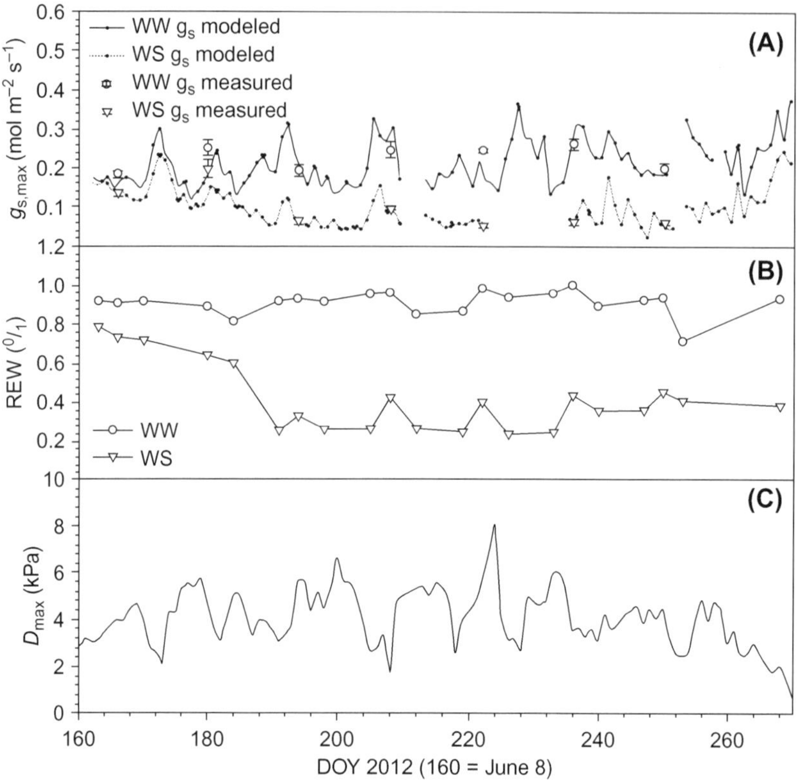

Figure 9.4 Time courses of (A) maximum stomatal conductance ($g_{s,max}$), (B) relative extractable water (REW), and (C) daily maximum vapor pressure deficit (D_{max}) recorded at the hedgerow olive orchard at the Sanabria farm. Lines in graph A depict $g_{s,max}$ estimated from sap flux density measurements applying the kind of relationships shown in Fig. 9.3. *Circles* are the average of eight $g_{s,max}$ measurements at about 8:30 GMT. *DOY*, day of the year. *(After Hernandez-Santana, V., Fernández, J.E., Rodriguez-Dominguez, C.M., Romero, R., Diaz-Espejo, A., 2016. The dynamics of radial sap flux density reflects changes in stomatal conductance in response to soil and air water deficit. Agricultural and Forest Meteorology 218–219, 92–101.)*

absorbing water from the soil and accumulating it in storage tissues, such that the diameter of the trunk increases. Early in the morning the stomata open and water is lost by transpiration, which causes the trunk diameter to decrease. In a day cycle the trunk diameter may vary from a few tens of microns to a few hundred microns. The dendrometers are sensors able to register these variations continuously and automatically. The recorded data have a high potential for irrigation scheduling, as proven by Fernández and Cuevas (2010), Moriana et al. (2010), Girón et al. (2015), and Fernández (2014b, 2017), among others.

Several water stress indicators can be derived from TDV records (Fernández and Cuevas, 2010). Both the maximum daily shrinkage (MDS) and trunk growth rate are recommended for a variety of fruit tree species (Fernández, 2017). The reliability of MDS is compromised, however, because, for different species, including olive, its value tends to increase as water stress progresses, but after a certain level the opposite is observed (Ortuño et al., 2010). For olive, that level corresponds to midday values of stem water potential (Ψ_{stem}) around -1.5 MPa. Among the indicators recommended for olive, the D_{MXTD} indicator proposed by Fernández et al. (2011a,b) yielded good results. Values of D_{MXTD} are calculated as the difference for daily maximum trunk diameter between a representative tree of the orchard, normally under DI conditions, and a reference tree, i.e., a representative tree of the orchard growing under nonlimiting soil water conditions. The D_{MXTD} indicator therefore is based on the signal intensity approach proposed by Goldhamer and Fereres (2001). Fig. 9.5 shows data from an experiment carried out by Fernández et al. (2011b), in which MDS values were compared to D_{MXTD} values. Measurements were made in trees from La Nava, a commercial orchard with 12-year-old 'Arbequina' trees planted at 7 m × 6 m (238 trees ha^{-1}), under two irrigation treatments: T1, trees irrigated twice a week with enough water to replace two-thirds of ET_c, and T2, trees without irrigation from June 25 until fruit shriveling. Then, a recovery irrigation amounting to the total water supplied in T1 for that period was applied. They also had control trees irrigated at 150% ET_c, to ensure nonlimiting soil water conditions. The orchard had deep soil with high water retention capacity. The D_{MXTD} values calculated from TDV records were more closely related to soil water content than MDS values (Fig. 9.5). Cuevas et al. (2013) tested the potential of D_{MXTD} values for assessing water stress in trees of the hedgerow olive orchard at the Sanabria farm, which had 1667 trees ha^{-1}. An alternative indicator is the time course of the maximum trunk diameter (MXTD), also shown in Fig. 9.5. It provides similar information to D_{MXTD}, but it does not require either data processing or reference trees (Fernández et al., 2011b). Fernández (2017) discussed an approach alternative to the signal intensity approach, for which MXTD values can also be used.

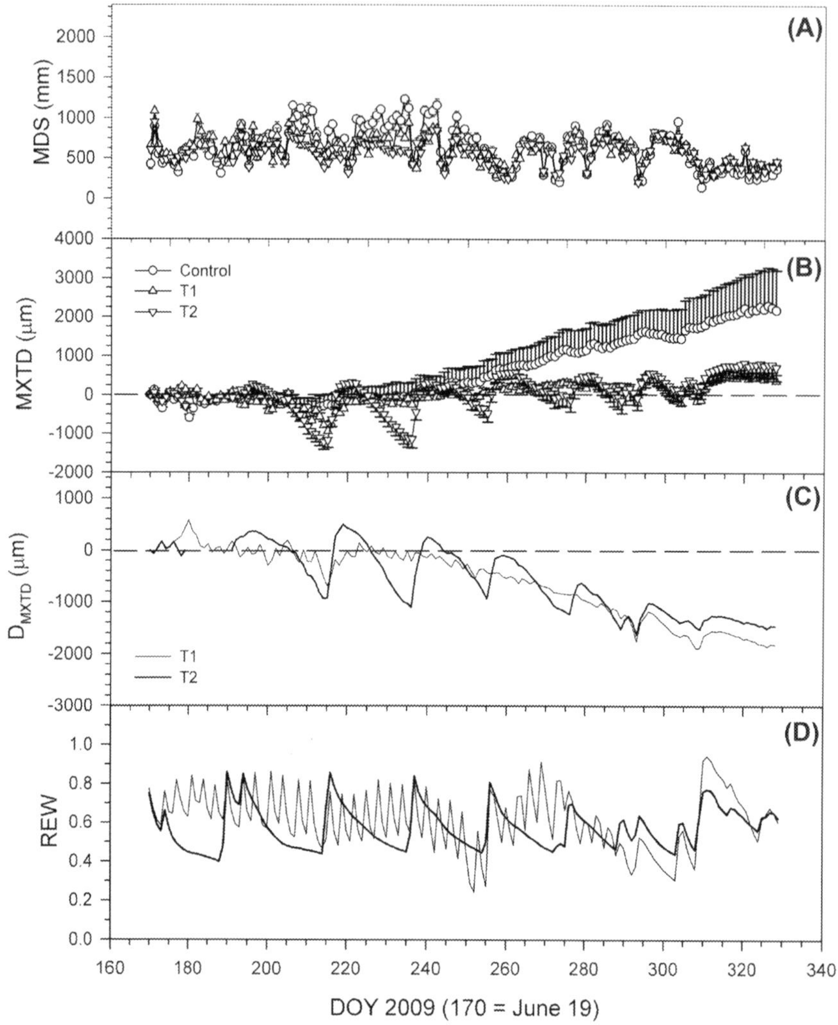

Figure 9.5 Time courses of (A) maximum daily shrinkage (MDS), (B) maximum trunk diameter (MXTD), (C) the difference among maximum trunk diameter of stressed and full irrigation trees (D_{MXTD}), and (D) the relative extractable water (REW) in the soil, for the different irrigation treatments applied at La Nava olive orchard. See Fernández et al. (2011b) for details. *DOY*, day of the year.

4.3 Leaf Turgor

The leaf is one of the first organs of the plant to lose water during dehydration. The loss of water causes a decrease in the turgor pressure of leaf cells. Zimmermann et al. (2008) developed the leaf patch clamp pressure

(LPCP) probe, a sensor to register relative changes in LTP (or P_c) in plants under field conditions. The sensor has a highly sensitive pressure sensor that senses the so-called output pressure (P_p), which is inversely proportional to P_c. The LPCP probe is noninvasive, easy to install, and capable of operating automatically and continuously under field conditions for months. Records by LPCP probes are sent by mobile network to an internet server, where the data are stored and visualized in terms of graphics or tables in nearly real time. The LPCP probe and related system has therefore a high potential for precision irrigation. The performance of the LPCP probe, both to assess plant water stress and to schedule irrigation, has been tested for a variety of species, including olive (Ben-Gal et al., 2010; Fernández et al., 2011c; Padilla-Díaz et al., 2016). A tight correlation between the shape of the daily P_p curves and the olive tree water status, expressed in values of midday Ψ_{stem} measured with a Scholander-type pressure chamber, was first reported by Fernández et al. (2011c) and Ehrenberger et al. (2012), and confirmed by Padilla-Díaz et al. (2016) (Fig. 9.6). Similar analyses made by other authors suggest that the ranges of Ψ_{stem} values for States 1—3 depend on the cultivar and orchard location (Table 9.1). Thus Marino et al. (2016) worked with 'Nocellara del Belice' and 'Olivo di Mandanici' olive trees in the peninsula of Sicily, and found lower Ψ_{stem} thresholds than those reported by Padilla-Díaz et al. (2016). They suggested differences being due to factors affecting the crop response to water stress, such as genotype, management, orchard characteristics, and location. Data in Table 9.1 suggest the risk of using the published ranges of Ψ_{stem} values corresponding to each state of the daily P_p curve in orchards with different cultivars and conditions.

Padilla-Díaz et al. (2016) developed an approach for scheduling irrigation based on the occurrence of States 1—3 in daily P_p curves recorded by LPCP probes in 'Arbequina' trees of the hedgerow olive orchard at the Sanabria farm. They validated their approach by comparing the derived IA with those calculated with the crop coefficient approach. In their approach, they combined the occurrence of the onset of water stress (as indicated by a shift from State 1 to State 2) in the orchard zone most sensitive to water stress with a 3-day weather forecast. IA were increased by 15% in case of an increase in the tree's water stress and a prediction for increasing atmospheric demand. Similarly, a 15% decrease in IA was scheduled when the tree's water stress decreased. Despite the empirical aspects of this approach (e.g., the 15% variation in IA), results showed a close agreement between the IA values determined with this new approach and those calculated with the

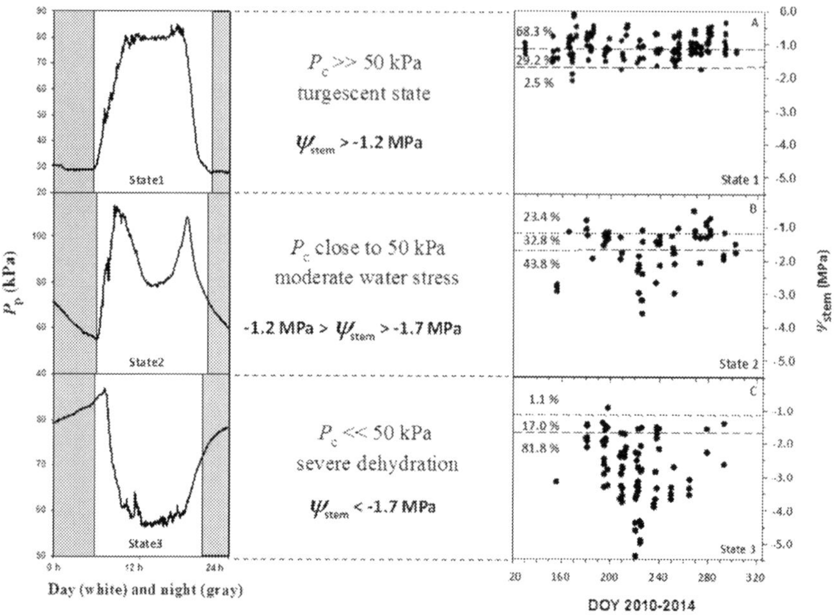

Figure 9.6 *Left panels*: Daily leaf patch clamp pressure (LPCP) (P_p) curves recorded in 2010 with LPCP probes in 'Arbequina' olive trees of the hedgerow olive orchard at the Sanabria farm under different levels of water stress. The shown curve shapes are typical of three levels of water stress identified in the figure by values both of leaf turgor pressure (P_c) and midday stem water potential (Ψ_{stem}) (*central panels*). *Right panels*: Midday Ψ_{stem} values measured in representative trees of all the irrigation treatments applied at the hedgerow olive orchard at the Sanabria farm during the irrigation seasons of 2010–14. Each data point corresponds to a single measurement with a Scholander-type pressure chamber. *Dashed lines* represent the Ψ_{stem} values identified by Fernández et al. (2011c) and Ehrenberger et al. (2012) as typical of State 1 (A), State 2 (B), and State 3 (C). *DOY*, day of the year. (Left panels: *First published in Fernández, J.E., 2014b. Plant-based sensing to monitor water stress: applicability to commercial orchards. Agricultural Water Management 142, 99–109, after Fernández, J.E., Rodriguez-Dominguez, C.M., Perez-Martin, A., Zimmermann, U., Rüger, S., Martín-Palomo, M.J., Torres-Ruiz, J.M., Cuevas, M.V., Sann, C., Ehrenberger, W., Diaz-Espejo, A., 2011c. Online-monitoring of tree water stress in a hedgerow olive orchard using the leaf patch clamp pressure probe. Agricultural Water Management 100, 25–35 and Ehrenberger, W., Rüger, S., Rodríguez-Domínguez, C., Díaz-Espejo, A., Fernández, J.E., Moreno, J., Zimmermann, D., Sukhorukov, V.L., Zimmermann, U., 2012. Leaf patch clamp pressure probe measurements on olive leaves in a nearly turgorless state. Plant Biology 14, 666–674.* Right panels: *After Padilla-Díaz, C.M., Rodriguez-Dominguez, C.M., Hernandez-Santana, V., Perez-Martin, A., Fernández, J.E., 2016. Scheduling regulated deficit irrigation in a hedgerow olive orchard from leaf turgor pressure related measurements. Agricultural Water Management 164, 28–37.)*

Table 9.1 Threshold values of midday stem water potential (Ψ_{stem}) for the three states, described in Fig. 9.6, of daily curves recorded with leaf patch clamp pressure (LPCP) probes

Authors	Species	Orchard type	Years	State 1	State 2	State 3
				Ψ_{stem} thresholds		
Padilla-Díaz et al. (2016)	*Olea europaea* L., cv. Arbequina	SHD	2010 to 2014	$\Psi_{stem} > -1.2$ MPa	-1.2 MPa $> \Psi_{stem} > -1.7$ MPa	$\Psi_{stem} < -1.7$ MPa
Marino et al. (2016)	*O. europaea* L., cv. Nocellara del Belice	SHD	2014	-1.8 MPa $> \Psi_{stem} > -2.5$ MPa	-2.2 MPa $> \Psi_{stem} > -3.2$ MPa	$\Psi_{stem} < -3$ MPa
	cv. Olivo di Mandanici	TO	2015	-1.5 MPa $> \Psi_{stem} > -2.1$ MPa	-1.8 MPa $> \Psi_{stem} > -2.8$ MPa	-2.1 MPa $> \Psi_{stem} > -3.7$ MPa
Martínez-Gimeno et al. (2017)	*Diospyros kaki* L.f., cv. *Rojo Brillante* Rootstock *Diospyros lotus*	CO	2014	$\Psi_{stem} = -0.66 \pm 0.1$ MPa	$\Psi_{stem} = -1.03 \pm 0.18$ MPa	$\Psi_{stem} = -1.41 \pm 0.23$ MPa
	L. Rootstock *Diospyros virginiana* L.			$\Psi_{stem} = -0.74 \pm 0.12$ MPa	$\Psi_{stem} = -0.91 \pm 0.10$ MPa	$\Psi_{stem} = -1.02 \pm 0.29$ MPa

The shown values were recorded in olive orchards with different cultivars and locations, and in persimmon trees of three different cultivars: *CO*, commercial orchard; *SHD*, super-high density; and *TO*, traditional orchard.

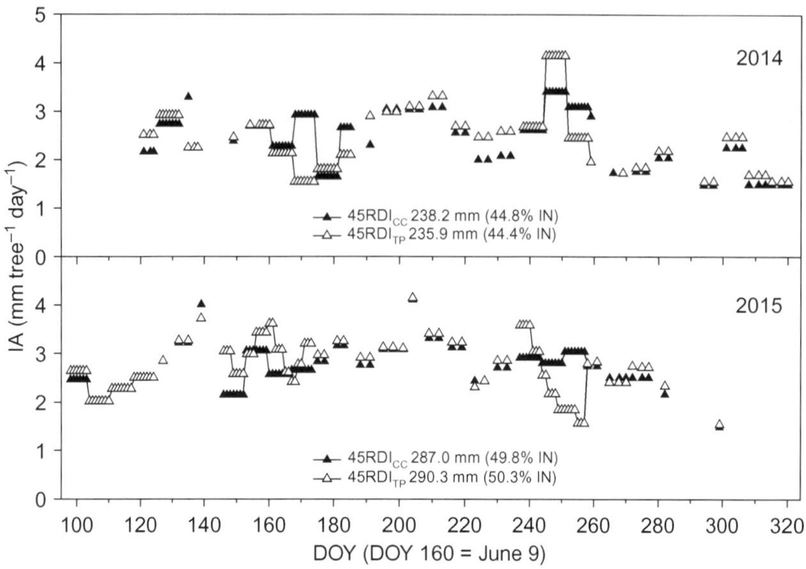

Figure 9.7 Seasonal courses of the irrigation amounts (IA) supplied to trees of the hedgerow olive orchard at the Sanabria farm under the 45% regulated deficit irrigation (45RDI) strategy shown in Fig. 9.2. Measurements were made on the irrigation seasons of 2014 and 2015. Also shown are the total amounts of water supplied by irrigation to each treatment, expressed both in millimeters and as a percentage of the irrigation needs (IN) for full irrigation. The $45RDI_{CC}$ treatment was scheduled with the crop coefficient approach and the $45RDI_{TP}$ treatment was scheduled from leaf turgor-related measurements (see Padilla-Díaz et al., 2016, for details). *DOY*, day of the year.

crop coefficient approach (Fig. 9.7). Furthermore, no differences either in irrigation WP or crop performance were found between the two approaches. These results prove the usefulness of the LPCP system for farmers with little training to schedule irrigation in commercial olive orchards from just the visualization of the daily P_p curves (no data processing required) and from weather predictions downloaded from the internet. The potential of other water stress indicators derived from LPCP records is being investigated (Fernández, 2014b; Marino et al., 2016).

5. TELEMATICS AND IRRIGATION

Precision irrigation began in the 1960s with the use of the geographic information system (GIS). In the late 1980s and early 1990s the development of the Global Positioning System (GPS) meant a significant step forward in precision farming. Since then, both increasing knowledge of crop physiology related to water stress and advances in electronics, data processing, and data transmission have favored precision irrigation. However, most systems with a potential for automatically and continuously assessing crop water stress and for irrigation scheduling, such as those based on SF-, TDV-, and LTP-related measurements, are being produced at a reduced scale, some of them even for research purposes only, which makes them expensive and difficult to use. These two limitations have curtailed the wide adoption of those systems by farmers and orchardists (Fernández, 2014b). However, solutions to overcome those limitations can be easily adopted. For instance, the user can hire companies that, for an affordable fee, install and maintain the systems and provide useful information through user-friendly applications. Also the use of miniaturized cameras mounted on unmanned aerial vehicles, also known as remote-piloted aerial systems (Gago et al., 2015), has lowered the cost of applying remote imagery to the point of making zoning or zonification, i.e., the identification of zones with different sensitivities to water stress within commercial farms and orchards, feasible, a key step for precision irrigation. This greatly helps to lower the cost of applying SF-, TDV-, and LTP-related measurements for scheduling irrigation in commercial orchards, since zonification allows the user to determine the right number of trees to be instrumented, thus reducing investment and maintenance costs. Examples of the use of remote imagery for managing irrigation in olive orchards are given by Gonzalez-Dugo et al. (2013, 2015), Agam et al. (2014), and Egea et al. (2017b), among others. In addition, electronics and communications companies are joining forces to facilitate a wider use of telematics in agriculture, by providing cheaper sensors and related systems, more efficient communication networks, and more capable storage information systems (the cloud). All these advances can contribute to the increasing popularity of precision irrigation.

6. IRRIGATION AND CROP PERFORMANCE

6.1 Production

Relationships between water consumed and both fruit and oil production have been derived for olive (Moriana et al., 2003; Grattan et al., 2006; Martín-Vertedor et al., 2011), and WP values, for both fruit and oil, have been determined (see Fernández et al., 2013, for details). It is well known that high soil water availability favors the number of fruits per tree, fruit fresh weight, fruit volume, and pulp:stone ratio and, as a consequence, increases fruit and oil yields (Goldhamer et al., 1994; d'Andria et al., 2004; Gucci et al., 2007). However, determining the right values of irrigation for the most advantageous soil water status is far from straightforward. First, relationships between water consumed by olive trees and the achieved fruit and oil yields are curvilinear, and the point at which the best balance between water supplied and orchard profitability is achieved (Fig. 9.1) is not clear. Second, results depend to a great extent on cultivar, orchard characteristics, local conditions, and management, among other factors, so caution must be taken when applying the findings reported by different authors to a particular orchard. It is known that DI has a greater impact on fresh fruit yield than on oil yield (Lavee et al., 2007). This can be due, at least in part, to the fact that irrigation increases the mesocarp water content, which penalizes physical oil extraction, as is discussed in the next section. For oil yield, consistent results on oil yield reductions of c.20% with c.50% DI strategies have been reported by different authors (Moriana et al., 2003; Iniesta et al., 2009; Caruso et al., 2013). More details on the impact of DI on olive production are given in Fernández (2014a) and Gucci et al. (2007, 2012). The additional benefits of DI in areas where vegetative growth in olive is often excessive, such as in northern Argentina, were evaluated by Correa-Tedesco et al. (2010) and Searles et al. (2011). Apart from the amount of water in the soil, its distribution within the root zone affects water consumption, growth, and production performance of the olive tree. Thus Fernández et al. (2003) observed that the reduced wetted soil volumes in olive trees under localized irrigation restricted E_p, even if the irrigation supplies were enough to replace the crop water needs. Pastor et al. (2005) irrigated mature 'Picual' olive trees with the same amount of water but with increasing number of drippers, which resulted in increasing fractions of the root zone wetted by irrigation. They found that yield increased with the volume of soil wetted by irrigation. This agrees with results reported by Gispert et al. (2013). The fraction of the root zone wetted by irrigation is related both to the effective soil water potential (Ψ_s) and to the triggering of root-to-shoot signaling mechanisms, which affect water status and water consumption by the tree (Fernández, 2014a).

The management of irrigation in hedgerow olive orchards with high plant densities, or SHD orchards, is particularly challenging. First, irrigation may have a negative impact on the quality of the olive oil. This subject is discussed in detail in the next section. Second, irrigation promotes production, but also growth. Excessive growth in SHD orchards leads to low radiation interception by the lower parts of the canopy, which decreases orchard productivity in the medium and long term (Connor et al., 2009, 2014). Also the vineyard-type straddle harvesters normally used in SHD orchards cannot work properly with large trees (León et al., 2007). We carried out a 6-year experiment in the SHD olive orchard at the Sanabria farm. The orchard had 'Arbequina' trees planted in 2007 at the top of 0.4 m high ridges, at 4 m × 1.5 m (1667 trees ha^{-1}). The aim of the experiment was to derive irrigation management practices for maximum profitability, i.e., to achieve the best balance between oil production and quality without reducing the orchard's productive life. In the first 3 years (2010–12) we evaluated an RDI strategy derived from state-of-the-art knowledge in 2010 at the beginning of the experiments, with two irrigation levels amounting to 60% and 30% of the irrigation needs for FI (IN) calculated with the crop coefficient approach (see Fernández et al., 2013, for details). We used a fully irrigated treatment as control. Main findings from those years suggested both slight changes in the RDI strategy and an irrigation level of 45% being the most appropriate for achieving the chosen production target under the orchard conditions. Thus we derived the 45% regulated deficit irrigation (45RDI) strategy depicted in Fig. 9.2, which we applied in the years 2013–15. In each experimental year we analyzed the response of main physiological variables related to water stress, as well as growth, production, and quality (Fernández et al., 2013; García et al., 2017; Hernández-Santana et al., 2017). The supplied IA for each irrigation treatment and year are detailed in Table 9.2, together with the yearly FAO-56 potential evapotranspiration calculated from meteorological measurements in the area, and the yearly precipitation. Table 9.2 shows the average yearly IN value to replace the crop water demand in this type of SHD olive orchard, and also year-to-year variations. With the exception of the year 2013, the supplied IA were close to the calculated IN. In 2013 the irrigation season started about 1 month later than required, because of problems with the irrigation system. The impact of the tested irrigation strategies and irrigation levels on fruit yield and oil yield is shown in Tables 9.3 and 9.4, respectively. Results from 2014–15 (results from 2013 are excluded because of the mentioned problem with the irrigation system) show that the derived irrigation strategy for SHD olive orchards (Fig. 9.2), with total IA amounting to 49% of IN, led to fruit and oil yields equal to 80% and 74%, respectively, of the yields obtained in fully irrigated trees. The WP values are described in Table 9.5. It has to be taken into account that pursuing the greatest WP values is not necessarily the best target when planning

Table 9.2 Irrigation amounts (IA) supplied to the trees of each irrigation treatment tested at the hedgerow olive orchard at the Sanabria farm over 6 years of the experiment described in Section 6.1

	Irrigation amount (mm)					
Year	**FI**	**60RDI**	**45RDI**	**30RDI**	**ET_o (mm)**	***P* (mm)**
2010	502.0 (101%)	308.0 (62%)		148.0 (30%)	1466.5	912.0
2011	477.2 (107%)	288.3 (65%)		127.3 (29%)	1473.6	477.0
2012	478.7 (102%)	291.3 (63%)		134.6 (29%)	1598.2	404.6
2013	367.6 (85%)[a]		197.9 (46%)		1464.3	476.4
2014	462.9 (92%)		238.2 (47%)		1437.6	549.4
2015	585.5 (101%)		287.0 (50%)		1581.3	237.6
AVG ± SE	479.0 ± 28.6	295.9 ± 6.1	241.0 ± 25.8	136.6 ± 6.1	1503.6 ± 27.8	509.5 ± 91.4

AVG ± *SE*, average ± standard error; ET_o, yearly value of the FAO-56 Penman–Monteith potential evapotranspiration calculated from data provided by a nearby standard weather station; *FI*, full irrigation; *P*, yearly precipitation collected in the orchard. Values in parentheses refer to the IA expressed as a percentage of the irrigation needs for full irrigation, calculated for each treatment with the crop coefficient approach; *RDI*, regulated deficit irrigation.

[a]In 2013 the irrigation season started about 3 weeks later than required, because of problems with the irrigation system.

Table 9.3 Fruit yields recorded for each irrigation treatment tested at the hedgerow olive orchard at the Sanabria farm over 6 years of the experiment described in Section 6.1

	Fruit yield (kg ha^{-1})			
Year	**FI**	**60RDI**	**45RDI**	**30RDI**
2010	15609.8	12680.9 (81%)		9185.2 (59%)
2011	19764.4	14480.0 (73%)		9731.1 (49%)
2012	23616.4	18140.3 (77%)		16229.9 (69%)
2013	22559.8		14952.1 (66%)	
2014	19283.0		16036.5 (83%)	
2015	19724.8		15073.8 (76%)	
AVG ± SE	20093.0 ± 1147.4	15100.4 ± 1606.2	15354.1 ± 343.0	11715.4 ± 2262.7

AVG ± SE, average ± standard error; *FI*, full irrigation; *RDI*, regulated deficit irrigation. Values in parentheses refer to the fruit yield expressed as a percentage of the FI fruit yield. See Fernández et al. (2013) for details of the sampling procedure.

Table 9.4 Oil yields recorded for each irrigation treatment tested at the hedgerow olive orchard at the Sanabria farm over 6 years of the experiment described in Section 6.1

	Oil yield (kg ha^{-1})			
Year	**FI**	**60RDI**	**45RDI**	**30RDI**
2010	999.0	912.1 (91%)		865.4 (86%)
2011	1264.8	861.2 (68%)		932.2 (74%)
2012	1291.2	733.3 (57%)		816.5 (63%)
2013	1557.1		867.6 (56%)	
2014	1875.8		1416.9 (76%)	
2015	1957.4		1387.4 (71%)	
AVG ± SE	1490.9 ± 153.1	835.5 ± 53.2	1224.0 ± 178.4	871.4 ± 33.5

AVG ± SE, average ± standard error; *FI*, full irrigation; *RDI*, regulated deficit irrigation. Values in parentheses refer to the oil yield expressed as a percentage of the FI oil yield. See Fernández et al. (2013) for details of the sampling and oil extraction procedures.

irrigation. The concepts of water use efficiency and WP, and the complexity of choosing the right production target in highly technified orchards, were addressed by Fernández (2017). Concerning the effect of DI on leaf area (LA), Fig. 9.8 shows seasonal courses of LA representative of those observed in the 6-year study made at the orchard. The trees were pruned every year, in December–January, to 4–6 m^2 LA. Data from 2014 to 2015 show that the average LA in the 45RDI trees at the end of the growing season was c.75% of that in the FI trees.

Table 9.5 Water productivity (WP) values in terms of fresh fruit and olive oil for each irrigation treatment tested at the hedgerow olive orchard at the Sanabria farm over 6 years of the experiment described in Section 6.1

Year	WP for fruit (kg fruit m^{-3} water supplied by irrigation)				WP for oil (kg oil m^{-3} water supplied by irrigation)			
	FI	60RDI	45RDI	30RDI	FI	60RDI	45RDI	30RDI
2010	3.11	4.12		6.21	0.20	0.30		0.58
2011	4.14	5.02		7.64	0.27	0.30		0.73
2012	4.93	6.23		12.05	0.27	0.25		0.61
2013	6.14		7.55		0.42		0.44	
2014	4.16		6.73		0.41		0.59	
2015	3.37		5.25		0.33		0.48	
AVG ± SE	4.31 ± 0.45	5.12 ± 0.61	6.51 ± 0.67	8.63 ± 1.76	0.32 ± 0.04	0.28 ± 0.02	0.50 ± 0.04	0.64 ± 0.05

AVG ± *SE*, average ± standard error; *FI*, full irrigation; *RDI*, regulated deficit irrigation.

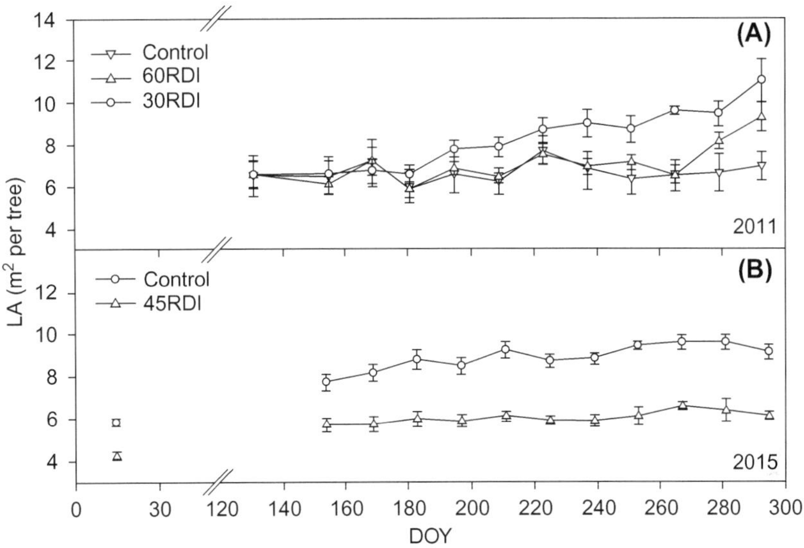

Figure 9.8 Seasonal courses of leaf area (LA, $n = 4$) in trees of the hedgerow olive orchard at the Sanabria farm under the tested irrigation treatments detailed in Section 6.1. Data from the experimental years 2011 (A) and 2015 (B) are shown. *Vertical bars* represent ± the standard errors. *DOY*, day of year. For details on the measurements, see Fernández et al. (2013).

6.2 Quality

There is a remarkable amount of literature on the effects of irrigation on the growth and function of both the root system and the canopy, and on the production and quality of olive fruits both for table consumption and for oil production. Doubts remain, however, regarding the impact of irrigation management on relevant aspects of olive tree production. Thus, and although the positive effect of irrigation in olive production is widely accepted (Gucci et al., 2007; Gomez-Rico et al., 2007), the literature reports apparently contradictory results. Ramos and Santos (2010), for instance, found no improvement in the production of 'Cordovil' trees under an SDI strategy supplying 60% of ET_c as compared to fully irrigated trees (100% ET_c). Furthermore, Grattan et al. (2006) studied the effect of increasing irrigation in virgin olive oil (VOO) production of 'Arbequina' trees. They found that the increase in the volume of irrigation reduced the physical extraction of the oil. This negative effect has also been reported by Gómez-del-Campo (2013), García et al. (2013), and Fernández et al. (2013), among others. Concerning quality, most authors agree on a negligible effect of irrigation levels on main physical and chemical parameters considered when assessing VOO quality, such as free acidity, peroxide value, and absorbance in the ultraviolet region (Patumi et al., 1999; Faci et al., 2002; Gómez-Rico et al., 2009; Stefanoudaki et al., 2009; García et al., 2013). Although it is true that the level of irrigation has an impact on several positive attributes of the olive oil (fruity, bitter, pungent, spicy, fresh, leaf, grass, etc.), it does not determine the development of negative attributes. Thus irrigation does not significantly affect the evaluation of the panel test, focused to detect negative attributes of VOO (García et al., 2017).

The lack of effect of irrigation on VOO fatty acid composition has been widely reported (Patumi et al., 1999; Tovar et al., 2002; Servili et al., 2007; Gómez-Rico et al., 2009; Caruso et al., 2014). Many authors, however, have found that the increase in the volume of irrigation decreased the oleic acid/linoleic acid ratio (García et al., 2013). In other cases, results showed a lack of consistency among seasons (Berenguer et al., 2006; Fernandes-Silva et al., 2013; García et al., 2017) and locations (Issaoui et al., 2013). Fatty acid composition depends on a set of enzymatic activities involved, first, in the synthesis and desaturation of fatty acids and, later, in the synthesis and accumulation of triglycerides (Hernández et al., 2009). The involved enzymes act during practically all the processes of fruit maturation (García and Mancha, 1992). In consequence, fatty acid composition is likely conditioned by the genetic characteristics of the cultivar and also by the growing characteristics in the orchard (García et al., 2017). For this reason we could assume that the cultivars with little capacity of desaturation, i.e., with higher oleic acid content, such as 'Frantoio', 'Coratina', and 'Picual', are little affected by differences in irrigation, while other cultivars such as 'Arbequina', which admit a wider range for the concentration of oleic and linoleic acids, may show a greater response to irrigation (Rondanini et al., 2014). On the other hand, it seems that high temperature during fruit ripening favors greater desaturation in the fatty acids of VOO due to an increase in the irrigation supplies (Caruso et al., 2014).

Most authors reported a reduction in VOO phenolic compounds with increasing water supply (Patumi et al., 1999; Romero et al., 2002; Berenguer et al., 2006; Caruso et al., 2014). This applies mainly to *O*-diphenols and secoiridoid derivatives of oleuropein and ligstroside. Other phenolic compounds remained unchanged (Servili et al., 2007; Gómez-Rico et al., 2009; Caruso et al., 2014). However, the opposite effect has also been reported (Baccouri et al., 2009; Dabbou et al., 2010). On the other hand, some authors have not found any effect of irrigation on

VOO phenolic content (Tognetti et al., 2007; Zeleke and Ayton, 2014; Dabbou et al., 2015). This lack of consistency in the reported findings could be due to the occurrence of rainfall events before harvesting, which have a marked impact on VOO phenolic compounds. Thus a rain event prior to harvesting increases the fruit water content, which reduces both the physical extraction of the oil and the dissolution of phenolic compounds in the oil. Part of the phenolic compounds can be lost, in fact, in the waste water (García et al., 2017). Caruso et al. (2014) worked with both irrigated and severely stressed 'Frantoio' trees. After a humid summer they found similar concentrations of O-diphenols in VOO from both treatments, while in the two, less rainy, previous seasons the VOO from stressed trees contained 14% more of this group of phenolic compounds. Moriana et al. (2007) found in 'Cornicabra' trees grown in a season with a rainy spring and a dry autumn that oils from rain-fed trees suffered an unexpectedly sharp reduction in VOO total phenol content. Likely, the biosynthesis of the phenolic compounds was affected in this case. Machado et al. (2013) observed that fruits from rain-fed 'Cobrançosa' trees exhibited higher activity of the enzyme L-phenylalanine ammonia-lyase, which belongs to the phenol biosynthetic pathway and produces VOO with higher phenolic content than those from irrigated trees. In summary, the seasonal course of water supply during fruit development may affect the phenolic content of VOO. Thus spring rains could inhibit the biosynthesis of phenolic compounds in the fruit and autumn rains may prevent access of these compounds to VOO.

Other factors significantly influencing oil quality are fruit load and harvesting date. The impact of water stress on yield and quality increases with crop load (Martín-Vertedor et al., 2011). Dag et al. (2011) observed contrasting effects of harvesting date on the amount and quality of oil produced by 'Barnea' and 'Souri', 'Barnea' required late harvest and advanced fruit maturity to get the maximum amount of good quality oil. In 'Souri', however, late harvest and advanced maturation caused an increase in fatty acids combined with a decline in polyphenol content resulting in loss of oil quality. Concerning maturation and harvesting date, there is a general agreement on irrigation delaying maturation (Motilva et al., 2000; Gómez-Rico et al., 2006; Fernández-Silva et al., 2013; Zeleke and Ayton, 2014). Strikingly, however, a decrease in polyphenol content, an increase in the fatty acid unsaturation, and a decline of photosynthetic pigments, three characteristics of VOO favored by irrigation, are usually found in oils extracted from fruits with a greater degree of maturation (Gutierrez et al., 1999; Gómez-Rico et al., 2008; Bengana et al., 2013). A possible explanation is that the increase in irrigation delays the change of color of the fruit skin, but accelerates other processes occurring in the fruit during ripening. To test this hypothesis, a study was carried out by our research group in the hedgerow 'Arbequina' orchard at the Sanabria farm, from 2010 to 2013, when the trees were under the FI, 60RDI, and 30RDI irrigation treatments. The effect of irrigation was tested on four variables related to the progress of fruit maturation: epidermis color and the contents of polyunsaturated fatty acids, total phenols, and chlorophylls. In each season, fruits were harvested at three different dates (Fig. 9.9). For all harvesting dates and seasons, fruits from the less-irrigated trees showed the lowest values of ripening index (Fig. 9.9A). Olives from the most-irrigated trees showed the highest

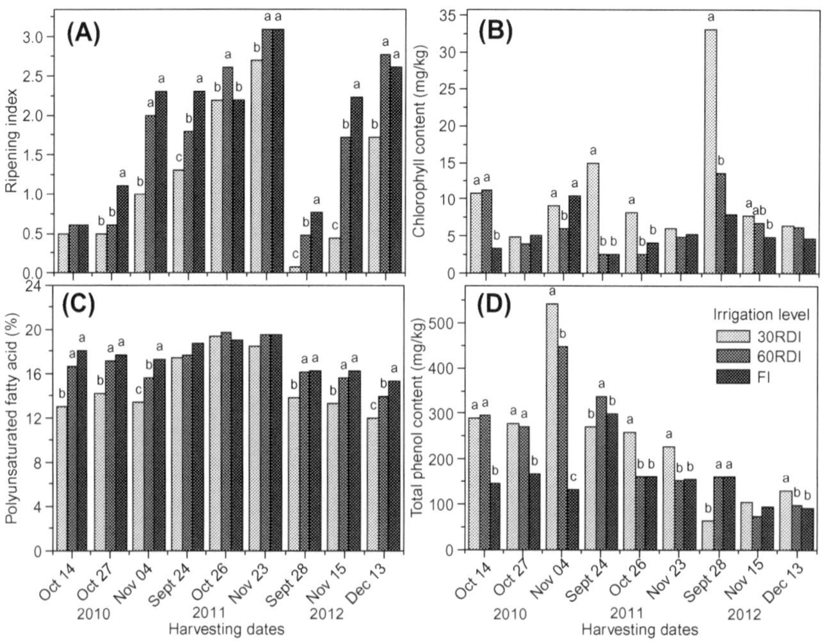

Figure 9.9 Changes of ripening index (A), chlorophyll content (B), polyunsaturated fatty acids content (C), and total phenol content (D) of 'Arbequina' fruits and virgin oils from the hedgerow olive orchard at the Sanabria farm. Values were recorded from 2010 to 2013 in trees of the FI, 60RDI, and 30RDI treatments. See Section 6.1 for details on the treatments. Each season, harvesting was made on three different dates. For each harvesting date, letters show differences among irrigation treatments at $P \leq .05$ (Tukey test). No letter means no significant effect of the irrigation treatments on the considered variable.

ripening index in seven of the nine harvesting dates studied. These results contradict those normally reported on the ripening index decreasing as a result of irrigation increase. It is known that factors such as the crop load or the temperature of the fruit during its development can inhibit the synthesis of anthocyanins in the olive fruit (Roca and Mínguez-Mosquera, 2001). It is possible that irrigation is another factor that can intervene in this process, causing an abnormal development of the change in the epidermis color of the olive.

Returning to our experiment, the chlorophyll contents of the oils extracted from the less-watered fruits, harvested early in the season, were significantly higher than those from the most-irrigated trees in the initial harvesting dates of the three seasons (Fig. 9.9B). Differences among treatments were lower when the harvesting date was delayed. Perhaps the effectiveness of the chloroplast break during the grinding of the fruit is a determining factor for recovering this compound in VOO or for losing it in the olive pomace. The reason why different levels of irrigation affect these organelles remains unknown. On the other hand, the water content of the fruit may hinder the dissolution of chlorophyll in the oil, given its amphiphilic nature. Finally, chlorophyll can be destroyed during the extraction process by the action of enzymes such as chlorophylases, peroxidases, or lipoxygenases, whose activities could be modified by the level of irrigation (Gallardo-Guerrero et al., 2002). In any case, and despite seasonal differences in our results, it is clear that the application of a low level of watering induced a greater recovery of chlorophylls in VOO, especially in the earliest harvesting, when the oils had the highest quality, i.e., the best sensory attributes and the largest number of nutritionally important minority compounds (Morrone et al., 2017). For the first and third seasons, the content of polyunsaturated fatty acid of the oils changed inversely to the amount of water received by the trees. In the second season, no significant differences were found (Fig. 9.9C). As desaturation increased with the progress of fruit ripening, the higher level of irrigation often induced a significant increase in the maturity level (Köseoglu et al., 2016).

In nearly all the harvesting dates, the content of phenols in the oils decreased with irrigation (Fig. 9.9D). The oils had significantly higher value for this parameter only in the first harvest of the third season. However, this harvest was carried out immediately after an 81.4 mm rainfall event. This agrees with the greater amount of these antioxidants in VOO being linked to low water supplies. It may happen, however, that after a whole season under DI, heavy rain prior to harvesting lowers the antioxidant content of the oil. García et al. (2017) also reported a reduction in the phenol content in VOO from fruits harvested after abundant autumn rain. In summary, and apart from the effect of occasional precipitation, results obtained in the experiment related to Fig. 9.9 indicate that the increase of irrigation induced both an early change of skin pigmentation and changes in the chemical composition of the oils, which is typical of an advanced state of fruit maturation.

7. CONCLUSIONS

In the current agricultural context, crop production per unit of cropped land and of water consumed by irrigation must increase to meet the demand of food, fiber, and fuel by the growing world population. The scientific community has responded to this challenge with advances in irrigation systems, the irrigation strategy, and the method for irrigation scheduling, the three main columns on which precision irrigation is based. Precision agriculture originated in the 1960s with the development of GIS and progressed in the late 1980s and early 1990s with the appearance of GPS. Precision irrigation requires both enough knowledge of crop response to water stress for the development of effective irrigation strategies and reliable, inexpensive, and user friendly methods for irrigation scheduling. For olive, these two conditions are close to being accomplished to a reasonable level. The current level of information on the response of the species to water stress is enough for the development of effective DI strategies, allowing for substantial water savings with reduced impacts on crop performance. On the other hand, technological advances for the continuous and automatic assessment of tree water stress, and progress on data transmission and storage, as well as on data processing, have placed precision irrigation among the most suitable options for irrigation management in olive orchards. Still, the cost of sensors and related systems for data recording and transmission has to decrease, and applications for the delivery of processed information to the users must improve. It seems that for the first time in history, electronics and data transmission companies are joining forces to overcome those two limitations, such that precision irrigation can soon be widely adopted in olive orchards. It is time for the user to become familiar with new advances in the three mentioned main components of precision irrigation, and to choose the best options for orchard conditions.

ACKNOWLEDGMENTS

Results from our group reported in this work are from experiments carried out in the frame of research projects AGL2009-11310/AGR (Spanish Ministry of Science and Innovation), AGL2012-34544 and AGL2015-71585-R (Spanish Ministry of Economy and Competitiveness), AGR-6456-2010 (*Junta de Andalucía*), and the Operational Programme of the European Social Fund 2007–13 "Andalusía is moving with Europe." Other funds were from the FEDER program. Thanks to the owners of *Internacional Olivarera, S.A. (Interoliva)* for allowing us to make the experiments at the Sanabria farm.

REFERENCES

Abd-El-Rahman, A.A., El-Sharkawi, H.M., 1974. Response of olive and almonds orchards to partial irrigation under dry-farming practices in semi-arid regions: II. Plant-soil water relations in olive during the growing season. Plant Soil 41, 13–31.

Agam, N., Segal, E., Peeters, A., Levi, A., Dag, A., Yermiyahu, U., Ben-Gal, A., 2014. Spatial distribution of water status in irrigated olive orchards by thermal imaging. Precision Agriculture 15, 346–359.

Agüero Alcaras, L.M., Rousseaux, M.C., Searles, P.S., 2016. Responses of several soil and plant indicators to post-harvest regulated deficit irrigation in olive trees and their potential for irrigation scheduling. Agricultural Water Management 171, 10–20.

Allen, R.G., Pereira, L.S., Raes, D., Smith, M., 1998. Crop evapotranspiration. Guidelines for computing crop water requirements. In: Irrigation and Drainage Paper 56. FAO, Rome, Italy.

Allen, R.G., Pereira, L.S., Smith, M., Raes, D., Wright, J., 2005. FAO-56 dual crop coefficient method for estimating evaporation from soil and application extensions. Journal of Irrigation and Drainage Engineering 131, 2–13.

Baccouri, O., Guerfel, M., Bonoli-Carbognin, M., Cerretani, L., Bendini, A., Zarrouk, M., Daoud, D., 2009. Influence of irrigation and site of cultivation on qualitative and sensory characteristics of a Tunisian minor olive variety (cv. Marsaline). La Rivista Italiana Delle Sostanze Grasse 86, 173–180.

Ben-Gal, A., Kool, D., Agam, N., van Halsema, G.E., Yermiyahu, U., Yafe, A., Presnov, E., Erel, R., Majdop, A., Zipori, I., Segal, E., Rüger, S., Zimmermann, U., Cohen, Y., Alchanatis, V., Dag, A., 2010. Whole-tree water balance and indicators for short-term drought stress in non-bearing 'Barnea' olives. Agricultural Water Management 98, 124–133.

Bengana, M., Bakhouchec, A., Lozano-Sánchez, J., Amir, Y., Youyou, A., Segura-Carretero, A., Fernández-Gutiérrez, A., 2013. Influence of olive ripeness on chemical properties and phenolic composition of Chemlal extra-virgin olive oil. Food Research International 54, 1868–1875.

Berenguer, M.J., Vossen, P.M., Grattan, S.R., Connell, J.H., Polito, V.S., 2006. Tree irrigation levels for optimum chemical and sensory properties of olive oil. Horticultural Science 41, 427–432.

Burgess, S.S.O., Adams, M.A., Turner, N.C., Beverly, C.R., Ong, C.K., Khan, A.A.H., Bleby, T.M., 2001. An improved heat pulse method to measure low and reverse rates of sap flow in woody plants. Tree Physiology 21, 589–598.

Cammalleri, C., Ciracolo, G., Minacapilli, M., Rallo, G., 2013. Evapotranspiration from an olive orchard using remote sensing-based dual crop coefficient approach. Water Resources Management 27, 4877–4895.

Carr, M.K.V., 2013. The water relations and irrigation requirements of olive (*Olea europaea* L.): a review. Experimental Agriculture 49, 597–639.

Caruso, G., Gucci, R., Urbani, S., Esposto, S., Taticchi, A., Di Maio, I., Selvaggini, R., Servili, M., 2014. Effect of different irrigation volumes during fruit development on quality of virgin olive oil of cv. Frantoio. Agricultural Water Management 134, 94–103.

Caruso, G., Rapoport, H.F., Gucci, R., 2013. Long-term evaluation of yield components of young olive trees during the onset of fruit production under different irrigation regimes. Irrigation Science 31, 37–47.

Chalmers, D.J., Mitchell, P.D., van Heek, L., 1981. Control of peach tree growth and productivity by regulated water supply, tree density and summer pruning. Journal of the American Society for Horticultural Science 106, 307–312.

Connor, D.J., Centeno, A., Gómez-del-Campo, M., 2009. Yield determination in olive hedgerow orchards. II. Analysis of radiation and fruiting profiles. Crop and Pasture Science 60, 443–452.

Connor, D.J., Fereres, E., 2005. The physiology of adaptation and yield expression in olive. Horticultural Reviews 34, 155–229.

Connor, D.J., Gómez-del-Campo, M., Rousseaux, M.C., Searles, P.S., 2014. Structure, management and productivity of hedgerow olive orchards: a review. Scientia Horticulturae 169, 71–93.

Correa-Tedesco, G., Rousseaux, M.C., Searles, P.S., 2010. Plant growth and yield responses in olive (*Olea europaea*) to different irrigation levels in an arid region of Argentina. Agricultural Water Management 97, 1829–1837.

Cuevas, M.V., Martín-Palomo, M.J., Diaz-Espejo, A., Torres-Ruiz, J.M., Rodriguez-Dominguez, C.M., Perez-Martin, A., Pino-Mejías, R., Fernández, J.E., 2013. Assessing water stress in a hedgerow olive orchard from sap flow and trunk diameter measurements. Irrigation Science 31, 729–746.

Cuevas, M.V., Torres-Ruiz, J.M., Álvarez, R., Jiménez, M.D., Cuerva, J., Fernández, J.E., 2010. Assessment of trunk diameter variation derived indices as water stress indicators in mature olive trees. Agricultural Water Management 97, 1293–1302.

Dabbou, S., Chehab, H., Faten, B., Dabbou, S., Esposto, S., Selvaggini, R., Taticchi, A., Servili, M., Montedoro, G.F., Hammami, M., 2010. Effect of three irrigation regimes on Arbequina olive oil produced under Tunisian growing conditions. Agricultural Water Management 97, 763–768.

Dabbou, S., Dabbou, S., Chehab, H., Taticchi, A., Servili, M., Hammami, M., 2015. Content of fatty acids and phenolics in coratina olive oil from Tunisia: influence of irrigation and ripening. Chemistry and Biodiversity 12, 397–406.

Dag, A., Kerem, Z., Yogev, N., Zipori, I., Lavee, S., Ben-David, E., 2011. Influence of time of harvest and maturity index on olive oil yield and quality. Scientia Horticulturae 127, 358–366.

d'Andria, R., Lavini, A., Morelli, G., Patumi, M., Terenziani, S., Calandrelli, D., Fragnito, F., 2004. Effect of water regimes on five picking and double aptitude olive cultivars (*Olea europaea* L.). Journal of Horticultural Science and Biotechnology 78, 15–23.

Dry, P.R., Loveys, B.R., Botting, D., During, H., 1996. Effects of partial rootzone drying on grapevine vigour, yield, composition of fruit and use of water. In: Stockley, C.S., Sas, A.N., Johnstone, R.S., Lee, T.H. (Eds.), Proceedings of the 9th Australian Wine Industry Technical Conference, pp. 126–131. Winetitles, Adelaide, Australia.

Egea, G., Fernández, J.E., Alcon, F., 2017a. Financial assessment of adopting irrigation technology for plant-based regulated deficit irrigation scheduling in super high-density olive orchards. Agricultural Water Management 187, 47–56.

Egea, G., Padilla-Díaz, C.M., Martínez, J., Fernández, J.E., Pérez-Ruiz, M., 2017b. Assessing a crop water stress index derived from aerial thermal imaging and infrared thermometry in super-high density olive orchards. Agricultural Water Management 187, 210–221.

Ehrenberger, W., Rüger, S., Rodríguez-Domínguez, C., Díaz-Espejo, A., Fernández, J.E., Moreno, J., Zimmermann, D., Sukhorukov, V.L., Zimmermann, U., 2012. Leaf patch clamp pressure probe measurements on olive leaves in a nearly turgorless state. Plant Biology 14, 666–674.

Faci, J.M., Berenguer, M.J., Espada, J.L., Gracia, S., 2002. Effect of variable water irrigation supply in olive (*Olea europaea* L.) cv. Arbequina in Aragon (Spain). II. Extra Virgin oil quality parameters. Acta Horticulturae 586, 649–652.

FAO, October 12–13, 2009. High Level Expert Forum – How to Feed the World in 2050. Rome, 4 p.

FAO, 2011. The State of the World's Land and Water Resources for Food and Agriculture (SOLAW) – Managing Systems at Risk. Food and Agriculture Organization of the United Nations, Rome and Earthscan, London.

Fereres, E., Soriano, M.A., 2007. Deficit irrigation for reducing agricultural water use. Journal of Experimental Botany 58 (2), 147–159.

Fernandes-Silva, A.A., Gouveia, J.B., Vasconcelos, P., Ferreira, T.C., Villalobos, F.J., 2013. Effect of different irrigation regimes on the quality attributes of monovarietal virgin olive oil from cv. 'Cobrançosa'. Grasas y Aceites 64, 41–49.

Fernández, J.E., 2014a. Understanding olive adaptation to abiotic stresses as a tool to increase crop performance. Environmental and Experimental Botany 103, 158–179.

Fernández, J.E., 2014b. Plant-based sensing to monitor water stress: applicability to commercial orchards. Agricultural Water Management 142, 99–109.

Fernández, J.E., 2017. Plant-based methods for irrigation scheduling of woody crops. Horticulturae 3, 35. https://doi.org/10.3390/horticulturae3020035.

Fernández, J.E., Cuevas, M.V., 2010. Irrigation scheduling from stem diameter variations: a review. Agricultural and Forest Meteorology 150, 135–151.

Fernández, J.E., Diaz-Espejo, A., d'Andria, R., Sebastiani, L., Tognetti, R., 2008a. Potential and limitations of improving olive orchard design and management through modelling. Plant Biosystems 142 (1), 130–137.

Fernández, J.E., Díaz-Espejo, A., Infante, J.M., Durán, P., Palomo, M.J., Chamorro, V., Girón, I.F., Villagarcía, L., 2006a. Water relations and gas exchange in olive trees under regulated deficit irrigation and partial rootzone drying. Plant and Soil 284, 273–291.

Fernández, J.E., Durán, P.J., Palomo, M.J., Díaz-Espejo, A., Chamorro, V., Girón, I.F., 2006b. Calibration of sap flow measurements by the compensation heat-pulse method in olive, plum and orange trees: relations with xylem anatomy. Tree Physiology 26, 719–728.

Fernández, J.E., Green, S.R., Caspari, H.W., Diaz-Espejo, A., Cuevas, M.V., 2008b. The use of sap flow measurements for scheduling irrigation in olive, apple and Asian pear trees and in grapevines. Plant and Soil 305, 91–104.

Fernández, J.E., Moreno, F., 1999. Water use by the olive tree. Journal of Crop Production 2, 101–162.

Fernández, J.E., Moreno, F., Martín-Palomo, M.J., Cuevas, M.V., Torres-Ruiz, J.M., Moriana, A., 2011a. Combining sap flow and trunk diameter measurements to assess water needs in mature olive orchards. Environmental and Experimental Botany 72, 330–338.

Fernández, J.E., Palomo, M.J., Díaz-Espejo, A., Clothier, B.E., Green, S.R., Girón, I.F., Moreno, F., 2001. Heat-pulse measurements of sap flow in olives for automating irrigation: tests, root flow and diagnostics of water stress. Agricultural Water Management 51, 99–123.

Fernández, J.E., Palomo, M.J., Diaz-Espejo, A., Girón, I.F., 2003. Influence of partial soil wetting on water relation parameters of the olive tree. Agronomie 23, 545–552.

Fernández, J.E., Perez-Martin, A., Torres-Ruiz, J.M., Cuevas, M.V., Rodriguez-Dominguez, C.M., Elsayed-Farag, S., Morales-Sillero, A., García, J.M., Hernandez-Santana, V., Diaz-Espejo, A., 2013. A regulated deficit irrigation strategy for hedgerow olive orchards with high plant density. Plant and Soil 372, 279–295.

Fernández, J.E., Rodriguez-Dominguez, C.M., Perez-Martin, A., Zimmermann, U., Rüger, S., Martín-Palomo, M.J., Torres-Ruiz, J.M., Cuevas, M.V., Sann, C., Ehrenberger, W., Diaz-Espejo, A., 2011c. Online-monitoring of tree water stress in a hedgerow olive orchard using the leaf patch clamp pressure probe. Agricultural Water Management 100, 25–35.

Fernández, J.E., Romero, R., Montaño, J.C., Diaz-Espejo, A., Muriel, J.L., Cuevas, M.V., Moreno, F., Girón, I.F., Palomo, M.J., 2008c. Design and testing of an automatic irrigation controller for fruit tree orchards, based on sap flow measurements. Australian Journal of Agricultural Research 59, 589–598.

Fernández, J.E., Torres-Ruiz, J.M., Diaz-Espejo, A., Montero, A., Álvarez, R., Jiménez, M.D., Cuerva, J., Cuevas, M.V., 2011b. Use of maximum trunk diameter measurements to detect water stress in mature 'Arbequina' olive trees under deficit irrigation. Agricultural Water Management 98, 1813–1821.

Ferreira, M.I., Conceição, N., David, T.S., Nadezhdina, N., 2013. Role of lignotuber versus roots in the water supply of rainfed olive. Acta Horticulturae 991, 181–188.

Flexas, J., Scoffoni, C., Gago, J., Sack, L., 2013. Leaf mesophyll conductance and leaf-hydraulic conductance: an introduction to their measurement and coordination. Journal of Experimental Botany 64, 3965–3981.

Gago, J., Douthe, C., Coopman, R.E., Gallego, P.P., Ribas-Carbo, M., Flexas, J., Escalona, J., Medrano, H., 2015. UAVs challenge to assess water stress for sustainable agriculture. Agricultural Water Management 153, 9–19.

Gallardo-Guerrero, L., Roca, M., Mínguez-Mosquera, M.I., 2002. Distribution of chlorophylls and carotenoids in ripening olives and between oil and alperujo when processed using a two-phase extraction system. Journal of the American Oil Chemists' Society 79 (1), 105–109.

García, J.M., Cuevas, M.V., Fernández, J.E., 2013. Production and oil quality in 'Arbequina' olive (*Olea europaea, L.*) trees under two deficit irrigation strategies. Irrigation Science 31 (3), 359–370.

García, J.M., Mancha, M., 1992. Evolution of the lipids biosynthesis during the maturation of the olive varieties 'Picual' and 'Gordal'. Grasas y Aceites 43, 277–280.

García, J.M., Morales-Sillero, A., Pérez-Rubio, A.G., Diaz-Espejo, A., Montero, A., Fernández, J.E., 2017. Virgin olive oil quality of hedgerow 'Arbequina'olive trees under deficit irrigation. Journal of the Science of Food and Agriculture 97, 1018–1026.

Girón, I.F., Corell, M., Martín-Palomo, M.J., Galindo, A., Torrecillas, A., Moreno, F., Moriana, A., 2015. Feasibility of trunk diameter fluctuations in the scheduling of regulated deficit irrigation for table olive trees without reference trees. Agricultural Water Management 161, 114–126.

Girón, I.F., Corell, M., Martín-Palomo, M.J., Galindo, A., Torrecillas, A., Moreno, F., Moriana, A., 2016. Limitations and usefulness of maximum daily shrinkage (MDS) and trunk growth rate (TGR) indicators in the irrigation scheduling of table olive trees. Agricultural Water Management 164, 38–45.

Gispert, J.R., Ramírez de Cartagena, F., Villar, J.M., Girona, J., 2013. Wet soil volume and strategy effects on drip-irrigated olive trees (cv. 'Arbequina'). Irrigation Science 31 (3), 479–489.

Goldhamer, D.A., 1999. Regulated deficit irrigation for California canning olives. Acta Horticulturae 474, 369–372.

Goldhamer, D.A., Fereres, E., 2001. Irrigation scheduling protocols using continuously recorded trunk diameter measurements. Irrigation Science 20, 115–125.

Goldhamer, D.A., Ferguson, L., Dunai, J., 1994. Irrigation requirements of olive trees and responses to sustained deficit irrigation. Acta Horticulturae 356, 172–175.

Gómez-del-Campo, M., 2013. Summer deficit–irrigation strategies in a hedgerow olive orchard cv. 'Arbequina': effect on fruit characteristics and yield. Irrigation Science 31 (3), 259–269.

Gómez-Rico, A., Fregapane, G., Salvador, M.D., 2008. Effect of cultivar and ripening on minor components in Spanish olive fruits and their corresponding virgin olive oils. Food Research International 41, 433–440.

Gómez-Rico, A., Salvador, M.D., Fregapane, G., 2009. Virgin olive oil and olive fruit minor constituents as affected by irrigation management based on SWP and TDF as compared to ETc in medium-density young olive orchards (*Olea europaea* L. cv. Cornicabra and Morisca). Food Research International 42, 1067–1076.

Gómez-Rico, A., Salvador, M.D., La Greca, M., Fregapane, G., 2006. Phenolic and volatile compounds of extra virgin olive oil (*Olea europaea* L. cv. Cornicabra) with regard to fruit ripening and irrigation management. Journal of the Agricultural and Food Chemistry 54, 7130–7136.

Gomez-Rico, A., Salvador, M.D., Moriana, A., Pérez, D., Olmedilla, N., Ribas, F., Fregapane, G., 2007. Influence of different irrigation strategies in a traditional Cornicabra cv. olive orchard on virgin olive oil composition and quality. Food Chemistry 100, 568–578.

Gonzalez-Dugo, V., Goldhamer, D.A., Zarco-Tejada, P.J., Fereres, E., 2015. Improving the precision of irrigation in a pistachio farm using an unmanned airborne thermal system. Irrigation Science 33, 43–52.

Gonzalez-Dugo, V., Zarco-Tejada, P., Nicolás, E., Nortes, P.A., Alarcón, J.J., Intrigliolo, D.S., Fereres, E., 2013. Using high resolution UAV thermal imagery to assess the variability in the water status of five fruit tree species within a commercial orchard. Precision Agriculture 14, 660–678.

Grattan, S.R., Berenguer, M.J., Connell, J.H., Polito, V.S., Vossen, P.M., 2006. Olive oil production as influenced by different quantities of applied water. Agricultural Water Management 85, 133–140.

Green, S.R., Clothier, B.E., Jardine, B.J., 2003. Theory and practical application of heat pulse to measure sap flow. Agronomy Journal 95, 1371–1379.

Green, S., Clothier, B., Perie, E., 2009. A Re-analysis of heat pulse theory across a wide range of sap flows. Acta Horticulturae 846, 95–114.

Green, S.R., Romero, R., 2012. Can we improve heat-pulse to measure low and reverse flows? Acta Horticulturae 951, 19–22.

Gucci, R., Goldhamer, D.A., Fereres, E., 2012. Olive. In: Steduto, P., Hsiao, T.C., Fereres, E., Raes, D. (Eds.), Crop Yield Response to Water. Irrigation & Drainage Paper No. 66. FAO, Rome, Italy, pp. 298–313.

Gucci, R., Lodolini, E., Rapoport, H.F., 2007. Productivity of olive trees with different water status and crop load. Journal of Horticultural Science and Biotechnology 82, 648–656.

Gutierrez, F., Jimenez, B., Ruiz, A., Albi, M.A., 1999. Effect of olive ripeness on the oxidative stability of virgin olive oil extracted from the varieties Picual and Hojiblanca and the different compounds involved. Journal of Agricultural and Food Chemistry 47, 121–127.

Hernandez, M.L., Padilla, M.N., Mancha, M., Martinez-Rivas, J.M., 2009. Expression analysis identifies FAD2-2 as the olive oleate desaturase gene mainly responsible for the linoleic acid content in virgin olive oil. Journal of Agricultural and Food Chemistry 57, 6199–6206.

Hernandez-Santana, V., Fernández, J.E., Cuevas, M.V., Perez-Martin, A., Diaz-Espejo, A., 2017. Photosynthetic limitations by water deficit: effect on fruit and olive oil yield, leaf area and trunk diameter and its potential use to control vegetative growth of super-high density olive orchards. Agricultural Water Management 184, 9–18.

Hernandez-Santana, V., Fernández, J.E., Rodriguez-Dominguez, C.M., Romero, R., Diaz-Espejo, A., 2016. The dynamics of radial sap flux density reflects changes in stomatal conductance in response to soil and air water deficit. Agricultural and Forest Meteorology 218–219, 92–101.

Hernandez-Santana, V., Hernandez-Hernandez, A., Vadeboncoeur, M., Asbjornsen, H., 2015. Scaling from single-point sap velocity measurements to stand transpiration in a multispecies deciduous forest: uncertainty sources, stand structure effect, and future scenarios. Canadian Journal of Forest Research 45, 1489–1497.

Iniesta, F., Testi, L., Orgaz, F., Villalobos, F.J., 2009. The effects of regulated and continuous deficit irrigation on the water use, growth and yield of olive trees. European Journal of Agronomy 30, 258–265.

IOC (International Olive Council), 2016. Market Newsletter, No 104, May 2016.

Issaoui, M., Flamini, G., Chehab, H., Cioni, P.L., Braham, M., Hammami, M., 2013. Irrigation regimes and growing area effects on 'Chétoui' olive oil quality. Journal of Food Biochemistry 37, 185–192.

Jones, H.G., 2004. Irrigation scheduling: advantages and pitfalls of plant-based methods. Journal of Experimental Botany 55, 2427–2436.

Jones, H.G., 2007. Monitoring plant and soil water status: established and novel methods revisited and their relevance to studies of drought tolerance. Journal of Experimental Botany 58, 119–130.

Köseoğlu, O., Sevim, D., Kadiroğlu, P., 2016. Quality characteristics and antioxidant properties of Turkish monovarietal olive oils regarding stages of olive ripening. Food Chemistry 212, 628–634.

Lavee, S., 1996. Biology and physiology of the olive tree. In: Lavee, S., Barranco, D., Bongi, G., Jardak, T., Loussert, R., Martin, G.C., Trigui, International Olive Council (Eds.), A World Olive Encyclopedia, pp. 61–110.

Lavee, S., Hanoch, E., Wodner, M., Abramowich, H., 2007. The effect of predetermined deficit irrigation on the performance of cv. Muhasan olives (*Olea europaea* L.) in the eastern coastal plain of Israel. Scientia Horticulturae 112, 156–163.

Lavee, S., Nashef, M., Wodner, M., Harshemesh, H., 1990. The effect of complementary irrigation added to old olive trees (*Olea europaea* L.) cv. Souri on fruit characteristics, yield and oil production. Advances in Horticultural Science 4, 135–138.

Lavee, S., Wodner, M., 1991. Factors affecting the nature of oil accumulation in fruit of olive (*Olea europaea* L.) cultivars. Journal of Horticultural Science 66 (5), 583–591.

León, L., de la Rosa, R., Rallo, L., Guerrero, N., Barranco, D., 2007. Influence of spacing on the initial production of hedgerow 'Arbequina' olive orchards. Spanish Journal of Agricultural Research 5 (4), 554–558.

López-Bernal, Á., García-Tejera, O., Vega, V.A., Hidalgo, J.C., Testi, L., Orgaz, F., Villalobos, F.J., 2015. Using sap flow measurements to estimate net assimilation in olive trees under different irrigation regimes. Irrigation Science 33, 357–366.

Machado, M., Felizardo, C., Fernandes-Silva, A.A., Nunes, F.M., Barros, A., 2013. Polyphenolic compounds, antioxidant activity and l-phenylalanine ammonia-lyase activity during ripening of olive cv. "Cobrançosa" under different irrigation regimes. Food Research International 51 (1), 412–421.

Marino, G., Pernice, F., Marra, F.P., Caruso, T., 2016. Validation of an online system for the continuous monitoring of tree water status for sustainable irrigation managements in olive (*Olea europaea* L.). Agricultural Water Management 177, 298–307.

Martín-Vertedor, A.I., Pérez-Rodríguez, J.M., Prieto-Losada, E., Fereres-Castiel, E., 2011. Interactive responses to water deficits and crop load in olive (*Olea europaea* L., cv. Morisca). II: water use, fruit and oil yield. Agricultural Water Management 98, 950–958.

Martínez-Gimeno, M.A., Castiella, M., Rüger, S., Intrigliolo, D.S., Ballester, C., 2017. Evaluating the usefulness of continuous leaf turgor pressure measurements for the assessment of Persimmon tree water status. Irrigation Science 35 (2), 159–167.

Molden, D.A., 2003. Water-productivity framework for understanding and action. In: Kijne, J.W., Barker, R., Molden, D. (Eds.), Water Productivity in Agriculture: Limits and Opportunities for Improvement. International Water Management Institute, Colombo, Sri Lanka, pp. 1–18.

Moreno, F., Fernández, J.E., Clothier, B., Green, S., 1996. Transpiration and root water uptake by olives. Plant and Soil 184, 85–96.

Moriana, A., Corell, M., Girón, I.F., Conejero, W., Morales, D., Torrecillas, A., Moreno, F., 2013. Regulated deficit irrigation based on threshold values of trunk diameter fluctuation indicators in table olive trees. Scientia Horticulturae 164, 102–111.

Moriana, A., Girón, I.F., Martín-Palomo, M.J., Conejero, W., Ortuño, M.F., Torrecillas, A., Moreno, F., 2010. New approach for olive trees irrigation scheduling using trunk diameter sensors. Agricultural Water Management 97, 1822–1828.

Moriana, A., Orgaz, F., Pastor, M., Fereres, E., 2003. Yield responses of a mature olive orchard to water deficits. Journal of the American Society for Horticultural Science 128 (3), 425–431.

Moriana, A., Perez-Lopez, D., Gomez-Rico, A., Salvador, M., Olmedilla, N., Ribas, F., Fregapane, G., 2007. Irrigation scheduling for traditional, low-density olive orchards: water relations and influence on oil characteristics. Agricultural Water Management 8, 171–179.

Morrone, L., Pupillo, S., Neri, L., Bertazza, G., Magli, M., Rotondi, A., 2017. Influence of olive ripening degree and crusher typology on chemical and sensory characteristics of Correggiolo virgin olive oil. Journal of the Science of Food and Agriculture 97 (5), 1443–1450.

Motilva, M.J., Tovar, M.J., Romero, M.P., Alegre, S., Girona, J., 2000. Influence of regulated deficit irrigation strategies applied to olive trees (Arbequina cultivar) on oil yield and oil composition during the fruit ripening period. Journal of the Science of Food and Agriculture 80, 2037–2043.

Nadezhdina, N., Čermák, J., Nadezhdin, V., 1998. Heat field deformation method for sap flow measurements. In: Čermak, J., Nadezhdina, N. (Eds.), Measuring Sap Flow in Intact Plants. IUFRO Publications, Publishing House of Mendel University, Brno, Czech Republic, pp. 72–92.

Nadezhdina, N., David, T.S., David, J.S., Nadezhdin, V., Cermak, J., Gebauer, R., Ferreira, M.I., Conceição, N., Dohnal, M., Tesar, M., Gartner, K., Ceulemans, R., 2012. Root function: in situ studies through sap flow research. In: Mancuso, S. (Ed.), Measuring Roots: An Updated Approach. Springer, pp. 267–290 (Chapter 14).

Orgaz, F., Fereres, E., Testi, L., 2017. Riego. In: Barranco, et al. (Eds.), El Cultivo del Olivo. Ediciones Mundi-Prensa, Madrid, pp. 461–490 (Chapter 12).

Orgaz, F., Testi, L., Villalobos, F.J., Fereres, E., 2006. Water requirements of olive orchards-II: determination of crop coefficients for irrigation scheduling. Irrigation Science 24, 77–84.

Ortuño, M.F., Conejero, W., Moreno, F., Moriana, A., Intrigliolo, D.S., Biel, C., Mellisho, C.D., Pérez-Pastor, A., Domingo, R., Ruiz-Sánchez, M.C., Casadesus, J., Bonany, J., Torrecillas, A., 2010. Could trunk diameter sensors be used in woody crops for irrigation scheduling? A review of current knowledge and future perspectives. Agricultural Water Management 97, 1–11.

Paço, T.A., Pôças, I., Cunha, M., Silvestre, J.C., Santos, F.L., Paredes, P., Pereira, L.S., 2014. Evapotranspiration and crop coefficients for a super intensive olive orchard. An application of SIMDualKc and METRIC models using ground and satellite observations. Journal of Hydrology 519, 2067–2080.

Padilla-Díaz, C.M., Rodriguez-Dominguez, C.M., Hernandez-Santana, V., Perez-Martin, A., Fernández, J.E., 2016. Scheduling regulated deficit irrigation in a hedgerow olive orchard from leaf turgor pressure related measurements. Agricultural Water Management 164, 28–37.

Pastor, M., Vega, V., Hidalgo, J.C., 2005. Ensayos en plantaciones de olivar superintensivas e intensivas. Vida Rural 218, 30–40.

Patumi, M., d'Andria, R., Fontanazza, G., Morelli, G., Giorio, P., Sorrentino, G., 1999. Yield and oil quality of intensively trained trees of three cultivars of olive (*Olea europaea* L.) under different irrigation regimes. Journal of Horticultural Science and Biotechnology 74, 729–737.

Pereira, L.S., Cordery, I., Iacovides, I., 2012. Improved indicators of water use performance and productivity for sustainable water conservation and saving. Agricultural Water Management 108, 39–51.

Proietti, P., Nasini, L., Ilarioni, L., 2012. Photosynthetic behavior of Spanish Arbequina and Italian Maurino olive (*Olea europaea* L.) cultivars under super-intensive grove conditions. Photosynthetica 50 (2), 239–246.

Rallo, L., Caruso, T., Díez, C.M., Campisi, G., 2016. Olive growing in a time of change: from empiricism to genomics. In: Rugini, et al. (Eds.), The Olive Tree Genome, Compendium of Plant Genomes. Springer, pp. 55–64 (Chapter 4).

Ramos, A.F., Santos, F.L., 2009. Water use, transpiration, and crop coefficients for olives (cv. Cordovil), grown in orchards in Southern Portugal. Biosystems Engineering 102, 321–333.

Ramos, A.F., Santos, F.L., 2010. Yield and olive oil characteristics of a low–density orchard (cv. Cordovil) subjected to different irrigation regimes. Agricultural Water Management 97, 363–373.

Rapoport, H.F., Fabbri, A., Sebastiani, L., 2016. Olive biology. In: Rugini, et al. (Eds.), The Olive Tree Genome, Compendium of Plant Genomes. Springer, pp. 13–25 (Chapter 2).

Rius, X., Lacarte, J.M., 2010. La revolución del olivar. El cultivo en seto, 340 p.

Roca, M., Mínguez-Mosquera, M.I., 2001. Unusual carotenogenesis in fruits with pronounced anthocyanic ripening (*Olea europea* var.'Arbequina'). Journal of Agricultural and Food Chemistry 49, 4414–4419.

Rodriguez-Dominguez, C.M., Ehrenberger, W., Sann, C., Rüger, S., Sukhorukov, V., Martín-Palomo, M.J., Diaz-Espejo, A., Cuevas, M.V., Torres-Ruiz, J.M., Perez-Martin, A., Zimmermann, U., Fernández, J.E., 2012. Concomitant measurements of stem sap flow and leaf turgor pressure in olive trees using the leaf patch clamp pressure probe. Agricultural Water Management 114, 50–58.

Romero, R., Muriel, J.L., Garcia, I., 2009. Automatic irrigation system in almonds and walnuts trees based on sap flow measurements. Acta Horticulturae 846, 135–142.

Romero, R., Muriel, J.L., García, I., Green, S.R., Clothier, B.E., 2012b. Improving heat-pulse methods to extend the measurement range including reverse flows. Acta Horticulturae 951, 31–38.

Romero, R., Muriel, J.L., García, I., Muñoz de la Peña, D., 2012a. Research on automatic irrigation control: state of the art and recent results. Agricultural Water Management 114, 59–66.

Romero, M.P., Tovar, M.J., Girona, J., Motilva, M.J., 2002. Changes in the HPLC phenolic profile of virgin olive oil from young trees (*Olea europaea* L. cv. Arbequina) grown under different deficit irrigation strategies. Journal of Agricultural and Food Chemistry 50, 5349–5354.

Rondanini, D.P., Castro, D.N., Searles, P.S., Rousseaux, M.C., 2014. Contrasting patterns of fatty acid composition and oil accumulation during fruit growth in several olive varieties and locations in a non-Mediterranean region. European Journal of Agronomy 52, 237–246.

Searles, P.S., Agüero-Alcarás, M., Rousseaux, M.C., 2011. El consumo del agua por el cultivo de olivo (*Olea europaea* L.) en el noroeste de Argentina: una comparación con la Cuenca Mediterránea. Ecología Austral 21, 15–28.

Sebastiani, L., Gucci, R., Kerem, Z., Fernández, J.E., 2016. Physiological responses to abiotic stresses. In: Rugini, et al. (Eds.), The Olive Tree Genome, Compendium of Plant Genomes. Springer, pp. 99–122 (Chapter 7).

Servili, M., Esposto, S., Lodolini, E., Selvaggini, R., Taticchi, A., Urbani, S., Montedoro, G., Serravalle, M., Gucci, R., 2007. Irrigation effects on quality, phenolic composition, and selected volatiles of virgin olive oils cv. Leccino. Journal of Agricultural and Food Chemistry 55, 6609–6618.

Smith, R.J., Baillie, J.N., 2009. Defining precision irrigation: a new approach to irrigation management. In: Proceedings of the Irrigation and Drainage Conference, Irrigation Australia Ltd., Swan Hill, Vic, Australia, 18–24 October 2009, pp. 1–6.

Stefanoudaki, E., Williams, M., Chartzoulakis, K., Harwood, J., 2009. Effect of irrigation on quality attributes of olive oil. Journal of Agricultural and Food Chemistry 57, 7048–7055.

Swanson, R.H., Whitfield, D.W.A., 1981. A numerical analysis of heat pulse velocity and practice. Journal of Experimental Botany 32, 221–239.

Testi, L., Villalobos, L., 2009. New approach for measuring low sap velocities in trees. Agriculture and Forest Meteorology 149 (3–4), 730–734.

Testi, L., Villalobos, F.J., Orgaz, F., Fereres, E., 2006. Water requirements of olive orchards: I. Simulation of daily evapotranspiration for scenario analysis. Irrigation Science 24, 69–76.

Tognetti, R., d'Andria, R., Sacchi, R., Lavini, A., Morelli, G., Alvino, A., 2007. Deficit irrigation affects seasonal changes in leaf physiology and oil quality of *Olea europaea* (cultivars Frantoio and Leccino). Annals of Applied Biology 150, 169–186.

Tognetti, R., Giovannelli, A., Lavini, A., Morelli, G., Fragnito, F., d'Andria, R., 2009. Assessing environmental controls over conductances through the soil–plant–atmosphere continuum in an experimental olive tree plantation of southern Italy. Agricultural and Forest Meteorology 149, 1229–1243.

Tovar, M.J., Romero, M.P., Alegre, S., Girona, J., Motilva, M.J., 2002. Composition and organoleptic characteristics of oil from Arbequina olive (*Olea europaea* L) trees under deficit irrigation. Journal of the Science of Food and Agriculture 82, 1755–1763.

Xiloyannis, C., Dichio, B., Nuzzo, V., Celano, G., 1996. 'L'olivo: pianta esempio per la sua capacità di resistenza in condizioni di estrema siccità. In: Seminari di Olivicoltura, Spoleto, 7 and 28 June, pp. 79–111.

Zeleke, K.T., Ayton, J., 2014. Fruit and oil quality of olive (*Olea europaea* L.) under different irrigation regimes and harvest times in south eastern Australia. Journal of Food, Agriculture and Environment 12 (2), 458–464.

Zimmermann, D., Reuss, R., Westhoff, M., Geßner, P., Bauer, W., Bamberg, E., Bentrup, F.-W., Zimmermann, U., 2008. A novel, non-invasive, online-monitoring, versatile and easy plant-based probe for measuring leaf water status. Journal of Experimental Botany 59, 3157–3167.

CHAPTER 10

Water and Heat Fluxes in Mediterranean Vineyards: Indicators and Relevance for Management

Joaquim M. Costa[1,2], Ricardo Egipto[1,3], José Silvestre[1,3], Carlos M. Lopes[1], Maria M. Chaves[2]

[1]Universidade de Lisboa, Lisboa, Portugal; [2]Universidade NOVA de Lisboa, Oeiras, Portugal; [3]INIAV, I.P., Pólo de Dois Portos, Dois Portos, Portugal

Contents

1. INTRODUCTION: DROUGHT AND HEAT STRESS IN MEDITERRANEAN VINEYARDS

Alterations in temperature and rainfall patterns due to climate changes are forcing farmers to rethink crop management strategies (EIP-AGRI Focus Group, 2016). Future performance and sustainability of Mediterranean

Water Scarcity and Sustainable Agriculture in Semiarid Environment
ISBN 978-0-12-813164-0
https://doi.org/10.1016/B978-0-12-813164-0.00010-7

viticulture can be limited by water scarcity, more frequent heat waves, and increased average air temperature (T_{air}). T_{air} is predicted to increase from 0.4 to 2.6°C in certain regions of the globe such as Australia and the Mediterranean Basin countries (Sadras and Moran, 2012; CIHEAM, 2015; Lotze-Campen and Schellnhuber, 2009; Fraga et al., 2016; IPCC, 2014). This tendency for warmer and drier conditions could be negative for grape yield and berry composition and for vine's longevity (Greer and Weedon, 2013; Rebetzke et al., 2016; Zarrouk et al., 2016).

In regions with a typical Mediterranean climate, T_{air} commonly exceeds supraoptimal values for plant performance (e.g., $T_{air} > 40°C$) and often co-occurs with dry air and soil conditions. Supraoptimal T_{air}, combined with high vapor pressure deficit (VPD), high radiation fluxes, and soil water deficits, reduces photosynthesis and promotes leaf senescence (Ferrini et al., 1995; Greer and Weedon, 2013; Zsófi et al., 2009), and ultimately causes defoliation, mainly at the shoot basal part (cluster zone), favoring berry sunburn (Chaves et al., 2016; Webb et al., 2009; Gibberd and Proffitt, 2012).

Soil temperature (T_s) is one of the most important transient soil physical properties influencing plant physiology and crop growth. However, the impact of climate change on T_s is less described in the literature (Bai et al., 2014). Climate change may encompass smaller, but still significant, changes in T_s (van Gestel et al., 2011). Higher T_s can be negative for soil structure and fertility (e.g., nitrogen soil dynamics) and soil microbiological activity (Wang et al., 2006; Frazer, 2009; Shvaleva et al., 2014).

High T_{air} during spring is known to stimulate carbohydrate mobilization in roots and speed up shoot and reproductive development in grapevine (Rogiers et al., 2013). The negative impact of supraoptimal T_{air} becomes more evident under soil water deficit conditions and when dealing with row crops such as grapevines, which are more exposed to radiative soil fluxes, increasing their vulnerability to heat stress (Ham et al., 1991; Nobel, 2005a; Taghvaeian et al., 2013).

Irrigation helps to mitigate the effects of drought and high T_{air} (e.g., heat waves) (Costa et al., 2007; Retallack, 2012) in grapevines, by promoting evaporative cooling that decreases canopy temperature (T_c) and promotes photosynthetic activity as compared to nonirrigated vines. Irrigation also contributes to decrease T_s, which can be relevant under heat wave conditions. As a result, irrigation has expanded in warm and dry wine regions worldwide (e.g., California, Australia, Mediterranean Basin countries) to mitigate climate limitations and guarantee grape yield and quality (Chaves et al., 2010; Retallack, 2012; Webb et al., 2012; Costa et al. 2016).

In the Mediterranean regions, climate is also characterized by large seasonal and interannual variation, which puts pressure on the highly variable soil water reserves and therefore on irrigation requirements. Improved monitoring, enabling the prediction of diurnal/seasonal water and heat fluxes in vineyards, can support more efficient forecasts of irrigation water needs and water management in Mediterranean viticulture (see Table 10.1).

The major aim of this chapter is to review the present understanding on water and heat fluxes in irrigated vineyards and to explain how crop and soil management strategies can influence these fluxes. Additionally, we aim to address the relevance of thermography to monitor temperature patterns in the soil and canopy, to detect heat and water stress to improve the forecast of irrigation needs, and finally to save irrigation water. We will focus on the impact of soil and irrigation management on water and heat fluxes, although the role of canopy management is acknowledged also to be important for the regulation of vine's water relations and temperature (for a review of this topic, see Keller, 2010; Retallack, 2012).

2. HEAT AND WATER FLUXES IN THE VINEYARD

Plant water relations and their capacity to regulate temperature below identified thresholds are crucial for optimal performance and survival under stress conditions. Water and heat fluxes are closely related. Indeed, transpiration is the major process influencing maintenance of plant temperature and water content (Nobel, 2005a; Jones and Rotenberg, 2011; Schymanski et al., 2013). Moreover, the literature emphasizes the relevance of a compromise between optimal temperature regulation and high water use efficiency (WUE) as targets for crop selection/breeding for dry and warm climates (Nicholas and Durham, 2012; Chaves et al., 2016). At the agronomic level, both crop and soil management strategies, e.g., irrigation, floor, and canopy management, strongly influence leaf temperature (T_l), T_c, T_s, and the diurnal and seasonal fluxes of water and heat.

2.1 Energy Balance and Heat Fluxes

Improved comprehension of sensible and latent heat fluxes, during the day and season, can help to optimize crop irrigation management under drought conditions (Majozi et al., 2017). The temperature of an individual

Table 10.1 Nonexhaustive list of parameters and strategies (environmental, plant, and anthropic) influencing water and heat fluxes in an irrigated vineyard and its practical relevance for vineyard and water management

Level	Parameters and strategies	How they influence fluxes	How they can be used
Air	Solar radiation, T_{air}, T_s, rainfall, relative humidity, wind speed and patterns, VPD	Leaf gas exchange rates, water loss via transpiration, soil evaporation, plant water uptake	Climate models, irrigation management models Predict risks related to drought and heat stress, forecast water needs
Soil	Physical traits (profile depth, type/texture, soil organic matter content, porosity, soil hydraulic conductivity, water-holding capacity, slope, color, thermal properties, available soil water, soil salinity)	Plant water uptake (water flow from soil to plant), rainfall water storage/runoff/deep percolation, irrigation needs	Soil water erosion models Irrigation management models Predict risks related to drought and heat stress, forecast irrigation needs
Plant	Genotype (scion and rootstock) traits, e.g., vigor, root density/depth, stem and leaf hydraulics, leaf gas exchange, leaf morphology (shape, thickness, size)	Water uptake and transport capacity, sensible and latent heat fluxes (evaporative cooling via transpiration)	Plant growth models Forecasting yield and quality Predict risks related to drought and heat stress, forecast irrigation needs
Agronomic strategies	Cultivar, rootstock, planting density, row orientation, training system, canopy and soil management, irrigation scheduling	Volume of the exploited soil by roots, canopy leaf area, intercepted sun radiation by canopy and soil, available soil water, T_s, soil evaporation	Mitigate climate extremes (heat waves), save water Support decision-making on crop and soil management and irrigation
Education	Knowhow and skills of employees on crop water relations, irrigation water management, crop and soil monitoring strategies and technologies	Perception on water and sustainability issues, capacity to innovate and adopt new strategies for precise crop/soil monitoring and water metrics	Support decision-making on crop and soil management and on irrigation Promote water metrics and water savings

Based on Heilman et al., 1994; Costa et al., 2007, 2012, 2016; Chaves et al., 2010; CWSA 2012; Nicholas and Durham 2012; Kool et al., 2014, 2016; Radke et al., 2015; Ashenfelter and Storchmann 2016; Bota et al., 2001, 2016; Simonneau et al., 2017.

leaf depends on the balance of the energy fluxes between the leaf and its surroundings (Jones, 1992; Nobel, 2005a). Solar radiation (shortwave) normally dominates energy inputs in sunlit leaves, whereas energy loss occurs via emission of longwave radiation, loss of sensible heat, and latent heat (evaporative cooling) (Jones, 1992, 1999; Lambers et al., 2008). When considering a crop, the energy balance is expressed by the following equation (Monteith and Unsworth, 1990): $Rn + G + H + LE + P + M = 0$. Rn is the net radiation and represents the energy gain by the surface from radiation, G is the soil heat flux (energy loss via heat conduction through soil), H is the sensible heat flux (energy loss by the surface by heat transfer to the atmosphere), LE is the latent heat flux (evaporative energy loss from the surface), P is the energy fixed in plants by photosynthesis, and M is the energy related to miscellaneous processes (e.g., canopy heat storage and respiration). Due to the small value of P and M, these components are often ignored making the equation simpler: $Rn + G = H + LE$, in which the term $Rn + G$ is known as the available energy and $H + LE$ as turbulent fluxes (latent and sensible heat fluxes) (Wilson et al., 2002). Sensible heat fluxes are proportional to the difference between T_{air} and the effective surface temperature and vary inversely with the aerodynamic resistance to heat transfer (Monteith and Unsworth, 1990). Latent heat flux is proportional to the difference in vapor pressure of air (e_a) and the saturated vapor pressure of the surface ($e \times [Ts]$), and it is inversely related to aerodynamic resistance to water transfer (Monteith and Unsworth, 1990).

Turbulent fluxes are important in meteorology, hydrology, and ecology. They influence climate and environmental parameters that alter mass and energy exchanges occurring between ground and atmosphere (Falge et al., 2005).

At the crop level, we must account for the effect of meteorological variables (T_{air}, VPD, Rn, rain, wind speed) on soil moisture availability and on the dynamics of energy partitioning at the vegetation level. The Bowen ratio is a useful approach to assess the partition of available energy between H and LE indicating whether most of the available energy is used to evaporate water or to warm up the environment. The Bowen ratio depends on the water potential gradient and surface resistance (Baldocchi et al., 2001), which in turn depends on rainfall dynamics, irrigation, and canopy development.

Partitioning of latent heat flux between the soil and the plant's canopy is particularly relevant in row crops, such as grapevine, to gain a mechanistic understanding of the effects of vegetation on water use (Dubbert et al., 2014; Kool et al., 2016). In parallel, soil can influence a plant's energy balance by reflecting shortwave irradiation that becomes a source of longwave radiation for the plant's canopy (Nobel, 2005b).

2.2 Water Fluxes and Water Balances

2.2.1 Leaf and Canopy Water Balances

Leaf water balance is regulated by stomatal behavior and plant hydraulics. At the canopy scale, water balance relates to the evaporative demand (VPD_{air}), canopy architecture influencing the exposed leaf area, and the available soil water (Table 10.1). Contrary to dense and closed canopies in row crops such as vineyards, soil surface has a determinant impact on overall energy balance and on a crop's water status. In addition, vine's morphological structure varies during the season, ranging from an empty canopy during dormancy (fall—winter) to foliage of variable compactness according to the training system, from flowering to harvest (spring—summer). Hence soil evaporation and canopy transpiration must be considered separately when predicting a vineyard's water needs. Moreover, sunlit and shadow areas of grapevine canopy and soil show different thermal patterns during the day and season (Colaizzi et al., 2016; García-Tejero et al., 2016; Kool et al., 2016).

Some models allow independent calculation of canopy transpiration and soil evaporation. This is the case in the Shuttleworth and Wallace model (Shuttleworth and Wallace, 1985), a two-source model often used in row crops, e.g., vineyards (Sene, 1994; Ortega-Farias et al., 2007). In turn, the Clumped model (Brenner and Incoll, 1997), a three-source model, separately estimates canopy transpiration and soil evaporation under the canopy and between rows (Poblete-Echeverria and Ortega-Farias, 2009). Nevertheless, the FAO-56 method (Allen et al., 1998) is still the most practical method to estimate canopy transpiration and canopy water fluxes, using the dual crop coefficient approach (crop coefficient [K_c] = basal crop coefficient [K_{cb}] + soil water evaporation coefficient [K_e]). Under water stress conditions, accurate K_{cb} and stress functions are needed. K_{cb} has been correlated with grapevine traits such as leaf area index and ground cover (Williams and Ayars, 2005; Netzer et al., 2009; Ferreira et al., 2012).

2.2.2 Soil Water Balance

Rainfall and irrigation are the two major sources of water for field-grown crops. However, part of these water inputs is lost via bare soil evaporation, by runoff when a marked slope exists, and by deep percolation.

The remaining soil water is potentially available to roots for crop growth and transpiration. Basic equations to describe soil water balance of an irrigated crop are described by Allen et al. (1998). Soil water balance is used to estimate total available soil water (TAW) and, indirectly, a crop's water use. The amount of available soil water varies during the season and can be evaluated at different time steps by accounting it at the beginning of the season (TAW_0), adding eventual water inputs from rainfall and/or irrigation, and finally subtracting the outputs related to evapotranspiration (ET) (plant transpiration + evaporation at soil surface) (Pellegrino et al., 2006; van Leeuwen et al., 2009). However, it is difficult to estimate TAW and TAW_0 due to a vineyard's spatial variability (e.g., rooting depth, soil physical characteristics) (van Leeuwen et al., 2009). For irrigation purposes, it is crucial to assess the degree of water stress by estimating the water stress coefficient (K_s), a dimensionless transpiration reduction factor dependent on the available soil water. K_s is calculated by considering the root zone depletion (D_r), i.e., water shortage relative to field capacity, and the TAW, the readily available soil water in the root zone ($RAW = (1 - p) \times TAW$), p being the fraction of TAW that a crop can extract from the root zone without suffering water stress. For $D_r < RAW$, K_s is 1. For $D_r > RAW$, $K_s = (TAW - D_r)/((1 - p) \times TAW)$. The actual ET and irrigation depths can be estimated by $ET_o \times (K_{cb} \times K_s + K_e)$ (Allen et al., 1998).

2.3 Diurnal and Seasonal Heat Exchanges in the Vineyard

Diurnal and seasonal variation of environmental variables influences plant condition, generating adaptive strategies for temperature regulation. Surface fluxes vary over the course of a day. As more sun radiation is received, the soil warms up and more energy is dissipated as sensible and latent heat. In addition, plant stomata will open allowing leaf gas exchange and favoring transpiration and related latent heat exchange (Lambers et al., 2008). Diurnal variations of T_l or T_c and berry temperature (T_{berry}) tend to follow diurnal changes of T_{air} (Figs. 10.1–10.3) with T_l or T_c being sometimes significantly higher (>10°C) than T_{air} due to strong sunlight irradiance and water stress, or cooler (often by 2–3°C) due to the leaf evaporative cooling effect (Davidson, 1969; Lambers et al., 2008). The magnitude of the difference between T_l and T_{air} depends on factors influencing aerodynamic resistances to heat and vapor, such as leaf morphoanatomical traits (e.g., size) and wind speed and on the degree of water stress experienced by plants (Jones, 1992). At the canopy level, it may also depend on canopy geometry and degree of compactness, which varies with the variety. In grapevine, and under Mediterranean summer conditions, T_l and T_c are often several degrees above optimal T_l for photosynthesis during most of the day (11:00–17:00 h), and especially under more stressful atmospheric conditions (high VPD and high T_{air}) combined with mild to severe water stress conditions (Fig. 10.1) (Egipto et al., 2016; García-Tejero et al., 2016). This reveals the degree of heat stress exposure of grapevine under Mediterranean conditions and shows the great resilience of grapevine species.

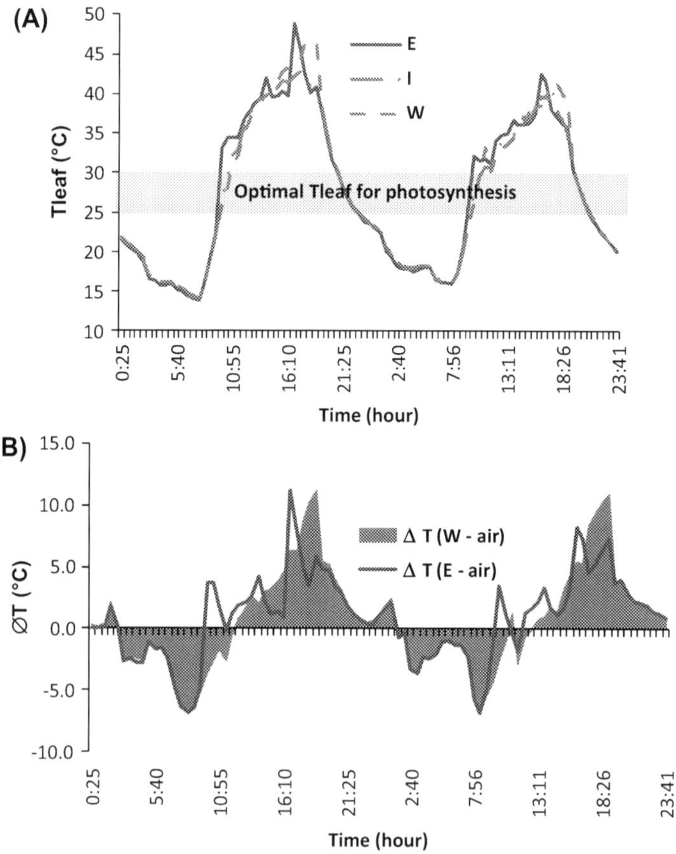

Figure 10.1 (A) Variation of leaf temperature (T_l) patterns observed in plants of the cv. Touriga Nacional, grown under deficit irrigation conditions in Alentejo (southern Portugal) in 2 consecutive days of the 2014 growing cycle. (B) Thermal difference between T_l and T_{air} (ΔT) in the same period. East (E) and west (W) canopy face exposed leaves and canopy interior leaves (I). *(Adapted from Egipto et al. (2016).)*

During the day, the absorbed thermal energy by a leaf/canopy is partly dissipated by emission of longwave radiation, preventing overheating. At night, the situation reverses; leaves gain heat from the surroundings (air and soil surface). The rate at which heat is gained/lost depends on the difference between T_{air} and T (ΔT). In turn, T_s is, in general, closely related to T_{air}. Usually, higher T_{air} will result in higher T_s, in particular at the most superficial soil layers (van Gestel et al., 2011).

As for sunlight reflectance from the soil surface, bare soil can disperse incoming energy in different ways: (1) conduct energy into the soil, (2) warm the air above it by convection, or (3) reflect energy as longwave radiation (Cierniewski and Courault, 1993; Matthias et al., 2000; Coakley, 2003).

Depending on soil type, soil humidity, and also on the degree of soil compactness, bare soils store more energy during the day period and have higher T_s than vegetative surfaces, which can convert a large part of the incoming light energy into latent heat via transpiration (Jones, 1992). Our own thermal imaging measurements done in an irrigated vineyard at Alentejo (southern Portugal) during the summer season of 2015 (from pea-size berry stage to harvest) showed that T_s was on average about 10–15°C higher than T_c. In the same trial we observed as well that soil tillage influences the diurnal variation of T_s. Maximum temperature difference between the two soil management systems (tillage vs. sown cover crops) occurred at early noon with maximum peak temperatures occurring in the tilled soil at 12:00–14:00 h (Fig. 10.2). In turn, soils with sown cover crops tend to keep higher T_s during cooler periods (e.g., predawn and before sunset) (Fig. 10.2). The seasonal variations in daily total energy fluxes are influenced by changes in sun angle, day length, weather conditions, and variation in the surface reflectance (albedo) as vine canopy leaves and floor vegetation grow and change color (either green foliage plants, dried mulch, or bare soil occur during the season and as a function of soil management strategies) (Evett et al., 2012).

Seasonal variations of water and heat fluxes also depend on soil water availability and air VPD. Under Mediterranean conditions, the second half of spring (corresponding to flowering/fruit set phenological phases) and summer periods (ripening stage) co-occur with maximum T_{air} and VPD, minimal precipitation, and soil water deficit, resulting in more severe water/heat stress experienced by vines.

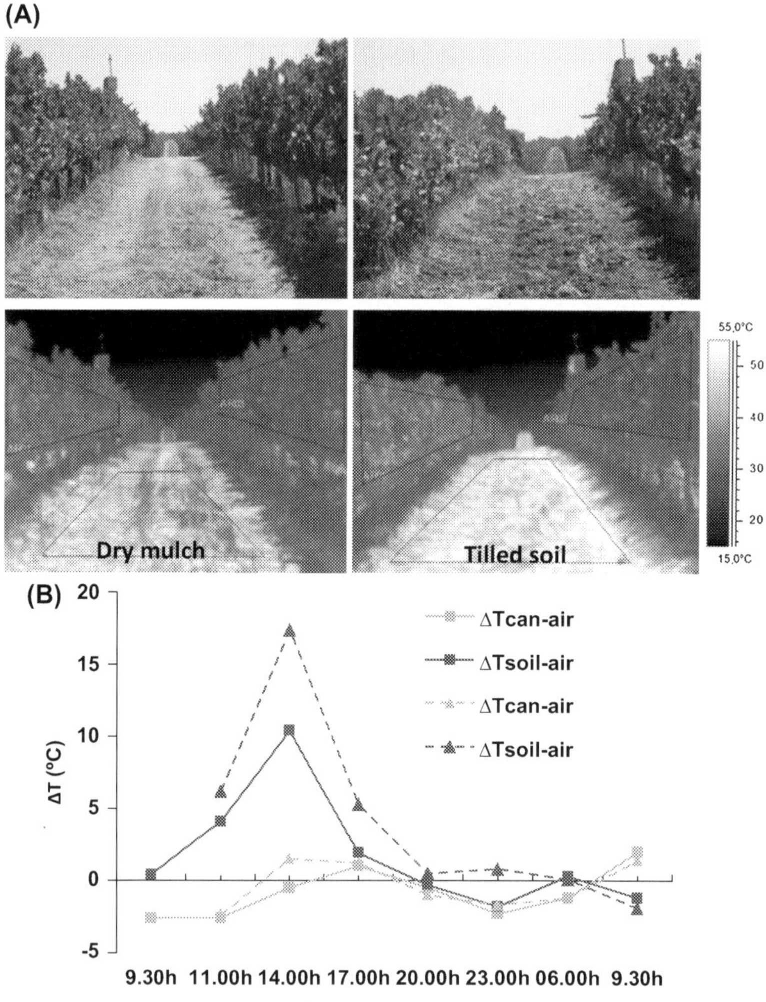

Figure 10.2 (A) Visible (top) and infrared thermal images (bottom) of a deficit-irrigated vineyard (*Vitis vinifera* cv. Touriga Nacional) with dry mulch from resident vegetation (left) and tilled bare soil (right) on the interrow. (B) ΔT patterns for the sunlit side of the canopy ([light gray in print version]) and soil (— [dark gray in print version]), under dry mulch (*continuous line*) and tilled bare soil (*dotted lines*), measured on August 19–20, 2014, in Alentejo, southern Portugal. Thermal imaging measurements were done with a Flir ThermaCAM B20, 240 × 320 pixels, $\varepsilon = 0.97$). Environmental conditions along measurements: max/min $T_{air} = 30.9°C/14.0°C$; max/min VPD = 3.47 KPa/0.07 KPa; max/min radiation = 1023 Wm^{-2}/0 Wm^{-2}.

In row crops such as grapevine, with a large soil surface exposition, part of the heat emanating from the soil surface is transported to the canopy influencing leaf and canopy temperature (Hicks, 1973; Heilman et al., 1994; Nobel, 2005a) and also T_{berry} (Pogue, 2008; Pieri, 2010a,b) due to microadvection. The available energy is also affected by floor surface shading, which is not uniform across the interrow and varies during the day and season (Horton, 1989; Pieri, 2010a; Kool et al., 2016). Hence despite the important contribution of soil-to-system fluxes, partitioning of energy balance components between soil and canopy is not easy to predict (Sene, 1994; Spano et al., 2000; Ortega-Farias et al., 2010; Kool et al., 2016). Consequently, the use of more wide-spaced rows and/or training systems with higher trunk and canopy height will result in larger diurnal changes of vines and soil exposure to the sun. In vineyards, wider row spacing favors soil evaporation and consequently its relevance and effect on vineyards' water balance (Heilman et al., 1994). In irrigated vineyards, wet soil near drippers and bare strips between vine rows must be considered separately when evaluating thermal patterns (Poblete-Echeverria and Ortega-Farias, 2009; Kool et al., 2014). However, microclimate variables, such as wind speed below the canopy in widely spaced vineyards with relatively sparse vegetation, are still not yet well understood (Poblete-Echeverria and Ortega-Farias, 2009). T_s is influenced by soil color, specific heat (influenced by soil water, soil organic matter), slope row orientation, slope, and type of soil-covering materials (vegetation vs. coarse material or mulching) (Davidoff and Selim, 1988; Pogue, 2008; Stoll et al. 2008).

As soil color influences albedo, darker soils reflect less solar radiation and warm up more than a light-colored soil. Soil color influences both quality and quantity of the reflected light (Jones, 1992). During the night, darker soils have higher irradiance in the long wave length, whereas during the day, light-colored soils reflect more radiation, heating up more of the basal canopy zone. Stoll et al. (2008) showed that, up to 0.6 m above the soil, there was no effect of soil reflection on T_{berry}, but Pogue (2008) found an increase in T_{berry} due to soil basalt covering material in the vineyards of Columbia Valley (USA) and for clusters selected to be about 0.5 m above the ground surface and shaded by leaves from direct sunlight. Under Mediterranean conditions where T_s can easily be above 50°C in summer and T_{air} can reach 40–45°C (Fig. 10.2), soil- or mulch-reflected radiation or emitted infrared can thus influence T_{berry}, as shown in Fig. 10.3.

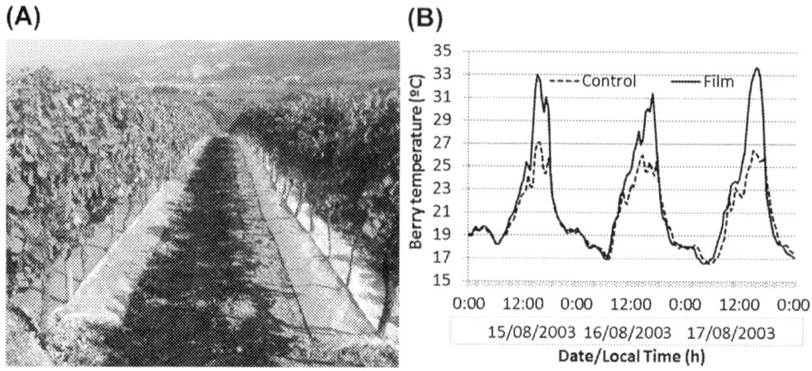

Figure 10.3 View of a field experiment testing the use of a reflective mulch (Vitexsol—aluminized polystyrene sheeting film) under nonirrigated vines (A) and the corresponding effects on T_{berry} measured with two-junction, fine-wire, copper–constantan thermocouples inserted into the berries of the red cv. Touriga Nacional, at the Lisbon wine-growing region, for control and film cover situation (B). T_{berry} data are means of three west-exposed berries during 3 consecutive ripening days of the summer of 2003. *(From Martins (2004).)*

3. ADAPTATION STRATEGIES TO MITIGATE WATER AND HEAT STRESS AND SAVE WATER IN THE VINEYARD

3.1 Genotypes

Grapevine species have large biodiversity in terms of growth morphology (canopy and roots) and physiology (Chaves et al., 2010; Keller, 2010). A better characterization of this biodiversity is required for selection and breeding purposes, as well as to improve crop management (Costa et al., 2012; Bota et al., 2016; Chaves et al., 2016; Coupel-Ledru et al., 2016; Martínez-Lüscher et al., 2016). Selection of varieties and rootstocks, which combine high WUE and high thermal plasticity for physiological activity, is required. On the other hand, selection of specific leaf morphological (e.g., leaf size, leaf specific weight, leaf shape, venation), biophysical (leaf optics, e.g., leaf color), or stay green character is needed for improved adaptation to heat stress and longer and severe drought periods (Costa et al., 2012; Medrano et al., 2015a,b; Chaves et al., 2016). At the canopy level, the morphology and architecture of the canopy (e.g., total and exposed leaf area, number and shape of vegetation plans, shoot orientation, etc.) can influence transpiration and related evaporative cooling (Lovisolo and Schubert, 2000). Root and stem hydraulics are also relevant traits for rootstocks because certain genotypes are more susceptible than others to xylem cavitation under water stress (Keller, 2010). Rootstocks have a key role in adaptation to environmental stress (Ollat et al., 2016). They confer tolerance to a large range of abiotic and biotic stresses, including drought and heat stress, salinity, or nutrient deficiency (e.g., limestone-induced iron deficiency). Rootstocks have a determinant influence on water and nutrient uptake/transport, and on water balance and WUE of the scion in grafted plants (Lovisolo et al., 2010; Corso and Bonghi, 2014; Ollat et al., 2016; Simonneau et al., 2017). A proper combination of rootstock and variety is thus crucial for adaptation to adverse climate change and water scarcity scenarios in Mediterranean regions.

3.2 Soil Management

Soil management aims to control weeds, guarantee soil conservation/improvement, improve nutrient/water management, and positively influence biodiversity in the vineyard (Guerra and Steenwerth, 2012). Common practices include soil tillage, keeping bare soil with chemical weed control, and cover cropping and mulching, used either in row or interrow, separated, combined, or alternated (Monteiro and Lopes, 2007; Stoll et al., 2008; Medrano et al., 2015a) (e.g., Figs. 10.2 and 10.3). Soil management influences water infiltration by changing surface roughness and minimizing runoff and ET by modifying the available energy, soil water, and water vapor exchange rate between soil and the atmosphere (Hatfield et al., 2001, 2011). Bare soil conditions favor runoff and reduce infiltration and water storage as compared to floor cover (Prichard, 1998). Soil tillage influences soil water fluxes by suppressing weed competition for water (Lipecki and Berbeć, 1997) and by eliminating surface crusts and hardpans that favor evaporation rates, runoff, infiltration, and water redistribution among soil layers (Schwartz et al., 2010). However, tillage promotes soil compaction and loss of structure (Merwin et al., 1994; Nicholas, 2004), which will reduce water infiltration and storage at deeper layers (Van Huyssteen and Weber, 1980; Gulick et al., 1994; Buckerfield and Webster, 2002). In tilled soils, surface microtopography and subsurface soil thermal properties are affected, and the changes in surface microrelief can influence local radiation characteristics and hence the heat balance due to changes in surface orientation and surface slope (Benjamin et al., 1990).

Interrow cover cropping is a soil management practice used in many of the world's viticultural areas and is being adopted to promote environmental sustainability and to deal with climate change in vineyards (Schultz and Stoll, 2010). The use of cover crops has, however, limitations under

Mediterranean-type climate conditions related to net loss of soil moisture due to cover crop transpiration, which can markedly contribute to a vineyard's ET (0.71−4.45 mm/day) (Yunusa et al., 1997; Lopes et al., 2004; Uliarte et al., 2013). Rainfall amount and distribution along the growing cycle will determine the potential contribution of cover crops for seasonal ET. In Mediterranean-type climates the more intense water uptake by cover crops occurs mainly during spring (Monteiro and Lopes, 2007; Celette et al., 2008; Lopes et al., 2011; Celette and Gary, 2013). In most parts of the seasons, the swards stop growing or even dry out during the summer, thus creating a dead mulch that will reduce evaporative soil water loss (Skrotch and Shribbs, 1986; Prichard, 1998). Plasticity of the grapevine root system and its high sensitivity to soil water content may influence root distribution, as well as changes in the soil water profile. Indeed, after some seasons of competition the number of roots in the interrow tends to be reduced but is counterbalanced by the development of deeper roots in the row, increasing the capacity for water extraction from deeper soil layers (Morlat and Jacquet, 2003; Celette et al., 2005, 2008). This behavior was also inferred by tracking changes in volumetric soil water content as shown in three Portuguese case studies reported in Lopes (2016); they showed that vineyard cover cropping can help to manipulate soil water availability in Mediterranean-type climates and to enhance the exploitation of soil water reserves into deeper layers.

The use of organic and inorganic mulches in vineyards conserves soil moisture by minimizing evaporation and improving soil infiltration (Pinamonti, 1998; Chan et al., 2010; Medrano et al., 2015b). In addition, organic mulches increase soil organic matter and improve soil structure and soil water-holding capacity (Proffitt, 2013). The use of mulches also has a regulatory effect on T_s (Figs. 10.2 and 10.3). Such an effect depends on the type of material used and its color (Pinamonti, 1998; Fourie and Freitag, 2010; Chan et al., 2010). In the case of dry grass mulching, our results show that it dampened T_s over time and avoided T_s extremes contrary to a bare cultivated soil situation (Fig. 10.2). Contrary to the use of mulches, a clean tilled soil surface promotes evaporation on top layers and does not conserve soil moisture effectively, especially when done after rain (Van Huyssteen and Weber, 1980). In turn, in the case of a bare soil with chemical weed control, the dry vegetation surface layer can have a mulch effect but soil cracking may occur (mainly in clay soils), which favors evaporation from deeper soil layers.

3.3 Irrigation Management

Irrigation influences soil water fluxes and also T_s. Under Mediterranean climate conditions some important vine developmental phases (shoot growth, flowering, and berry ripening) may take place under mild to severe water stress conditions due to limited soil water availability. Besides, Mediterranean soils are also poor in organic matter (e.g., in Portugal, the majority of soils have less than 1% organic matter) and very shallow (Brito et al., 2012) limiting soil water-holding capacity.

Irrigation solutions used in modern Mediterranean viticulture are mainly based on drip irrigation systems, namely, the deficit irrigation strategies (irrigation below ET losses) that can save water and improve berry quality (Chaves et al., 2010; Medrano et al. 2015a,b). However, irrigation must be adjusted to the terroir (climate, soil, plant) to avoid excessive vigor with negative effects on berry composition (Chaves et al., 2010). Precise irrigation scheduling should be based on soil and crop water status monitoring, the relevance of those tools being more important when using deficit irrigation strategies (Costa et al., 2007). The increasingly wider range of phytomonitoring tools combining point and imaging approaches, ground and aerial monitoring, image processing, and modeling can support decision-making in irrigation management in viticulture (see, for example, Jones et al., 2002; Bellvert et al., 2014; Costa et al., 2010, 2016; Fuentes et al., 2012; Jones and Grant, 2016; García-Tejero et al., 2016; Egipto et al., 2016). Alternative irrigation strategies such as subsurface drip irrigation are also being tested, making possible the use of lower quality water (Levidow et al., 2014; Costa et al., 2016). Nevertheless, subsurface irrigation remains an expensive alternative and it has risks related to dripper clogging and limitations of detecting and repairing problems (Levidow et al., 2014; Jacoby et al., 2015).

3.4 Canopy Management

The geometrical structure of the vineyard's canopy also has a determinant effect on vineyard energy and water fluxes and therefore crop productivity and berry composition. Indeed, distribution of leaf and branches (canopy architecture) influences light interception and consequently carbon assimilation and transpirational water loss (Schultz et al., 2009). Geometry may also influence canopy porosity to air movements. Examples of this role are shown by the training system components, such as row space and orientation, canopy dimensions, number and architecture of the vegetation plans, or trunk height. Furthermore, for the same training system, canopy management practices (desuckering, shoot thinning, shoot positioning, shoot trimming, leaf removal), by modifying the relative position and/or the amount of leaves, induce important changes in the radiation exposure of the remaining leaves influencing vine's water use and canopy temperature. Furthermore, the analysis will become more complex if we take into account the interactions between canopy management and genotypes. In fact, for the same training system, the large biodiversity of grapevine regarding vigor, growth habits (erect vs. prostrated), and leaf morphology will also influence plant water relations and temperature regulation.

3.5 Precision Monitoring to Support Decision-Making

Nowadays, the available precision viticulture tools and phytomonitoring techniques allow the acquisition of a big pool of soil—plant—weather data in real time, making possible a more precise estimation of vineyard water fluxes. Moreover, remote and georeferenced measurements establish robust relationships between spatial information on soil and crops (e.g., soil physical properties vs. yield or water and fertilizer contents and growth). Remote sensing approaches, e.g., ground-based and aerial thermography, made it possible to assess vineyards' soil and plant variability in space and time (Costa et al., 2013; Jones and Grant, 2016), with gains in terms of efficiency and localized application, or limited water losses. Thermography emerged as a feasible approach to assess crop water stress and plant traits (e.g., transpiration) (Jones et al., 2002; Fuentes et al., 2012; Costa et al., 2013; Grant et al., 2016). Leaf and canopy temperature correlated well with leaf gas exchange (stomatal conductance to water vapor and photosynthesis) (Egipto et al., 2016; García-Tejero et al., 2016) validating the robustness of thermal imaging data as a tool to interpret grapevine ecophysiology and vineyards' micrometeorology. Katurji and Zawar-Reza (2016) showed that thermography performed at a relatively high temporal rate of 10 s could be used to monitor vineyard-scale micrometeorology.

Several thermal indicators are used to support stress detection and irrigation management. They vary in their simplicity of calculation and robustness as a function of the thermal indicators used and the time of measurement (Costa et al., 2013; García-Tejero et al., 2016). Simpler thermal indicators such as T_c and T_c-T_{air} correlated well with vines' water potential and leaf gas exchange, which supports their practical value for irrigation decision-making and to feed models based on thermal data.

Thermography has also been used to monitor T_s and indirectly soil moisture near the surface. Higher soil moisture content resulted in lower T_s during the day and warmer T_s at night, with the diurnal range of surface T_s being correlated with near-surface soil moisture and soil evaporation (Heilman and Moore, 1980; Schmugge et al., 1980; Kerridge et al., 2013). This allows the use of T_s, assessed by thermal infrared remote sensing (3−14 μm), as a proxy of near-surface soil moisture content (Curran, 1985; Hain et al., 2009; Rahimzadeh-Bajgiran et al., 2013). Soliman et al. (2011) also demonstrated the potential of airborne thermography to monitor soil compaction problems in vineyards. The combination of ground and aerial thermography enlarged the practical usage and benefits of thermography for agricultural uses (Bellvert et al., 2014; Gago et al., 2015; Jones and Grant, 2016) and ecological studies (e.g., heat and water fluxes, evaporation patterns at a larger scale), as well as to feed hydrological models that aim to estimate crop water needs and water fluxes (Lambin et al., 1993; Ortega-Farias et al., 2010, 2016; Anderson et al., 2012; Lakshmi, 2013; González-Piqueras et al., 2014; Hoffmann et al., 2016; Kool et al., 2016; Brenner et al., 2017). The operational use of remote-sensing thermal data can be optimized by using it to feed decision support systems (DSS). DSS enable the use of models/indexes to convert acquired data into valuable information for crop and vineyard management, namely, irrigation and crop protection (Rossi et al., 2012; Pérez-Expósito et al., 2017). Empirical models able to predict leaf water potential were developed for grapevine using a machine-learning approach (Brillante et al., 2016). These authors found that T_{air} can predict drought stress in grapevine, especially at midday, and that T_{air} has a direct effect on stem water potential, independently from rainfall and soil water. T_{air} can therefore be used as a proxy for grapevine's evaporation demand. Moreover, novel algorithms to support irrigation scheduling based on thermal data (T_l and T_s) could support more precise water management and forecast irrigation needs (García-Tejero et al., 2016).

4. CONCLUSIONS AND FUTURE PROSPECTS

Scenarios of increasingly adverse climate conditions, including scarcer water resources, higher temperatures, and more extreme events, are expected for Mediterranean countries. This is a major threat and a challenge for wine production in these areas and demands solutions to mitigate the problem along with more sustainable production in terms of water use. Improved knowledge on heat fluxes in vineyards can be an important approach to optimize irrigation and to feed models to estimate crop water needs (Ortega-Farias et al., 2007, 2016; Anderson et al., 2012; Hoffmann et al., 2016; Kool et al., 2016).

In parallel, the use of better adapted varieties and rootstocks offers new opportunities to viticulture (Ollat et al., 2016) in combination with novel approaches to soil and canopy management (e.g., plant density, row orientation, canopy management adjusted to each variety, soil management, etc.). The future of Mediterranean viticulture encompasses optimized agronomic approaches to improve adaptation to climate change and improved heat and drought tolerance by vine canopies and berries. For example, in dry conditions, preservation of leaf area at the cluster zone will reduce the likelihood of berry damage by heat. However, since transpiration cooling is essential to reduce leaf damage, varieties that close their stomata in response to heat stress may be more susceptible to leaf damage in soil water deficit conditions (heat tolerance) (Costa et al., 2012; Chaves et al., 2016). Therefore a compromise between canopy WUE and temperature regulation is crucial to mitigate the increased risks posed by severe drought and heat waves and must be considered when selecting the planting material.

In the same way, improved knowledge on the role of soil traits and management strategies (tillage, no tillage, cover cropping, mulching) on heat and water fluxes in the vineyard can support decision-making in terms of irrigation and soil and canopy management to face extreme climate events such as heat waves. At the vineyard management level we may consider adaptation strategies based on canopy architecture and management. For example, manipulation of trunk height should be better studied, regarding the possibilities to mitigate the negative effects of high T_s on leaves and berries at the cluster zone.

Advances in seasonal climate forecasting, phytomonitoring, data retrieval, and modeling are creating new opportunities to respond to within-season climate-related risks in vineyard management (Egipto et al., 2016) and will help to predict water needs under the highly variable climate conditions of Mediterranean viticulture. Moreover, cheaper sensors, open access image analysis software solutions for crop monitoring, and phenotyping are being studied by researchers and industry (IPPN, 2017). The use of thermography and related thermal data can be part of the solution for future grapevine phenotyping and breeding, as well as for crop and soil monitoring. Image analysis and processing software are also advancing rapidly, making it easier to retrieve information from thermal data. In parallel, more efficient use of irrigation water in wine production involves improved water metrics. This will ultimately contribute to identify major problems concerning inefficient water flows and save irrigation water.

ACKNOWLEDGMENTS

J. Miguel Costa was sponsored with a scholarship of FCT (SFRH/BPD/93334/2013). This research has received funding from the European Community's Seventh Framework Programme (FP7/2007–2013), grant agreement no. 311775, Project INNOVINE. We also wish to thank Fundação para a Ciência e Tecnologia through the research unit UID/AGR/04129/2013 (LEAF) for the financial support and GREEN-it "Bioresources for Sustainability" (UID/Multi/04551/2013).

REFERENCES

Allen, R.G., Pereira, L.S., Raes, D., Smith, M., 1998. Crop evapotranspiration: guidelines for computing crop water requirements. In: FAO Irrigation and Drainage Paper 56. FAO, Rome.

Anderson, M.C., Allen, R.G., Morse, A., Kustas, W.P., 2012. Use of Landsat thermal imagery in monitoring evapotranspiration and managing water resources. Remote Sensing of Environment 122, 50–65.

Ashenfelter, O., Storchmann, K., 2016. Climate change and wine: a review of the economic implications. The Journal of Wine Economics 11, 105–138.

Bai, Y., Scott, T.A., Min, Q., 2014. Climate change implications of soil temperature in the Mojave Desert, USA. Frontiers in Earth Science 8, 302–308.

Baldocchi, D., Falge, E., Gu, L., Olson, R., Hollinger, D., Running, S., Anthoni, P., Bernhofer, C., Davis, K., Evans, R., Fuentes, J., Goldstein, A., Katul, G., Law, B., Lee, X., Malhi, Y., Meyers, T., Munger, W., Oechel, W., Paw, U.K.T., Pilegaard, K., Schmid, H.P., Valentini, R., Verma, S., Vesala, T., Wilson, K., Wofsy, S., 2001. FLUXNET: a new tool to study the temporal and spatial variability of ecosystem-scale carbon dioxide, water vapor, and energy flux densities. Bulletin of the American Meteorological Society 82, 2415–2434.

Bellvert, J., Zarco-Tejada, P.J., Girona, J., Fereres, E., 2014. Mapping crop water stress index in a 'Pinot-noir' vineyard: comparing ground measurements with thermal remote sensing imagery from an unmanned aerial vehicle. Precision Agriculture 15, 361–376.

Benjamin, J.G., Blaylock, A.D., Brown, H.J., Cruse, R.M., 1990. Ridge tillage effects on simulated water and heat transport. Soil and Tillage Research 18, 167–180.

Bota, J., Flexas, J., Medrano, H., 2001. Genetic variability of photosynthesis and water use in Balearic grapevine cultivars. Annals of Applied Biology 138, 353–361.

Bota, J., Tomás, M., Flexas, J., Medrano, H., Escalona, J.M., 2016. Differences among grapevine cultivars in their stomatal behavior and water use efficiency under progressive water stress. Agricultural Water Management 164, 91–99.

Brenner, A.J., Incoll, L.D., 1997. The effect of clumping and stomatal response on evaporation from sparsely vegetation shrublands. Agricultural and Forest Meteorology 84, 187–205.

Brenner, C., Thiem, C.E., Wizemann, H.-D., Bernhardt, M., Schulz, K., 2017. Estimating spatially distributed turbulent heat fluxes from high resolution thermal imagery acquired with a UAV system. International Journal of Remote Sensing 38 (8–10), 3003–3026.

Brillante, L., Mathieu, O., Lévêque, J., Bois, B., 2016. Ecophysiological modeling of grapevine water stress in Burgundy terroirs by a machine-learning approach. Frontiers in Plant Science 7, 796.

Brito, I., Goss, M.J., de Carvalho, M., Chatagnier, O., van Tuinen, D., 2012. Impact of tillage system on arbuscular mycorrhiza fungal communities in the soil under Mediterranean conditions. Soil and Tillage Research 121, 63–67.

Buckerfield, J.C., Webster, K.A., 2002. Organic matter management in vineyards: mulches for soil maintenance. The Australian and New Zealand Grapegrower and Winemaker 61, 26–33.

Celette, F., Gary, C., 2013. Dynamics of water and nitrogen stress along the grapevine cycle as affected by cover cropping. European Journal of Agronomy 45, 142–152.

Celette, F., Gaudin, R., Gary, C., 2008. Spatial and temporal changes in the water regime of a Mediterranean vineyard due to the adoption of cover cropping. European Journal of Agronomy 29, 153–162.

Celette, F., Wery, J., Chantelot, E., Cellete, J., Gary, C., 2005. Belowground interactions in a vine *Vitis vinifera* L.-tall fescue *Festuca arundinacea* Shreb. Intercropping system: water relations and growth. Plant and Soil 276, 205–217.

Chan, K.Y., Fahey, D.J., Newell, M., Barchia, I., 2010. Using composted mulch in vineyards – effects on grape yield and quality. International Journal of Fruit Science 10, 441–453.

Chaves, M.M., Costa, J.M., Zarrouk, O., Pinheiro, C., Lopes, C.M., Pereira, J.S., 2016. Controlling stomatal aperture in semi-arid regions—the dilemma of saving water or being cool? Plant Science 251, 54–64.

Chaves, M.M., Zarrouk, O., Francisco, R., Costa, J.M., Santos, T., Regalado, A.P., Rodrigues, M.L., Lopes, C.M., 2010. Grapevine under deficit irrigation: hints from physiological and molecular data. Annals of Botany 105, 661–676.

Cierniewski, J., Courault, D., 1993. Bidirectional reflectance of bare soil surfaces in the visible and near-infrared range. Remote Sensing of Environment 7, 321–339.

CIHEAM, 2015. Climate change and Mediterranean agriculture expected impacts, possible solutions and the way forward. Watch Letter 37. September 2016 ref. https://www.ciheam.org/uploads/attachments/249/05_Todorovic_WL_37.pdf.

Coakley, J.A., 2003. Encyclopedia of atmospheric sciences. In: Holton, J.R., Pyle, J., Curry, J.A. (Eds.), first ed. Academic Press, pp. 1914–1923.

Colaizzi, P.D., Evett, S.R., Agam, N., Schwartz, R.C., Kustas, W.P., 2016. Soil heat flux calculation for sunlit and shaded surfaces under row crops: 1. Model development and sensitivity analysis. Agricultural and Forest Meteorology 216, 115–128.

Corso, M., Bonghi, C., 2014. Grapevine rootstock effects on abiotic stress tolerance. Plant Science Today 1, 108–113.

Costa, J.M., Ortuño, M.F., Chaves, M.M., 2007. Deficit irrigation as a strategy to save water: physiology and potential application to horticulture. International Journal of Plant Biology 49, 1421–1434.

Costa, J.M., Grant, O., Chaves, M.M., 2010. Use of thermal imaging in viticulture: current application and future prospects. In: Delrot, S., Medrano, H., Or, E., Bavaresco, L., Grando, S. (Eds.), Methodologies and Results in Grapevine Research. Springer, pp. 135–150.

Costa, J.M., Ortuño, M.F., Lopes, C.M., Chaves, M.M., 2012. Grapevine varieties exhibiting differences in stomatal response to water deficit. Functional Plant Biology 39, 179.

Costa, J.M., Vaz, M., Escalona, J., Egipto, R., Lopes, C., Medrano, H., Chaves, M.M., 2016. Modern viticulture in southern Europe: vulnerabilities and strategies for adaptation to water scarcity. Agricultural Water Management 164, 5–18.

Costa, J.M., Grant, O.M., Chaves, M.M., 2013. Thermography to explore plant–environment interactions. Journal of Experimental Botany 64, 3937–3949.

Coupel-Ledru, A., Lebon, E., Christophe, A., Gallo, A., Gago, P., Pantin, F., Doligez, A., Simonneau, T., 2016. Reduced nighttime transpiration is a relevant breeding target for high water-use efficiency in grapevine. Proceedings of the National Academy of Sciences of the United States of America 113, 8963–8968.

Curran, P.J., 1985. Principles of remote sensing. International Journal of Remote Sensing 6 (11), 1765.

CWSA, 2012. Performance Metrics and the California Sustainable Winegrowing Program. California Sustainable Winegrowing Aliance. https://metrics.sustainablewinegrowing.org/docs/Performance%20Metrics%20and%20the%20SWP Final 3.15.12.pdf.

Davidoff, B., Selim, H.M., 1988. Correlation between spatially variable soil moisture content and soil temperature. Soil Science 145, 1–10.

Davidson, R.L., 1969. Effect of root/leaf temperature differentials on root/shoot ratios in some pasture grasses and clover. Annals of Botany 33, 561–569.

Dubbert, M., Cuntz, M., Piayda, A., Werner, C., 2014. Oxygen isotope signatures of transpired water vapor: the role of isotopic non-steady-state transpiration under natural conditions. The New Phytologist 203, 1242–1252.

Egipto, R., Barriguinha, A., Costa, J.M., Carvalho, N., Neto, M., Silvestre, J., Lopes, C.M., 2016. Phytomonitoring as a decision support tool for grapevine deficit irrigation management – a case study at a Mediterranean vineyard. In: Book of Abstracts of the INNOVINE Final International Symposium, Toulouse, South-West of France, 16 and 17 November, 2016.

EIP-AGRI Focus Group, 2016. Water & Agriculture: Adaptive Strategies at Farm Level. Final Report. 65 p.

Evett, S.R., Prueger, J.H., Tolk, J.A., 2012. Water and energy balances in the soil-plant-atmosphere continuum. In: Huang, P.M., Li, Y., Sumner, M.E. (Eds.), Handbook of Soil Sciences: Properties and Processes, second ed. CRC Press, Florida, USA, ISBN 978-1-4398-0305-9, pp. 6-1–6-44.

Falge, E., Reth, S., Bruggemann, N., Butterbach-Bahl, K., Goldberg, V., Oltchev, A., Schaaf, S., Spindler, G., Stiller, B., Queck, R., Koostner, B., Bernhofer, C., 2005. Comparison of surface energy exchange models with eddy flux data in forest and grassland ecosystems of Germany. Ecological Modelling 188, 174–216.

Ferreira, M.I., Silvestre, J., Conceição, N., Malheiro, A.C., 2012. Crop and stress coefficients in rainfed and deficit irrigation vineyards using sap flow techniques. Irrigation Science 30, 433–447.

Ferrini, F., Mattii, G.B., Nicese, F.P., 1995. Effect of temperature on key physiological responses of grapevine leaf. American Journal of Enology and Viticulture 46, 375–379.

Fourie, J.C., Freitag, K., 2010. Soil management in the Breede River Valley wine grape region, South Africa. 2. Soil temperature. South African Journal of Enology and Viticulture 31, 165–168.

Fraga, H., Garcia, C., Malheiro, A., Santos, J., 2016. Modelling climate change impacts on viticultural yield, phenology and stress conditions in Europe. Global Change Biology 22, 3774–3788.

Frazer, L., 2009. Climate change. Will warmer soil be as fertile? Environmental Health Perspectives 117, A59.

Fuentes, S., Bei, R., Pech, J., Tyerman, S., 2012. Computational water stress indices obtained from thermal image analysis of grapevine canopies. Irrigation Science 30, 523–536.

Gago, J., Douthe, C., Coopman, R.E., Gallego, P.P., Ribas-Carbo, M., Flexas, J., Escalona, J.M., Medrano, H., 2015. UAVs challenge to assess water stress for sustainable agriculture. Agricultural Water Management 153, 9–19.

García-Tejero, I.F., Costa, J.M., Egipto, R., Durán-Zuazo, V.H., Lima, R.S., Lopes, C.M., Chaves, M.M., 2016. Thermal data to monitor crop-water status in irrigated Mediterranean viticulture. Agricultural Water Management 176, 80–90.

Gibberd, M., Proffitt, T., 2012. Vineyards for 2025 – Improved Industry Returns from Soil and Water Research –Communication and Adoption of Research Outcomes. Curtin University of Technology, Australia, 169pp.

González-Piqueras, J., Villodre, J., Campos, I., Calera, A., Balbontín, C., 2014. Seguimiento de los flujos de calor sensible y calor latente en vid mediante la aplicación del balance de energía METRIC. Revista de Teledetección 43, 43–54.

Grant, O.M., Ochagavía, H., Baluja, J., Diago, M.P., Tardáguila, J., 2016. Thermal imaging to detect spatial and temporal variation in the water status of grapevine (*Vitis vinifera* L.). The Journal of Horticultural Science and Biotechnology 91, 43–54.

Greer, D.H., Weedon, M.M., 2013. The impact of high temperatures on *Vitis vinifera* cv. Semillon grapevine performance and berry ripening. Frontiers in Plant Science 4, 491.

Guerra, B., Steenwerth, K., 2012. Influence of floor management technique on grapevine growth, disease pressure, and juice and wine composition: a review. American Journal of Enology and Viticulture 63, 149–164.

Gulick, S.H., Grimes, D.W., Munk, D.S., Goldhamer, D.A., 1994. Cover crop-enhanced water infiltration of a slowly permeable fine sandy loam soil. Soil Science Society of America Journal 58, 1539–1546.

Hain, C.R., Mecikalski, J.R., Anderson, M.C., 2009. Retrieval of an available water-based soil moisture proxy from thermal infrared remote sensing. Part I: methodology and validation. Journal of Hydrometeorology 10, 665–683.

Ham, J.M., Heilman, J.L., Lascano, R.J., 1991. Soil and canopy energy balances of a row crop at partial cover. Agronomy Journal 83, 744–753.

Hatfield, J.L., Sauer, T.J., Prueger, J.H., 2001. Managing soils for greater water use efficiency: a review. Agronomy Journal 93, 271–280.

Hatfield, J.L., Boote, K.J., Kimball, B.A., Ziska, L.H., Izaurralde, R.C., Ort, D., Thomson, A.M., Wolfe, D., 2011. Climate impacts on agriculture: implications for crop production. In: Publications from USDA-ARS/UNL Faculty Paper 1350. http://digitalcommons.unl.edu/usdaarsfacpub/1350.

Heilman, J.L., Moore, D.G., 1980. Thermography for estimating near-surface soil moisture under developing crop canopies. Journal of Applied Meteorology 19, 324–328.

Heilman, J.L., McInnes, K.J., Savage, M.J., Gesch, R.W., Lascano, R.J., 1994. Soil and canopy energy balances in a west Texas vineyard. Agricultural and Forest Meteorology 71, 99–114.

Hicks, B.B., 1973. Eddy fluxes over a vineyard. Agricultural and Forest Meteorology 2, 203–215.

Hoffmann, H., Nieto, H., Jensen, R., Guzinski, R., Zarco-Tejada, P., Friborg, T., 2016. Estimating evaporation with thermal UAV data and two-source energy balance models. Hydrology and Earth System Sciences 20, 697–713.

Horton, R., 1989. Canopy shading effects on soil heat and water flow. Soil Science Society of America Journal 53, 669–679.

IPCC, 2014. In: Core Writing Team, Pachauri, R.K., Meyer, L.A. (Eds.), Climate Change 2014: Synthesis Report. Contribution of Working Groups I, II and III to the Fifth Assessment Report of the Intergovernmental Panel on Climate Change. Geneva, Switzerland, 151 pp.

IPPN, 2017. International Plant Phenonotyping Network. https://www.plant-phenotyping.org/.

Jacoby, P.W., Peters, T., Sankaran, S., Khot, L.R., 2015. Advancing water use efficiency in vineyards with sub-surface micro-irrigation techniques. In: 2015 ASABE Irrigation Symposium, pp. 486–493.

Jones, H.G., 1992. Plants and Microclimate. A Quantitative Approach to Environmental Plant Physiology, second ed. Cambridge University Press. 428 pp.

Jones, H.G., 1999. Use of thermography for quantitative studies of spatial and temporal variation of stomatal conductance over leaf surfaces. Plant, Cell and Environment 22, 1043–1055.

Jones, H.G., Stoll, M., Santos, T., Sousa, C., Chaves, M.M., Grant, O.M., 2002. Use of infra-red thermography for monitoring stomatal closure in the field: application to grapevine. Journal of Experimental Botany 53, 2249–2260.

Jones, H.G., Rotenberg, E., 2011. Energy Radiation and Temperature Regulation in Plants. eLS. John Wiley & Sons Ltd., Chichester. http://www.els.net.

Jones, H.G., Grant, O.M., 2016. Remote sensing and other imaging technologies to monitor grapevine performance. In: Gerós, H., Chaves, M., Medrano, H., Delrot, S. (Eds.), Grapevine under Environmental Stress: From Ecophysiology to Molecular Mechanisms. Wiley-Blackwell, Hoboken, New Jersey, pp. 179–201.

Katurji, M., Zawar-Reza, P., 2016. Forward-looking infrared cameras for micrometeorological applications within vineyards. Sensors 16, 1518.

Keller, M., 2010. The Science of Grapevines: Anatomy and Physiology. Academic Press Elsevier, New York.

Kerridge, B.L., Hornbuckle, J.W., Christen, E.W., Faulkner, R.D., 2013. Using soil surface temperature to assess soil evaporation in a drip irrigated vineyard. Agricultural Water Management 116, 128–141.

Kool, D., Ben-Gal, A., Agam, N., Simunek, J., Heitman, J.L., Sauer, T.J., Lazarovitch, N., 2014. Spatial and diurnal below canopy evaporation in a desert vineyard: measurements and modeling. Water Resources Research 50, 7035–7049.

Kool, D., Kustas, W.P., Ben-Gal, A., Lazarovitcha, N., Heitmand, J.L., Sauer, T.J., Agama, N., 2016. Energy and evapotranspiration partitioning in a desert vineyard. Agricultural and Forest Meteorology 218–219, 277–287.

Lakshmi, V., 2013. Remote sensing of soil moisture. ISRN Soil Science 2013, 424178. https://doi.org/10.1155/2013/424178, 33 pp.

Lambers, H., Chapin, F.S., Pons, T.L., 2008. Plant Physiological Ecology. Springer.

Lambin, E.F., Cashman, P., Moody, A., Parkhurst, B.H., Pax, M.H., 1993. Agricultural production monitoring in the Sahel using remote sensing: present possibilities and research needs. Journal of Environmental Management 38, 301–322.

Levidow, L., Lindgaard-Jørgensen, P., Nilsson, Å., Skenhall, S.A., Assimacopoulos, D., 2014. Eco-efficiency improvements in industrial water-service systems: assessing options with stakeholders. Water Science and Technology: a Journal of the International Association on Water Pollution Research 69 (10), 2113–2121.

Lipecki, J., Berbeć, S., 1997. Soil management in perennial crops: orchards and hop gardens. Soil and Tillage Research 43, 169–184.

Lopes, C.M., Monteiro, A., Rückert, F.E., Gruber, B., Steinberg, B., Schultz, H.R., 2004. Transpiration of grapevines and co-habitating cover crop and weed species in a vineyard. A "snapshot" at diurnal trends. Vitis 43, 111–117.

Lopes, C.M., Santos, T., Monteiro, A., Rodrigues, M.L., Costa, J.M., Chaves, M.M., 2011. Combining cover cropping with deficit irrigation in a Mediterranean low vigor vineyard. Scientia Horticulturae 129, 603–612.

Lopes, C.M., 2016. Cover crops competition for water in vineyards: case studies in Mediterranean terroirs. In: Jones, G., Doran, N. (Eds.), Proceedings 11th Int. Terroir Congress. Southern Oregon University, Ashland, USA, pp. 117–123.

Lotze-Campen, H., Schellnhuber, H.J., 2009. Climate impacts and adaptation options in agriculture: what we know and what we don't know. Journal für Verbraucherschutz und Lebensmittelsicherheit 4, 145–150.

Lovisolo, C., Schubert, A., 2000. Downward shoot positioning affects water transport in field-grown grapevines. Vitis 39, 49–53.

Lovisolo, C., Perrone, I., Carra, A., Ferrandino, A., Flexas, J., Medrano, H., Schubert, A., 2010. Drought-induced changes in development and function of grapevine (Vitis spp.) organs and in their hydraulic and non-hydraulic interactions at the whole-plant level: a physiological and molecular update. Functional Plant Biology 37, 98–116.

Majozi, N.P., Mannaerts, C.M., Ramoelo, A., Mathieu, R., Nickless, A., Verhoef, W., 2017. Analysing surface energy balance closure and partitioning over a semi-arid savanna FLUXNET site in Skukuza, Kruger National Park, South Africa. Hydrology and Earth System Sciences (Discussion). https://doi.org/10.5194/hess-2016-685.

Martínez-Lüscher, J., Kizildeniz, T., Vucetic, V., Dai, Z., Luedeling, E., van Leeuwen, C., Gomès, E., Pascual, I., Irigoyen, J.J., Morales, F., Delrot, S., 2016. Sensitivity of grapevine phenology to water availability, temperature and CO_2 concentration. Frontiers in Environmental Science 4, 48.

Martins, R., 2004. Influência da utilização de filmes reflectores na qualidade da uva, casta 'Touriga Nacional' (MSc thesis). Enga agronómica, ISA/UTL, p. 57.

Matthias, A.D., Fimbres, A., Sano, E.E., Post, D.F., Accioly, L., Batchily, A.K., Ferreira, L.G., 2000. Surface roughness effects on soil albedo. Soil Science Society of America Journal 64, 1035–1041.

Medrano, H., Tomás, M., Martorell, S., Escalona, J.M., Pou, A., Fuentes, S., Flexas, J., Bota, J., 2015a. Improving water use efficiency of vineyards in semi-arid regions. A review. Agronomy for Sustainable Development 35, 499–517.

Medrano, H., Tomás, M., Martorell, S., Flexas, J., Hernández, E., Rosselló, J., Pou, A., Escalona, J.-M., Bota, J., 2015b. From leaf to whole-plant water use efficiency (WUE) in complex canopies: limitations of leaf WUE as a selection target. The Crop Journal 3, 220–228.

Merwin, I.A., Stiles, W., van Es, H., 1994. Orchard groundcover management impacts on soil physical properties. Journal of the American Society for Horticultural Science 119, 216–222.

Monteiro, A., Lopes, C.M., 2007. Influence of cover crop on water use and performance of vineyard in Mediterranean Portugal. Agriculture, Ecosystems and Environment 121, 336–342.

Monteith, J.L., Unsworth, M.H., 1990. Principles of Environmental Physics. E.A. Arnold.

Morlat, R., Jacquet, A., 2003. Grapevine root system and soil characteristics in a vineyard maintained long-term with or without interrow sward. American Journal of Enology and Viticulture 54, 1–7.

Netzer, Y., Yao, C., Shenker, M., Bravdo, B.-A., Schwartz, A., 2009. Water use and the development of seasonal crop coefficients or superior seedless grapevines trained to an open-gable trellis system. Irrigation Science 27, 109–120.

Nicholas, K.A., Durham, W.H., 2012. Farm-scale adaptation and vulnerability to environmental stresses: insights from winegrowing in Northern California. Global Environmental Change 22, 483–494.

Nicholas, P., 2004. Soil, irrigation and nutrition. In: Grape Production Series, vol. 2. South Australian Research and Development Institute, Adelaide, Australia.

Nobel, P.S., 2005a. Physicochemical and Environmental Plant Physiology, second ed. Academic Press, San Diego, CA.

Nobel, P., 2005b. Physicochemical and Environmental Plant Physiology, third ed. Academic Press, ISBN 9780125200264. 540 p.

Ollat, N., Touzard, J.-M., van Leeuwen, C., 2016. Climate change impacts and adaptations: new challenges for the wine industry. Journal of Wine Economics 11 (1), 139–149.

Ortega-Farias, S., Carrasco, M., Olioso, A., Acevedo, C., Poblete, C., 2007. Latent heat flux over Cabernet Sauvignon vineyard using the Shuttleworth and Wallace model. Irrigation Science 25, 161–170.

Ortega-Farias, S., Poblete-Echeverría, C., Brisson, N., 2010. Parameterization of a two-layer model for estimating vineyard evapotranspiration using meteorological measurements. Agricultural and Forest Meteorology 150, 276–286.

Ortega-Farias, S., Ortega-Salazar, S., Poblete, T., Kilic, A., Allen, R., Poblete-Echeverría, C., Ahumada-Orellana, L., Zuñiga, M., Sepúlveda, D., 2016. Estimation of energy balance components over a drip-irrigated olive orchard using thermal and multispectral cameras placed on a helicopter-based unmanned aerial vehicle (UAV). Remote Sensing 8, 638.

Pellegrino, A., Goze, E., Lebon, E., Wery, J., 2006. A model-based diagnosis tool to evaluate the water stress experienced by grapevine in field sites. European Journal of Agronomy 25, 49–59.

Pérez-Expósito, J.P., Fernández-Caramés, T.M., Fraga-Lamas, P., Castedo, L., 2017. VineSens: an eco-smart decision-support viticulture system. Sensors (Basel) 17 (3) pii: E465.

Pieri, P., 2010a. Modeling radiative balance in a row-crop canopy: row-soil surface net radiation partition. Ecological Modelling 221, 791–801.

Pieri, P., 2010b. Modeling radiative balance in a row-crop canopy: cross-row distribution of net radiation at the soil surface and energy available to clusters in a vineyard. Ecological Modelling 221, 802–811.

Pinamonti, F., 1998. Compost mulch effects on soil fertility, nutritional status and performance of grapevine. Nutrient Cycling in Agroecossystems 51, 239–248.

Poblete-Echeverria, C., Ortega-Farias, S., 2009. Estimation of actual evapotranspiration for a drip-irrigated Merlot vineyard using a three-source model. Irrigation Science 28, 65–78.

Pogue, K.R., 2008. Influence of basalt on the terroir of the Columbia Valley American viticultural area. In: Proceedings of the 8th International Terroir Congress, Centro di Ricerca per la Viticoltura, Soave, Italy, vol. 1, pp. 4-50–4-55. http://people.whitman.edu/~pogue/soave.pdf.

Prichard, T.L., 1998. Water use and infiltration. In: Ingels, C.A., McGourty, G.T., Christensen, L.P. (Eds.), Cover Cropping in Vineyards: A Grower's Handbook. University of California, Oakland, USA, pp. 85–90. Publication 3338.

Proffitt, T., 2013. The Use of Mulch and Compost in Vineyards. www.winewa.asn.au.

Radke, J., Pinto, P., Lachhwani, K., Kondolf, G.M., Rocha, J., Llobet, A.S., Edwards, D., Francella, V., Jurich, K., McKnight, K., Alex, R.A., Eng, T., Harrell, B., Uennatornwaranggoon, F., Wolfson, E., Alfaro, P.J., Ding, E., Marzion, R., 2015. Alqueva Changing Ecologies of the Montado Landscape, Alentejo, Portugal. http://ced.berkeley.edu/downloads/research/AlquevaReportLA205-2015.pdf.

Rahimzadeh-Bajgiran, P., Berg, A.A., Champagne, C., Omasa, K., 2013. Estimation of soil moisture using optical/thermal infrared remote sensing in the Canadian Prairies. ISPRS Journal of Photogrammetry and Remote Sensing 83, 94–103.

Rebetzke, G.J., Zheng, B., Chapman, S.C., 2016. Do wheat breeders have suitable genetic variation to overcome short coleoptiles and poor establishment in the warmer soils of future climates? Functional Plant Biology 43, 961–972.

Retallack, M., 2012. What can be done in the vineyard to manage risk in difficult seasons? The Australian and New Zealand Grapegrower and Winemaker 586, 30–37.

Rogiers, S.Y., Clarke, S.J., Schmidtke, L.M., 2013. Elevated root-zone temperature hastens vegetative and reproductive development in Shiraz grapevines. Australian Journal of Grape and Wine Research 20, 123–133.

Rossi, V., Caffi, T., Salinari, F., 2012. Helping farmers face the increasing complexity of decision-making for crop protection. Phytopathologia Mediterranea 51, 480–491.

Sadras, V.O., Moran, M.A., 2012. Elevated temperature decouples anthocyanins and sugars in berries of Shiraz and Cabernet Franc. Australian Journal of Grape and Wine Research 18, 115–122.

Schmugge, T.J., Jackson, T.J., McKim, H.L., 1980. Survey of soil moisture determination. Water Resources Research 16, 961–979.

Schultz, H.R., Stoll, M., 2010. Some critical issues in environmental physiology of grapevines: future challenges and current limitations. Australian Journal of Grape and Wine Research 16, 4–24.

Schwartz, R.C., Baumhardt, R.L., Evett, S.R., 2010. Tillage effects on soil water redistribution and bare soil evaporation throughout a season. Soil and Tillage Research 110, 221–229.

Schymanski, S., Or, D., Zwieniecki, M., 2013. Somatal control and leaf thermal and hydraulic capacitances under rapid environmental fluctuations. PLoS One 8, e54231.

Sene, K.J., 1994. Parameterizations for energy transfers from a sparse vine crop. Agricultural and Forest Meteorology 71, 1–18.

Schultz, H.R., Pieri, P., Poni, S., Lebon, E., 2009. The ecophysiology of grapevine canopy systems – learning from models. In: Proceedings of Recent Advances in Grapevine Canopy Management: An Int. Symposium in Honor of Dr. Mark Kliewer, U. California, Davis, USA, pp. 7–12.

Shuttleworth, W.J., Wallace, J.S., 1985. Evaporation from sparse crops-an energy combination theory. Quarterly Journal of the Royal Meteorological Society 111, 839–855.

Shvaleva, A., Costa e Silva, F., Costa, J.M., Correia, A., Anderson, M., Lobo-do-Vale, R., Fangueiro, D., Bicho, C., Pereira, J.S., Chaves, M.M., Skiba, U., Cruz, C., 2014. Comparison of methane, nitrous oxide fluxes and CO_2 respiration rates from a Mediterranean cork oak ecosystem and improved pasture. Plant and Soil 374, 883–898.

Simonneau, T., Lebon, E., Coupel-Ledru, A., Marguerit, E., Rossdeutsch, L., Ollat, N., 2017. Adapting plant material to face water stress in vineyards: which physiological targets for an optimal control of plant water status? OENO One 51 (2), 167–179.

Skrotch, W., Shribbs, J., 1986. Orchard floor management – an overview. HortScience 21, 390–394.

Soliman, A., Brown, R., Hecka, R.J., 2011. Separating near surface thermal inertia signals from a thermal time series by standardized principal component analysis. International Journal of Applied Earth Observation and Geoinformation 13, 607–615.

Spano, D., Snyder, R.L., Duce, P., Paw, U.K.T., 2000. Estimating sensible and latent heat flux densities from grapevine canopies using surface renewal. Agricultural and Forest Meteorology 104, 171–183.

Stoll, M., Stuebinger, M., Lafontaine, M., Schultz, H.R., 2008. Radiative and thermal effects on fruit ripening induced by differences in soil colour. In: VIIe Congrès International des terroirs viticoles/VIIth Int. Terroir Congress. ACW, Agroscope Changins-Wädenswil.

Taghvaeian, S., Chávez, J.L., Altenhofen, J., Trout, T., DeJonge, K., 2013. Remote sensing for evaluating crop water stress at field scale using infrared thermography: potential and limitations. Hydrology Days 73–83.

Uliarte, M., Schultz, H.R., Frings, C., Pfister, M., Parera, C.A., del Monte, R.F., 2013. Seasonal dynamics of CO_2 balance and water consumption of C_3 and C_4-type cover crops compared to bare soil in a suitability study for their use in vineyards in Germany and Argentina. Agricultural and Forest Meteorology 181, 1–16.

van Gestel, N.C., Schwilk, D.W., Tissue, D.T., Zak, J.C., 2011. Reductions in daily soil temperature variability increase soil microbial biomass C and decrease soil N availability in the Chihuahuan desert: potential implications for ecosystem C and N fluxes. Global Change Biology 17, 3564–3576.

Van Huyssteen, L., Weber, H.W., 1980. The effect of conventional and minimum tillage practices on some soil properties in a dryland vineyard. South African Journal of Enology and Viticulture 1, 35–45.

van Leeuwen, C., Trégoat, O., Choné, X., Bois, B., Pernet, D., Gaudillère, J.-P., 2009. Vine water status is a key factor in grape ripening and vintage quality for red Bordeaux wine. How can it be assessed for vineyard management purposes? Journal International des Sciences de la Vigne et du Vin 43, 121–134.

Wang, C., Wan, S., Xing, X., Zhang, L., Han, X., 2006. Temperature and soil moisture interactively affected soil net N mineralization in temperate grassland in Northern China. Soil Biology and Biochemistry 38, 1101–1110.

Webb, L.B., Watt, A., Hill, T., Whiting, J., Wigg, F., Dunn, G., Needs, S., Barlow, S., 2009. Extreme heat: managing grapevine response. In: Documenting Regional and Inter-regional Variation of Viticultural Impact and Management Input Relating to the 2009 Heatwave in South-Eastern Australia. www.landfood.unimelb.edu.au/vitum.

Webb, L.B., Whetton, P.H., Bhend, J., Darbyshire, R., Briggs, P.R., Barlow, E.W.R., 2012. Earlier winegrape ripening driven by climatic warming and drying and management practices. Nature Climate Change 2, 259–264.

Williams, L.E., Ayars, J.E., 2005. Grapevine water use and crop coefficient are linear functions of the shaded area measured beneath the canopy. Agricultural and Forest Meteorology 132, 201–211.

Wilson, K., Goldstein, A., Falge, E., Aubinet, M., Baldocchi, D., Berbigier, P., Bernhofer, C., Ceulemans, R., Dolman, H., Field, C., Grelle, A., Ibrom, A., 2002. Energy balance closure at FLUXNET sites. Agricultural and Forest Meteorology 113, 223–243.

Yunusa, I.A.M., Walker, R.R., Guy, J.R., 1997. Partitioning of seasonal evapotranspiration from a commercial furrow-irrigated Sultana vineyard. Irrigation Science 18, 45–54.

Zarrouk, O., Costa, J.M., Francisco, R., Lopes, C.M., Chaves, M.M., 2016. Drought and water management in Mediterranean vineyards. In: Delrot, S., Chaves, M.M., Gerós, H., Medrano, H. (Eds.), Grapevine in a Changing Environment. A Molecular and Ecophysiological Perspective. Wiley-Blackwell, UK, pp. 38–67.

Zsófi, Z., Gál, L., Szilágy, Z., Szücs, E., Marschall, M., Nagy, Z., Bálo, B., 2009. Use of stomatal conductance and pre-dawn water potential to classify terroir for the grape variety Kékfrankos. Australian Journal of Grape and Wine Research 15, 36–47.

CHAPTER 11

Irrigation of Pistachios: Strategies to Confront Water Scarcity

David Pérez-López[1], Houssem Memmi[1,2], María del Carmen Gijón-López[2], Marta Maria Moreno[3], José Francisco Couceiro[2], Ana Centeno[1], María J. Martín-Palomo[4], Mireia Corell[4], Luis Noguera-Artiaga[5], Alejandro Galindo[6], Arturo Torrecillas[7], Alfonso Moriana[4]

[1]Technical University of Madrid, Madrid, Spain; [2]IRIAF, Ciudad Real, Spain; [3]Universidad de Castilla la Mancha, Ciudad Real, Spain; [4]Universidad de Sevilla, Seville, Spain; [5]Universidad Miguel Hernández de Elche (UMH), Alicante, Spain; [6]University of Twente, Enschede, The Netherlands; [7]Centro de Edafología y Biología Aplicada del Segura (CEBAS-CSIC), Murcia, Spain

Contents

1. INTRODUCTION

Pistachios (*Pistacia vera* L.) are native to western Asia and Asia Minor, from Syria to the Caucasus and Afghanistan. It is a member of the Anacardiaceae (cashew and mango family), being the only commercially edible nut among

Water Scarcity and Sustainable Agriculture in Semiarid Environment
ISBN 978-0-12-813164-0
https://doi.org/10.1016/B978-0-12-813164-0.00011-9

the 11 species in the genus *Pistacia* and by far the most economically important. The pistachio was introduced to Italy from Syria early in the first century C.E., and subsequently its cultivation spread to other Mediterranean countries. Iran, Turkey, and Syria hold 80% of the area dedicated to its cultivation, while only two more countries (Tunisia and the United States) bring this percentage to 94% of the planted area. This area dedicated to pistachio cultivation was almost doubled in the 14 years from 2000 to 2014, increasing from 421,000 to 827,000 ha in 2014. This was mainly due to increases in Turkey (from 36,000 to 281,000 ha) and the United States (from 30,000 to 82,000 ha) (FAOSTAT, 2017). Three facts help explain this increase. First, harvesting can be fully mechanized; second, few inputs are needed for pistachio cultivation, both aspects meaning that little investment is needed; and third, the price paid to producers tripled during the same period (FAOSTAT, 2017). In summary, pistachio cultivation has become a very profitable business. Such an increase in harvested area and the tripling of producer prices at the same time can only be explained by consumers' appreciation of its taste and nutritional value. This trend could even be stronger in the future due to recent studies that support the healthy effects of pistachio consumption (Kay et al., 2010; Tomaino et al., 2010; Aldemir et al., 2011).

Cultivated for a centuries in Mediterranean semiarid and arid areas, the pistachio has been considered as a drought-tolerant species (Spiegel-Roy et al., 1977; Behboudian et al., 1986; Rieger, 1995), mainly for its ability to survive under extreme conditions of water stress (Kanber et al., 1990; Goldhamer, 1995). In Tunisia, for example, the pistachio has been considered of great importance for valorizing drylands in the center and south of the country under rain-fed conditions (Ghrab et al., 2008). This area is characterized by an annual rainfall of around 200 mm and a reference evapotranspiration (ETo) of more than 1300 mm (Elloumi et al., 2013). However, the average yield in Tunisia is about 0.9 ton ha^{-1} (FAOSTAT, 2017) or even lower (Ghrab et al., 2008), since more than 95% are nonirrigated orchards (Abidi, 2016). By contrast, in the United States average yields, under irrigation, are about 2.6 ton ha^{-1}. This statistic confirms the fact that even though the pistachio is considered a drought-tolerant species, the term refers to its capacity to survive and produce rather than its ability to provide high profitable yields. In actual fact, the water demand of pistachio trees is not particularly low (Goldhamer et al., 1985), and the potential evapotranspiration rate at the maximum seasonal evaporative demand is equal to or exceeds that of many other deciduous species (Goldhamer et al., 1983).

Fruit culture under water-scarcity scenarios should be directed toward the use of plant materials that are less water demanding or able to withstand deficit irrigation with minimum impact on yield and fruit quality. In this sense, the pistachio plants can be considered an ideal alternative for arid and semiarid agrosystems because they can increase farmer revenues by saving important irrigation water amounts, leading to important collateral advantages such as the enrichment of agricultural biodiversity.

Taking into consideration all of the previous discussion, our emphasis in this chapter was to review the state of knowledge on the pistachio response to irrigation, paying special attention to (1) irrigation water requirements, (2) the effect of water deficit on vegetative and fruit growth, (3) the effect of water deficit on tree water relations on different rootstocks, (4) the response of pistachio to regulated deficit irrigation, and (5) the use of plant sensors for pistachio irrigation scheduling.

2. VEGETATIVE AND FRUIT GROWTH

The first point to be considered is that yield is determined by the previous year. Therefore, the yield of the present year modulates shoot growth, which determines the number of potential floral buds for the following year. In addition, the yield of the present year regulates the number of floral buds retained in "on" years (high fruit load), trees have a high percentage of dropping fruit buds, while in "off" years (low fruit load) this percentage is smaller. Water stress increases this percentage and will reduce yield in the next year (Fig. 11.1). Usually shoot growth is not related to regulation, and, in fact, pistachio trees are pruned frequently without causing reductions in yield because of its specific vegetative growth habit. This behavior is mainly because pistachio produces two types of shoots, the preformed or neoformed, depending on the moment of their differentiation. These two types of shoot have different reproductive capacities. Preformed shoots, commonly shorter than neoformed shoots, are essentially productive, while neoformed shoots are mostly vegetative and undesirable. Hence, they are usually removed by pruning without affecting yield.

Flowering is a process that occurs in two phases in temperate fruit trees. During winter, buds need to accumulate chilling, the exact amount depending on the variety. When the temperature increases in spring, the

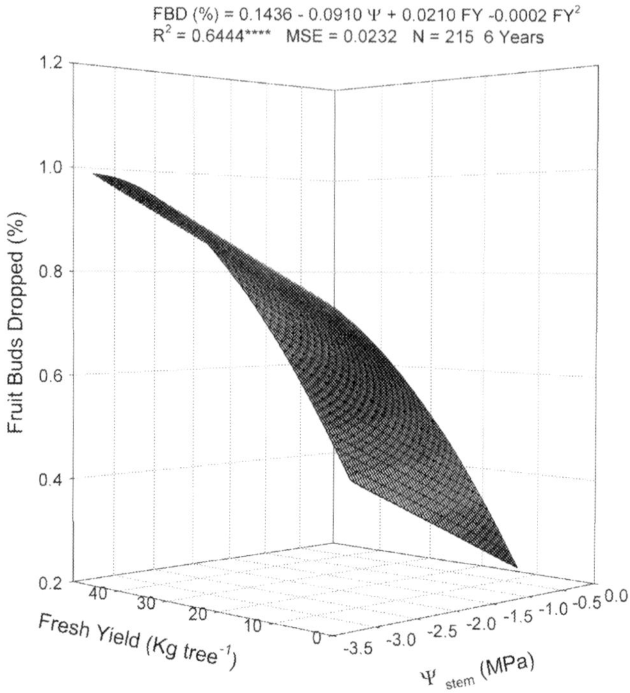

Figure 11.1 Fruit bud retention (FBP) in relation to fresh yield (FY) and minimum year stem water potential at midday (Ψ_{stem}). N = 215. Data from 6 years.

buds accumulate degree-days until flowering or leaf out takes place. The experimental data presented here for pistachio were obtained in Ciudad Real (central Southern Spain). Flowering in pistachio trees, var. Kerman, occurs during mid-April in Ciudad Real, similar to pistachio trees in San Joaquin Valley, California, USA (Goldhamer and Kamber, 2012). Since leaf out takes place until the second half of June, shoot growth occurs in two or three flushes. At the same time, buds differentiate into flower buds. Simultaneously there is a strong degree of fruit abscission, and the number of fruits per cluster changes from more than 200 to less than 30, depending on the current yield (Fig. 11.2A).

The first step in fruit development is rapid growth of the hull and the shell until they reach the final size. This process is easily monitored by measuring the transversal diameter of the fruit. This stage involves a strong increase in fresh fruit weight (Fig. 11.2A and B).

Then, the shell hardening starts for more than a month this is the only process that occurs in the fruit. In fact, the fresh fruit weight keeps more or

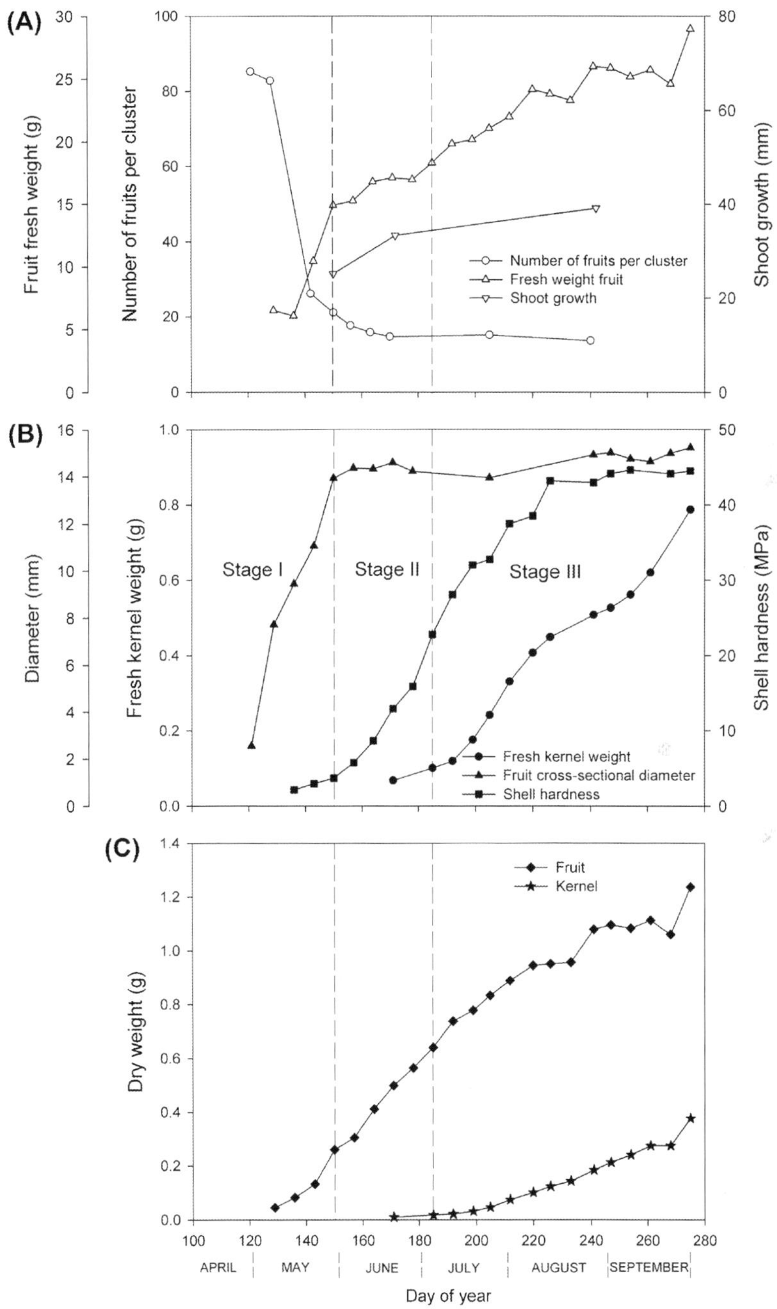

Figure 11.2 Pistachio fruit development and shoot growth in a mature pistachio orchard (cv. Kerman on *P. terebinthus*). (A) *White circles*, number of fruits per cluster; *white up triangle*, fresh weight fruit; *white down triangle*, shoot growth; (B) *black circle*, fresh kernel weight; *black up triangle*, fruit cross-sectional diameter; *black square*, shell hardness; (C) *black diamond*, dry weight fruit; *black star*, dry weight kernel.

less stable during this period. At the end of it the fruit buds' abscission occurs. This period finishes when the kernel growth begins, which implies an increase of fresh fruit weight, although the shell hardening continues for another month. The kernel grows during the last 3 months of the season, and then trees are harvested at the end of September or in early October. Goldhamer and Beede (2004) divided this nut development into three stages for irrigation scheduling purposes and named them Stage I, from leaf out to full shell expansion; Stage II, from full shell expansion to the onset of rapid kernel growth; and Stage III, from rapid growth to harvest.

In summary, three different phenological stages can be considered within the season: Stage I, when there is strong shoot growth, widespread abscission of fruit and the shell and hull grow until they reach the final size; Stage II, which sees shell hardening, the end of abscission of fruit until the final number of fruits per rachis is reached, accompanied occasionally by, slight shoot growth; and Stage III, during which shell hardening finishes and kernel growth occurs.

3. PLANT WATER RELATIONS

Pistachio trees are considered one of the most drought-resistant fruit species because they exhibit a high resistance to severe water stress conditions (Spiegel-Roy et al., 1977; Behboudian et al., 1986; Rieger, 1995) and can be cultivated in rain-fed and irrigated conditions. In this sense, pistachio vegetative growth is the earliest indicator of water stress conditions because it decreases in response to a reduction in water availability, even when water stress is not detected by conventional plant water status indicators. Lin et al. (1984) and Germana (1997) also suggested that pistachio drought resistance mainly depends on the extensive root system, because, although it is commonly believed to be a xerophyte, it does not present the morphologic characteristics of such in the leaves, which show high values of Pn and g_{leaf}.

Gijón et al. (2011) showed that different mechanisms of drought resistance operate in the different phenological stages in pistachio trees: (1) the relationships between stem water potential and gas exchange parameters were different during stage II and almost the same in stages I and III; (2) a change in the elastic modulus of the leaf cells in stage II was observed; and (3) different degrees of osmotic adjustment in response to water stress were observed in each phenological stage.

In rain-fed trees, stem water potential (Ψ_{stem}), and the water potential in covered leaves, had clearly lower values than those of fully irrigated trees, even though diurnal time course values were qualitatively similar, gradually decreasing from early morning, stabilizing at minimum values at around midday, and recovering in the afternoon (Fig. 11.3). For this reason, midday Ψ_{stem} values are commonly used to diagnose plant water status and in irrigation scheduling.

The diurnal time course of leaf conductance (g_{leaf}) in fully irrigated pistachio plants is characterized by values that increase from sunrise to reach high and maximum values between 11:00 and 13:00 h, after which they decrease (Fig. 11.4). In contrast, rain-fed plants show a gradual decrease from sunrise to midday when a certain stabilization of g_{leaf} values is observed, before they fall from 18:00 h onward (Fig. 11.4). Moreover, it is important to consider that Ψ_{stem} values were seen to decrease in response to water deficit earlier than g_{leaf} values (Figs. 11.3 and 11.4). However, when irrigation was restarted in water-stressed pistachio plants, Ψ_{stem} values increased faster than g_{leaf} values, which were lower than those in full-irrigated trees 32 days after plant recovery (H. Memmi, unpublished data). This behavior indicated that stomatal closure in pistachio plants is not a simple passive response to water deficit; the pattern of midday g_{leaf} under

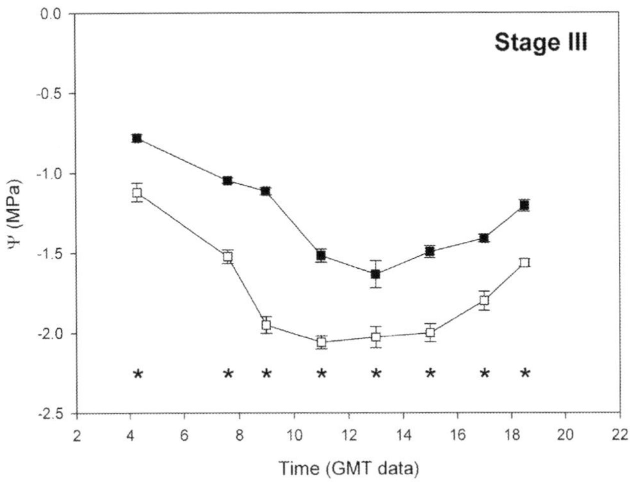

Figure 11.3 Daily pattern of stem water potential (Ψ_{stem}) in a mature pistachio orchard (cv. Kerman on *P. terebinthus*) during stage III of nut development. Symbols: ■, fully irrigated trees; □, rain-fed trees. *Asterisks* indicate significant statistical differences.

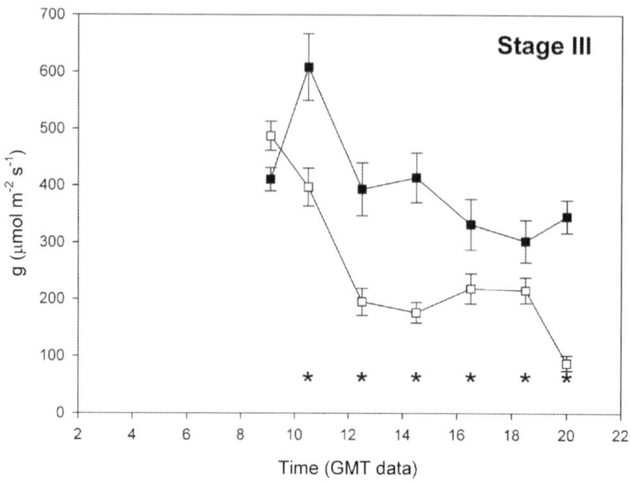

Figure 11.4 Daily pattern of leaf conductance (g_{leaf}) in a mature pistachio orchard (cv. Kerman on *P. terebinthus*) during phase III of nut development. Symbols as in Fig. 11.3.

water stress may be related to hormonal changes within the leaf, such as an increase in abscisic acid and/or a decrease in the cytokinin content (Mansfield, 1987; Davies and Zhang, 1991).

4. IRRIGATION WATER REQUIREMENTS

A crop is fully irrigated when irrigation avoids water stress due to low water levels in the soil. Crop irrigation requirements (ETc) are determined according to daily crop reference evapotranspiration (ETo), calculated using the Penman–Monteith equation (Allen et al., 1998) and a crop factor based on the time of the year. In woody crops, a third coefficient (Kr) is used, a reducing coefficient used to consider tree density and size, reaching a maximum value of one. Three studies calculate the Kc for pistachio (Fig. 11.5) (Goldhamer, 1995; Testi et al., 2008; Memmi et al., 2016). All the values are deemed to change as the season progresses, two according to the calendar date and the last one as a function of fruit growth stage. The Kc published by Goldhamer in 1995 is the most widely used because it is the oldest, but also it is the most contrasted. Goldhamer (1995) and Testi et al. (2008) did their studies in California, so dates for other places must be adjusted to plant development. Memmi et al. (2016) made their calculations in Ciudad Real with drip irrigation, instead of the microsprinklers used in previous assays, which may help explain the differences between their study and others.

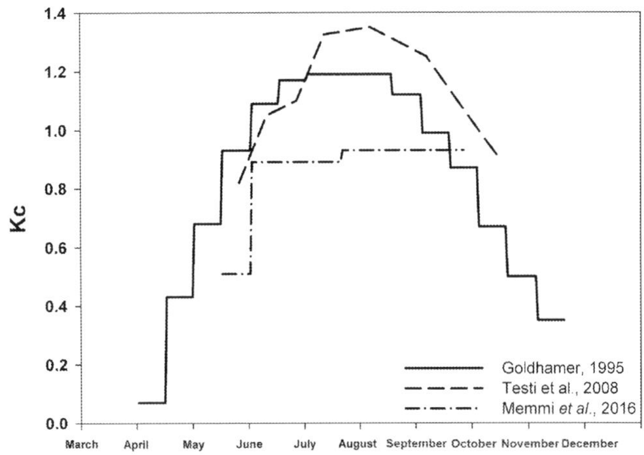

Figure 11.5 Crop coefficients (Kc) calculated for pistachio trees in three studies (Goldhamer, 1995; Testi et al., 2008; Memmi et al., 2016).

ETc can vary from 547 mm, calculated as described by Memmi et al. (2016), to 1000 mm calculated according to Goldhamer (1995). Crop coefficients proposed by Testi et al. (2008) estimate ETc values at around 840 mm season^{-1}. These substantial differences among estimated ETc values could be due to the effect of the irrigation system and tree size. In this sense, Kermani and Salehi (2006), in Iran, concluded that seasonal irrigation requirements oscillate between 600 and 1200 mm when trees are drip or flood irrigated, respectively. Therefore although pistachio trees can suffer a severe water stress and maintain a certain level of production, irrigation water requirements that avoid any water stress situation are quite large in excess of 600 mm season^{-1}. Frequently pistachio trees are grown in arid and semiarid regions, so these irrigation requirements are greater than the water available for use by pistachio farmers. Therefore water stress avoidance during the whole seasons is a luxury that farmers frequently cannot afford.

5. REGULATED DEFICIT IRRIGATION

Water scarcity around the world and increasing agricultural costs, such as those associated with pruning, have forced growers to reduce the amount of water supplied to fruit trees. In vegetables and crops, any reduction in irrigation is commonly associated with a reduction in yield and, therefore, profits. However, in fruit trees subjected to accurate irrigation scheduling, the yield reduction in deficit conditions is null or, at least, lower than in other crops. Deficit irrigation is defined as an irrigation strategy that provides water below the full needs and that produces periods of water stress.

Several deficit irrigation strategies may be followed, although they are all focused on improving water productivity (WP), i.e., the marketable yield per unit of water used (Kijne et al., 2003), rather than maximum yields (Zhang and Oweis, 1999), in other words increasing the efficiency of the irrigation water applied (Zhang, 2003). Regulated deficit irrigation (RDI) is based on the fact that fruit tree sensitivity to water deficit is not constant throughout the growing season and that water deficit during particular periods may benefit WP by increasing irrigation water savings, while minimizing or eliminating negative impacts on yield and crop revenue and even improving harvest quality (Chalmers et al., 1981; McCarthy et al., 2002; Domingo et al., 1996; Goldhamer, 1999; Loveys et al., 2004; Cameron et al., 2006). In RDI, full irrigation is applied during drought-sensitive phenological stages (critical periods) of fruit trees, and irrigation is limited or even unnecessary if rainfall provides a minimum supply of water during the drought-tolerant phenological stages (noncritical periods) (Chalmers et al., 1981; Mitchell and Chalmers, 1982; Geerts and Raes, 2009). RDI is therefore deemed suitable for fruit trees if the following criteria are satisfied:

- The fruit species have to recover from the stress period as soon as possible in order to avoid water stress during a sensitive phenological stage.
- They should be drought-resistant species in order to permit significant water saving with a low yield reduction.
- Vegetative growth and fruit growth in the growing season should be clearly distinguishable or, at least, have a short overlap that allows growers to select the process to be affected.

In this respect, while pistachio trees are extremely resistant to drought conditions, they do not recover rapidly from water stress and neither is there a clear separation between vegetative and nut growth. Therefore, RDI needs to be scheduled as accurately as possible. The effect of water stress applied in the different phenological stages of nut development in pistachio trees is still a subject of study because of its complexity. However, depending on the water stress periods applied, the main effects are:

Stage I. In this phenological stage, which takes place at the beginning of the season, water stress is uncommon due to the low number of transpiring leaves and the low evaporative demand; moreover, soil moisture is usually high. Water stress in this period reduces the final nut size at harvest because most of the nut growth takes place at this time. Experimental works reported a slight increase of split nuts with water stress during this stage but at the expense of nut size. Nevertheless, water stress during this stage has no effect on crop load since no reduction in preformed shoots was reported in response to water stress. Preformed growth occurs early in the season, well before the beginning of stage II, when water stress is infrequent due to winter and spring rain, being sufficient to guarantee maximum yield potential, since fruiting buds are located mostly on these short shoots.

Stage II. This phenological stage is considered the most tolerant to drought. The effect of water stress is low or even null during this period for the subsequent amount and quality of the yield. Severe water stress conditions would reduce the number of nuts and therefore the current yield. During this phenological stage, buds for the next year are being produced and there are natural bud abscissions. Severe water stress during this period could also increase this abscission and reduce the next yield. The influence of water stress in this process is complex because bud abscission is also related to yield. Fig. 11.1 presents a fitted surface from experimental data in which bud abscission is related to yield and midday stem water potential. In conditions of high fruit load (around 40 kg tree^{-1}) a decrease in water potential from −1.5 MPa to values of around −3 MPa would increase bud abscission from 70% to 90%. When yield is moderate, the influence of water stress is lower. In a 20 kg tree^{-1}, the same decrease in water potential (−1.5 to −3 MPa) would increase the bud abscission from 65% to around 80%.

Stage III. This phenological stage is considered the most sensitive in relation to the final yield. Moderate water stress conditions reduce final nut size and the quality of the nuts, a double effect that strongly advises against any reduction in the irrigation water supplied in this period.

The quality of pistachio nuts in the field is related to the splitting of the nut. In the total yield three different kinds of nuts can be identified: splitting, nonsplitting, and blank. Nonsplitting pistachios are those where the shell remains closed, while blank pistachios are those without kernel. The economic value of splitting pistachios is greater than the value of nonsplitting pistachios. The process of splitting could be related to water stress, although this process is still not completely understood. Water stress during stage III decreases the percentage of splitting pistachios, which, some authors suggest, may be related with slow kernel development. However, moderate water stress conditions during stage I improve the percentage of splitting nuts at harvest. The most accepted hypothesis is that a decreased nut size promotes easy nut splitting.

There is another factor to consider in nut splitting, which is early splitting. This involves rupture of the hull before the harvest and may lead to mold infections of the kernel with a subsequent decrease in the value. Goldhamer and Beede (2004) reported that moderate water stress conditions during stage I are related to an increase in early splitting. However, a 3-year experiment performed in central Spain suggested that part of this increase could be linked to a minimum daily temperature under 13°C (Gijón et al., 2009).

Vegetative growth in pistachio trees takes place throughout the season, although it is higher during stage I. Therefore any water stress in this period will be the first to affect the increase of the tree size. In mature pistachio orchards, growth may partly occur outside the water stress period; the seasonal growth is not, apparently, a limitation to the following year's yield.

Postharvest water stress did not affect the yield in the following year, a result that suggests that the carbohydrate reservoir is not limiting. Stevenson and Shackel (1998) similarly reported that alternate bearing is not caused by differences in the carbohydrate reservoir between the "on" and the "off" years. These authors suggest that in the "off" years water and fertilizer could be reduced without any effect on the following yield.

The capacity of recovery is another important issue. In some studies recovery from water stress in stage II was slow and, for a short time during stage III, trees suffered water stress conditions. However, apparently, this water stress did not affect the yield. Such results suggest that the level of water stress could be an important additional factor regulating the response of pistachio to drought conditions.

In summary, RDI conditions during stage II or postharvest does not reduce yield, whereas water stress during stage I and III reduces the quantity and quality of the yield. The level of water stress is a factor whose effects have not been sufficiently studied. The use of biosensors that increase the accuracy of water status measurements is likely to increase the water-saving capacity.

6. ROOTSTOCK RESPONSE TO DROUGHT

Rootstocks can affect the water status and, consequently, irrigation sensitivity in two main ways. Firstly, root distribution and the capacity to take up water determine the water usage efficiency. In this respect, the nutrient content of leaves has been reported to be related to the rootstock, suggesting a differential bioelement uptake capacity. If rootstocks have different capacities to obtain bioelements, they may behave differently regarding water. Such differences could modify the irrigation frequency, although studies in this field to date are insufficient for any conclusions to be reached.

The other way in which rootstocks affect water relations is by sending signals from the roots to the shoots. Different substances in the roots might induce a reduction in the gas exchange that takes place in the leaves. Therefore, the more sensitive the rootstock is to drought conditions, the more sensitive the cultivar will be to the same. For instance, greater leaf transpiration and photosynthetic rates have been recorded in *Pistacia atlantica* Desf. than in *Pistacia terebinthus* L.; a cultivar on *P. atlantica* rootstock would present more transpiration than the same cultivar on *P. terebinthus*.

The selection of the most effective rootstock for different environmental conditions is crucial for improving pistachio crop performance under water stress conditions and for developing the best deficit irrigation strategy. In this sense, Gijón et al. (2010) showed that the responses of cv. Kerman depended on the rootstock onto which it had been grafted. The hybrid *P. atlantica* Desf. × *P. vera* L. rootstock was associated with a higher degree of stomatal control and reduced leaf senescence compared to *P. atlantica* and *P. terebinthus*, despite being associated with the most vigorous shoot growth. *P. terebinthus* enabled very effective stomatal control but was also associated with the most rapid leaf senescence. *P. atlantica* was associated with less vigorous shoot growth and similar levels of water stress as occurred with other rootstocks under conditions of high evaporative demand, which was associated with lower stomatal control. Memmi et al. (2016) observed that pistachios of cv. Kerman on *Pistacia integerrima* L. usually have a higher Ψ_{stem} than on *P. atlantica* and *P. terebinthus* at the beginning of the irrigation season, i.e., the end of May and beginning of June. However, at the end of the irrigation season, pistachios on *P. terebinthus* have the highest Ψ_{stem} values. Damage due to freezing has been observed in pistachios on *P. integerrima* when exposed to spring frost in Spain. Pistachios on *P. integerrima* seem to come out of dormancy before the other rootstocks, which makes these trees more sensitive to frost in locations with a cold spring but provides them with a longer growth season in warmer locations. On the other hand, at the end of summer, pistachio trees with *P. terebinthus* are likely to have a better water status, which could be related to higher drought tolerance.

Interactions between rootstocks and RDI are not consistent, so no particular type of rootstock can be recommended.

7. PLANT BIOSENSORS FOR IRRIGATION SCHEDULING

RDI is a technique that "combines" the moment of application with the level of water stress that the withholding of irrigation water will produce. The influence of the phenological stage (the moment) on RDI was described previously and it is widely known, while the level of water stress has been less studied despite its importance for the accurate control of the drought period. Water stress conditions have different effects depending on the phenological stage when they occur. Moreover, the level of water stress may reduce the response if it is mild or change it completely if it is too severe. For instance, severe water restriction during stage II (shell hardening) is likely to reduce the yield even though this phenological stage has been described as drought tolerant.

There are several methodologies that could be used to quantify water stress under field conditions. These are grouped based on measurements of the amount of water in the soil and estimations of the plant water status. Although soil water measurements are most commonly used in orchards, plant water status measurements are considered more sensitive for measuring water stress conditions in trees. Irrigation scheduling based on plant water status measurements is still poorly understood in the case of most fruit trees, and even less so in the case of pistachio trees because there are a few research groups working with this tree worldwide. A number of considerations may be mentioned regarding the use of biosensors:

1. Relationship with the environment. These methodologies measure the plant physiology and are strongly affected by other conditions in addition to water stress, mainly fruit load and evaporative demand.
2. Threshold values. When the environmental effect has been eliminated, threshold values that limit the effect of water stress must be obtained.
3. Sampling methodology. The target of these methodologies is irrigation scheduling in commercial orchards. As some of these measurements are performed on individual trees, a minimum number of trees must be studied and their distribution throughout the orchard should be considered. In addition, some measurements are not continuous and the periodicity must be established.

7.1 Stem Water Potential

Water potential is an easy, traditional way to determine water status in plants. It has been used widely as a water status indicator in research and is considered the reference methodology. To illustrate how the water potential varies throughout the day (Fig. 11.3), two specific moments have been considered as a reference: before dawn and at noon. Furthermore, the measurement at noon can be determined by two methods: using a leaf from the tree, or using a leaf that has been previously covered with aluminum foil at least an hour before for avoiding irradiation to reach stomatal closure. The second method is called stem water potential because a balance of the stem water status is reached. This measurement has less variability around the tree and is therefore the preferred method in research literature.

To determine the water potential, blotting paper must be placed at the cut end of the petiole because in pistachio leaves at the onset of gas injection into the chamber exudates appear, presumably from the phloem, at the cut end of the petiole, and this fluid may interfere with the xylem fluids (Goldhamer and Kamber, 2012).

Leaf water potential and stem water potential are influenced by factors other than water stress, such as crop load or meteorological conditions. However, the water potential is widely used for commercial irrigation scheduling in several species, including peach trees and vines.

Stem water potential can be used as an indicator in irrigation scheduling, and, if it is lower than a specific value (threshold value; TV), the amount of irrigation water must be increased. Theoretically, the stem water potential cannot be higher than TV, but this depends on how the TV is defined. When the stem water potential remains around or above TV, water irrigation must be reduced. However, the TV varies with the evaporative demand. In fully irrigated conditions, the vapor pressure deficit (VPD) is a good predictor of midday water potential values. Fig. 11.6 presents a baseline between stem water potential of pistachio trees fully irrigated for 7 years and VPD. The data included were measured for different fruit load conditions and with different rootstock. There were no significant differences between "ON" and "OFF," years neither with rootstocks in the years "OFF," but there were statistical differences in the years "ON" between intercept but not in the slope of the equation. As such, differences were small (less than 0.6 MPa), and we consider that the entire pool of data could be used. Our field experience suggests that if no meteorological data are available, a constant value of around −1.1 MPa could be used throughout the season.

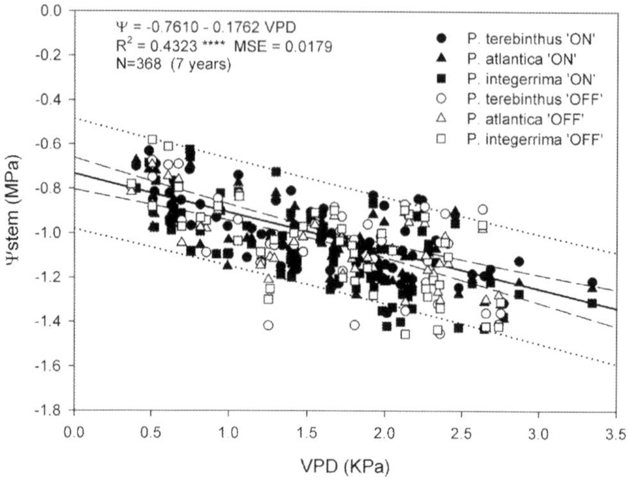

Figure 11.6 Relationship between stem water potential (MPa) and vapor pressure deficit (KPa) in well-irrigated mature pistachios on three rootstocks.

Experimental studies in central Spain using midday stem water potential in RDI are promising. Threshold values of -1.5 and -2 MPa during stage II, and keeping the plants well irrigated during the other stages, did not produce significant differences with the control trees (fully irrigated during the three stages) in terms of yield in a 3-year experiment in young pistachio trees. However, there were differences in vegetative growth, so it is very important to restrict this kind of irrigation when the canopy reaches its maximum volume. In addition, it was found that a higher number of buds dropped under water stress conditions (Fig. 11.1), so, in the long term, this could limit the yield if the water stress conditions are severe. Although further research is needed, the results suggest that a threshold value of stem water potential of -1.5 MPa for stage II could be sustainable in the long term, while saving a substantial amount of water.

7.2 Trunk Diameter Fluctuations

This methodology uses a promising indicator—trunk diameter fluctuations—because continuous and automatic measurements are possible. Although its usefulness may be limited by the cost and interference from environmental conditions, it has found commercial applications in some fruit trees. Trunks/stems in all plants present daily cycles of swelling and shrinking, which are caused by the partial hydration/rehydration of the bark. During the day the trunk shrinks, while during midafternoon it begins to swell and continues to do so until the following dawn. Two indicators are calculated from these daily curves: maximum daily trunk shrinkage (MDS) and trunk growth rate (TGR). MDS is the difference between the daily maximum (in the morning) and the minimum (afternoon). TGR is the difference between two consecutive maximum values.

Water stress in fruit trees increases MDS compared with what occurs in fully irrigated trees. However, MDS is also strongly associated with transpiration, and most of the daily variations are related to evaporative demand. An increase in evaporative demand also increases MDS. In other fruit trees, the relationships between MDS and an easy meteorological variable (such as temperature or vapor pressure deficit), the so-called baseline, have been used to estimate MDS in fully irrigated trees. The real indicator, the MDS signal, is the ratio between the measured MDS and that estimated from the baseline. There are no studies in pistachio trees that estimate the MDS, except one published work, in which the MDS presented higher values in water stress conditions than in fully irrigated trees, even after the beginning of the recovery. This result could indicate a high sensitivity to drought conditions.

TGR is commonly associated with growth, so a decrease in TGR is considered a representative water stress value. However, the seasonal pattern of TGR changes. In mature fruit trees, where the vegetative growth stops in favor of the fruit development, TGR is around zero even in conditions of full irrigation. Moreover, the fruit load could also affect the TGR seasonal pattern, with very high values in low fruit load conditions and around zero in high fruit load conditions. Therefore, the TGR could be a useful indicator because it shows changes in phenological stages although the threshold values are difficult to establish. In addition, in deciduous trees like pistachio, the effect of leaf drop should be considered.

7.3 Canopy Temperature

Water stress conditions affect gas exchange in fruit trees. The reduction of leaf transpiration produces an increase in the leaf temperature, which can usually be measured with an infrared thermometer. The main problem of this indicator is interference from the ambient temperature. This limitation is partially eliminated with the use of thermal images and equations that can reduce the error in the measurements. This technique is widely used in other fruit trees by flying over orchards. Imaging even allows identification of the plots with problems or poor irrigation. The two problems of this methodology are the cost and the absence of continuous measurements. However, flying costs have been reduced in recent years with the use of drones.

There is a study that uses this technique for pistachio trees (Testi et al., 2008). Although it did not use drones, the conclusions indicated that this methodology is very sensitive for identifying water stress, even in conditions of mild level. Therefore, although further research is necessary in pistachio trees, this technique will probably become commercially viable.

8. DEFICIT IRRIGATION AND FRUIT QUALITY

As already mentioned, RDI mainly focuses on reducing the water supply to the plant by imposing water deficits during specific phenological stages, while minimizing any impact on fruit quality. These changes in irrigation scheduling could be associated with variation in nut quality, understood as color flavor and consumer appreciation. The work described in this chapter (Carbonell-Barrachina et al., 2015) was performed with Kerman nuts in an RDI experiment in central Spain, as briefly described in the previous section, using the stem water potential as a plant biosensor. The influence of RDI on fruit quality can be studied from different points of view: (1) morphology, including fruit weight and size; (2) physicochemical properties, including instrumental color, mineral composition, fatty acid profile, and other parameters directly related to the quality of the nuts; and (3) sensory analysis, including descriptive tests using a trained panel specialized in nuts and/or affective tests conducted with a consumer panel.

8.1 Morphology

The morphology of fresh fruit is an important factor when making decisions on the quality of pistachios because a decrease in the size and/or weight leads to lower prices and, later, reduced consumer acceptance. The price of pistachios is determined by their size and edible yield (ratio of edible portion to total weight of nut).

To control the fruit morphology, the weight and size (length, width, and height) of pistachios subjected to RDI were analyzed and compared to those of the control samples (irrigated with enough water to avoid water stress). The morphological parameters were studied independently in the edible nut and shell to verify whether RDI affected the various portions of the pistachio nut differently.

The analyses showed that the application of RDI in stage II of phenological development, as described in this chapter, did not significantly affect the weight or size of the pistachios. The weights of the whole pistachio, edible nut, and shell were 1.52, 0.76, and 0.76 g, respectively; while the dimensions (length, width, and height) of the pistachios were 17.7, 9.8, and 11.2 mm, respectively (Carbonell-Barrachina et al., 2015).

This experimental observation is crucial because it means that by using RDI, farmers can claim to be environmentally friendly by saving irrigation water while the morphology of the pistachios is not compromised. This will ensure that the selling prices of RDI pistachios and later consumer acceptance will not, in principle, be affected.

Raw pistachios were salted and roasted, using standardized protocols, to simulate the key properties of the commercial products. Consequently, such "salted and roasted" nuts are the materials mentioned in the following evaluation.

8.2 Physical and Chemical Characteristics

In general, RDI conditions had no significant effects on the instrumental color (CIE$L^*a^*b^*$ color) and mineral composition of pistachios, although it significantly affected hardness, fatty acid profile, and volatile composition. Whatever the case, pistachios are a good source of minerals and contain high amounts of potassium (~ 10.5 g kg^{-1}), iron (~ 40 mg kg^{-1}), copper (~ 14 mg kg^{-1}), and zinc (~ 17 mg kg^{-1}). The typical green color of pistachios can be characterized by the following values of the main color coordinates: L^* (lightness) 61.3, a^* (green-red coordinate) −3.3, and b^* (blue-yellow coordinate) 27.3 (Carbonell-Barrachina et al., 2015).

Through an instrumental texture analysis (force needed to cut the nut), it has been observed that the hardness of the toasted nuts increase with the intensity of RDI. That is, the more severe the water stress, the harder and crunchier the pistachios.

The application of moderate RDI significantly increased the oil content of the nuts from 40.6% (control) to 42.1%, while more severe RDI conditions reduced the oil content to 38.3%. The fatty acid profile was also affected. The fatty acid profile of pistachios is dominated by three main compounds: oleic acid ($\sim 50\%$), linoleic acid ($\sim 33\%$), and palmitic acid ($\sim 13\%$). The application of a moderate or severe water stress triggered a significant increase in the content of linoleic acid, which is an essential fatty acid for humans.

The volatile compounds of a particular food are responsible for its smell and aroma, and the application of moderate RDI usually did not affect these attributes. However, the application of severe RDI conditions seemed to increase the contents of aldehydes (associated with green and vegetable notes) and reduce those of pyrazines (nut and toasted notes) and terpenes (citric notes).

8.3 Sensory Analysis

The descriptive analysis of pistachios showed that moderate RDI conditions lead to an intense "green pistachio" color, accompanied by higher intensities of nutty and pistachio notes in harder, crunchier nuts with a longer aftertaste. More severe RDI conditions are not reflected in higher intensities of any of the key sensory attributes.

An international consumer study was conducted in four countries (Spain, France, Poland, and the Slovak Republic) to gather information about the opinion of European consumers on pistachios obtained after RDI. This type of product was called "hydroSOStainable" because it contributes to saving a very scarce resource, water; this information was not provided to the consumers. Pistachios obtained under moderate RDI conditions obtained a higher satisfaction degree ("like them very much") than control and severe RDI pistachios ("like them moderately"). Moreover, a very high percentage of the European consumers (85%) were willing to buy the hydroSOStainable pistachios grown under moderate RDI conditions (Fig. 11.7).

For all the just-discussed products, we conclude that pistachios from trees irrigated using moderate RDI strategies have normal (not-reduced) weight and size, normal mineral contents, characteristic "pistachio" color, a high content of oil and linoleic acid, and high intensities of key sensory attributes (e.g., nutty and toasted pistachio notes). Besides, European

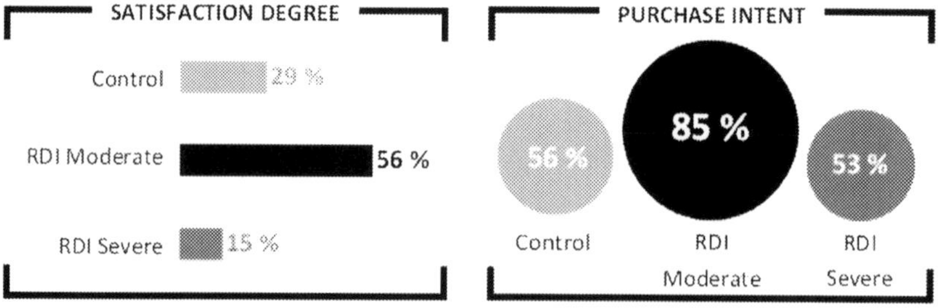

Figure 11.7 Results of an international affective study on consumer opinion of "hydroSOStainable" pistachios.

consumers indicated their high degree of satisfaction with the nuts, and a very high percentage (85%) said they were willing to buy the hydroSOStainable pistachios (Carbonell-Barrachina et al., 2015).

9. CONCLUSIONS

Growing pistachios in rain-fed conditions is profitable because the species exhibits a multitude of mechanisms of resistance to severe water stress conditions; however, production increases under irrigated conditions. On the other hand, complete avoidance of water stress situations requires a large amount of water, which growers rarely have. The use of RDI allows substantial water savings without affecting yield. Additionally, the use of plant biosensors improves irrigation efficiency when used in conjunction with RDI or other strategies. Pistachios grown under RDI, with a stem water potential threshold of −1.5 MPa during stage II of fruit growth, were much appreciated by consumers in an affective test.

ACKNOWLEDGMENTS

This research was supported by Ministerio de Economía y Competitividad (MINECO) (CICYT/FEDER AGL2013-45922-C2-1-R and AGL2013-45922-C2-2-R) grants to the authors. AG acknowledges the postdoctoral financial support received from Ramón Areces Fondation program. LNA was funded by a FPU grant from the Spanish government (FPU014/01874).

REFERENCES

Abidi, W., 2016. Pomological and physical attributes of pistachio (*Pistacia vera* L.) varieties grown in west-central Tunisia. Journal of New Sciences 28 (4), 1582–1588.

Aldemir, M., Okulo, E., Neselioglu, S., Erel, O., Kayigil, Ö., 2011. Pistachio diet improves erectile function parameters and serum lipid profiles in patients with erectile dysfunction. International Journal of Impotence Research 23 (1), 32–38.

Allen, R.G., Pereira, R.S., Raes, D., Smith, M., 1998. Crop Evapotranspiration-Guidelines for Computing Crop Water Requirements. Irrigation and Drainage. Food and Agriculture Organization of the United Nations, Roma.

Behboudian, M.H., Walker, R.R., Törökfalvy, E., 1986. Effects of water stress and salinity on photosynthesis of pistachio. Scientia Horticulturae 29, 251–261.

Cameron, R.W.F., Harrison-Murray, R.S., Atkinson, C.J., Judd, H.L., 2006. Regulated deficit irrigation – a means to control growth in woody ornamentals. The Journal of Horticultural Science and Biotechnology 81, 435–443.

Carbonell-Barrachina, A.A., Memmi, H., Noguera-Artiaga, L., Gijón-López, M.C., Ciapa, R., Pérez-López, D., 2015. Quality attributes of pistachio nuts as affected by rootstock and deficit irrigation. Journal of the Science of Food and Agriculture 95 (14), 2866–2873.

Chalmers, D.J., Mitchell, P.D., van Heek, L., 1981. Control of peach tree growth and productivity by regulated water supply, tree density, and summer pruning. Journal of the American Society for Horticultural Science 106, 307–312.

Davies, W.J., Zhang, J., 1991. Root signals and regulation of growth and development of plants in drying soils. Annual Review of Plant Physiology and Plant Molecular Biology 42, 55–76.

Domingo, R., Ruiz-Sanchez, M.C., Sanchez-Blanco, M.J., Torrecillas, A., 1996. Water relations, growth and yield of Fino lemon trees under regulated deficit irrigation. Irrigation Science 16, 115–123.

Elloumi, O., Ghrab, M., Ben Mimoun, M., 2013. Chilling accumulation effects on performance of pistachio trees cv. Mateur in dry and warm area climate. Scientia Horticulturae 159, 80–87.

FAOSTAT, February 28, 2017. Statistic Information. Available from: http://www.fao.org/faostat/en/#home.

Geerts, S., Raes, D., 2009. Deficit irrigation as an on-farm strategy to maximize crop water productivity in dry areas. Agricultural Water Management 96, 1275–1284.

Germana, C., 1997. The response of pistachio trees to water stress as affected by two different rootstock. Acta Horticulturae 449, 513–520.

Ghrab, M., Gargouri, K., Ben Mimoun, M., 2008. Long term effects of dry conditions and drought on fruit trees yield in dry land areas of Tunisia. Options Méditerranean 80, 107–111.

Gijón, M.C., Giménez, C., Pérez, D., Guerrero, J., Couceiro, J.F., Moriana, A., 2010. Rootstock influences the response of pistachio (*Pistacia vera* L. cv. Kerman) to water stress and rehydration. Scientia Horticulturae 125, 666–671.

Gijón, M.C., Gimenez, C., Pérez, D., Guerrero, J., Couceiro, J.F., Moriana, A., 2011. Water relations of pistachio (*Pistacia vera* L.) as affected by phenological stages and water regimes. Scientia Horticulturae 128, 415–422.

Gijón, M.C., Guerrero, J., Couceiro, J.F., Moriana, A., 2009. Deficit irrigation without reducing yield or nut splitting in pistachio (*Pistacia vera* cv. Kerman on *Pistacia terebinthus* L.). Agricultural Water Management 96, 12–22.

Goldhamer, D., Kamber, R., 2012. Pistachio. In: Steduto, P., Hsiao, T.C., Fereres, E., Raes, D. (Eds.), Crop Yield Response to Water. Food and Agriculture Organization of the United Nations, Roma, pp. 416–429.

Goldhamer, D.A., 1995. Irrigation management. In: Fergusson, L. (Ed.), Pistachio Production Manual. Center for Fruit and Nut Research and Information, Davis, pp. 71–81.

Goldhamer, D., 1999. Regulated deficit irrigation for California canning olives. Acta Horticulturae 474, 369–372.

Goldhamer, D., Beede, R., 2004. Regulated deficit irrigation effects on yield, nut quality and water-use efficiency of mature pistachio trees. Journal of Horticultural Science and Biotechnology 79, 538–545.

Goldhamer, D.A., Beede, R.H., Moore, J.M., Weinberger, G., Menezes, J.J., 1983. Water use requirements and physiological response to water stress in pistachio. In: Annual Report of the California Pistachio Commission, Crop Year 1982–1983, pp. 53–57.

Goldhamer, D.A., Kjelgren, R., Beede, R., Williams, L., Moore, M.J., Lane, J., Weinberger, G., Menezes, J., 1985. Water use requirements of pistachio tree sand response to water stress. In: Annual Report of the California Pistachio Commission, Crop Year 1984–1985, pp. 85–92.

Kanber, R., Eylen, M., Köksal, H., Yüksek, G., 1990. Güneydogu AnadoluKosullarinda Antepfistigi Verim ve Su tüketiminin Irdelenmesi. In: Türkiye Antepfistigi Simpozyumu. Gaziantep, pp. 45–160 (in Turkish).

Kay, C.D., Gebauer, S.K., West, S.G., Kris-Etherton, P.M., 2010. Pistachios increase serum antioxidants and lower serum oxidized-LDL in hypercholesterolemic adults. The Journal of Nutrition 140, 1093–1098.

Kermani, M.M., Salehi, F., 2006. Determination of pistachio crop evapotranspiration (ETc). Acta Horticulturae 726, 441–448.

Kijne, J.W., Barker, R., Molden, D., 2003. Improving water productivity in agriculture. In: Kijne, J.W., Barker, R.M.D. (Eds.), Water Productivity in Agriculture: Limits and Opportunities for Improvement. International Water Management Institute, Colombo, Sri Lanka, pp. xi–xix.

Lin, T.S., Crane, J.C., Ryugo, K., Polito, V.S., DeJong, T.M., 1984. Comparative study of leaf morphology, photosynthesis, and leaf conductance in selected Pistacia species. Journal of American Society of Horticultural Science 109, 325–330.

Loveys, B.R., Stoll, M., Davies, W.J., 2004. Physiological approaches to enhance water use efficiency in agriculture: exploiting plant signaling in novel irrigation practice. In: Bacon, M.A. (Ed.), Water Use Efficiency in Plant Biology. Blackwell Publishing Ltd., Oxford, UK, pp. 113–141.

Mansfield, T.A., 1987. Hormones as regulators of water balance. In: Davies, W. (Ed.), Plant Hormones and Their Role in Plant Growth and Development. Martinus Nijhoff Publishers, Dordrecht, pp. 411–430.

McCarthy, M.G., Loveys, B.R., Dry, P.R., 2002. Regulated deficit irrigation and partial rootzone drying as irrigation management techniques for grapevines. In: Deficit Irrigation Practices. Water Reports Publication n. 22, Food and Agriculture Organization of the United Nations, Rome, pp. 79–87.

Memmi, H., Gijón-López, M.C., Couceiro, J.F., Pérez-López, D., 2016. Water stress thresholds for regulated deficit irrigation in pistachio trees. Rootstock influence and effects on yield quality. Agricultural Water Management 164, 58–72.

Mitchell, P.D., Chalmers, D.J., 1982. The effect of reduced water supply on peach tree growth and yields. Journal of the American Society for Horticultural Science 107, 853–856.

Rieger, M., 1995. Offsetting effects of reduced root hydraulic conductivity and osmotic adjustment following drought. Tree Physiology 15, 379–385.

Spiegel-Roy, P., Mazigh, D., Evenari, M., 1977. Response of pistachio to low soil moisture conditions. Journal of the American Society for Horticultural Science 102, 470–473.

Stevenson, M.T., Shackel, K.A., 1998. Alternate bearing in pistachio as a masting phenomenon: construction cost of reproduction versus vegetative growth and storage. Journal of the American Society for Horticultural Science 123, 1069–1075.

Testi, L., Goldhamer, D.A., Iniesta, F., Salinas, M., 2008. Crop water stress index is a sensitive water stress indicator in pistachio trees. Irrigation Science 26, 395–405.

Tomaino, A., Martorana, M., Arcoraci, T., Monteleone, D., Giovinazzo, C., Saija, A., 2010. Antioxidant activity and phenolic profile of pistachio (*Pistacia vera* L., variety Bronte) seeds and skins. Biochimie 92, 1115–1122.

Zhang, H.P., Oweis, T., 1999. Water yield relations and optimal irrigation scheduling of wheat in the Mediterranean region. Agricultural Water Management 38, 195–211.

Zhang, H.P., 2003. Improving water productivity through deficit irrigation: examples from Syria, the North China Plain and Oregon, USA. In: Kijne, J.W., Barker, R., Molden, D. (Eds.), Water Productivity in Agriculture: Limits and Opportunities for Improvement. Centre for Agriculture and Biosciences International, pp. 301–309.

CHAPTER 12

Sustainable Deficit-Irrigation Management in Almonds (*Prunus dulcis* L.): Different Strategies to Assess the Crop Water Status

Iván F. García Tejero[1], Alfonso Moriana[2], Carmen R. Rodríguez Pleguezuelo[3], Víctor H. Durán Zuazo[1], Gregorio Egea[2]
[1]IFAPA Centro "Las Torres-Tomejil", Seville, Spain; [2]Universidad de Sevilla, Seville, Spain; [3]IFAPA Centro "Camino de Purchil", Granada, Spain

Contents

1. INTRODUCTION

Irrigated agriculture in southern Europe, and more specifically in arid and semiarid areas, is crucial for economic development, especially in rural regions with a lower income potential. For the case of Andalusia, irrigated crops generate more than 60% of rural employment and represent 64% of agricultural production. Currently, 1,176,000 ha are under irrigation,

Water Scarcity and Sustainable Agriculture in Semiarid Environment
ISBN 978-0-12-813164-0
https://doi.org/10.1016/B978-0-12-813164-0.00012-0

representing 24% of total Andalusian agricultural surface area and constituting 33% of the irrigated cropland in Spain (ARA, 2011). These data are crucial to the farming sector for economic and social development of regions such as Andalusia, with regard to irrigated agriculture (García-Tejero et al., 2014a).

The climate of southern Europe is characterized by the scarcity and irregularity of rainfall, the dry period coinciding with the season of highest evapotranspiration. Moreover, the latest studies on the European Mediterranean countries predict significant water-resource depletion, with major declines in the soil-water reserves and more severe periods of lower rainfall during the wet periods (IPCC, 2014). Also, different climate change models predict a significant depletion of groundwater reserves for the river basins of Mediterranean countries, and, apart from sharp reductions in the annual rainfall, average temperatures are projected to rise. In addition, crop water demand is expected to increase (30% compared to current consumption), causing an imbalance between the irrigation-water demand and the available resources (IPCC, 2014).

This new paradigm will compromise the profitability of rain-fed agricultural systems and will strain the supply of water required for the irrigated crops (Trnka et al., 2011; Daccache et al., 2012; Savé et al., 2012).

An overview of the current and future conditions indicates that it is crucial to establish the most profitable methods to boost water productivity by defining the best strategies for efficient and sustainable water management throughout society, this being even more determinant in irrigated agriculture (García-Tejero et al., 2011a). Within the different options, the introduction of alternative crops that can maximize the profitability of irrigated agriculture and the implementation of different strategies to improve the agricultural water management would be the key factors in reaching an equilibrium between farming and the water resources available (COM, 2012). In this context, it is vital to consider that the future scenarios of climate change will require major irrigation restrictions that will hamper the irrigation of traditional crops, making it necessary to consider not only a change in the regional crop distribution but also to increase the irrigation water-use efficiency, such as deficit irrigation (DI) programs (Fereres and Soriano, 2007; García-Tejero et al., 2011b), especially in arid and semiarid areas as the southern regions of Europe (Trnka et al., 2009).

Almond (*Prunus dulcis* Mill) is the third crop in terms of surface area in Spain. The United States, specifically California, is the world's largest producer of this crop, representing 80% of the global market. In Spain, this crop represents almost 40% of the overall cultivated area, and 84% of the almond crop within the European Union, although in terms of production, only 5% of world production (FAOSTAT, 2016).

Notably, the United States with 350,000 ha of almond crop produces more than 800,000 t, whereas Spain, with a total surface to 530,000 ha has an average production of around 211,000 t per year (FAOSTAT, 2016). This low production (in terms of yield per unit of surface) is part of the reason why this crop has traditionally been associated with marginal areas, being cultivated under very limiting conditions and thus producing only $\sim$150 kg ha^{-1}, while in other countries this crop is cultivated under maximum irrigation-water requirement, as in the United States ($\sim$2600 kg ha^{-1}) or Australia ($\sim$2200 kg ha^{-1}). Therefore, the profit margin is relatively low in Spain compared with more productive countries (Almond Board of California, 2015; CAPDR, 2016).

In terms of yield and nut quality, irrigation can be considered the main limiting factor for this crop (Goldhamer and Fereres, 2016). In works related to annual crop water demand, authors such as Goldhamer and Girona (2012) have reported water requirements of between 900 and 1350 mm, depending on the cultivar, rootstock, or tree density, among other factors. Goldhamer and Fereres (2016) reported different production functions for almond crop, reaching the maximum yield values close to 4000 kg ha^{-1} (depending on the season) for irrigation rates of around 1250 mm. By the contrast, when these rates were decreased to 1000 mm, yield values fell 14% with respect to the maximum yield. However, these authors were unable to define the maximum threshold value for the kernel unit weight, which increased progressively from 1.0 to 1.2 g between irrigation amounts of 1000 and 1350 mm, respectively.

Nevertheless, almond is considered a drought-tolerant crop because of its xeromorphic properties (Torrecillas et al., 1996), and, for decades many authors have reported different results related to the effects of strategies of DI applied to this crop, with the aim of improving the management of water scarcity and minimizing the effects in terms of yield and nut quality (e.g., Fereres et al., 1981; Romero et al., 2004, 2006; Girona et al., 2005; Goldhamer et al., 2006 or Egea et al., 2010). Given the maximum irrigation rates required by this tree, its consideration as an alternative crop in arid and semiarid areas of Mediterranean countries would be justified in terms of sustainable development only if DI strategies are considered on the basis of an equilibrium between agricultural and environmental sustainability (García-Tejero et al., 2014a). With the aim of improving its productivity under water-scarcity conditions, the irrigation of this crop requires proper knowledge concerning its phenological development as well as its agronomical and physiological response under different degrees of water stress and DI strategies (Spinelli et al., 2016), especially when water stress can be applied at different crop stages.

2. ALMOND WATER REQUIREMENTS

2.1 Crop Coefficient (K_c)

Crop water requirements of almond orchards have been studied for over three decades (Fereres et al., 1982; Goldhamer and Fereres, 2016). The studies that have determined K_c values for this species concern: (1) soil—water balance approaches, which give nonstressed crop evapotranspiration (ET_c) values as a residual term of the soil—water balance equation and K_c as the ratio of ET_c to reference evapotranspiration (ET_o) (Fereres et al., 1982; García-Tejero et al., 2015; Girona, 2006; Goldhamer, 1989); (2) weighing lysimeters (Espadafor et al., 2015); and (3) the eddy covariance technique, a micrometeorological method that may also provide accurate and representative estimations of crop water requirements (Sanden et al., 2012; Stevens et al., 2012). Table 12.1 summarizes the monthly K_c values determined for almond trees worldwide. Despite the relatively high number of studies performed, these works were conducted in only three countries, i.e., the United States (California), Spain, and Australia. The only study that used weighing lysimetry to estimate almond water use (Espadafor et al., 2015) is not included in Table 12.1 because the experiment was conducted with young trees and because transpiration coefficients (K_T = Crop transpiration/ET_o) instead of K_c values were reported.

By comparing the K_c values recommended by FAO I&D Papers No. 24 and 56 (Allen et al., 1998; Doorenbos and Pruitt, 1977) and by studies conducted three to four decades ago (Goldhamer, 1989) with the more recent estimates of K_c, it can be seen that maximum (midseason) K_c values progressively increased with time. While Goldhamer (1989) and FAO I&D Papers No. 24 and 56 recommended midseason (summer) K_c values of around 0.9—0.95 for almond orchards without a cover crop, Girona (2006) found midseason K_c values of 1.05, García-Tejero et al. (2014b, 2015) and Sanden et al. (2012) of around 1.15, Goldhamer and Fereres (2016) of 1.17 and Stevens et al. (2012) of 1.21. Whereas Girona (2006) maintained an unattended cover crop in summer, the rest of the studies were conducted without a cover crop.

Table 12.1 Estimates of the monthly crop coefficients (K_c) determined for almond orchards

Month	FAO24[a]	FAO56[b]	GOL89[c]	GIR06[d]	SAN12[e]	GT15[f]	GF16[g]	Month	STE12[h]
Mar	0.5	0.40	0.54	0.4	0.88	—	0.2	Sep	0.93
Apr	0.7	0.57	0.63	0.65	1.01	0.55	0.75	Oct	1.11
May	0.85	0.84	0.76	0.8	1.02	0.9	0.95	Nov	1.24
Jun	0.9	0.90	0.85	0.92	1.09	1.05	1.1	Dec	1.3
Jul	0.9	0.90	0.93	0.96	1.14	1.15	1.15	Jan	1.16
Aug	0.9	0.90	0.94	1.05	>1.08	1.05	1.17	Feb	1.21
Sep	0.8	0.90	0.92	0.82	>1.08	0.8	1.12	Mar	>1.03
Oct	0.75	0.75	0.83	0.6	0.96	0.7	0.87	Apr	>0.61
Nov	0.65	0.65	0.7	0.4	0.85	—	—	May	>0.55

[a]Doorenbos and Pruitt (1977): Values for orchards without ground cover crop, from arid regions with light to moderate wind.
[b]Allen et al. (1998): Values for orchards without ground cover crop using the reported lengths of crop development stages for California and Mediterranean regions.
[c]Goldhamer (1989): Values for mature trees (kernel yield of 2–2.5 t ha^{-1}) without cover crop and irrigated with microsprinklers.
[d]Girona (2006): Values derived for mature trees with average kernel yield of 1.8 t ha^{-1}, irrigated with microsprinklers and with an unattended ground cover crop during the summer season. Method of ET_c calculation: soil–water balance.
[e]Sanden et al. (2012): Values for mature, high-yielding (~4.5 t ha^{-1}) almond trees irrigated with microsprinklers without ground cover crop. Method of ET_c calculation: eddy covariance technique. K_c value of Aug and Sep are somewhat underestimated because of the stress of harvest cutoff prior to shaking.
[f]García-Tejero et al. (2014b, 2015): Values derived for drip-irrigated 4-year-old trees without ground cover crop. Method of ET_c calculation: soil–water balance using drainage lysimeters.
[g]Goldhamer and Fereres (2016): Values derived for mature, high-yielding (~4.0 t ha^{-1}) trees irrigated with microsprinklers without ground cover crop. These values are not exactly K_c, but irrigation coefficients (ICs) calculated as the ratio of water applied to ET_o. The very limited effective rainfall (about 60 mm/year) and negligible deep percolation losses makes the calculated ICs for midseason to be very close to actual K_c values.
[h]Stevens et al. (2012): Values derived for mature, high-yielding (4.3–4.6 t ha^{-1}) trees irrigated with microsprinklers with ground cover crop from late winter to late November. Method of ET_c calculation: eddy covariance technique. K_c values of Mar, Apr, and May are somewhat underestimated because of a water stress coefficient, $K_s < 0.97$.

The significantly higher maximum K_c over time, as compared to FAO I&D Papers No. 24 and 56 recommendations and former studies performed in California, indicates that recommendations of almond water requirements have notably increased during recent decades. This marked increase in almond water requirements over the last four decades may be due to the changes implemented during the intensification of the almond productive systems, with higher tree densities, larger tree canopies, with minimal or no pruning, better nutrient management, and much higher fruit loads (Goldhamer and Fereres, 2016; Steduto et al., 2012). It is also important to highlight that the first estimations of almond water use were made in orchards with surface irrigation (border strip). In recent studies, because drip or microsprinkler irrigation was used, the most frequent soil-wetting patterns may have increased surface evaporation, this being another potential factor to explain the higher K_c values (Steduto et al., 2012). Given that K_c values include not only transpiration but also soil evaporation, which might vary markedly depending on the irrigation method and frequency, the K_c values shown in Table 12.1 are reliable only if used under conditions similar to those determined experimentally.

Notably, the two studies that used the eddy covariance technique to derive almond K_c values (Sanden et al., 2012; Stevens et al., 2012) reported, as compared to the maximum midseason values, higher K_c values during the spring season than in the rest of studies. This fact may reveal the limitations of soil-water balance and lysimetry-based methods to estimate soil evaporation with accuracy. This problem may be exacerbated in orchards with large tree spacings, where the differences between lysimeter surface area and tree spacing are so great that the evaporation component of ET_c should be estimated by other means (Lorite et al., 2012).

An important issue to consider when determining orchard water requirements is the year-to-year variation in K_c, since this value will change radically from young to mature trees. When the coverage of the ground by tree canopy is incomplete, there is considerable uncertainty about how to adjust K_c for these situations (Espadafor et al., 2015). In an experiment conducted in California nearly four decades ago, Fereres et al. (1982) found a relation between the percentage of area shaded by young almond trees (S, %) in July and the percentage of ET_c of mature trees:

$$ET_c\ (\%) = -0.0194S^2 + 2.8119S - 1.0080 \tag{12.1}$$

This relationship is valid for S values within 8%–65%. Above 65%, the ET_c (100%) of mature trees is assumed. Eq. (12.1), formulated by

Girona (2006) using Fereres et al. (1982) data, has been widely used since then to determine crop water requirements for many fruit-tree species with incomplete ground covers. More recently, the original relation derived by Fereres et al. (1982) has been completed with data from experiments of several other tree species and readjusted (Steduto et al., 2012):

$$ET_c\ (\%) = -0.012S^2 + 2.26S \tag{12.2}$$

This equation is valid for S varying within the range 0%–70%, assuming 100% of the ET_c of mature trees at higher ground covers.

More complex procedures to estimate K_c as a function of fraction of ground cover and crop height can be found in the FAO I&D Paper No. 56 (Allen et al., 1998) and Allen and Pereira (2009), the latter being a modification of the former. In any case, the differences in the K_c values calculated with these two methods reflect the existing uncertainty to estimate accurately the K_c of developing tree crops (Espadafor et al., 2015).

2.2 Phenological Stages of Almond

Crop phenology in almond can be characterized by different phenological stages (dormant, bloom, fruit development, harvest, and postharvest), which encompasses bud development, flowering, pollination, vegetative growth (leaves and shoots) and kernel growth, the kernel-filling period (dry-matter accumulation), the harvest, and the postharvest period, with the accumulation of reserves prior to leaf fall and bud differentiation (Doll, 2009).

As in other stone fruits, the phenological development in almond is determined by the cold temperatures, although the nut crop does not require long chilling hours, although this condition is determinant for the bud development. Therefore, almond development (vegetative growth and flowering) begins very early (from late January to March, depending on the crop variety), and ends in mid-August to September with the harvest period (Goldhamer and Girona, 2012).

Flowering and leaf budding are the first processes, which occur almost simultaneously (depending on the cultivar, these processes can occur at the same time or within a few days of each other), when the necessary cold hours have been reached and the temperatures start to increase toward the middle or end of winter. These processes, known as the "bloom" stage, depend heavily on the carbohydrate reserves accumulated by the tree during the previous season. Because of this, after the harvest some important processes determine the crop development and final yield for the following seasons. In this sense, after the harvest the accumulation process continues, bolstering the carbohydrate reserves, and exerting a long-term impact on almond-yield components (Goldhamer and Viveros, 2000; Esparza et al., 2001a).

In this context, carbohydrate accumulations during the kernel-filling period and after the harvest, before leaf fall, could determine the fruit growth in the following years (Esparza et al., 2001b). This is consistent with the results pointed out by Girona et al. (2005), showing that in a long-term experiment, in which different degrees of water stress were applied during the kernel-filling period, the kernel dry mass diminished during the last seasons of their experiment. Thus, transport of carbohydrates, and ultimately their availability for the following seasons, can be affected by different water management (e.g., restrictions) applied during the different phenological periods.

Depending on the cultivar, after the necessary cold hours, flowering buds start to develop (bloom period) from an initial stage (dormant spur) to fruit ripening, when the period of rapid fruit growth starts. Before this point, the different flowering stages include: (1) green-tip stage, when the calyx starts to be visible; (2) pink-bud stage, when the corolla starts to be visible, (3) popcorn stage, when the stamens are visible; (4) full bloom; (5) petal fall; and (6) fruit set (Doll, 2009) (Fig. 12.1).

According to Arquero (2013), in southern Spain the bloom period depends on the cultivar and the weather conditions. Thus, for a same cultivar, this period varies according to the area and its climatic characteristics. Without the climate effect being taken into account, and in terms of flowering development, almond cultivars can be classified into three groups: very early varieties (e.g., Desmayo largueta or Garrigues); early cultivars (e.g., Marcona, Cartayera, or Pajarera); and late cultivars (Guara, Marta, Masbovera, Antoñeta, Ferragnès, Ferraduel, or Lauranne).

Whereas proper bloom development is directly affected by crop management during the previous season (since both the accumulation of reserves and bud differentiation for the next season are determined during the preceding one), for this second stage the crop starts to produce new metabolites, this being determinant for healthy vegetative and fruit growth (Doll, 2009).

After fruit set, three different stages of fruit growth can be defined. Stage I is rapid growth, during which the hull, shell, and integuments develop, and the fruit reaches the final size. During this stage, both the hull and shell are easily distinguished, but the integument is a white enclosure that forms

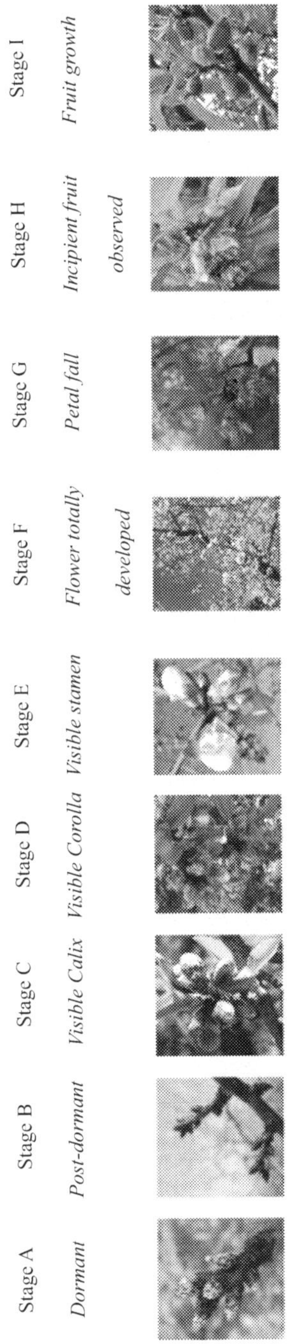

Figure 12.1 Stages of almond flower development.

around the embryo (Doll, 2013). During this first stage (8–10 weeks after fruit set), there is a phase of rapid cell division and afterwards its expansion. At the end of this period (April to May, depending on the cultivar, crop management, and weather), there is an incipient development of the kernel, this being a soft structure, with a gel consistency, even though the hull, shell, and kernel will have reached their final size (Goldhamer and Girona, 2012). Once the fruit reaches its final size, the processes of shell hardening and kernel growth begin (stage II). At the end of this stage (1 month after finishing the stage I; late May to mid-June), the shell will complete the hardening process, and embryo growth prompts changes in the kernel, changing from a watery tissue to an opaque white with a jelly texture (Doll, 2013).

Finally, stage III starts, with the gain in kernel dry weight. During this phase, dry matter accumulates in the kernel until the end of the maturity period. In this stage, the cell development of hull, shell, and kernel finishes, and the dry weight of pericarp does not change. The kernel carbohydrates stored in the fruit change to fats and proteins and the ripening period begins. In this process, two changes occur at the end of the fruit development: endocarp dehiscence and the formation of an abscission layer between the fruit and its peduncle. Although once fruit development has ended, the hull-splitting stage limits nut processing by the farmer, this being a quality parameter (the ease of separating the hull from the shell). Thus, a severe water-stress situation during this stage would promote the formation of tight hulls (Goldhamer and Girona, 2012).

2.3 Different Tools to Assess the Almond Water Status

In a world facing increasingly scarce supplies of irrigation water, it is essential for farmers to adopt irrigation technologies and management practices that contribute to produce more yield with less water. This is especially relevant for almond growers due to the high water requirements of this species (Goldhamer and Fereres, 2016). Optimizing irrigation scheduling, especially DI, has been demonstrated to substantially improve the harvest while saving water (Jones, 2004). In this regard, during the past two decades a number of studies on field-grown almond trees have assessed the suitability of several plant-sensing techniques.

2.3.1 Plant Water Potential

The potential of estimating plant water stress through plant water potential lies in its integrative effect of soil, plant, and atmospheric conditions on the measure of water availability within the plant itself (Shackel, 2011). Leaf water potential (Ψ_{leaf}) of fully exposed leaves, which has been the typical measure of plant water potential during decades, has proven to be unreliable as a water-stress indicator under field conditions (Nortes et al., 2005), mainly due to the effect that the leaf microclimate exerts on Ψ_{leaf} through the interaction with stomatal conductance. The finding that stem water potential (Ψ_{stem}) could better indicate differences among irrigation regimes in *Prunus* species (Garnier and Berger, 1985) opened the possibility of extending the use of this water-status indicator for irrigation control of almond trees and other species (Mirás-Avalos et al., 2016; Shackel et al., 1997). The Ψ_{stem} is measured indirectly through attached leaves that have been prevented from transpiring by enclosure in a darkened plastic bag, sometimes overlaid with silver foil (Fereres and Goldhamer, 2003), to allow equilibration with Ψ_{stem} (Shackel et al., 1997).

The time required for the covered leaf to equilibrate with Ψ_{stem} prior to measurement is somewhat uncertain, as the reported times vary from periods as brief as 10 min (Shackel et al., 1997) to periods of up to 2 h (Fereres and Goldhamer, 2003). Sometimes, even uncovered leaves are used as a surrogate of Ψ_{stem} (Goldhamer and Fereres, 2016), and Goldhamer and Fereres (2001b) demonstrated that water potential of shaded leaves (Ψ_{shaded}) covered with a damp cloth immediately prior to excision and quickly transported to the pressure chamber were well correlated with Ψ_{stem} in almond [Ψ_{shaded} (MPa) = 0.902·SWP (MPa) − 0.112; $R^2 = 0.942$].

Concerning the optimal time of the day for Ψ_{stem} measurement, there is general consensus that Ψ_{stem} measured at midday under maximum evaporative demand is a reliable and consistent measure of almond tree water stress for irrigation control (Fereres and Goldhamer, 2003; Shackel, 2011; Shackel et al., 1997). To determine whether a given Ψ_{stem} value denotes well-irrigated or water-stressed conditions, a reference Ψ_{stem} representing well-irrigated conditions must be used. These reference values are not constant but they vary with the evaporative demand. McCutchan and Shackel (1992), studying well-irrigated prune trees, found a strong relationship (Eq. 12.3) between midday Ψ_{stem} and midday vapor pressure deficit (VPD).

$$\Psi_{stem}\ (\text{MPa}) = -0.12 \cdot \text{VPD}\ (\text{kPa}) - 0.41\ (R^2 = 0.81) \tag{12.3}$$

This relationship has proven also to be a consistent predictor of Ψ_{stem} of well-irrigated almond trees (Dhillon et al., 2014; Shackel, 2011; Shackel et al., 1997).

2.3.2 Trunk-Diameter Variations

Measurements of trunk-diameter variations (TDV) have proven to be highly sensitive to variations in almond tree water status. Several plant water-stress indicators can be determined from TDV measurements (Goldhamer and Fereres, 2001a), of which maximum daily trunk shrinkage (MDS) has been demonstrated to be the most reliable parameter for irrigation scheduling of almond trees (Egea et al., 2009a; Goldhamer and Fereres, 2004; Puerto et al., 2013).

As occurs with Ψ_{stem}, the MDS of well-irrigated trees is also affected by atmospheric conditions. Thus, reference MDS values for nonstressed trees need to be calculated for the prevailing atmospheric conditions on a given day in order to be used as threshold values for irrigation scheduling. Unlike what has been observed for midday Ψ_{stem}, whose reference baseline has proven to be consistent under different locations, years, and even different *Prunus* species (McCutchan and Shackel, 1992; Shackel et al., 1997), MDS baselines derived for almond differed among almond orchards (Egea et al., 2009a; Puerto et al., 2013). The reason for this lack of consistency of the MDS baselines lies in the variety of factors that have been documented to affect MDS, such as phenological stages, tree size, trunk diameter, or crop load (Egea et al., 2009a; Fereres and Goldhamer, 2003; Intrigliolo and Castel, 2007, 2006).

To overcome this drawback while taking advantage of the negligible effect that almond phenology plays on the MDS baseline, Egea et al. (2009a) proposed a practical procedure based on using 15–20 days of the early growth stages II–III to determine the MDS baselines periodically (every 1 or 2 years) for use as reference MDS for the subsequent growth stages. Like Ψ_{stem} baselines, the MDS of well-irrigated almond trees also show a linear response with VPD. Egea et al. (2009a) and Puerto et al. (2013) found that mean VPD of the period 10–15 h GMT was the best predictor of MDS, whereas Goldhamer and Fereres (2004) reported an MDS baseline calculated with daily average VPD.

2.3.3 Canopy Temperature

The use of canopy temperature to monitor plant water status was proposed in the 1960s (Fuchs and Tanner, 1966) and has been extensively used ever since for irrigation decision making in multiple crops. With the advent of modern high-resolution thermal cameras and unmanned aerial vehicles, the technique is being revived today due to its potential to be used in remote-sensing applications that allow, for example, mapping spatial variability in crop water status (Bellvert et al., 2016).

In almond trees, the technique has been evaluated at three levels: measuring temperature of single leaves with an infrared thermometer (Dhillon et al., 2014); measuring canopy temperature with high-resolution thermal cameras placed at proximal distances from the canopy (proximal sensing) (García-Tejero et al., 2012); and measuring canopy temperature with thermal cameras onboard aircrafts (remote sensing) (Gonzalez-Dugo et al., 2012). In all cases, the results for irrigation management were promising. Dhillon et al. (2014) classified trees into water stressed and unstressed using the leaf-temperature measurements along with microclimatic variables. García-Tejero et al. (2012) developed a methodological protocol for analyzing thermal images and established sound relationships between the canopy-to-air temperature differences with Ψ_{stem} and stomatal conductance. González-Dugo et al. (2012) found that intracrown temperature variation may be a useful indicator of the onset of tree water stress, since it increased from fully irrigated trees to intermediate irrigation levels, diminishing afterwards in the most stressed treatments.

3. DEFICIT-IRRIGATION STRATEGIES FOR WATER-SCARCITY CONDITIONS

In the previous sections, we have defined the main phenological stages of almond, these being flowering, fruit development, bud formation and differentiation, and, finally, the accumulation of reserves. Moreover, we have established how to estimate the almond water requirements with no water restrictions. However, as we have discussed during this chapter, despite that irrigation is considered the most important factor in terms of yield and fruit quality (Egea et al., 2009b), almond has a well-known capacity to withstand water stress (Fereres and Goldhamer, 1990; Hutmacher et al., 1994). In this sense, this species is considered resistant to drought (Castel and Fereres, 1982) and salinity (Franco et al., 2000), and, therefore, although almond is cultivated in irrigated lands, it is also grown worldwide under rain-fed conditions. Thus, DI strategies offer a suitable opportunity to achieve good commercial results, with significant water savings and without compromising yield or nut quality (Fereres and Soriano, 2007). As in other tree crops, a moderate water-stress strategy can be imposed with a positive impact, i.e., to control an excessive vegetative growth, to save water, or to ease the harvest and reduce fungal diseases (Chalmers et al., 1981; Goldhamer et al., 2006; Teviotdale et al., 2001). In this context, it bears distinguishing between the need to define a proper DI strategy, which will be closely related to the phenology as well as to the physiology of a particular crop. That is, knowledge is needed concerning the physiological characteristics that make this crop a good alternative for arid and semiarid areas with water-limiting conditions (Spinelli et al., 2016).

3.1 Physiological Response of Almond to Water Stress

In physiological terms, when a crop undergoes mild-to-moderate water stress, the first response is related to the stomatal conductance (g_S) reduction (Hsiao, 1990; Torrecillas et al., 1996), i.e., minimizing the water loss through the stomata. Moreover, almond is not considered as an isohydric crop; that is, its capability of regulating g_S when subjected to water stress is limited (and strongly dependent on the cultivar), as has been discussed by authors such as Wartinger et al. (1990), Egea et al. (2011), or Eichi (2013).

In short, the drought responses of this tree involve different resistance processes to avoid water stress. Thus, prior to a significant reduction in the g_S, this crop responds with sharp decreases in Ψ_{stem} or Ψ_{leaf}. As reflected in Fig. 12.2, whereas significant differences in terms of Ψ_{leaf} appear between two irrigation treatments in almond trees (cv. Guara), this contrast may not be as evident for stomatal conductance.

However, when the water stress applied is severe and lasts for a long period, the almond tree commonly undergoes leaf senescence, promoting a leaf-area reduction and thus canopy conductance (Spinelli et al., 2016). In this situation, the almond crop could maintain acceptable levels of g_S (promoting significant decreases in the crop water potential), while keeping optimum values of carbon assimilation, photosynthetic rate, and hence

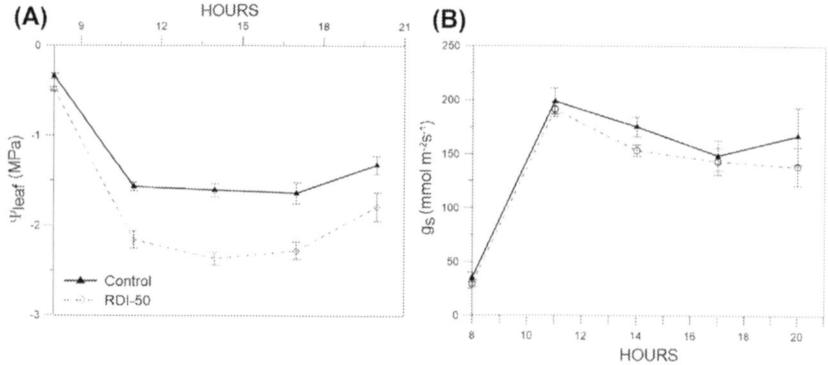

Figure 12.2 Leaf water potential (Ψ_{leaf}) (A) and stomatal conductance (g_S) (B) dynamics: control trees irrigated at 100% ET_C versus RDI-50 irrigated at 50% ET_C.

increasing intrinsic water-use efficiency (WUE) (McCutchan and Shackel, 1992; Rouhi et al., 2007).

Nevertheless, Spinelli et al. (2016) found highly significant relationships between the g_S values ($R^2 = 0.74$) and photosynthetic rate (A) ($R^2 = 0.61$) with Ψ_{stem} values ranging between −0.5 and −2.0 MPa, depending on the irrigation schedule. Also, this pattern depends on the cultivar, as argued by Gomes-Laranjo et al. (2006), according to whom the optimum range of predawn (Ψ_{pd}), Ψ_{leaf} at midday, g_S and A for different cultivars could not be considered similar (Table 12.2) when these were properly irrigated.

Goldhamer and Viveros (2000) reported minimum Ψ_{pd} values of around −4.0 MPa, although these values were associated with highly significant reductions in terms of yield during the current and next season. Moreover, the relationship between midday Ψ_{stem} and MDS showed a threshold value of −2 MPa (Puerto et al., 2013), which is comparable with those found in olive trees (Moriana et al., 2000) and higher than in other fruit trees such as lemons or peaches (Ortuño et al., 2010).

Under fully irrigated conditions, the daily cycle of gas exchange is almost constant when radiation is not limited (Torrecillas et al., 1988; Klein et al., 2001; Romero and Botia, 2006) and the vapor pressure deficit is not higher than 2 KPa (Romero and Botia, 2006). Values of leaf conductance vary over the season according to the phenological stage, producing a hysteresis loop between water potential and leaf conductance with maximum values around May (Torrecillas et al., 1988) and maximum net photosynthesis during spring (Nortes et al., 2009). The decrease in net photosynthesis, even under full irrigation, has been reported to be significant from the second part of the kernel-filling stage (Nortes et al., 2009).

Water-stress conditions trigger partial stomatal closure in the daily cycle (Torrecillas et al., 1988), this becoming more intense at midday than in the morning (Klein et al., 2001) and reducing the plant's sensitivity to

Table 12.2 Optimum values for predawn, leaf water potential at midday, stomatal conductance, and photosynthesis rate for different cultivars (Gómes-Laranjo et al., 2006)

Cultivar	Ψ_{pd} (MPa)	Ψ_{leaf} (MPa)	g_S (mmol m^{-2} s^{-1})	A (μmol m^{-2} s^{-1})
Glorieta	−0.6 to −0.9	−1.7 to −2.6	280−458	7.1−9.6
Ferragnes	−0.7 to −0.8	−1.7 to −2.4	281−418	7.9−10.7
Masbovera	−0.7 to −0.8	−1.9 to −2.4	223−360	7.2−10.1
Lauranne	−0.9 to −1.0	−2.0 to −2.3	291−405	8.7−9.8

evaporative demand (Romero and Botia, 2006). The net photosynthesis rate is limited under moderate water stress due to stomatal closure, but the effect on the photosynthetic system has been reported when Ψ_{pd} was lower than -1.5 MPa (Romero et al., 2004). Although the sensitivity of net photosynthesis to water deficit differs between cultivars, a 50% reduction was reported in a range between -2.5 and -3.5 MPa of Ψ_{leaf} (Gomes-Laranjo et al., 2006).

DI strategies are based on a great capacity of recovery after a period of water stress. Almond trees quickly recover their water potential, but lag in the gas-exchange values (Torrecillas et al., 1996; Romero et al., 2004; Nortes et al., 2009), a delay that could become persistent with repeated deficit treatments (Nortes etal., 2009). However, although significant recovery has been reported, water-stress conditions before harvest are commonly related to greater leaf abscission (Klein et al., 2001; Romero et al., 2004) and higher percentages of full hull splitting (Goldhamer and Viveros, 2000; Puerto et al., 2013), indicating a drought-avoidance response.

3.2 Nut Yield and Quality Responses to Water Stress

The results of DI scheduling in terms of nut yield and quality are affected mainly by the intensity of the water stress and the phenological period when water deficit is applied (Fig. 12.3). Nortes et al. (2009) suggested six phenological periods from flowering to postharvest. To date, the main experiments based on DI strategies in almond are concentrated exclusively during kernel filling (stage IV) and postharvest (stage V) periods. However, as discussed earlier, the period of water stress determines whether the effects on yield occur in the current or in the next season.

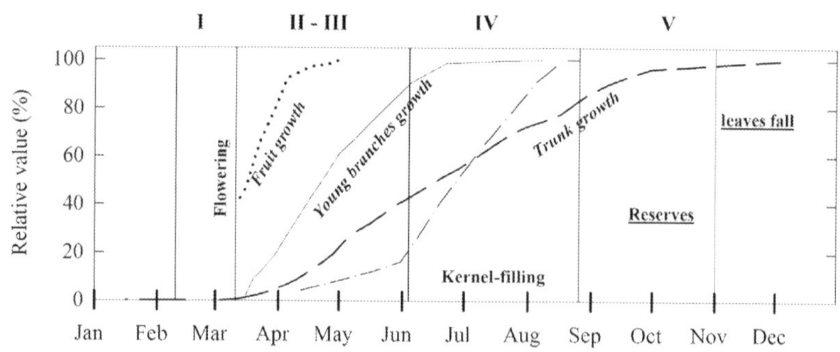

Figure 12.3 Phenological stages for almond cv. Marta in Murcia, Spain (Nortes et al., 2009).

However, several irrigation studies performed over at least three seasons reported that the final current yield was reduced when the water stress was applied during the kernel-filling period (Hutmacher et al., 1994; Goldhamer and Viveros, 2000; Girona et al., 2005; Egea et al., 2013; Alcón et al., 2013; Mañas et al., 2014). In contrast, other works have reported no significant differences in yield when water stress occurred in this phenological stage (Goldhamer and Fereres, 2004; Romero et al., 2004; Egea et al., 2009b, 2010; Cornacchia et al., 2010; Puerto et al., 2013).

This controversy could be related not only to the timing but also the water-stress level imposed during the phenological stage considered, suggesting threshold values for irrigation scheduling. Fig. 12.4 shows information on

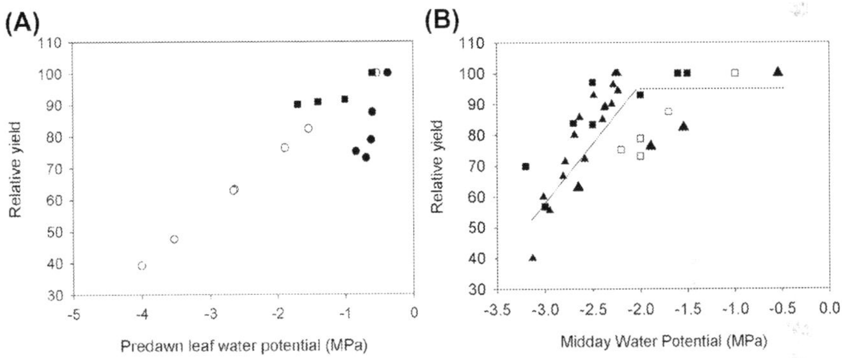

Figure 12.4 Relationship between relative yield and predawn (A) and midday (B) leaf water potential. *((A) Based on Girona, J., Mata, M., Marsal, J., 2005. Regulated deficit irrigation during the kernel-filling period and optimal irrigation rates in almond. Agricultural Water Management 75, 152–167,* solid square symbols, *and Egea, G., González-Real, M.M., Baille, A., Nortes, P.A., Sánchez-Bel, P., Domingo, R., 2009b. The effects of contrasted deficit irrigation strategies on the fruit growth and kernel quality of mature almond trees. Agricultural Water Management 96, 1605–1614,* solid circle symbols *and Romero, P., Botía, P., García, F., 2004. Effects of regulated deficit irrigation under subsurface drip irrigation conditions on vegetative development and yield of mature almond trees. Plant and Soil 260, 169–181,* blank circle. *(B) Based on Puerto, P., Domingo, R., Torres, R., Pérez-Pastor, A., García-Riquelme, M., 2013. Remote management of deficit irrigation in almond trees based on maximum daily trunk shrinkage. Water relations and yield. Agricultural Water Management 126, 33–45,* solid square symbols, *and Egea, G., González-Real, M.M., Baille, A., Nortes, P.A., Sánchez-Bel, P., Domingo, R., 2009b. The effects of contrasted deficit irrigation strategies on the fruit growth and kernel quality of mature almond trees. Agricultural Water Management 96, 1605–1614,* blank square symbols, *and Hutmacher et al. (1994),* solid circle. *Values of Puerto et al. (2013) and Egea et al. (2009b) were for covered leaves while values of Hutmacher et al. (1994) were for uncovered leaves (authors' own preparation).)*

different experiments where midday or predawn leaf water potential during kernel filling is related to relative yield (considering the maximum yield in each season as 100%). The predawn leaf water potential (Fig. 12.4A) presented an almost lineal relationship with yield but with different slopes. Data of Girona et al. (2005) and the regression equation formulated by Puerto et al. (2013) were very close, while data of Egea et al. (2009b) showed a steeper slope and a narrow range of variations. According to this figure a suggestion based on this indicator is not easy, though Romero et al. (2004) suggested −2 MPa of predawn leaf water potential as possible threshold value.

On the other hand, Fig. 12.4B presented data of midday leaf water potential, all the data corresponding to different orchards. This indicator presented a pattern which could be almost constant until values of around −2 MPa, with a lineal decrease from this point onward. A similar threshold (−2.5 MPa) was suggested in Hutmacher et al. (1994) and Puerto et al. (2013).

Girona et al. (2005) suggested that kernel-weight variation was the main factor involved in yield reduction when DI was imposed during kernel filling. In line with this, the base of the regulated DI during kernel filling is that during this stage the almond is more drought resistant than during other stages, such as vegetative growth (Girona et al., 2005; Egea et al., 2013). Fig. 12.5 represents the relationship between kernel weight and water potential in three different works. Note that although each work was performed in different cultivars (Nonpareil, Marta, and Guara) the weight values are comparable.

The variation pattern is similar to those reported in Fig. 12.4. While Ψ_{pd} presented a lineal decrease with values higher than −1 MPa, this threshold clearly changed in the slope. The pattern of midday Ψ_{leaf} was clearly different, again with a lineal decrease from values of around −2 MPa.

A postharvest water deficit diminishes the fruit load in the next season (Goldhamer and Viveros, 2000; Esparza et al., 2001a; Girona et al., 2005; Goldhamer et al., 2006). The few works that compare kernel filling and postharvest water stress conclude that the latter is more sensitive than the former (Goldhamer and Viveros, 2000; Girona et al., 2005; Goldhamer et al., 2006). Esparza et al. (2001a) suggested that this reduction would occur with stem water potential lower than −1.2 MPa and with a rate around −7.7 kg MPa^{-1}. According to this threshold value, some of the yield loss reported when the water stress is imposed during the kernel-filling period could be related to a delay in rehydration during postharvest (i.e., Mañas et al. (2014) with similar yield values under irrigated and rain-fed conditions). The yield reduction in the next season appears to be related

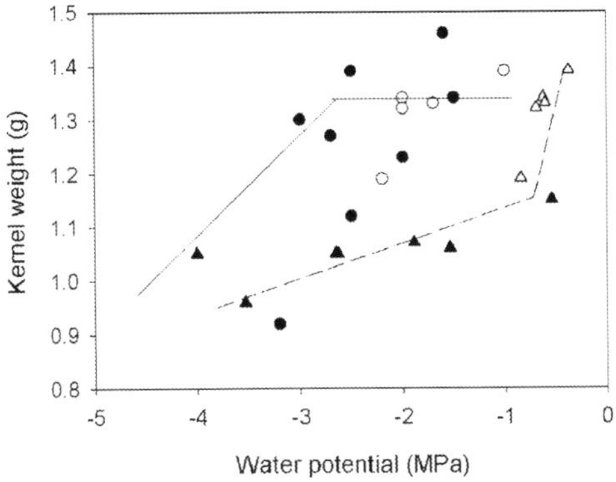

Figure 12.5 Relationship between kernel weight and predawn leaf water potential (*triangles*) and midday stem water potential (*circles*). *(Based on Goldhamer, D.A., Viveros, M., 2000. Effects of preharvest irrigation cutoff durations and postharvest water deprivation on almond tree performance. Irrigation Science 19, 125–131,* solid triangles, *and Puerto, P., Domingo, R., Torres, R., Pérez-Pastor, A., García-Riquelme, M., 2013. Remote management of deficit irrigation in almond trees based on maximum daily trunk shrinkage. Water relations and yield. Agricultural Water Management 126, 33–45,* solid circles, *and Egea, G., González-Real, M.M., Baille, A., Nortes, P.A., Sánchez-Bel, P., Domingo, R., 2009b. The effects of contrasted deficit irrigation strategies on the fruit growth and kernel quality of mature almond trees. Agricultural Water Management 96, 1605–1614,* blank triangles, blank circles *(authors' own preparation).)*

to the decline in carbohydrates (Esparza et al., 2001b) and likely in nitrogen reserves (Nortes et al., 2009) in the whole tree.

Water-deficit conditions are applied in the almond literature in three different ways: regulated deficit irrigation (RDI), sustained deficit irrigation (SDI), and partial root drying (PRD). According to the yield results, differences between RDI during kernel filling and SDI are not clear. Some works reported greater yield in RDI than in SDI when similar amounts of water are applied (Goldhamer et al., 2006) and even similar to fully irrigated conditions (Girona et al., 2005). However, in other works, these differences were not significant, though RDI tended to values lower than SDI (Girona et al., 2005; Alcón et al., 2013; Egea et al., 2013).

Comparing RDI strategies with PRD, Egea et al. (2009b, 2011) reported that a PRD treatment irrigated at 50% ET_c reported yield values similar to those of an RDI treatment irrigated at 70% ET_c. Moreover, only when this PRD was irrigated at 30% ET_c were these yield reductions significantly lower than under RDI (Egea et al., 2009b). However, neither water potential nor gas exchange significantly differed between RDI and PRD 50% (Egea et al., 2011), suggesting no effect of the root signal.

Nut quality is usually not considered in horticultural works, with hull splitting and kernel weight being the common indicators for evaluating the yield quality. Greater hull splitting on a given date would indicate an avoidance strategy of the almond. However, such a response is not commonly reported, and only works with moderate water-stress levels present such results (Goldhamer and Fereres, 2004; Puerto et al., 2013). Most works conclude that water deficit reduces hull splitting, though this is usually very high (Goldhamer and Viveros, 2000; Romero et al., 2004; Goldhamer et al., 2006; Egea et al., 2009b). Other effects, such as hull rot (Goldhamer and Viveros, 2000) and sticktights (Goldhamer et al., 2006), were less notable at more severe water-deficit levels. No clear effect was found in the percentage of empty and double nuts (Egea et al., 2009b).

The composition of almonds has been studied in relation with irrigation management, though such features are more affected by cultivar than by water management (Sánchez-Bel et al., 2008a). The main changes in the almond composition occur up to 100 days after fruit set, and later changes are related to the variation in kernel moisture (Sánchez-Bel et al., 2008b). Egea et al. (2009b) provided a clear description of the almond water status. In this work, no effect was found between irrigation treatments in the quantity and composition of carbohydrates and organic acids, lipids, and protein content, either during ripening or at harvest (Egea et al., 2009b). However, in comparisons between irrigated and nonirrigated orchards in experiments where the water status was not reported, irrigation was not found to affect lipid or carbohydrate quantity but it changed composition (Nanos et al., 2002; Sánchez-Bel et al., 2008a). By contrast, irrigation increased oleic acid and sucrose but decreased palmitic and linoleic acid (Nanos et al., 2002; Sánchez-Bel et al., 2008a). Cornacchia et al. (2010) in an experiment of storage at room temperature concluded that full irrigation and deficit irrigation did not affect the composition of the nuts, and only nonirrigated almonds lowered antioxidant and phenol contents and raised sugar and sucrose values.

The adoption of modern water-saving technologies and irrigation-management strategies (e.g., RDI) that maintain, or only slightly reduce, production levels while lowering input requirements could be key to guaranteeing irrigation sustainability in arid and drought-prone areas. However, despite that a great number of RDI experiments have been conducted in multiple herbaceous and tree crops, very few studies have analyzed the economic feasibility of adopting RDI strategies by growers, which represents an important drawback in terms of accelerating the adoption of environmentally friendly agricultural practices. In any case, almond is an exception because three studies have evaluated the economic performance of almond orchards under RDI, although the weak point is that the three case studies were conducted in the same region (southeast Spain). García et al. (2004), for example, carried out a cost—benefit analysis to evaluate the economic impact of implementing RDI in a 13-year-old almond orchard. The authors demonstrated that profit was higher in RDI and the production costs lower than under full irrigation. Two years later, Romero et al. (2006) assessed the economic feasibility of implementing jointly RDI and subsurface-drip irrigation in an adult almond orchard. Their data revealed that an RDI strategy based on reducing 80% of ET_c during kernel filling and 50% during postharvest under subsurface-drip irrigation conditions could be even more economically profitable than full irrigation in regions where the water price is high. More recently, Alcón et al. (2013) evaluated the long-term (6 years) economic feasibility of implementing several DI strategies in almond orchards from plantation. They established four irrigation treatments: full irrigation, RDI (mean water saving of c.45%), sustained DI with moderate water stress, SDI_m (mean water saving of c.35%), sustained DI with moderate to severe water stress, SDI_s (mean water saving of c.63%). Alcón et al. (2013) pointed out that RDI and SDI_m were the most economically feasible treatments over the long term, whereas FI and SDI_s presented lower, albeit similar, economic profitability.

4. CONCLUSIONS

Increasing water scarcity and drought indicate the urgency of achieving a more sustainable approach to water-resource management in agriculture. Agricultural water-use efficiency and water productivity needs to be improved urgently in water-scarce areas. Today many strategies are being implemented to improve water productivity, starting with the irrigation system, followed by the application of the proper scheduling in terms of both timing and quantity of irrigation water applied, and concluding with the choice of the best crop management with regard to the environmental conditions.

In this context, a promising approach is the use of deficit-irrigation strategies, which can both save water and induce plant physiological regulation such as stomatal opening, reproductive development, and growth rate. There is a special interest in whether crop water use can be regulated as a function of managing the physiological aspects in water-shortage scenarios. If so, then agricultural water resources can be reallocated to benefit situations of water scarcity.

Consequently, the main mission will be to build an integrative system to control crop water use during different phenological periods and actively regulate the plant's growth, productivity, and development based on physiological responses in a given crop under adverse environmental conditions such as water scarcity. In addition, as it has been demonstrated, the knowledge of the crop's physiological behavior (tolerance thresholds and baseline information) is crucial in order to manage the water stress, avoiding excessive deficit that could provoke irreversible damage. In this sense, the long-term effects of deficit irrigation should be examined, because the long-term plant responses could be different from short-term responses. Therefore, an understanding of the periods of plant sensitivity to water stress is essential so that the deficit is applied at times when the negative impact on yield and nut quality is lessened.

In particular, this chapter explores the proposition that deficit irrigation can provide reasonable yield in almond trees compared with full irrigation, representing higher water productivity. On the other hand, deficit irrigation demands more control over the amount and timing of irrigation application than does full irrigation. However, with crop water production functions and the sensitive stages of almond to water stress, optimal deficit irrigation can be scheduled with a minimum yield reduction compared to full irrigation. In addition, the crop production functions could solve the problem of allocation of limited water resources in crops competing for limited available water.

Thus, deficit irrigation can be considered a key strategy for boosting the water-use efficiency and water productivity, and in encouraging sustainable water-saving strategies for almond orchards in zones with reduced agricultural water resources.

ACKNOWLEDGMENTS

Part of this work was sponsored by the research project "Integrated management of the cultivation of almond and other nuts (INNOVA-Nuts)" (AVA.AVA201601.18), within the Operational Program FEDER 2014—20 "Andalusia moves with Europe." The author I. Garcia-Tejero has a contract cofinanced by the Operational Program of the European Social Fund (ESF) 2007—13 "Andalusia moves with Europe."

REFERENCES

Alcón, F., Egea, G., Nortes, P.A., 2013. Financial feasibility of implementing regulated and sustained deficit irrigation in almond orchards. Irrigation Science 31, 931–941.

Almond Board of California, 2015. Almond Industry Position Report. Available from: http://www.almonds.com/sites/default/files/content/newsletters/attachments/2016.07posrpt.pdf.

Allen, R.G., Pereira, L.S., 2009. Estimating crop coefficients from fraction of ground cover and height. Irrigation Science 28, 17–34.

Allen, R.G., Pereira, L.S., Raes, D., Smith, M., 1998. Crop evapotranspiration. In: FAO Irrigation and Drainage Paper 56. FAO, Rome.

ARA, 2011. Agenda del Regadío Andaluz, Horizonte 2015, Consejería de Agricultura y Pesca, Junta de Andalucía, 127 p.

Arquero, A., 2013. Manual del Almendro, Consejería de Agricultura, Pesca y Desarrollo Rural. Junta de Andalucía, 78 p.

Bellvert, J., Marsal, J., Girona, J., Gonzalez-Dugo, V., Fereres, E., Ustin, S.L., Zarco-Tejada, P.J., 2016. Airborne thermal imagery to detect the seasonal evolution of crop water status in peach, nectarine and saturn peach orchards. Remote Sensing 8, 1–17.

CAPDR, 2016. Caracterización del sector de la almendra en Andalucía. Secretaría General de Agricultura y Alimentación, Consejería de Agricultura, Pesca y Desarrollo Rural. Junta de Andalucía, 34 p. Available from: http://www.juntadeandalucia.es/agriculturaypesca/observatorio/servlet/FrontController?action=RecordContent&table=12030&element=1654785.

Castel, J.R., Fereres, E., 1982. Responses of young almond trees to two drought periods in the field'. Journal of Horticultural Science and Biotechnology 57, 175–187.

Chalmers, D.J., Mitchell, P.D., Van Heek, L.A.G., 1981. Control of peach tree growth and productivity by regulated water supply, tree density, and summer pruning. Journal of American Society of Horticultural Science 106, 307–312.

COM, 2012. Comunicación de la Comisión al Parlamento Europeo y al Consejo sobre la cooperación de innovación europea. In: Productividad y sostenibilidad agrícolas, Bruselas, 29.2.2012. Available from: http://ec.europa.eu/agriculture/eip/pdf/com 2012-79_es.pdf.

Cornacchia, R., Amodio, M.L., Colelli, G., Tortosa, P.A.N., 2010. Effect of irrigation water reduction strategies on quality at harvest and during storage of in-shell almonds. Acta Horticulturae 887, 251–260.

Daccache, A., Keay, C., Jones, R.J.A., Weatherhead, E.K., Stalham, M.A., Knox, J.W., 2012. Climate change and land suitability for potato production in England and Wales: impacts and adaptation. Journal of Agricultural Science 150, 161–177.

Dhillon, R., Udompetaikul, V., Rojo, F., Roach, J., 2014. Detection of plant water stress using leaf temperature and microclimatic measurements in almond, walnut, and grape crops. Trans ASABE 57, 297–304.

Doll, D., 2009. The Seasonal Patterns of Almond Production. University of California Cooperative Extension. Available from: http://thealmonddoctor.com/2009/06/22/the-seasonal-patterns-of-almond-production/.

Doll, D., 2013. Many factors can affect almond size and quality. In: Growing Produce. Available from: http://www.growingproduce.com/nuts/many-factors-can-affect-almond-size-and-quality/.

Doorenbos, J., Pruitt, W.O., 1977. Crop water requirements. In: Irrigation and Drainage. Paper 24. FAO, Rome.

Eichi, V.R., 2013. Water Use Efficiency in Almonds (*Prunus dulcis* (Mill)D.A. Web) (Ph.D. thesis). School of Agriculture, Food and Wine. Faculty of Science. Univ. of Adelaide, 77 p. Available from: https://digital.library.adelaide.edu.au/dspace/bitstream/2440/87112/8/02whole.pdf.

Egea, G., Pagán, E., Baille, A., Domingo, R., Nortes, P.A., Pérez-Pastor, A., 2009a. Usefulness of establishing trunk diameter based reference lines for irrigation scheduling in almond trees. Irrigation Science 27, 431–441.

Egea, G., González-Real, M.M., Baille, A., Nortes, P.A., Sánchez-Bel, P., Domingo, R., 2009b. The effects of contrasted deficit irrigation strategies on the fruit growth and kernel quality of mature almond trees. Agricultural Water Management 96, 1605–1614.

Egea, G., Nortes, P.A., Gonzalez-Real, M.M., Baille, A., Domingo, R., 2010. Agronomic response and water productivity of almond trees under contrasted deficit irrigation regimes. Agricultural Water Management 97, 171–181.

Egea, G., Dodd, I.C., González-Real, M.M., Domingo, R., Baile, A., 2011. Partial rootzone drying improves almond tree leaf-level water use efficiency and afternoon water status compared with regulated deficit irrigation. Functional Plant Biology 38, 372–385.

Egea, G., Nortes, P.A., Domingo, R., Baille, A., Pérez-Pastor, A., Gónzalez-Real, M.M., 2013. Almond agronomic response to long term deficit irrigation applied since orchard establishment. Irrigation Science 31, 445–454.

Espadafor, M., Orgaz, F., Testi, L., Lorite, I.J., Villalobos, F.J., 2015. Transpiration of young almond trees in relation to intercepted radiation. Irrigation Science 33, 265–275.

Esparza, G., DeJong, T.M., Weinbaum, S.A., Klein, I., 2001a. Effects of irrigation deprivation during harvest period on yield determinants in mature almond trees. Tree Physiology 21, 1073–1079.

Esparza, G., DeJong, T.M., Weinbaum, S.A., 2001b. Effects of irrigation deprivation during the harvest period on nonstructural carbohydrate and nitrogen contents of dormant, mature almond trees. Tree Physiology 21, 1081–1086.

FAOSTAT, 2016. Food and Agriculture Organisation of the United Nations. Available from: http://www.fao.org/faostat/en/#data/QC.

Fereres, E., Soriano, M.A., 2007. Deficit irrigation for reducing agricultural water use. Journal of Experimental Botany 58, 147–159.

Fereres, E., Goldhamer, D.A., 2003. Suitability of stem diameter variations and water potential as indicators for irrigation scheduling of almond trees. Journal of Horticultural Science and Biotechnology 78, 139–144.

Fereres, E., Goldhamer, D.A., 1990. Deciduous fruits and nuts. In: Stewart, B.A., Nielsen, D.R. (Eds.), Irrigation of Agricultural Crops, ASA Monograph, vol. 30, pp. 987–1017. Madison, WI.

Fereres, E., Martinich, D.A., Aldrich, T.M., Castel, J.R., Holzapfel, E., Schulbach, H., 1982. Drip irrigation saves money in young almond orchards. California Agriculture 36, 12–13.

Fereres, E., Aldrich, T.M., Schulbach, H., Martinich, D.A., 1981. Responses of young almond trees to late season drought. California Agriculture 35, 11–12.

Franco, J.A., Abrisqueta, J.M., Hernansaez, A., Moreno, F., 2000. Water balance in a young almond orchard under drip irrigation with water of low quality. Agricultural Water Management 43, 75–98.

Fuchs, M., Tanner, C.B., 1966. Infrared thermometry of vegetation. Agronomy Journal 58, 597–601.

García-Tejero, I.F., Durán-Zuazo, V.H., Rodríguez, C.R., Muriel-Fernández, J.L., 2011a. Water and Sustainable Agriculture. Springerbriefs in Agriculture, 94 p.

García-Tejero, I.F., Durán-Zuazo, V.H., Jiménez-Bocanegra, J.A., Muriel, J.L., 2011b. Improved water use efficiency by deficit irrigation programs: implications for saving water in citrus orchards. Scientia Horticulturae 124, 278–282.

García-Tejero, I., Durán-Zuazo, V.H., Arriaga, J., Hernández, A., Vélez, L.M., Muriel-Fernández, J.L., 2012. Approach to assess infrared thermal imaging of almond trees under water-stress conditions. Fruits 67, 463–474.

García-Tejero, I.F., Durán-Zuazo, V.H., Muriel-Fernández, J.L., 2014a. Towards sustainable irrigated Mediterranean agriculture: implications for Water conservation in semi-arid environments. Water International 39, 635–648.

García-Tejero, I.F., Hernández, A., Viñuela, I., Apolo, O.E., Montes, C., Muriel, J.L., Durán-Zuazo, V.H., 2014b. Gestión sostenible de los recursos hídricos en el cultivo del almendro: determinación de los coeficientes de cultivo y efectos del riego deficitario. Fruticultura 36, 2–11.

García-Tejero, I.F., Hernández, A., Rodríguez, V.M., Ponce, J.R., Ramos, V., Muriel, J.L., Durán-Zuazo, V.H., 2015. Estimating almond crop coefficients and physiological response to water stress in semiarid environments (SW Spain). Journal of Agricultural Science and Technology 17, 1255–1266.

García, J., Romero, P., Botía, P., García, F., 2004. Cost-benefit analysis of almond orchard under regulated deficit irrigation (RDI) in SE Spain. Spanish Journal of Agricultural Research 2, 157–165.

Garnier, E., Berger, A., 1985. Testing water potential in peach trees as an indicator of water stress. Journal of Horticultural Science 60, 47–56.

Girona, J., Mata, M., Marsal, J., 2005. Regulated deficit irrigation during the kernel-filling period and optimal irrigation rates in almond. Agricultural Water Management 75, 152–167.

Girona, J., 2006. La respuesta del cultivo del almendro al riego. Vida Rural 234, 12–16.

Goldhamer, D.A., 1989. Drought Irrigation Strategies for Deciduous Orchards. University of California Division of Agriculture and Natural Resources. Publication 21453.

Goldhamer, D.A., Fereres, E., 2016. Establishing an almond water production function for California using long-term yield response to variable irrigation. Irrigation Science 1–11. https://doi.org/10.1007/s00271-016-0528-2.

Goldhamer, D.A., Girona, J., 2012. Crop yield response to water: almond. In: Steduto, P., Hsiao, T.C., Fereres, E., Raes, D. (Eds.), FAO Irrigation and Drainage Paper No. 66. Food and Agriculture Organization of the United Nations, Rome, pp. 358–373.

Goldhamer, D.A., Viveros, M., Salinas, M., 2006. Regulated deficit irrigation in almonds: effects of variation in applied water and stress timing on yield and yield components. Irrigation Science 24, 101–114.

Goldhamer, D.A., Fereres, E., 2004. Irrigation scheduling of almond trees with trunk diameter sensors. Irrigation Science 23, 11–19.

Goldhamer, D.A., Fereres, E., 2001a. Irrigation scheduling protocols using continuously recorded trunk diameter measurements. Irrigation Science 20, 115–125.

Goldhamer, D., Fereres, E., 2001b. Simplified tree water status measurements can aid almond irrigation. California Agriculture 55, 32–37.

Goldhamer, D.A., Viveros, M., 2000. Effects of preharvest irrigation cutoff durations and postharvest water deprivation on almond tree performance. Irrigation Science 19, 125–131.

Gomes-Laranjo, J., Coutinho, J.P., Galhano, V., Cordeiro, V., 2006. Responses of five almond cultivars to irrigation: photosynthesis and leaf water potential. Agricultural Water Management 83, 261–265.

Gonzalez-Dugo, V., Zarco-Tejada, P., Berni, J.A.J., Suárez, L., Goldhamer, D., Fereres, E., 2012. Almond tree canopy temperature reveals intra-crown variability that is water stress-dependent. Agriculture. Forest and Meteorology 154, 156–165. https://doi.org/10.1016/j.agrformet.2011.11.004.

Hsiao, T.C., 1990. Measurements of plant water status. In: Steward, B.A., Nielsen, D.R. (Eds.), Irrigation of Agricultural Crops. Agronomy Monograph No. 30. Am. Soc. Agro., Madison, WI, pp. 243–279.

Hutmacher, R.B., Nightingale, H.I., Rolston, D.E., Biggar, J.W., Dale, F., Vail, S.S., Peters, D., 1994. Growth and yield responses of almond (Prunus amygdalus) to trickle irrigation. Irrigation Science 14, 117–126.

Intrigliolo, D.S., Castel, J.R., 2007. Crop load affects maximum daily trunk shrinkage of plum trees. Tree Physiology 27, 89–96.

Intrigliolo, D.S., Castel, J.R., 2006. Usefulness of diurnal trunk shrinkage as a water stress indicator in plum trees. Tree Physiolohy 26, 303–311.

IPCC, 2014. Climate change 2014. In: Barros, V., Field, C.B. (Eds.), Impacts, Adaptations and Vulnerabilities. Part B. Regional Aspects. Cambridge University Press, 688 p.

Jones, H.G., 2004. Irrigation scheduling: advantages and pitfalls of plant-based methods. Journal of Experimental Botany 55, 2427–2436.

Klein, I., Esparza, G., Weinbaum, S.A., DeJong, T.M., 2001. Effects of irrigation deprivation during the harvest period on leaf persistence and function in mature almond trees. Tree Physiology 21, 1063–1072.

Lorite, I.J., Santos, C., Testi, L., Fereres, E., 2012. Design and construction of a large weighing lysimeter in an almond orchard. Spanish Journal of Agricultural Research 10, 238–250.

Mañas, F., López-Fuster, P., López-Urrea, R., 2014. Effects of different regulated and sustained deficit irrigation strategies in almond production. Acta Horticulturae 1028, 391–394.

McCutchan, H., Shackel, K.A., 1992. Stem-water potential as a sensitive indicator of water stress in prune trees (*Prunus domestica* L. cv. French). Journal of American Society of Horticultural Science 117, 607–611.

Mirás-Avalos, J.M., Pérez-Sarmiento, F., Alcobendas, R., Alarcón, J.J., Mounzer, O., Nicolás, E., 2016. Using midday stem water potential for scheduling deficit irrigation in mid–late maturing peach trees under Mediterranean conditions. Irrigation Science 34, 161–173.

Moriana, A., Fereres, E., Orgaz, F., Castro, J., Humanes, M.D., Pastor, M., 2000. The relations between trunk diameter fluctuations and tree water status in olive tree (*Olea europea* L.). Acta Horticulturae 537, 293–297.

Nanos, G.D., Kazantzis, I., Kefalas, P., 2002. Irrigation and harvest time affect almond kernel quality and composition. Scientia Horticulturae 96, 249–256.

Nortes, P.A., Pérez-Pastor, A., Egea, G., Conejero, W., Domingo, R., 2005. Comparison of changes in stem diameter and water potential values for detecting water stress in young almond trees. Agricultural Water Management 77, 296–307.

Nortes, P., González-Real, M.M., Egea, G., Baille, A., 2009. Seasonal effects of deficit irrigation on leaf photosynthesis traits of fruiting and non-fruiting shoots in almond trees. Tree Physiology 29, 375–388.

Ortuño, M.F., Conejero, W., Moreno, F., Moriana, A., Intrigliolo, D.S., Biel, C., Mellisho, C.D., Pérez-Pastor, A., Domingo, R., Ruiz-Sánchez, M.C., Casadesus, J., Bonany, J., Torrecillas, A., 2010. Could trunk diameter sensors be used in woody crops for irrigation scheduling?. A review of current knowledge and future perspectives. Agricultural Water Management 97, 1–11.

Puerto, P., Domingo, R., Torres, R., Pérez-Pastor, A., García-Riquelme, M., 2013. Remote management of deficit irrigation in almond trees based on maximum daily trunk shrinkage. Water relations and yield. Agricultural Water Management 126, 33–45.

Romero, P., Botía, P., García, F., 2004. Effects of regulated deficit irrigation under subsurface drip irrigation conditions on vegetative development and yield of mature almond trees. Plant and Soil 260, 169–181.

Romero, P., Botía, P., 2006. Daily and seasonal patterns of leaf water relations and gas exchange of regulated deficit-irrigated almond trees under semiarid conditions. Environmental and Experimental Botany 56, 158–173. https://doi.org/10.1016/j.envexpbot.2005.01.012.

Romero, P., García, J., Botía, P., 2006. Cost–benefit analysis of a regulated deficit-irrigated almond orchard under subsurface drip irrigation conditions in Southeastern Spain. Irrigation Science 24, 175–184.

Rouhi, V., Samson, R., Lemeur, R., Van Damme, P., 2007. Photosynthetic gas exchange characteristics in three different almond species during drought stress and subsequent recovery. Environmental and Experimental Botany 59, 117–129.

Sánchez-Bel, P., Egea, I., Martinez-Madrid, M.C., Flores, B., Romojaro, F., 2008a. Influence of irrigation and organic/inorganic fertilization on chemical quality of almond (Prunus amygdalus cv Guara). Journal of Agricultural and Food Chemistry 56, 10056–10062.

Sánchez Bel, P., Egea, I., Martínez-Madrid, M.C., Flores, B., Romojaro, F., 2008b. Influence of irrigation and organic/inorganic fertilization on chemical quality of almond (Prunus amygdalus cv. Guara). Journal of Agricultural and Food Chemistry 56, 10056–10062.

Sanden, B., Brown, P., Snyder, R., 2012. New insights on water management in almonds. In: Regulatory Issues Impacting California Agriculture. Proceedings: American Society of Agronomy. Calif. Chap., Visalia, California, 88–93.

Savé, R., de Herralde, F., Aranda, X., Pla, E., Pascual, D., Funes, I., Biel, C., 2012. Potential changes in irrigation requirements and phenology of maize, apple trees and alfalfa under global change conditions in Fluvià watershed during XXI st century: results from a modeling approximation to watershed-level water balance. Agricultural Water Management 114, 78–87.

Shackel, K., 2011. A plant-based approach to deficit irrigation in trees and vines. HortScience 46, 173–177.

Shackel, K.A., Ahmadi, H., Biasi, W., Buchner, R., Goldhamer, D., Gurusinghe, S., Hasey, J., Kester, D., Krueger, B., Lampinen, B., McGourty, G., Micke, W., Mitcham, E., Olson, B., Pelletrau, K., Philips, H., Ramos, D., Schwankl, L., Sibbett, S., Snyder, R., Southwick, S., Stevenson, M., Thorpe, M., Weinbaum, S., Yeager, J., 1997. Plant water status as an index of irrigation need in deciduous fruit tree. Horttechnology 7, 23–29.

Steduto, P., Hsiao, T.C., Fereres, E., Raes, D., 2012. Crop yield response to water. In: FAO Irrigation and Drainage Paper 66. FAO, Rome.

Stevens, R.M., Ewenz, C.M., Grigson, G., Conner, S.M., 2012. Water use by an irrigated almond orchard. Irrigation Science 30, 189–200.

Spinelli, G.M., Snyder, R.L., Sanden, B.L., Shackel, K.A., 2016. Water stress causes stomatal closure but does not reduce canopy evapotranspiration in almond. Agricultural Water Management 168, 11–22.

Teviotdale, B., Goldhamer, D., Viveros, M., 2001. Effects of deficit irrigation on hullrot disease of almond trees caused by *Monilinia fructicola* and *Rhizopus stolonifer*. Plant Disease 85, 399–403.

Torrecillas, A., Ruiz-Sanchez, M.C., del Amor, F., León, A., 1988. Seasonal variations on water relations of *Amygdalus communis* L. under drip irrigated and non irrigated conditions. Plant and Soil 106, 215–220.

Torrecillas, A., Alarcón, J.J., Domingo, R., Planes, J., Sánchez-Blanco, M.J., 1996. Strategies for drought resistance in leaves of two almond cultivars. Plant Science 118, 135–143.

Trnka, M., Eitzinger, J., Hlavinka, P., Dubrovska, M., Semerádová, D., Åtapanek, P., Thaler, S., Åsalud, Z., Molna, M., Formayer, H., 2009. Climate-driven changes of production regions in Central Europe. Plant and Soil 521, 257–266.

Trnka, M., Olesen, J.E., Kersebaum, K.C., Skjelvåg, A.O., Eitzinger, J., Seguin, B., Peltonen-Sainio, P., Rötter, R., Iglesias, A., Orlandini, S., Dubrovský, M., Hlavinka, P., Balek, J., Eckersten, H., Cloppet, E., Calanca, P., Gobin, A., Vučetić, V., Nejedlik, P., Kumar, S., Lalic, B., Mestre, A., Rossi, F., Kozyra, J., Alexandrov, V., Semerádová, D., Žalud, Z., 2011. Agroclimatic conditions in Europe under climate change. Global Change Biology 17, 2298–2318.

Wartinger, A., Heilmeier, H., Hartung, W., Schulze, E.D., 1990. Daily and seasonal courses of leaf conductance and abscisic acid in the xylem sap of almond trees (*Prunus dulcis* (Miller) D.A.Webb) under desert conditions. New Phytologist 116, 581–587.

CHAPTER 13

Irrigation Strategies for Mango (*Mangifera indica* L.) Under Water-Scarcity Scenario in the Mediterranean Subtropical Environment

Carmen R. Rodríguez Pleguezuelo[1], Belén Cárceles Rodríguez[1], Iván F. García Tejero[2], Baltasar Gálvez Ruíz[1], Dionisio Franco Tarifa[3], José R. Francia Martínez[1], Víctor H. Durán Zuazo[2]

[1]IFAPA Centro "Camino de Purchil", Granada, Spain; [2]IFAPA Centro "Las Torres-Tomejil", Seville, Spain; [3]Finca Experimental "El Zahorí", Granada, Spain

Contents

1. INTRODUCTION

The declining availability of fresh water is becoming a chronic water-scarcity scenario, worsened by climate change, mainly in the Mediterranean basin, where the climate is characterized by hot, dry summers with high evapotranspiration rates and precipitation commonly concentrated in autumn and winter but largely unpredictable in amount. These conditions are significantly affecting the biotic system (Gasith and Resh, 1999).

Water Scarcity and Sustainable Agriculture in Semiarid Environment
ISBN 978-0-12-813164-0
https://doi.org/10.1016/B978-0-12-813164-0.00013-2

The estimated increase in the global population will inevitably increase the food demand, with a heavy impact on agricultural water use. Therefore, assessing the irrigation needs is essential for sustainable water-resource use in order to avoid excessive water consumption as well as to produce more food with less water (Kang et al., 2017).

Research progress and practical applications of highly efficient agricultural water use are urgently needed. Consequently, the adoption of water-saving strategies by agriculture is becoming increasingly critical, especially under shortage scenarios. In many areas where the lack of water is not yet critical, crops are currently irrigated in excess to avoid water shortage and promote plant growth; however, this practice may be detrimental to crop quality and obviously to water resources.

In this context, boosting water-use efficiency (WUE) in agriculture is expected to play a central role in ensuring food and water security. The present state and future trend of water scarcity and food security need special attention, since the importance of irrigation in ensuring food security has been highlighted based on the analysis of the time course of irrigation-water productivity.

Mango (*Mangifera indica* L.) is one of the most widely consumed fruits worldwide due to its bright color, characteristic texture and taste, and nutritional value (Akinyele and Keshinro, 1980; Septembre-Malaterre et al., 2016). Native to the Indo-Burmese region, mangoes are now cultivated in more than 90 countries, with production estimated to be over 42.7 Mt yr^{-1}. India produces 42%, China 10%, Thailand 7%, Indonesia 4.8%, and Mexico 4.4% of the world yield (FAOSTAT, 2013). Although India and China are the countries with the highest mango production, this latter is solely for their own local market, and therefore the largest exporters to Europe are Mexico (13.5%), the Netherlands (12.2%), Thailand (9.6%), India (9.4%) and Brazil (7.9%).

This tree is well adapted to tropical and subtropical environmental conditions, being cultivated until up to 1300 m a.s.l. However, most commercial cultivation is limited to areas below 600 m a.s.l. Mango is a typical fruit of southern subtropical climates in many areas worldwide, where the annual average temperature is above 20°C, as in Mediterranean Europe. In particular, the major mango planting areas within southeastern (SE) Spain, including the provinces of Granada and Malaga, have an annual average temperature over 21°C (Durán et al., 2013). Under such temperatures, mangos usually mature between September and November.

In this context, in SE Spain, this productive and profitable farming has been developed based on subtropical crops occupying recently terraced slopes, since Spain is the only producer country of these fruits in Europe (Durán et al., 2003, 2005; Barea et al., 2016). According to the area cultivated, as well as to the economic impact, the leading crop is avocado (*Persea americana* Mill.), with 14,938 ha in Spain, from which 13,393 ha are located in Andalusia (90%), and with a production of 49,807 t yr^{-1}, followed by mango (*Mangifera indica* L.), with 3763 ha in Spain and 3276 of them in Andalusia (87%), with a production of 20,579 t yr^{-1} (ESYRCE, 2015). That is, the expansion area of irrigated mango orchards (more than 90%) in the last 5 years in Andalusia was important as well as the price of the fruits (Fig. 13.1). In addition, there is an area of mango orchards in the Canary Islands of 487 ha that represents 13% of the total of Spanish mangoes (ESYRCE, 2015).

Nevertheless, few studies have examined the optimal water supply for subtropical farming, in particular for a Mediterranean climate. The primary problems of cultivation on terraces, as for the rest of the crops in the area, are the shortage of water resources and the excessive cost of energy required for pumping irrigation water to high levels. Thus, the traditional amount of irrigation water applied in terraced mango orchards in the study area needs to be adjusted to the real requirements by promoting water-savings strategies (García-Tejero et al., 2011).

Deficit irrigation signifies that the supplementary water applied is reduced to only a fraction of potential evapotranspiration of a well-watered reference crop (ET_C). In other words, deficit irrigation (DI) is an optimization strategy under which crops are deliberately subjected to some degree of water stress and yield reduction (English and Raja, 1996). This situation according to Fereres and Soriano (2007) reduces biomass production in canopy size and

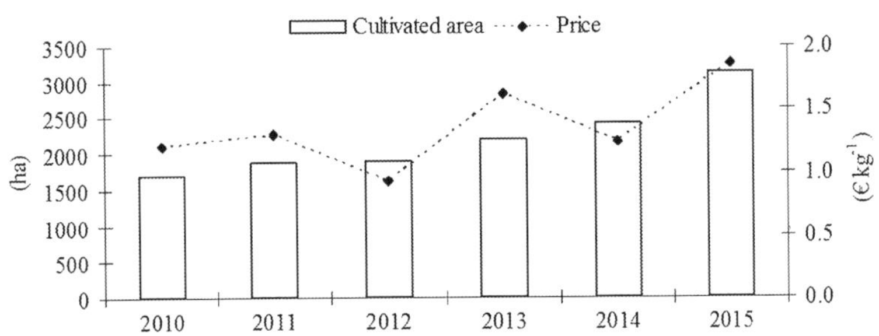

Figure 13.1 Progression of cultivated area of mango orchards and price per kilogram of fruits in south-eastern Spain.

radiation interception. Particularly, the DI technique based on a uniform water restriction has been successfully applied in many fruit crops such as pear (Kang et al., 2002), plum (Intrigliolo et al., 2013), nectarine (De la Rosa et al., 2015), peach (Faci et al., 2014), orange (García-Tejero et al., 2012), and apple (Van Hoojdonk et al., 2004). This strategy allows the crop to adapt to the stressful situation gradually as water becomes scarce.

Therefore in this chapter we present an experiment in which a shortage of agricultural water was simulated through the application of DI techniques in a mature terraced mango plantation, with the aim of assessing the impact on fruit yield and quality, and tree growth under subtropical Mediterranean climate.

2. EXPERIMENTAL CONDITIONS

The field study was conducted over three seasons at the experimental farm in Almuñécar (Granada) (Andalusia, SE Spain) (36° 48′ 00″ N, 3° 38′ 0″ W) at an elevation of 195 m (a.s.l.) (Fig. 13.2). The orchard, located in terraces, is representative of those commonly found in the area (reverse-sloped bench-terrace type), averaging 190 m long (Fig. 13.3). Each platform had a single row of 12-year-old mango trees (*Mangifera indica* L. cv. "Osteen" grafted onto "Gomera-1"), healthy and uniform in size, and spaced 3 m apart (Fig. 13.4).

Figure 13.2 Location of study area in south-eastern Spain.

Figure 13.3 Terraced mango orchards in the study area (Almuñécar, Granada, SE Spain).

Figure 13.4 Mango (*Mangifera indica* L.) tree cv. "Osteen" grafted onto rootstock Gomera-1.

The Florida (USA) cv. "Osteen" among other cultivars (Olano et al., 2005) is one of the scions most commonly used in SE Spain due to its commercial appeal. This cultivar has an oblong, ovoid fruit, reddish purple when ripe, often blotched with yellow and greenish coloring. The pulp has an exceptional quality, yellowish orange in color, very juicy, and nonfibrous.

In the study area, a cultivated hectare of mango trees on steeply sloped lands (65 degrees) would have 18 terraces (spaced about 5 m) 100 m long. The average annual rainfall in the area is 449.0 mm, and temperature is subtropical to semihot within the Mediterranean climatic category (average temperature is 20.8°C).

The soils in the study area are typical xerorthent according to the Soil Survey Staff (1999), with clay, sand, and silt of 79, 680, and 241 $g\ kg^{-1}$, respectively, containing 6.5 $g\ kg^{-1}$ of organic carbon, and 0.7 $g\ kg^{-1}$ of total N, with 15.0 $mg\ kg^{-1}$ P and 179.0 $mg\ kg^{-1}$ assimilable K.

2.1 Deficit Irrigation Regimes

The deficit irrigation regimes included three water-shortage types based on reductions of the crop evapotranspiration (ET_C): DI_{33}, DI_{50}, and DI_{75} each received 33%, 50%, and 75% of ET_C, respectively; with full irrigation (FI_{100}) as a control treatment that received 100% of ET_C (irrigation required to meet the crops' evapotranspiration demand). The experiment was a completely randomized block design with three replications per irrigation treatment. Each plot had eight trees per row. The four central trees of the rows were used for fruit yield and tree-size measurements, and the other four trees served as border trees.

The reference evapotranspiration (ET_0) was calculated using the Penman–Monteith method (Allen et al., 1998), and crop coefficients K_C with the adjustment of tree size were estimated from drainage lysimeters located in the same orchard (K_C values of 0.51, 0.72, and 0.60 at flowering, fruit set, and fruit growth, respectively). The irrigation regimes were applied by a combination of several self-regulating emitters ([4 and 8] $L\ h^{-1}$) in a double-line system. Irrigation was controlled automatically by a head unit programmer and electrohydraulic valves.

The experimental orchard was managed with the same fertilization (240 g N, 71 g P_2O_5 and 212 g K_2O) and routine cultivation techniques for diseases and insect control as used in the study area.

2.2 Field and Laboratory Measurements

The height, canopy diameter, and trunk circumference were measured 15 cm above the bud union in grafted trees. The canopy volume was calculated using the equation for one-half of a prolate spheroid (Vihotogbé et al., 2013):

$$C_V = \frac{2}{3}\pi r^2 H \tag{13.1}$$

where C_V is canopy volume, r is canopy radius, and H is canopy height. Trunk circumference was converted into trunk cross-sectional area (Dalal and Brar, 2012):

$$TCSA = \frac{C^2}{4\pi} \tag{13.2}$$

where C is the trunk circumference. Yield efficiency was calculated by dividing fruit yield by C_V and by $TCSA$. WUE was calculated as fresh mango yield divided by total seasonal irrigation water applied.

During the harvest period, the fruit yield per tree was recorded for each irrigation treatment. In addition, fruits per tree were collected to measure vertical and horizontal diameters with a Vernier caliper. The titratable acidity (TA) from fruit juice titrating against NaOH 0.05 N using phenolphthalein as the indicator was determined (AOAC, 1995). Total soluble solids (TSS) (°Brix) were measured by direct reading in a refractometer (Eclipse, Bellingham and Stanley, Ltd.).

3. FRUIT YIELD UNDER DEFICIT-IRRIGATION TREATMENTS

The mango fruit yield during the three monitoring seasons for DI_{33}, DI_{50}, DI_{75}, and control averaged of 16.0, 30.7, 22.5, and 24.1 kg tree^{-1}, with applied irrigation water of 2.8, 4.3, 5.8, and 7.9 m^3 tree^{-1}, respectively. Mango trees subjected to DI_{50} significantly yielded more than the rest of DI treatments and more than the control (FI_{100}) (Fig. 13.5). This means that a 50% reduction in the irrigation volume required to meet the crop evaporation demand (ET_C) could produce more in relation to DI_{33}, DI_{75}, and FI_{100} by 48%, 27%, and 21%, respectively.

In our study the mango trees covered 600 ha^{-1} (3 m apart with 18 terraces spaced about 5 m), and therefore the fruit yield for DI_{33}, DI_{50}, DI_{75}, and control with increasing amounts of irrigation water was 9.6, 18.4, 13.5, and 14.5 t ha^{-1}, respectively (Fig. 13.5). DI_{50}, in terms of fruit yield,

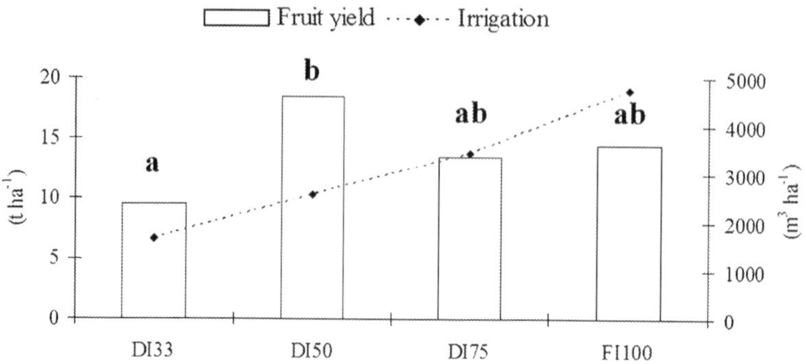

Figure 13.5 Fruit yield response to different deficit irrigation (DI) strategies. Different letters within the same column are statistically different by LSD analysis ($P < .05$).

saved about 2160 m^3 ha^{-1} (~56%) of irrigation water in relation to the fully irrigated mango trees at 100% of ET_C. An intermediate situation between the minimum and maximum production (DI_{33} heavily deficient and DI_{50}) was found for DI_{75}, and control trees FI_{100} with no significant differences.

The mango fruit yield per tree in our marginal area under subtropical Mediterranean climate proved to be much lower than those reported in Pakistan by Iqbal et al. (2012), ranging from 85.5 to 140.0 kg $tree^{-1}$. Similarly, Smith et al. (2008) in Australia reported 36 and 181 kg $tree^{-1}$ for the lowest yielding and more productive trees, respectively, in mango orchard with about 125 trees ha^{-1} (8 m × 10 m). Far more mango fruit production was found in Venezuela for cvs. Kent and Smith of 378 and 868 kg $tree^{-1}$, respectively (Avilán, 1974). In this context, for the coast of Granada (SE Spain), it should be taken into consideration that this area represents the climatic limit for commercially viable mango cultivation and, therefore, tree sizes and yields are far smaller, but this singularity enables the cultivation at high-density planting on terraces as well as in flat areas.

Due to the different planting patterns, the yields reached in the present study are in a good range in relation to those from typical mango-producing zones. Bhriguvanshi et al. (2012) reported mango yield of between 9.0 and 11.9 t ha^{-1} in subtropical zones in India, Pavel and Villiers (2004) 5.0–9.0 t ha^{-1} and Spreer et al. (2007, 2009) 4.0 and 20.0 t ha^{-1} in northern Thailand. In Brazil dos Santos et al. (2014) recorded the fruit yield for mango with 3450 m^3 ha^{-1} at 24.0 t ha^{-1}, which contrasts with 11,976 m^3 ha^{-1} for the total seasonal water requirement of mature mango trees in South

Africa (Mostert and Hoffman, 1997). For Australian conditions, Smith et al. (2008) stated the fruit yield ranging between 4.5 and 22.6 t ha^{-1}.

On the other hand, in the tropics of India, according to Sharma et al. (2001), high-density planting of 2.5 m × 2.5 m is standardized for intensive mango cultivation, but, after a profitable yield for up to 14–15 years, yield progressively declines due to overcrowding of canopies. In our subtropical area, as stated before, the climate significantly limits tree growth, and the canopy volume does not become crowded with appropriate pruning every 2 years (Durán et al., 2003). Therefore, the feasibility of densely planted mango orchards could be profitable for longer periods than in the tropics.

The calculated WUE in terms of applied water and productivity was significantly ($P < .05$) higher for DI_{50} than for the rest of deficit treatments and FI_{100}. Specifically, DI_{50} amounted to 7.1 kg m^{-3} in contrast with control trees with 3.1 kg m^{-3}. Thus, the WUE by reducing the irrigation at 50% ET_C was improved by 56% with respect to fully irrigated trees (100% ET_C). Therefore, increasing the water amount in mango trees did not result in higher yields or improvement of WUE, as reported by many experimental studies (da Campos et al., 2008; Pavel and Villiers, 2004; Spreer et al., 2007). Recently, in semiarid areas of Brazil, dos Santos et al. (2014), by applying regulated deficit irrigation, registered the best fruit yield at 50% ET_C.

In mango trees, water deficit could provoke deficient nutrient uptake, inhibited growth, a disturbed phenologic stage, a depressed photosynthesis rate, and finally lower fruit yield. This prospect is corroborated in the present study, since the trees under SDI_{33} presented lower yield, which is presumably associated with a lower gas-exchange rate in mango, as was stated by Luvaha et al. (2008).

The analysis of overall fruit yield and water consumption from irrigation gave a good fit to a binomial yield function for each season, showing high correlation coefficients, especially for the second and third season ($R^2 = 0.954$ and $R^2 = 0.991$, respectively) (Fig. 13.6). However, when the three seasons were taken into account, the fit to a binomial function was also good ($R^2 = 0.849$) (Fig. 13.6). Thus, from Fig. 13.6 the optimum level of irrigation water was approximately 4.5–5.0 m^3 $tree^{-1}$, this corresponding mainly to the SDI_{50} irrigation strategy. The binomial functions relating yield and irrigation water have also been used for mango by Spreer et al. (2009) as well as for other types of woody fruit crops such as olive (Gattan et al., 2006), in order to characterize the impact of the volume of irrigation water on fruit yield.

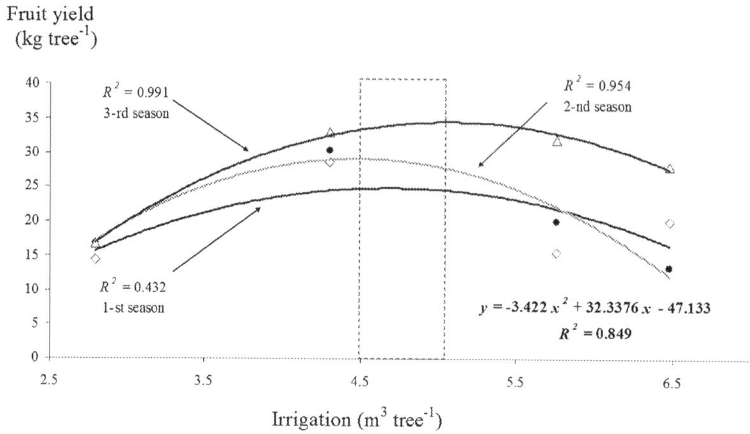

Figure 13.6 Production functions for three monitoring seasons relating mango fruit yield to irrigation amount. Each point represents the average of 16 experimental trees under different deficit irrigation strategies.

3.1 Pomological and Quality Characteristics of Mango Fruits

Fig. 13.7 displays the number, weight, length, and width for fruits cv. Osteen in relation to deficit irrigation strategies used. In terms of the average number of fruits per tree, the DI_{50} consistently resulted in a higher amount (54) than did the rest of the deficit treatments and control. However, the average fruit weight reached the highest in control trees, which differed especially from the DI_{33} and DI_{50} deficit treatments ($P < .05$). Consequently, differences in fruit yield were altered by the number as well as by the fruit weight. In addition, fruit yield was correlated lineally with the number of fruits harvested ($R^2 = 0.91$) and not with the average fruit weight.

On the other hand, the average fruit weight for DI_{33} and FI_{100} ranged from 536.9 to 648.4 g, respectively, showing that the mean fruit weight increased with higher irrigation amounts. That is, control trees produced the heaviest fruits (greater dry matter) and differed significantly from the remaining deficit treatments (Fig. 13.8). Thus, control trees produced the greatest fruit length and width, differing significantly from the deficit irrigation treatments, in accordance with the fruit weight. On the contrary, the DI_{33} produced the smallest fruits.

As mango is a fleshy fruit containing more than 80% water (Lakshimnarayana et al., 1970), the size of the fruit depends on the accumulation of water and dry matter in the various compartments during the

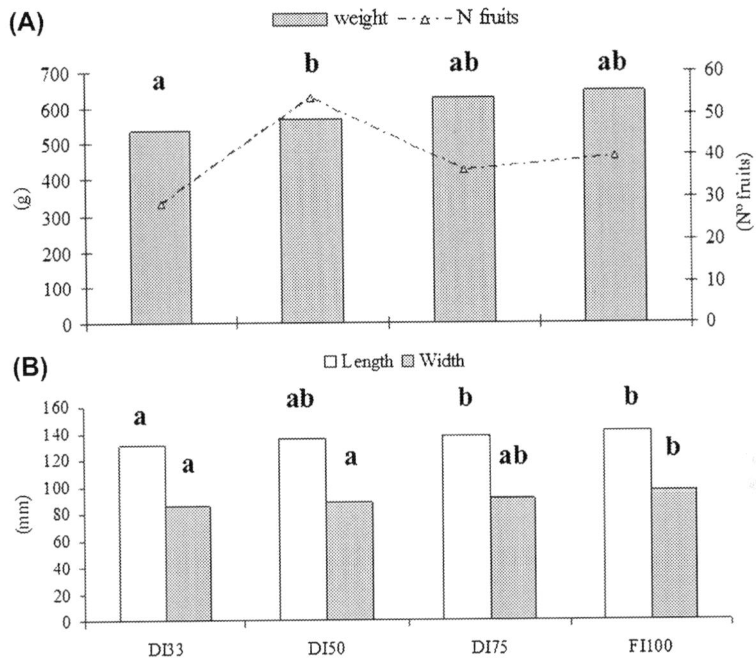

Figure 13.7 Impact of different deficit irrigation (DI) strategies on number, weight (A), length, and width (B) of mango fruits. Different letters within the same column are statistically different by LSD analysis ($P < .05$).

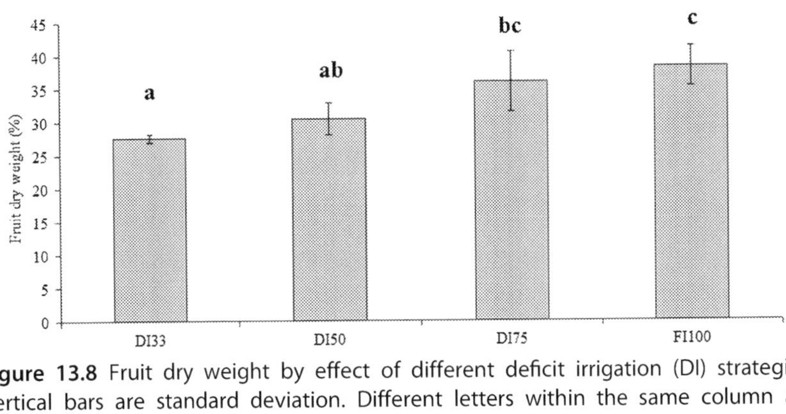

Figure 13.8 Fruit dry weight by effect of different deficit irrigation (DI) strategies. Vertical bars are standard deviation. Different letters within the same column are statistically different by LSD analysis ($P < .05$).

fruit-growth stage. The skin, pulp, and stone have specific compositions that appear to accumulate water and dry matter at different rates, depending on environmental conditions (Léchaudel et al., 2002). Fruit growth after cell division consists of the enlargement of fruit cells characterized by a large accumulation of water that results from the balance between incoming fluxes, such as phloem and xylem, and outgoing fluxes, such as transpiration (Ho et al., 1987). Therefore, changing the water supply by reducing the irrigation leads to significant variations in fruit volume, which was revealed with the size of fruits from stressed and nonstressed trees. Similarly, Simmons et al. (1998) observed that water stress affected fruit growth rate and final fruit size through an impact on the cell number.

Fig. 13.9 shows the TSS and TA for the experimental period. The TSS was significantly higher ($P < .05$) in fruits under DI_{33}, especially with control trees, indicating a possible active sugar accumulation due to the lower amount of irrigation (Fig. 13.9A). In general, the progressive trend in TSS under DI depended on the amount of irrigation water applied.

Higher TSS content is an indicator of good quality of fruits. In the present study, the amount of TSS was found to be significantly higher

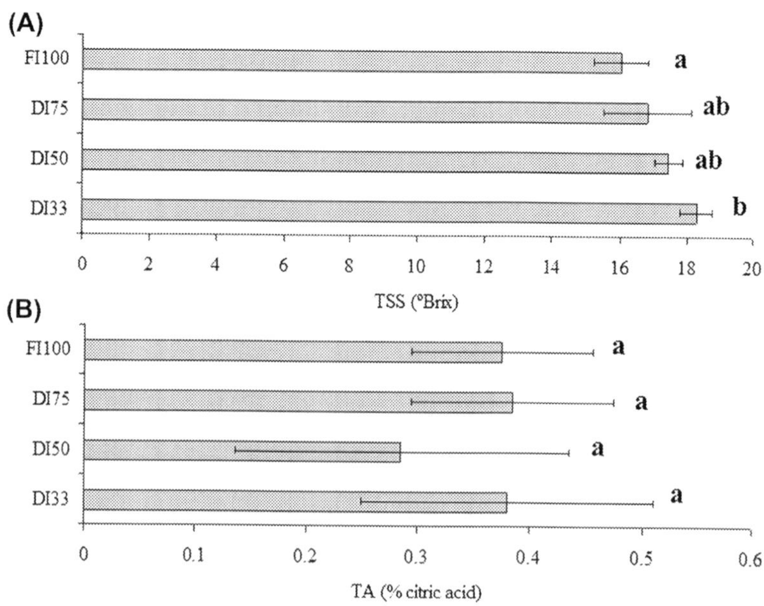

Figure 13.9 Average of total soluble solids (TTS) (A) and titratable acidity (TA) (B) in mango fruits due to effect of different deficit irrigation (DI) strategies. Horizontal bars are standard deviation. Different letters within the same bar are statistically different by LSD ($P < .05$).

(18.2%) for the most deficit-irrigated trees than in the rest of the treatments, especially control trees. Ripening was associated with increasing soluble solids and total sugars (Andrew and Anthony, 1985), and generally taste and particularly sweetness of the fruits depend on the percentage of TSS content (Shafique et al., 2006). An increase in TSS appears to be due to accumulation of more sugars in the fruits due to hydrolysis of starch from increased amylase activity during ripening (Fuchs et al., 1980; Tandon and Kalra, 1983). Similar findings were reported for mango fruits (Spreer et al., 2007), as well as for oranges (Aguado et al., 2012), peaches (Crisostol et al., 1994), and pear-jujube (Cui et al., 2008).

In addition, the values for TSS in our study are consistent with those reported by Dick et al. (2009) for cv. Kent, which varied from 14.2 to 20°Brix. On the other hand, the titratable acidity was not significantly affected by the irrigation treatment, although the trend was to be lower in DI_{50} (Fig. 13.9B).

4. TREE GROWTH UNDER DEFICIT-IRRIGATION TREATMENTS

From the findings of the present experiment the control trees (FI_{100}) had the highest trunk cross-sectional area (136.0 cm^2) in relation to the three deficit-irrigation treatments (Fig. 13.10). Similarly, tree size subjected to 100% ET_C had the highest tree height and canopy volume ($P < .05$) (Fig. 13.11). Consequently, the higher irrigation volume in FI_{100} was invested in vegetative growth rather than in fruit yield. By contrast, trees with DI_{33} reached the smallest tree size. The yield efficiency in terms of tree size of irrigated control trees was the least efficient with 3.1 kg m^{-3} in contrast with most deficient irrigated trees DI_{33} with 7.1 kg m^{-3}. Similarly,

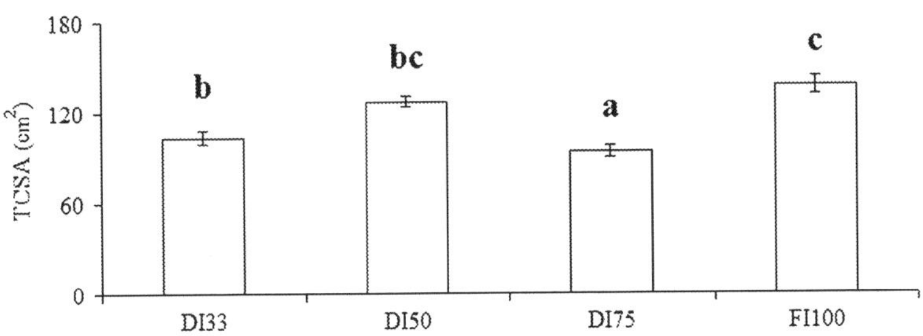

Figure 13.10 Trunk cross-sectional area (TCSA) by effect of different deficit irrigation (DI) strategies. Vertical bars are standard deviation. Different letters within the same column are statistically different by LSD analysis ($P < .05$).

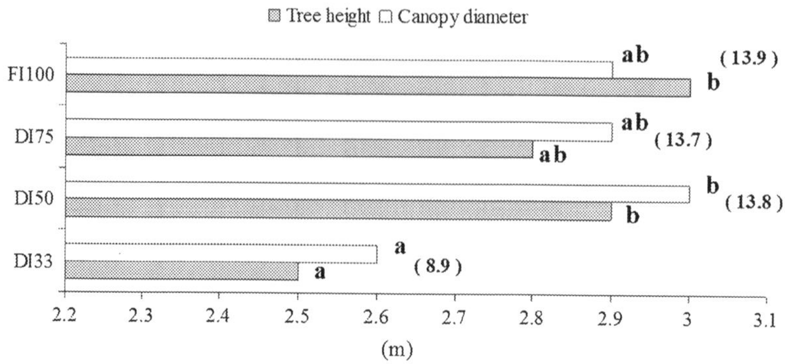

Figure 13.11 Canopy diameter and tree height under different deficit irrigation (DI) strategies. Different letters within the same bar are statistically different by LSD analysis ($P < .05$). Values in parenthesis are the average of canopy volume of eight monitored trees in m^3.

the efficiency in terms of area (TCSA) in our study, were the values for DI_{33} and FI_{100}, 219.8 and 171.2 g m^{-2}, respectively.

The vegetative growth of woody fruit trees is recognized as being the process most sensitive to water shortage. Reductions in shoot elongation and trunk cross-sectional area in response to water deficits lead to declines in tree size and smaller canopies (Intrigliolo and Castel, 2005; Marsal et al., 2008; Pérez et al., 2009). According to Pavel and Villiers (2004), water deficit significantly reduces the vegetative growth of mature mango trees. The results for tree size in SE Spain revealed that trees in terms of height, canopy diameter, canopy volume, as well as in trunk cross-sectional area values are much lower than trees in other traditional mango-producing countries. Mirjat et al. (2011) stated that mango trees with canopy diameters from 5.6 to 6.4 m in Pakistan contrasted with values in our study. According to Dayal et al. (2016), considerable differences in scion canopy volume were observed due to influence of rootstocks in different mango cultivars from 5.7 to 21.2 m^3 in subtropical climatic conditions in India. On the other hand, the yield efficiency in terms of tree growth was in line with the findings of Smith et al. (2008), who reported in Australia values ranging from 174.0 to 503.0 g cm^{-2}.

Finally, as was stated before, these characteristics regarding the size of mango trees in our subtropical climate in SE Spain provided the opportunity to test high-density planting, which gave similar fruit yields as in tropical and subtropical areas worldwide.

5. CONCLUSIONS

Water shortage is a primary global dilemma, and, according to climate change projections, it will become more critical in the coming decades. Current and future questions related to water scarcity need to be debated and focused on the urgency to implement more sustainable approaches to water-resource management for crop production. As a repercussion of augmenting water scarcity and drought, resulting from climate change, considerable water use for irrigation is expected in the context of harsh competition between farmers and other sectors of the economy.

The findings indicate that the volume of irrigation in mango trees is crucial in order to maximize the water-saving strategies under a scarcity scenario. In this context, in the present experiment the highest yield and water-use efficiency were attained by reducing use to 50% of ET_C, and thus the greatest volumes of water did not result in the highest yield as reflected by the control trees at 100% ET_C. However, control trees yielded the greatest fruit length and width, contrasting significantly from the deficit-irrigation treatments, in accordance with the fruit weight. By contrast, the smallest fruits were produced under the most severe deficit irrigation treatment (33% of ET_C). Tree size proved greater in control trees than in the deficit treatments, indicating that the higher irrigation volume was invested in vegetative growth rather than in fruit yield.

Thus, the DI program with 50% of ET_C should be adopted for achieving sustainable water management in terraced mango plantations under a scarcity scenario.

ACKNOWLEDGMENTS

Part of this publication was sponsored by the research project "Impact of deficit irrigation on productivity of subtropical fruit crops: tools for sustainable water stress management" (PP.AVA.AVA201601.8), and cofinanced by the European Regional Development Fund (ERDF) within the Operational Programme Andalusia 2014–20, "Andalucía is moving with Europe."

REFERENCES

Aguado, A., Frías, J., García, T.I.F., Romero, R., Muriel, F.J.L., Capote, N., 2012. Towards the improvement of fruit-quality parameters in citrus under deficit irrigation strategies. ISRN Agronomy. https://doi.org/10.5402/2012/940896.

Akinyele, I.O., Keshinro, O.O., 1980. Tropical fruits as sources of vitamin C. Food Chemistry 5, 163–167.

Allen, R.G., Pereira, L.S., Raes, D., Smith, M., 1998. Crop Evapotranspiration Guidelines for Computing Crop Water Requirements, Irrigation and Drain. Paper no. 56. FAO, Rome, Italy.

Andrew, P.M., Anthony, K.T., 1985. Analysis of sugars and organic acids in ripening mango fruits (*Mangifera indica* L. var Keitt) by high performance liquid chromatography. Journal of the Science of Food and Agriculture 36, 561–566.

AOAC, 1995. Official Methods of Analysis of Association of Official Analytical Chemists, sixteenth ed. Washington, DC, USA.

Avilán, R.L., 1974. Cuatro años de fertilización nitrogenada en mango (*Mangifera indica* L.) en suelos de la serie Maracay. Agronomia Tropical 24, 97–106.

Barea, A.M., Delgado, A.C., Haro, A., Olalla, M., Seiquer, I., Rufián, H.J.A., 2016. Subtropical fruits grown in Spain and elsewhere: a comparison of mineral profiles. Journal of Food Composition and Analysis 48, 34–40.

Bhriguvanshi, S.R., Adak, T., Kumar, K., Singh, V.K., Singh, A., Singh, V.K., 2012. Impact of fertigation regimes on yield and water use efficiency of mango (*Mangifera indica* L.) under subtropical condition. Indian Journal of Soil Conservation 3, 252–256.

Crisostol, C.H., Johnson, R.S., Luza, J.G., Crisosto, G.M., 1994. Irrigation regimes affect fruit soluble solids concentration and rate of water loss of 'O'Henry' peaches. Hortscience 29, 1169–1171.

Cui, N., Du, T., Kang, S., Li, F., Zhang, J., Wang, M., Li, Z., 2008. Regulated deficit irrigation improved fruit quality and water use efficiency of pear-jujube trees. Agricultural Water Management 95, 489–497.

da Campos, J.H.B., da Silva, V.P.R., de Azevedo, P.V., Borges, C.J.R., Soares, J.M., de Moura, M.S.B., da Silva, B.B., 2008. Evapotranspiração e produtividade da mangueira sob diferentes tratamentos de irrigação. Revista Brasileira de Engenharia Agrícola e Ambiental 12, 150–156.

Dalal, R.P.S., Brar, J.S., 2012. Relationship of trunk cross-sectional area with growth, yield, quality and leaf nutrient status in Kinnov mandarin. Indian Journal of Horticulture 69, 111–113.

Dayal, V., Dubey, A.K., Singh, S.K., Sharma, R.M., Dahuja, A., Kaur, C., 2016. Growth, yield and physiology of mango (*Mangifera indica* L.) cultivars as affected by polyembryonic rootstocks. Scientia Horticulturae 199, 186–197.

De la Rosa, J.M., Domingo, R., Gómez, M.J., Pérez, P.A., 2015. Implementing deficit irrigation scheduling through plant water stress indicators in early nectarine trees. Agricultural Water Management 152, 207–216.

Dick, E., Adopo, N.A., Camara, B., Moudioh, E., 2009. Influence of maturity stage of mango at harvest on its ripening quality. Fruits 64, 13–18.

dos Santos, R.M., Martínez, M.A., Rodrigues, D.S.L., Coelho, E.F., 2014. Tommy Atkins mango yield and photosynthesis under water deficit in semiarid region of Bahia. Revista Brasileira de Engenharia Agrícola e Ambiental 9, 899–907.

Durán, Z.V.H., Martínez, R.A., Aguilar, R.J., Franco, T.D., 2003. El cultivo del mango (*Mangifera indica* L.) en la costa granadina. Granada, Spain.

Durán, Z.V.H., Aguilar, R.J., Martínez, R.A., Franco, T., 2005. Impact of erosion in the taluses of subtropical orchard terraces. Agriculture, Ecosystems and Environment 107, 199–210.

Durán, Z.V.H., Rodríguez, P.C.R., Francia, M.J.R., Martin, P.J.F., 2013. Land-use changes in a small watershed in the Mediterranean landscape (SE Spain): environmental implications of a shift towards subtropical crops. Journal of Land Use Science 1, 47–58.

English, M., Raja, S.N., 1996. Perspectives on deficit irrigation. Agricultural Water Management 32, 1–14.

ESYRCE, 2015. Ministerio de Agricultura, Alimentación y Medio Ambiente. Available at: http://www.magrama.gob.es/es/estadistica/temas/estadisticas-grarias/agricultura/esyrce/.

Faci, J.M., Medina, E.T., Martínez, C.A., Alonso, J.M., 2014. Fruit yield and quality response of a late season peach orchard to different irrigation regimes in a semi-arid environment. Agricultural Water Management 143, 102–112.

FAOSTAT, 2013. United Nations Food and Agriculture Organization, Statistics Division. Available at: http://www.fao.org/faostat/es/#home.

Fereres, E., Soriano, M.A., 2007. Deficit irrigation for reducing agricultural water use. Journal of Experimental Botany 58, 147–159.

Fuchs, Y., Pesis, E., Zuberman, G., 1980. Changes in amylase activity, starch, and sugar contents in mango fruit pulp. Scientia Horticulturae 13, 155–160.

García-Tejero, I.F., Durán, Z.V.H., Rodríguez, P.C.R., Muriel, F.J.L., 2011. Water and Sustainable Agriculture. Springer Briefs, Springer Science + Business Media BV, The Netherlands.

García-Tejero, I.F., Durán, Z.V.H., Arriaga, J., Muriel, F.J.L., 2012. Relationships between trunk- and fruit-diameter growths under deficit-irrigation programmes in orange trees. Scientia Horticulturae 133, 64–71.

Gasith, A., Resh, V.H., 1999. Streams in Mediterranean climate regions: abiotic influences and biotic responses to predictable seasonal events. Annual Review of Ecology and Systematics 30, 51–81.

Gattan, S.R., Berenguer, M.J., Connell, J.H., Polito, V.S., Vossen, P.M., 2006. Olive oil production as influenced by different quantities of applied water. Agricultural Water Management 85, 133–140.

Ho, L.C., Grange, R.I., Picken, A.J., 1987. An analysis of the accumulation of water and dry matter in tomato fruit. Plant, Cell and Environment 10, 157–162.

Van Hoojdonk, B.M., Dorji, K., Behboudian, M.H., 2004. Responses of 'Pacific Rose' apple to partial rootzone drying and deficit irrigation. European Journal of Horticultural Science 69, 104–110.

Intrigliolo, D.S., Castel, J.R., 2005. Effects of regulated deficit irrigation on growth and yield of young Japanese plum trees. Journal of Horticultural Science and Biotechnology 80, 177–182.

Intrigliolo, D.S., Ballester, C., Castel, J.R., 2013. Carry-over effects of deficit irrigation applied over seven seasons in a developing Japanese plum orchard. Agricultural Water Management 128, 13–18.

Iqbal, M., Niamatullah, M., Hussain, A., Munir, M., Khan, I., Khan, M.Q., 2012. Performance of selected parameters of mango cultivars in Muzaffargarh district (Punjab), Pakistan. Sarhad Journal of Agriculture 3, 395–398.

Kang, S., Hu, X., Goodwin, I., Jerie, P., 2002. Soil water distribution, water use and yield response to partial root zone drying under a shallow groundwater table condition in a pear orchard. Acta Horticulturae 92, 277–291.

Kang, S., Hao, X., Du, T., Tong, L., Su, X., Lu, H., Li, X., Huo, Z., Li, S., Ding, R., 2017. Improving agricultural water productivity to ensure food security in China under changing environment: from research to practice. Agricultural Water Management 179, 5–17.

Lakshimnarayana, S., Subhadra, N.V., Subramanyam, H., 1970. Some aspects of developmental physiology of mango fruit. Journal of Horticultural Sciences 45, 133–142.

Léchaudel, M., Génard, M., Lescourret, F., Urban, L., Jannoyer, M., 2002. Leaf-to-fruit ratio affects water and drymatter content of mango fruit. Journal of Horticultural Science and Biotechnology 77, 773–777.

Luvaha, E., Netondo, G.W., Ouma, G., 2008. Effect of water deficit on the physiological and morphological characteristics of mango (*Mangifera indica* L.) rootstock seedlings. American Journal of Plant Physiology 3, 1–15.

Marsal, J., López, G., Girona, J., 2008. Recent advances in regulated deficit irrigation (RDI) in woody perennials and future perspectives. Acta Horticulturae 792, 429–439.

Mirjat, M.S., Jiskani, M.M., Siyal, A.A., Mirjat, M.U., 2011. Mango production and fruit quality under properly managed drip irrigation system. Pakistan Journal of Agriculture, Agricultural Engineering and Veterinary Sciences 27, 1–12.

Mostert, P.G., Hoffman, J.E., 1997. Water requirements and irrigation of mature mango trees. Acta Horticulturae 455, 331–338.

Olano, C.T., Schnell, R.J., Quintanilla, W.B., Campbell, R.J., 2005. Pedigree analysis of Florida mango cultivars. Proceedings of the Florida State Horticultural Society 118, 192–197.

Pavel, E.W., Villiers, A.J.D., 2004. Responses of mango trees to reduced irrigation regimes. Acta Horticulturae 646, 63–68.

Pérez, P.A., Domingo, R., Torrecillas, A., Ruiz, S.M.C., 2009. Response of apricot trees to deficit irrigation strategies. Irrigation Science 27, 231–242.

Septembre-Malaterre, A., Stanislas, G., Douraguia, E., Gonthier, M.P., 2016. Evaluation of nutritional and antioxidant properties of the tropical fruits banana, litchi, mango, papaya, passion fruit and pineapple cultivated in Réunion French Island. Food Chemistry 212, 225–233.

Shafique, M.Z., Ibrahim, M., Helali, M.O.H., Biswas, S.K., 2006. Studies on the physiological and biochemical composition of different mango cultivars at various maturity levels. Bangladesh Journal of Scientific and Industrial Research 41, 101–108.

Sharma, R.R., Singh, C.N., Saxena, S.K., Pandey, S.N., Chhonkar, O.P., 2001. Cluster planting favours malformation and influences yield and fruit quality in mango. Annals of Agricultural Research 22, 48–51.

Simmons, S.L., Hofman, P.J., Whiley, A.W., Hetherington, S.E., 1998. Effects of leaf: fruit ratios on fruit growth, mineral concentration and quality of mango (*Mangifera indica* L. cv. Kensington Pride). Journal of Horticultural Science and Biotechnology 73, 367–374.

Smith, W.M., Bright, D.J., Hoult, D.M., Renfree, A.R., Maddern, T., 2008. Field evaluation of 64 rootstocks for growth and yield of "Kensigton Pride" mango. Hortscience 43, 1720–1725.

Soil Survey Staff, 1999. Soil taxonomy. A basic system of soil classification for making and interpreting soil surveys. In: Agric. Handbook N, vol. 436. USDA, USA.

Spreer, W., Nagle, M., Neidhart, S., Carle, R., Ongprasert, S., Müller, J., 2007. Effect of regulated deficit irrigation and partial rootzone drying on the quality of mango fruits (*Mangifera indica* L., cv. 'Chok Anan'). Agricultural Water Management 88, 173–180.

Spreer, W., Ongprasert, S., Hegele, M., Wünsche, J.N., Müller, J., 2009. Yield and fruit development in mango (*Mangifera indica* L. cv. Chok Anan) under different irrigation regimes. Agricultural Water Management 96, 574–584.

Tandon, D.K., Kalra, S.K., 1983. Changes in sugars, starch and amylase activity during development of mango fruit cv. Dashehari. Journal of Horticultural Sciences 58, 449–453.

Vihotogbé, R., van den Berg, R., Sosef, S.M.M., 2013. Morphological characterization of African bush mango trees in West Africa. Genetic Resources and Crop Evolution 60, 1597–1614.

CHAPTER 14

Avocado (*Persea americana* Mill.) Trends in Water-Saving Strategies and Production Potential in a Mediterranean Climate, the Study Case of SE Spain: A Review

Carmen R. Rodríguez Pleguezuelo[1], José R. Francia Martínez[1], Iván F. García Tejero[2], Baltasar Gálvez Ruíz[1], Dionisio Franco Tarifa[3], Víctor H. Durán Zuazo[2]
[1]IFAPA Centro "Camino de Purchil", Granada, Spain; [2]IFAPA Centro "Las Torres-Tomejil", Seville, Spain; [3]Finca Experimental "El Zahorí", Granada, Spain

Contents

1. INTRODUCTION

Avocado (*Persea americana* Mill.), a fruit with a very high nutritional content, is considered one of the most important tropical fruits due to the high contents of protein and fat-soluble vitamins not present in other fruits. It also contains high concentrations of vitamins A and B (Fonseca et al., 2016).

Water Scarcity and Sustainable Agriculture in Semiarid Environment
ISBN 978-0-12-813164-0
https://doi.org/10.1016/B978-0-12-813164-0.00014-4

In addition, due to its high oil concentration, it has been utilized by the pharmaceutical and cosmetic industries (Bleinroth and Castro, 1992), and it is highly valued for its contents of omega fatty acids, phytosterols, tocopherols, and squalene (Santos et al., 2014; Fonseca et al., 2016).

In 2014 the surface devoted to avocado worldwide was 547,849 ha (FAOSTAT, 2015), with an average growth of 3.9% annually, following a positive trend over recent years; in fact, global production has maintained an average annual growth of 4.5% since 2009. In 2014 global production reached 5,028,756 ha, which represents an increment of 28% over 2009. The main producing countries are Mexico, Dominican Republic, Colombia, Peru, and Chile, which represent 48.1% of total production (Table 14.1).

Indonesia and the United States are also major producers. In Europe, Spain had an irrigated area of 10,893 ha in 2014, and globally is 15th in area and 17th in production (1.54% of global yield; Table 14.2).

With respect to fruit imports and exports, Mexico is the main world supplier of avocado. With a share ratio of more than 40%, their main clients are the United States, Netherlands, France, and Japan (Table 14.3). The European Union contributes roughly 25% of world trade with production coming from Spain as the only producer in Europe, which is later redistributed in the European markets and in non-EU countries (Table 14.3).

Table 14.1 Main avocado producers worldwide

Country	Area (ha)
Mexico	153,771
Colombia	34,513
Chile	31,727
Peru	30,320
Indonesia	23,981
United States	23,876
South Africa	20,000
Rwanda	19,000
China, mainland	17,500
Haiti	16,500
Cameroon	15,929
Ethiopia	13,798
Dominican Republic	12,927
Kenya	11,583
Spain	10,893
World	547,849

Table 14.2 Avocado fruit production during 2014

Country	Production (t)
Mexico	1,520,695
Dominican Republic	428,301
Peru	349,317
Indonesia	307,326
Colombia	288,739
Kenya	218,692
United States	179,124
Rwanda	161,519
Chile	160,000
Brazil	156,699
China, mainland	116,000
South Africa	107,176
Venezuela	104,331
Guatemala	104,113
Haiti	97,078
Israel	91,035
Spain	77,401
World	5,028,756

The leading export countries are Peru (114,515 t), Chile (88,360 t), and Spain (65,273 t). The United States is the main country that imports avocado (571,793 t), representing 45% of world imports, with its main supplier being Mexico (Table 14.3). For the European Union, with a share of 38% of all imports, the main suppliers lie outside the European Union: Peru, South Africa, Israel, and Chile.

In Spain, according to the latest MAPAMA statistics (2016), subtropical fruits represent approximately 4.3% of the total surface devoted to fruit trees, and 2.2% of the total production of fruit production, not including banana. The requirements for the development of these types of crops are a subtropical climate and cultivation areas in the provinces of Granada and Malaga (Andalusia, SE Spain) and the Canary Islands (Fig. 14.1). The subtropical agricultural sector in Andalusia implies an important thrust for socioeconomic development due to the direct profit generated and the indirect benefits from auxiliary services and employment generation (Observatorio de Precios y Mercados, 2015).

In Andalusia the main commercialized, irrigated subtropical fruits are avocado (*P. americana* Mill.), cherimoya (*Anonna cherimolia* L.), mango (*Mangifera indica* L.), and loquat (*Eriobotrya japonica* L.) (Durán et al., 2013). This important sector development has been possible due to the subtropical

Table 14.3 Global imports and exports for avocado fruit during 2014. Main studies undertaken for avocado water-deficit irrigation techniques

Country	(t)	(1000 US$)
Imports		
United States	571,793	1,141,751
Netherlands	143,555	326,041
France	99,014	230,686
Japan	60,458	162,143
Canada	57,519	145,811
United Kingdom	41,391	90,739
Spain	41,034	83,550
Germany	31,085	89,692
Sweden	19,635	53,715
Belgium	14,240	29,815
World	1,283,314	2,776,159
Exports		
Mexico	563,492	1,086,935
Netherlands	119,885	317,737
Peru	114,515	185,019
Chile	88,360	184,767
Spain	65,273	155,549
South Africa	50,742	77,480
Israel	48,848	122,121
United States	43,211	114,178
Kenya	25,002	29,280
Dominican Republic	20,628	22,810
World	1,216,993	2,479,194

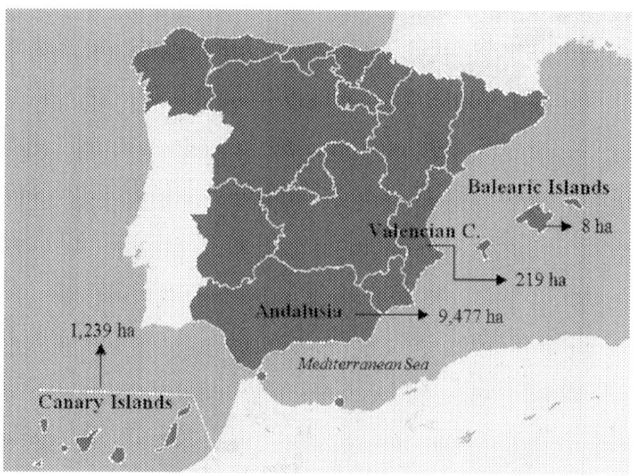

Figure 14.1 Spanish-irrigated areas with avocado orchards (ESYRCE, 2015).

Mediterranean climate, which is unique in the entire European continent. It is characterized by mild average temperatures (approximately 20°C throughout the year) with high humidity. In fact, this narrow strip of approximately 12 km wide parallel to the coast has a peculiar microclimate due to the special configuration of the intertropical valleys that are oriented north–south, and also protection against the north wind by the Penibetic mountain systems, which run along the coast from east to west.

In this context, in 2014 the surface area in Andalusia devoted to avocado was 9477 ha, which implied an increase of 10% compared to 2002 (Fig. 14.2). In this sense it is the most extensively grown subtropical crop in the area. Avocado is cultivated under irrigation in large areas of Malaga (6478 ha) and Granada (2612 ha) (CAPDR, 2014). Moreover, the surface area is expected to increase rapidly in the coming years since in the meadows of Almuñécar and Motril (Granada) loquat is being replaced by avocado (CAPDR, 2014).

Avocado yields in these irrigated areas are very variable, ranging from 6.26 to 10.06 t ha^{-1} and from 6.56 to 9.31 t ha^{-1} for Malaga and Granada, respectively.

On the other hand, water is the most important limiting factor for crop production in many regions worldwide and the Mediterranean basin is no exception. Water scarcity is a serious threat to food security in arid and semiarid areas, and therefore irrigation water needs to be used more efficiently (i.e., improved water productivity; Chartzoulakis and Bertaki, 2015). Moreover, climate change will power this situation by augmenting potential evapotranspiration, decreasing total rainfall, and increasing the frequency of droughts (IPCC, 2014). The amount of river water will

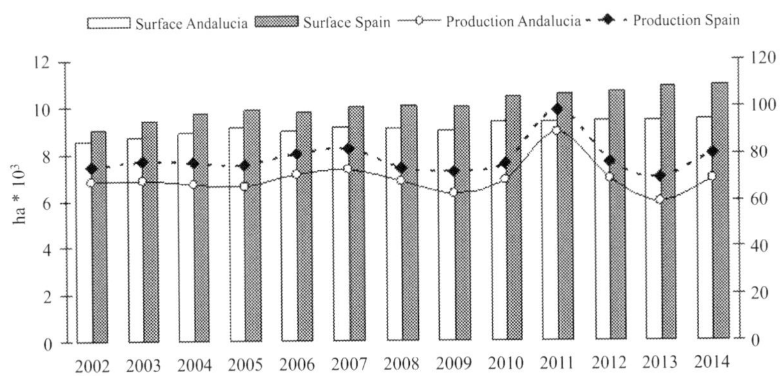

Figure 14.2 Area and avocado production in Andalusia and Spain (ESYRCE, 2015).

diminish by 40% by the middle of this century and agriculture will require greater amounts of water for irrigation (Forzieri et al., 2014). In addition, due to the expected larger global population in the coming years, there will undoubtedly be a strong debate concerning water allocation among the different economic sectors. In this sense, agriculture is the largest consumer of the world's water resources, representing 70% of all global freshwater withdrawals (Mbabazi et al., 2017).

In particular, in the Mediterranean Basin agriculture is seriously affected by this situation, since it is the sector with the highest water use (50% of total water withdrawal) (FAO, 2015). These problems are even more pressing and complex when groundwater is overexploited or salt-water intrusion occurs. Consequently, agriculture is expected to be the center of this debate, focusing on how to remove some of the available water for use in another economic sector, and how to utilize available water in a more sustainable and effective way. Measures related to efficient irrigation scheduling may palliate this problem of increasing demand of water supplies. This water scarcity context will affect avocado, since it is very sensitive to a lack of moisture, which depresses growth and yield and causes problems in the quality and postharvest life of the fruits.

An effective water management system in avocado requires proper irrigation scheduling. In this context, watering requirements for avocado in SE Spain are poorly understood. An inappropriate use of irrigation water (deficit or excess) can bring irregular production and consequently variable fruit yield. Also in this zone and by extension throughout Mediterranean climate areas, water is a limiting resource. Therefore this chapter reviews the different studies regarding the optimization of irrigation for avocado orchards and plant physiological parameters for assessing the water status with special attention given to the subtropical Mediterranean climate.

2. AVOCADO TREE AND WATER AVAILABILITY

Avocado, originally from Mexico and Central America, is part of the Lauraceae family, i.e., the genus *Persea*. This genus has two subgenera: *Persea* and *Eriodaphne*. The subgenus with commercial interest is *Persea*, which in turn has three botanical varieties: *P. americana* Mill. var. Drymifolia (Mexican species); *P. americana* Mill. var. American (West Indian species); and *Persea nubigena* Mill. var. Guatemalensis (Guatemalan species) (Fonseca et al., 2016).

The requirements of avocado vary according to the variety. The Antillean variety has a pyriform shape and is adapted to low temperatures. The Guatemalan variety has rounded fruits and is even more resistant to cold. Meanwhile, the Mexican variety has small fruits and is the most resistant to low temperatures (Fonseca et al., 2016).

The tree is evergreen, but normally it is considered as wintergreen since the leaves do not stay on the tree more than 12 months and the entire canopy is changed during the growth that follows the flowering period. Trunk and branches are monopodial and end in a vegetative bud. Leaves reach their final size in approximately 30 days (Scora et al., 2002). It has a vegetative preference in its growth pattern, and therefore assimilates preferentially to shoot growth and not to reproductive organs. This behaviour, together with the fast production of leaves and shading areas in the canopy, reduces the number of shoots available for flowering (Carr, 2013). In subtropical latitudes, flowering begins when shoot growth enters the inactive period. Flowering occurs normally during late winter and spring and anthesis lasts from 3 to 8 weeks (Whiley and Schaffer, 1994). Inflorescences are panicles of cymes. Avocados are protogynous and have synchronous dichogamy with the maturation of anthers and stigmas in different periods—that is, flowers open when the pistil is receptive and afterward close to open again on the second day when the pollen is dispersed and the pistil is no longer receptive (Carr, 2014). The timing of the opening and closing of the flowers changes according to the variety. A medium-flowering tree can have 1–2 million flowers, but produces only 200–300 fruits. In subtropical latitudes, the fruit load is decreased by leaf shed during approximately the first month after fruit set and then again at the beginning of summer (Schaffer and Whiley, 2002). The fruit is botanically a berry with a thick, fleshy mesocarp surrounding a single large seed (Carr, 2013). After flowering, a great number of fruits drop; this period is crucial since resources must be shared between fruit growth and new shoot growth.

Avocado has a potential photosynthetic capacity to produce more than 30 t ha^{-1} of fruits with 17% of oil (Wolstenholme, 1986). However, actual production is determined by high abscission rates of flowers and fruits, so that average avocado world production is 10.0 t ha^{-1} (Garner and Lovatt, 2008). Fruit and flower abscission has not been associated with tree growth or with other factors such as soil moisture or leaf nutritional status (Garner and Lovatt, 2008). Wolstenholme et al. (1990) suggested that the high need for carbohydrates during the fruit development period coinciding with high temperatures and often with high evaporative demands may be the main cause of fruit abscission.

On the other hand, the root distribution of avocado is very shallow, with more than 70% normally growing in the first 0.6 m layer of the soil (Lahav and Whiley, 2002). The reason for this particular distribution is still unknown; however, several authors have undertaken studies to monitor the influence of soil texture and irrigation management on this distribution (Cantuarias et al., 1995; Salgado and Cautdo, 2008). Therefore the capacity of soil exploitation is limited and as a consequence is highly sensitive to poor water availability periods. Moreover, there are periods where this lack of water availability is critical: flowering, fruit set, or seed development. In addition, higher temperatures are frequent in the period of avocado fruitlet development in the Mediterranean Basin and coincide with the period of lower water availability, which could eventually cause fruitlet abscission. Therefore the root system of avocado is an inherent characteristic, which is still more problematic in a context of water scarcity, an intrinsic problem of the Mediterranean Basin, due to the low amounts of total rainfall and to the irregular distribution throughout the year and between years. Also the intermittent periods of drought make the problem more pressing (Lionello et al., 2006).

In addition, during the flowering period, the canopy surface can potentially suffer significant water losses. Stomatal closure is a rapid indicator of water stress, which is associated with changes in leaf anatomy, restricting the diffusion of CO_2. Relatively few studies have been performed regarding water use in avocado trees. In Mediterranean climate environments, the peaks of water use (in summer) appear to be between 3 and 5 mm day^{-1}, according to Carr (2013). In agreement with this author, for mature avocado trees, the crop coefficient (K_c) is usually in the range of 0.4 and 0.6, with water use efficiency (WUE) being between 1 and 2 kg m^{-3}.

Thus due to the high energy costs related to the avocado growing conditions under the Mediterranean climate, with periods of drought that have occurred more frequently in recent decades, there is an urgent need to develop techniques to reduce water consumption without significantly affecting yield or quality. Understanding plant–water–soil relations is a major objective for developing an irrigation strategy to optimize avocado productivity.

3. IRRIGATION AND WATER USE EFFICIENCY

It is well known that the Mediterranean climate has scarce and erratic precipitation. In particular, the average annual rainfall in some areas in SE Spain is less than 300 mm per year. Proper management of irrigation is essential for achieving maximum efficiency of irrigation water use.

The efficiency of irrigation is normally defined as an economic income of yield per unit of utilized irrigation water. This concept is vital in arid and semiarid areas, and in particular in Mediterranean climate areas, where freshwater resources are overexploited due to the population increase, high climate (rainfall) variability, and intensive use of water for agricultural purposes, industry, and tourism (Blum, 2014).

The first step to achieve sustainable WUE is appropriate irrigation scheduling, which is the decision process concerning when to irrigate and the necessary amount to apply. This requires knowing the crop water requirements, the soil characteristics, the climate conditions, and the irrigation method.

3.1 Deficit-Irrigation Techniques

There are two main types of deficit irrigation: sustained deficit irrigation (SDI) and regulated deficit irrigation (RDI). In SDI, an equal proportion of crop evapotranspiration (ET_c) is applied throughout phenological development, resulting in constantly increasing water stress over the growing season (Fereres and Soriano, 2007). By contrast, RDI involves an alternation in the proportion of ET_c, causing more severe water stress during a specific phenological stage. The reasoning behind the RDI strategy is that crops respond differently to water stress during different phenological stages of development, and applying water stress at the appropriate stage will allow controlled vegetative growth, reduced fruit size, and a moderate yield loss. Thus the final task is to increase WUE by using these techniques.

On the other hand, partial root drying (PRD) consists of exposing half of the root system to dry conditions while the other part is well irrigated. The dry–wet cycles are normally 7–14 days. These systems provoke biochemical responses of plants to this stress, leading to a balance between vegetative and reproductive growth (Stoll et al., 2000; Chartzoulakis and Bertaki, 2015). The expected effects are, first, stress in root systems, which prompts a signal of abscisic acid to regulate stomatal opening and closing. The second is a change in plant growth and finally the prevention of physiological consequences related to severe drought periods (Tahi et al., 2007). These chemical signals are transported through the xylem from the roots to the shoots, reducing stomatal conductance (g_s) (Bravdo, 2005). This scenario has been demonstrated in several species. Deficit irrigation may be a sustainable technique in avocado under water scarcity situations in the Mediterranean subtropical climate.

3.1.1 Avocado Irrigation Approaches

At the plot level, crop water requirements are usually determined based on the reference evapotranspiration (ET_o) and K_c tabulated by the Food and Agriculture Organization (FAO) for different growing conditions (Allen et al., 1998). K_c can also be determined by lysimetry for either weighing or drainage (Durán et al., 2011; Lorite et al., 2012; García et al., 2015). This methodology is ideal for woody crops, as it is heavily influenced by local microclimate, the soil type, and the phenological stage, which allow the calculation of specific K_c for each of the different situations considered and therefore indicate the crop water needs during their development cycle.

Table 14.4 shows the main studies related to water use by avocado trees in many producer countries.

In this context, according to Neuhaus et al. (2007), under Mediterranean conditions, applying PRD could save water without considerably reducing fruit yield. These authors investigated whether drying half the root system could keep the axial conductivity of trunks and trigger a root signal to induce the reduction of water use and plant growth, compared with drying the entire root system. They found that for the treatment of drying the whole root system (DD), 34% of xylem vessels were occluded by tyloses (suberized outgrowths of ray or axial parenchymal cells protruding through the pit cavities into the xylem vessels, laticifers, or tracheary cells (Rioux et al., 1995)) in the lower trunk, even 84 days after rewatering. However, they found no tyloses in trees with half-watered root systems. DD induced the closing of stomata, reduced water use, and depressed tree growth, flower abortion, and fruit shedding. These researchers demonstrated that watering half of the root system used a similar amount of water as a complete water system.

Also Neuhaus et al. (2009) evaluated the effect of drying half the root system on fruit yield and fruit Ca concentration in Australia. Treatments were DD (drying the root zone beneath the whole canopy) and WD (half the canopy), and both compared with well-irrigated trees. Yields were almost the same in WD and in DD. According to the findings, yields were

Table 14.4 Studies on water use in avocado (*Persea americana* Mill.) trees

Authors	Objective	Main findings
Richards et al. (1962)	WUE	The experiment compared four irrigation intervals (7, 14, 21, and 28 days). Vegetative growth (trunk diameter, tree height, and volume) was reduced as the irrigation interval increased
Kurtz et al. (1992)	WUE	Treatments: 70%, 100%, and 130% of the annual amount water recommended for mature trees of the annual amount recommended for mature trees (700 mm). Fruit yields varied considerably but when totaled the only significant effects were the 32% yield advantage of the 130% treatment over the 100% (and 70%) treatments for cv. "Ettinger" (47 cf. 35 t ha^{-1}), and a corresponding 15% benefit for cv. "Fuerte" (40 cf. 32 t ha^{-1})
Lahav et al. (1992)	WUE	Treatments: 60%, 80%, 100%, and 120% of the commonly applied amount. Yields of fruit increased at 2.2 kg m^{-3} (cv. "Hass") and 1.6 kg m^{-3} (cv. "Fuerte") with the annual average water irrigation applied
Olalla et al. (1992)	WUE	Four treatments: three drippers per tree, with crop coefficient (K_c) = 0.44, 0.57, and 0.66, and five drippers per tree, $K_c = 0.57$. There were differences in vegetative growth but fruit yields were similar
Hoffman and du Plessis (1999)	WUE	Summer: water use reached 5 mm d^{-1} for cv. "Fuerte" and 4 mm d^{-1} for cv. "Hass." Winter: corresponding values were 1.5–2.0 mm d^{-1} and below 1.5 mm d^{-1}. The annual totals were 1020 and 890 mm, respectively

Continued

Table 14.4 Studies on water use in avocado (*Persea americana* Mill.) trees—cont'd

Authors	Objective	Main findings
Lahav and Whiley (2002)	Irrigation regimes	Midsummer water application rates recommended for young trees were: year 1, 4–8 L d^{-1} $tree^{-1}$; year 2, 8–15 L d^{-1} $tree^{-1}$; year 3, 30–50 L d^{-1} $tree^{-1}$, and year 4, 80–150 L d^{-1} $tree^{-1}$
Chartzoulakis et al. (2002)	Water deficit	Changes in leaf anatomy, reduction of 35%–45% in leaf porosity (decrease in CO_2 diffusion). Lower stomatal conductance (g_s) and photosynthesis rates. The cv. "Hass" was more susceptible than cv. "Fuerte" to water stress
Gardiazabal et al. (2003)	K_c	Treatments: 90%, 100%, 110%, and 130% of crop evapotranspiration (ET_c), 90% and 100% of ET_c: severe water stress. K_c with 110% and 130% of ET_c and they proposed a value of 0.72 of K_c. Economic analysis: 110% was the most productive treatment.
Lahav and Aycicegi-Lowengart (2006)	WUE	Confirmation of results of Lahav et al. (1992).
Neuhaus et al. (2009)	Drying root	Yields were reduced proportionally more than the reduction in water supply, compared to the well-watered treatment. Rewatering did not restore yields in the following season
Silber et al. (2011, 2012)	Sensitivity to water deficit	Avocados grown in lysimeters: three irrigation treatments and two soil volumes (100 and 200 L). Vegetative growth was higher in the 200 L volume in all irrigation frequencies. Differences between irrigation treatments statistically significant in the 2 years when fruit was harvested in the low volume (100 L). Fruit drop was greater in 100 L soil volume than in 200 L
Gil et al. (2011)	WUE and K_c	Four irrigation regimes: T_0 (110% reference evapotranspiration [ET_o]), T_1 (65% ET_o), T_2 (77% ET_o), and T_3 (132% ET_o). T_2 produced the highest yields and T_0 the lowest, while T_1 had the fruits of smaller size compared with T_3. Firmness of fruits after 45 days was higher for T_0

Kiggundu et al. (2012)	Water productivity	Combined irrigation and nutrient management treatments. Soil water status irrigation-based techniques reduced the volume of water applied by 87%, and the quantity of phosphorus leached by 74%, compared with irrigating to a local set schedule. Water productivity ranged from less than 1 kg m^{-3} (set traditional schedule) up to 12 kg m^{-3} (this study: water balance)
Kaneko (2016)	WUE, water deficit in flowering and fruit development	Monthly average K_c ranged from 0.45 to 0.60 for the mature trees, and from 0.25 to 0.30 for the young trees. In winter, K_c increased to 0.9–1.0 for the mature trees and 0.45–0.55 for the young trees. The control treatment registered a higher average yield at 36.4 ± 1.1 kg per tree than the stressed trees (27.8 ± 1.0 kg). Trees under deficit irrigation in spring and flowering: not very sensitive to water stress
Mbabazi et al. (2017)	Smartirrigation apps applied to avocado	Optimum number of previous days to calculate ET_c in avocado (6–7 days was enough). They recommended the use of a wireless connection for users to change the irrigation planning using the smartphone app to avoid water losses during the wet season
Holzapfel et al. (2017)	Water productivity, bearing	Production was lower in the "off-crop" years due to alternate bearing for all treatments. Yields in 25% and 50% ET_o were only 22% of the yield obtained in the two "on-crop" years, whereas the yields of 75% and 100% ET_o in the off-year reached 42% of the average yield of the 2 years of high production

K_c, crop coefficient; *WUE*, water use efficiency.

reduced by more than half and rewatering did not recover yield on DD or WD trees in the following season. No effects on Ca in the fruits were detected for WD or DD, and therefore quality was not affected. The reason in this case for the reduction of fruit yield was due to fruit abscission. Furthermore, inflorescences in avocado were more resistant to drying soil than were other vegetative structures (Neuhaus et al., 2007).

The literature is scarce regarding experiments with avocado tree deficit irrigation trials under a Mediterranean climate. Silber et al. (2012) studied the time course of avocado trees grown in containers under different irrigation regimes (pulsed irrigation, 10–20 min every 30 min) throughout the day (T_1), one daily irrigation event beginning at night and ending in the morning every day (T_2), and one irrigation event every 2 days (T_3) at two root volumes (100 and 200 L). The results indicated that water use by trees in T_1 corresponded to changes in meteorological conditions, whereas T_2 and T_3 were subjected to water stress. The treatments had a very low effect on flowering and fruit set, but the consequences on fruitlet abscission were significant, being higher in 100 L containers than in 200 L containers, and $T_3 > T_2 > T_1$. Fruit abscission is a key parameter to monitor because, in general, for high rates this parameter determines the overall yield for that season (Garner and Lovatt, 2008). In the same context and irrigation treatments, Silber et al. (2013) monitored the effects of the fruit sink on gas exchange and water uptake. To proceed, they compared fruited and defruited trees. Despite the higher vegetative growth of defruited plants, their water use was 40% lower than in fruited trees. Consequently, avocado fruit load is a major parameter for irrigation programs. These authors found no differences for g_s or for photosynthesis per unit leaf area among irrigation treatments. In defruited and fruited trees they detected lower carbohydrate concentrations before sunrise, concluding that leaf carbohydrate is an important parameter to take into account and further research in the process of stomata aperture is necessary (Silber et al., 2013).

On the other hand, Gil et al. (2011) conducted another experiment under Mediterranean conditions in Chile, comparing four irrigation regimes: T_0 (110% ET_o), T_1 (65% ET_o), T_2 (77% ET_o), and T_3 (132% ET_o). They found differences among treatments for trunk contraction and growth of the tree, but not for chlorophyll content or trunk transversal diameter. T_2 produced the highest yields and T_0 the lowest, while T_1 had the fruits of smaller size compared with T_3. The firmness of fruits after 45 days was higher for T_0. These authors concluded that T_2 (77% ET_o) was the best treatment under their conditions. Gardiazabal et al. (2003) conducted a similar experiment in California with cv. "Hass" to evaluate the recommended K_c. They applied irrigation corresponding to 90%, 100%, 110%, and 130% of ET_c, and found that with 90% and 100% of ET_c, trees were submitted to severe water stress. Under their climatic conditions, the K_c of avocado approached 110% and 130% of ET_c and they proposed a value of 0.72 of K_c for all the months of the year. They also made an economic analysis and identified 110% as the most productive treatment.

Also in Chile, Holzapfel et al. (2017) studied the effects of four irrigation treatments for Hass avocado (25%, 50%, 75%, and 100% of ET_o). They found that fruit size and production increased significantly for the treatment 75% ET_o. Furthermore, production was different among years. They discovered that 25% and 50% ET_o represented 22% of the yield obtained during the "on-crop" years, whereas yields of 75% and 100% ET_o in the off-year reached 42% of the average yield of the 2 years of high production.

Kaneko (2016) determined water requirements for avocado cv. "Hass" in New Zealand, monitoring the effect of water deficit on flowering and early fruit development in young avocado trees. ET_o was calculated by the FAO Penman–Monteith equation based on weather information at the orchard. ET_c of avocado was determined by measuring sap flow, and daily ET_c was also estimated from a soil–water balance. Therefore avocado tree water requirements were calculated as K_c. This researcher found that during summer, monthly average K_c ranged from 0.45 to 0.60 for the mature trees and from 0.25 to 0.30 for the young trees. In winter, K_c increased to 0.9–1.0 for the mature plants and 0.45–0.55 for the young plants. The same researcher also evaluated the consequences of deficit irrigation on flowering and fruit development, by setting rainout shelters under the young trees. Control plants were irrigated and fertilized, whereas the trees with deficit treatment had no irrigation, precipitation, or fertilizers. The rainout shelters were removed in early May 2015, and monitoring was continued until late October. The control treatment registered higher average yield at 36.4 ± 1.1 kg per tree than the stressed trees (27.8 ± 1.0 kg). Therefore flowering trees under deficit irrigation in spring did not appear to be very sensitive to water stress, but early fruit growth was in fact sensitive, resulting in a sharp reduction in fruit size.

In Malaga, Olalla et al. (1992) compared different treatments. The average annual depth of the water applied to each treatment over the 3 years ranged between 674 mm (drip, $K_c = 0.44$) and 1024 mm (five drippers, $K_c = 0.57$). Due to direct evaporation losses from wet soil, the larger the proportion of wetted soil surface, the higher the water application necessary, particularly with microsprinklers. Although there were differences in vegetative growth, fruit yields were similar (cv. Hass, 5–7 years old). In relation to this study, Carr (2013) reported that this study was relatively unclear, with the main conclusion that when estimating how much water to apply to a coarse textured soil, K_c should be higher than 0.55.

Hermoso et al. (2003) carried out a trial where water was applied to the canopy in periods of high evaporative demand with microsprinkler irrigation. This lowered the canopy temperature. Water consumption increased exponentially with the lowering of avocado leaf temperature. This water application on the canopy also lowered g_s in periods of high evaporative demand, enlarging the average fruit size. In the same study area, Hermoso et al. (2007) carried out an experiment with two cultivars (Fuerte and Hass) on two rootstocks (Topa Topa and Lula). The conventional irrigation system had two microsprinklers per tree. With alternate irrigation, every microsprinkler irrigated a larger area and water was applied alternatively to both sides of the tree using a double irrigation system. The irrigated side was changed when soil water potential reached −1 MPa at 50 cm. In the first season, differences between treatments were insignificant. In the second, yield and potential yield, as well as number and size of fruits, were slightly but not significantly higher with alternate irrigation.

Briefly, an increasing number of avocado farmers in Mediterranean areas are adopting techniques and practices to improve production. However, the rising water costs for irrigation together with water constraints due to severe droughts are forcing farmers to improve WUE. Consequently, when and how to irrigate requires detailed knowledge of soil properties and the use of monitoring devices to characterize the physiological response to water stress, avoiding water logging and maximizing water productivity. However, there is still a lack of information regarding to soil–water–plant relationship for avocado in subtropical Mediterranean areas.

3.2 Water–Soil–Plant Relationships in Avocado

Management and knowledge in relation to crop water status from the physiological parameters, dendrometry, and, more recently, infrared thermography, as well as soil moisture data are appropriate tools in irrigation-programming methods. That is, these technologies enable farmers to reduce the surplus water applied by using methods based exclusively on the empirical calculation of ET_c.

3.2.1 Soil Monitoring

The traditional procedures are the monitoring of soil moisture dynamics and readings of evapotranspiration by using standard class A pan evaporation data with the Penman–Monteith model. However, restrictions have been reported for the use of these techniques (Simonne et al., 1992). In fact, actual evapotranspiration based on the Penman–Monteith model or on class A pan evaporation data can severely underestimate avocado water requirements according to Arpaia and Hofshi (1999). Therefore further research is needed in plant-monitoring techniques in planning the irrigation schedule for avocado.

Several indicators and tools have been proposed to assess avocado water stress. Soil moisture sensors have the drawback of considerable variability of soil properties and the consequences on soil water distribution, and in some cases (with sprinkler irrigation systems) it is necessary to install a large number of probes to have a representative measure of soil moisture (Van Leeuwen et al., 2001).

Tensiometers have also been widely used to monitor water dynamics in avocado. In agreement with du Plessis (1991), the critical matric potentials for avocado at depths of 0.30 m were −30 kPa for sandy soils and −50 kPa for clay soils. Lahav et al. (1992) used tensiomenters as well in an experiment in Israel that compared irrigation treatments (60%, 80%, 100%, and 120% of the commonly applied amount), reporting higher yields with greater amounts of water. These authors confirmed these results in another series of experiments (Lahav and Aycicegi-Lowengart, 2006).

On the other hand, Chartzoulaki et al. (2002) used tensiometers in an experiment with avocado cv. "Fuerte" and cv. "Hass" grown in 50 L containers to evaluate physiological and anatomical changes by applying two moderate water deficits: water was administered when the soil water potential reached −0.03 and −0.5 MPa for the wet and dry treatments, respectively. They found that "Hass" seems to be more affected by moderate water stress.

In general terms, the use of ET devices to control ET is not always possible due to the lack of knowledge of K_c (Migliaccio et al., 2010), as well as the high cost and technical skills needed to gather the information (Mbabazi et al., 2017). Moreover, as stated earlier, soil water sensors can make serious errors because of intrinsic spatial and temporal variability of soil water conditions. Alternative scheduling devices have been proposed to help farmers schedule water irrigation, e.g., Irris Scheduler, Irrigation Scheduler Mobile, and KanSched (Mbabazi et al., 2017). These latter tools do not provide the necessary time for irrigation, but only the water depth to be applied. In this context, the University of Florida developed Smartirrigation apps that use real-time weather data and provide irrigation planning with information on amount and runtime. The Smartirrigation app was composed for avocado and uses real-time ET_c, irrigation system characteristics, root depth, and field capacity (Mbabazi et al., 2017). These latter authors carried out an experiment to determine the optimum number of previous days to calculate ET_c in avocado. They compared 3, 4, 6, and 7 previous days' ET_c, and compared them with 5 days (the current methodology), but found no significant differences, concluding that the use of the previous 6 and 7 days was enough. They recommended the use of a wireless connection for users to change the irrigation planning using the smartphone app to avoid water losses during the wet season (Mbabazi et al., 2017).

3.2.2 Plant Physiological Parameters

Plant-based water stress indicators have been widely utilized in research for many fruit crops: leaf water potential (Ψ_{leaf}) and stem water potential (Ψ_{stem}), trunk diameter, sap flow, canopy temperature, g_s, etc. With the use of all these indicators, it is assumed that the stomatal opening is regulated according to the plant water status and evaporative factors. Moreover, there is feedback from the end-product metabolite that is related to the control of the net photosynthesis rate (Paul and Pellny, 2003; Silber et al., 2013). However, it is important to point out that the thresholds of these indicators are related at the same time to the phenological state of the tree.

Measurements in the plant such as Ψ_{leaf}, Ψ_{stem}, and trunk diameter changes have the advantage of being direct information of water status (Table 14.5).

In this context, Celedón et al. (2012) evaluated the sensitivity of Ψ_{stem} and maximum daily trunk shrinkage (MDS) in avocado, by withholding irrigation and making comparisons with control trees. These authors also monitored leaf abscisic acid, g_s, soil moisture, and vapor pressure deficit, showing that these indicators differed from the control treatment after 3 days of retaining irrigation. They noted that MDS detected water stress in a more effective way than did Ψ_{stem}, but had higher variability. Therefore further research is necessary in the comparison of indicators of water stress in avocado. This high precision of MDS can be crucial for avocado scheduling where early detection of stress is needed. Along this line, other studies related to Ψ_{stem} in avocado reported values of -1.0 to -1.2 MPa as

Table 14.5 Main studies using different plant physiological parameters under water-deficit irrigation techniques in avocado trees

Authors	Physiological parameters	Main findings
Blanke and Lovatt (1993)	Stomatal conductance (g_s)	The floral stomata were functional. During flowering, there are more than 2 million transpiring flowers with an estimated surface area of 54 m^2 in the periphery of the tree; their contribution to the total water use of a tree is considerable
Barrientos and Sanchez (1987)	Stomatal density among cultivars	Fourfold difference in density extremes, higher stomatal density with reductions in tree height
Turner et al. (2001)	g_s, young leaf expansion, young fruit expansion	Water deficits caused the stomata to close before any other changes
Cantuarias et al. (1995)	Leaf water potential (Ψ_{leaf}) and water stress	Actual transpiration (T) in well-irrigated trees reached a maximum of 3 mm d^{-1}. The ratio T:ET_o was low and remained in the range 0.13–0.21. Ψ_{leaf} remained constant. Enlargement of the wetted soil volume from 25% to 75% improved tree water status and root growth
Whiley et al. (1988)	g_s	High contribution of flowers to water use
Neuhaus (2003); Schaffer and Whiley (2003)	g_s	g_s a reliable early indicator of water deficit
Sharon (1999); Sharon et al. (2001)	Trunk and leaf contractions	Demonstrated the sensitivity of trunk and leaf contractions to diurnal changes in Ψ_{leaf} (diurnal range −0.15 to −1.05 MPa)
Celedón et al. (2012)	Stem water potential (Ψ_{stem}) and maximum daily trunk shrinkage, leaf abscisic acid, g_s, soil moisture, and VPD	MDTS detected water stress in a more effective way than Ψ_{stem}, but had higher variability
Turner et al. (2001)	g_s to schedule irrigation	Scheduling by g_s reduced the water applied by one-third, leaf–gas exchange, and the number of fruits on the trees with respect to the water balance method. g_s was not recommended for scheduling

ET_o, reference evapotranspiration; *MDTS*, maximum daily trunk shrinkage; *VPD*, vapor pressure deficit.

water stress indicators (Scholefield et al., 1980). MDS has also been used as water status in other crops, such as in vineyards (Gurovich and Saggé, 2005) and peach (Naor and Cohen, 2003).

In a Mediterranean climate, stomatal closure (g_s) has also been used in avocado as a water deficit indicator (Turner et al., 2001). These authors found that water stress in avocado trees triggered the stomata to close before any other changes in other parameters. Other authors (Neuhaus, 2003; Schaffer and Whiley, 2003) stated that g_s is a reliable indicator, i.e., better than Ψ_{leaf}, Ψ_{stem}, or growth parameters.

Sharon (1999) and Sharon et al. (2001) confirmed the sensitivity of trunk and leaf changes to diurnal shifts in Ψ_{leaf}, and demonstrated how stomata remained opened during the day. They attributed the capacity of trees to maintain a fast transpiration rate to rapid hydraulic conductivity in the root.

It is crucial to advance the integration of different technologies for controlling water stress (physiological sensors, dendrometry, thermography, and soil water content) in avocado trees subjected to deficit irrigation strategies in a subtropical Mediterranean climate (Fig. 14.3).

The challenge is to find an alternative sustainable adaptation to face the growing scenario of shortages of irrigation water. Transversely, within the use of precision farming practices, a detailed analysis related to the effects on production and quality is needed, on the level of stress generated in the spatial variability and the characteristics of the environment, to adjust sustainable water use resources to real crop demand. Therefore these sources of information will be the practical support for estimating the different production functions and crop development, which will be the basis for a predictive tool relating to the response in water scarcity situations.

Consequently, the information generated by this set of variables and production functions will be integrated into scheduling and irrigation control systems, and its use will be essential for practical implementation of alternative irrigation strategies for avocado based on sustainable use and conservation of water resources.

4. SPANISH AVOCADO CULTIVATION

On the coast of Granada and Malaga it has been possible to establish subtropical fruit farming despite the high latitude (36° 44′). This is due to the presence of a high mountain system that stops the north winds, propitiating higher minimum average winter temperatures. Differences in altitude, wind

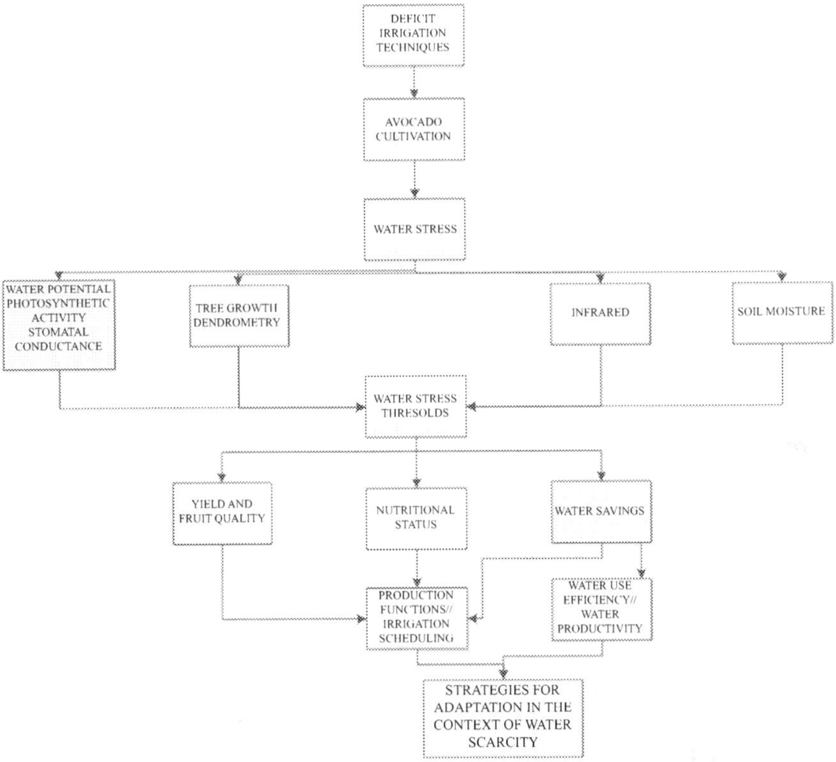

Figure 14.3 Deficit irrigation as a sustainable strategy under a water scarcity scenario.

exposition, and orientation can bring different conditions in short distances, and thus differences in frost risks (Olalla and Rodríguez, 1991).

Subtropical crops were introduced in Granada and Malaga in 1924, with the creation of the experimental research center "Estación de Agricultura Meridional," when in the same project there was the intention of introducing exotic fruits. A subsequent commercial interest promoted the plantation of approximately 2 ha in Almuñécar (Granada) (Olalla, 2012). Public initiatives converged with the high prices of avocado and many farmers assumed the investment of establishing irrigation systems in hillslope areas, since a great number of avocados are planted on orchard terraces (Rodríguez et al., 2009; Durán et al., 2013) (Fig. 14.4). During the 1970s, an exponential increase of plantations took place from several tens of hectares to thousands of hectares at the end of the decade (Olalla, 2012).

According to the latest data from the Spanish Ministry of Agriculture, Fisheries and Rural Development (CAPDR) (2014), during 2013 and 2014

Figure 14.4 Avocado orchards on terraces and hillslopes in SE Spain.

avocado production amounted to 59,066 and 68,885 t, respectively, with an average yield of 6.5 t ha^{-1}. And the prices in 2013 and 2014 were 1.35 and 1.55 € kg^{-1}, respectively. More than 50% of Andalusian avocado production is exported to other countries of the European market, reaching about 50,000 t, with France as the main importer.

Economically, according to the CAPDR (2014), the production costs of an avocado plantation for the 2012 season related to irrigation were 73.8 € ha^{-1} for "depreciation" and 472.8 € ha^{-1} for "irrigation association, repair, and maintenance." Taking into account that the total average indirect costs are 1162.8 € ha^{-1}, irrigation represents 47% of the total cost. In addition, the average price in the area amounted to 0.50 € m^{-3}, a substantial investment for avocado production.

Due to the economic importance of subtropical crops in Andalusia and its competitive advantage in the European market, as well as the environmental problems and water scarcity risk, an optimization of the water supply in avocado goes directly to a reduction in transpiration levels of the crop, and at the same time a lowering of production costs, especially concerning the cost of water for farmers. Knowing the limit of this reduction, maintaining the sustainability of productive systems and proposing adaptation strategies for future water scarcity scenarios will be main challenges for farmers.

The most common avocado variety in this area is "Hass," grafted on Mexican rootstocks. Flowering takes place in March and the harvest period is between January and February; however, fruits can be retained on the tree until July (Olalla, 2012). Table 14.6 shows the main avocado cultivars grown in SE Spain (Hueso and Cuevas, 2014; BROKAW, 2017).

According to Olalla (1997) and Farré et al. (1992) the irrigation needs in Mediterranean semiarid conditions amounted to 7500–8000 $m^3 ha^{-1}$. At the beginning of establishing plantations, trees were not pruned, but there was a problem of excessive vegetative growth (Fig. 14.4). Currently, there are two trends in pruning: one is the maintenance of a continuous edge with distance between rows of 8 m and tree height not higher than 4 m, while the other is the periodic pruning of either alternating trees or alternating different branches on the same tree.

On the other hand, the strategic geographical location gives Spain an advantage with respect to other suppliers. In this context, it would be desirable to make a major effort to inform European consumers of the advantages of the quality of Spanish avocado over other fruits from overseas. For this reason, it would be interesting to find out some other varieties that give production out of the typic period of cv. 'Hass', which is from July to December (Olalla, 2012), even if the final yields are lower. Furthermore, there is strong willingness from Central European consumers to pay, with a high average income per capita and at the same time an increasing demand for healthy, fresh produce. Therefore Spanish production needs to establish its market niche to compete with other countries, by knowing the needs

Table 14.6 Main varieties utilized on the coast of Granada and Malaga (SE Spain)

Cultivar	Growth pattern	Flower type	Fruit size (g)	Fruit color	Fruit skin	Harvest period
Bacon	Erect	B	170–510	Green	Thin	Oct–Dec
Fuerte	Extended	B	170–500	Green	Thin	Nov–Feb
Pinkerton	Extended	A	230–425	Green	Coarse	Nov–Feb
Hass	Rounded	A	140–400	Green-Black	Coarse	Dec–Jun
Reed	Erect	A	270–680	Green	Thin-coarse	Mar–Jul
Lamb Hass	Erect	A	283–510	Black	Coarse	May–Jul

Flower type: "A" varieties open as female on the morning of the first day. The flower closes in late morning or early afternoon. The flower will remain closed until the afternoon of the second day when it opens as male. "B" varieties open as female on the afternoon of the first day, close in late afternoon, and reopen in the male phase the following morning.

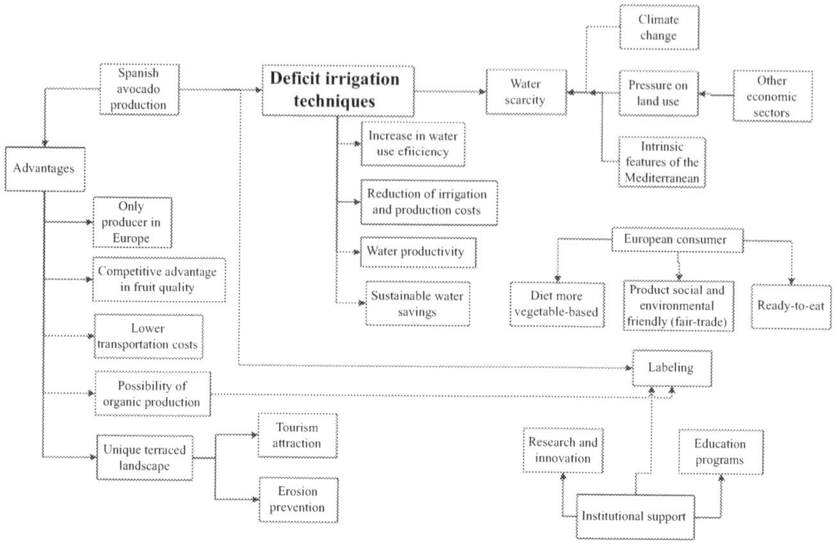

Figure 14.5 Spanish avocado cultivation in a context of a water scarcity scenario: socioeconomic and environmental implications.

and preferences of the European consumer and supporting and encouraging innovation to distinguish the Spanish avocado from others by strategies such as labeling, including water sustainability (Fig. 14.5). In this sense, water-saving strategies should be included in the added value of an avocado that is produced in an environmentally friendly way, and this facet should be included in the development of new labeling features for the consumer (Krishnajumar et al., 2014).

On the other hand, the pressure for territories with other economic sectors, together with climate change and the aforementioned characteristics of the Mediterranean climate (which will become more severe), will force growers to seek higher efficiency in water consumption for avocado irrigation in SE Spain (Fig. 14.5). Further research and economic support is of high priority since Spanish avocado, as stated earlier, has several strategic advantages in the European market: higher quality for longer time of maturation in tree, lower transportations costs, and the creation of a singular terraced landscape in SE Spain, which has its own inherent value of promoting green tourism.

In addition, for the coming years, the importance of a healthy diet will continue to grow, shifting from a carbohydrate- and meat-based diet toward one rich in fruits and vegetables. Furthermore, European consumers prefer products based on sustainability—that is, they have a strong desire for social and environmentally friendly products. New innovative labeling systems could involve water efficient systems for growing avocado, which should be included on labels related to the protection of the environment (Fig. 14.5). European consumers are increasingly demanding the so-called ready-to-eat avocado. As a result, naturally matured fruits from nearby locations will have an advantage. Also in SE Spain, organic avocado cultivation has promising possibilities of becoming an important sector, which could be an asset in marketing. However, the sector needs further institutional support for research, as well as innovation and educational programs for irrigation associations, farmers, and local governments from the area of SE Spain to promote more efficient water use in subtropical farming.

5. CONCLUSIONS

Under the climatic conditions of SE Spain and under a water scarcity scenario, considerable effort must be devoted to introduce strategies aimed at improving WUE based on the premise that higher yields can be achieved with less water. From the review carried out in this chapter, we conclude that sustainability, in particular for Spanish avocado orchards, can be achieved by the following means:

- Improving the efficiency of irrigation systems. It is necessary to complete research on proper irrigation scheduling according to actual requirements of avocado in a subtropical Mediterranean environment.
- Deficit irrigation programs have been adopted in some areas; however, due to the scarcity of trials in the area, further research is needed.
- More studies concerning the willingness to pay and the preferences of the European consumer are needed to promote the potential of the Spanish avocado, taking into account the geographical advantage.
- National and regional governments should invest in capital formation and productive assets. Spanish avocado agriculture will have to adapt to climate change and help to mitigate the effects by promoting policies of soil and water conservation by improving both the quality of yields and the efficient use of these resources.

ACKNOWLEDGMENTS

Part of this publication was sponsored by the following research project: "Impact of deficit irrigation on productivity of subtropical fruit crops: tools for sustainable water stress management" (PP.AVA.AVA201601.8), and was cofinanced by the European Regional Development Fund within the Operational Programme Andalusia 2014—20 "Andalucía is moving with Europe." The author I. Garcia-Tejero has a contract cofinanced by the Operational Programme of the European Social Fund 2007—13 "Andalucía is moving with Europe."

REFERENCES

Allen, R.G., Pereira, L.S., Raes, D., Smith, M., 1998. Crop Evapotranspiration – Guidelines for Computing Crop Water Requirements. FAO Irrigation and Drainage Paper 56. UN-FAO, Rome.

Arpaia, M.L., Hofshi, R., 1999. Irrigation frequency affects soil salinity of drip irrigated avocado. In: Proceedings of Avocado Brainstorming '99, Riverside USA, pp. 27–28.

Barrientos, P.A., Sanchez, C.S., 1987. In: Stomatal Density and its Relationship to Growth Habit in Avocado, vol. 10. South African Avocado Grower's Association Yearbook, pp. 66–67.

Blanke, M.M., Lovatt, C.J., 1993. Anatomy and transpiration of the avocado inflorescence. Annals of Botany 71, 543–547.

Bleinroth, E.W., Castro, E.W., 1992. 'Matéria prima' in ITAL. Abacate cultura, matéria-prima, processamento e aspectos econômicos. Série Frutas Tropicais, second ed. ITAL, Campinas, pp. 58–147.

Blum, W., 2014. Land degradation and security linkages in the Mediterranean region. In: Kapur, B.S., Ersahin, S. (Eds.), Soil security for Ecosystem Management. Springer Briefs in Environment, Security, Development and Peace, pp. 19–29.

Bravdo, B.A., 2005. Physiological mechanisms involved in the production of non-hydraulic root signals by partial rootzone drying – a review. Acta Horticulturae 689, 267–276.

BROKAW, 2017. Viveros BROKAW. Available from: http://www.viverosbrokaw.com/aguacate_variedades.html.

Cantuarias, T., Cohen, Y., Tomer, E., 1995. Improving avocado tree water status under severe climatic conditions by increasing wetted soil volume. In: Proceedings of the World Avocado Congress III, Israel, pp. 196–204.

CAPDR, 2014. Consejería de Agricultura, Pesca y Desarrollo Rural, Junta de Andalucía. Anuario de Estadísticas Agrarias y Pesqueras de Andalucía. Available from: http://www.juntadeandalucia.es/export/drupaljda/Anuario%202014_0.zip.

Carr, M.K., 2013. The water relations and irrigation requirements of avocado (*Persea Americana* Mill.): a review. Experimental Agriculture 49, 256–278.

Carr, M.K., 2014. Advances in Irrigation Agronomy. Cambridge Uiversity Press, UK.

Celedón, J.M., Gil, P.M., Ferreyra, R., Maldonado, P., Barrera, C., 2012. Sensivity and variability of two plant water stress indicators: exploring criteria of choosing a plant monitoring method for avocado irrigation management. Chilean Journal of Agricultural Research 72, 379–387.

Chartzoulakis, K., Bertaki, M., 2015. Sustainable water management in agriculture under climate change. Agriculture and Agricultural Science Procedia 4, 88–98.

Chartzoulakis, K., Patakas, A., Kofidis, G., Bosabalisis, A., Nastoou, A., 2002. Water stress affects leaf anatomy, gas exchange, water relations and growth of two avocado cultivars. Scientia Horticulturae 95, 39–50.

du Plessis, S.F., 1991. In: Factors Important for Optimal Irrigation Scheduling of Avocado Orchards, vol. 14. Yearbook -South-African-Avocado-Growers'-Association, pp. 91–93.

Durán, Z.V.H., Rodríguez, P.C.R., Muriel, F.J.L., Franco, T.D., 2011. Monitoring the pollution risk and water use in orchard terraces with mango and cherimoya trees by drainage lysimeters. Irrigation and Drainage Systems 922, 83–90.

Durán, Z.V.H., Rodríguez, P.C.R., Francia, M.J.R., Martin, P.J.F., 2013. Land-use changes in a small watershed in the Mediterranean landscape (SE Spain): environmental implications of a shift towards subtropical crops. Journal of Land Use Science 1, 47–58.

ESYRCE, 2015. Encuesta sobre Superficies y Rendimientos de Cultivos. Resultados Nacionales y Autonómicos. Ministerio de Agricultura, Alimentación y Medio Ambiente. Available from: http://www.mapama.gob.es/es/estadistica/temas/estadisticas-agrarias/espana2015web_tcm7-401244.pdf.

FAO, 2015. AQUASTAT Database. Availablr from: http://www.fao.org/nr/water/aquastat/data/query/index.html?lang=en.

FAOSTAT, 2015. Food and Agriculture Organization of the United Nations Statistics. Available from: http://www.fao.org/faostat/en/#data.

Farré, J.M., Hermoso, J.M., Jaime, M., 1992. Riego y nutrición del aguacate en el litoral oriental andaluz. Información Técnica 19/92 Junta de Andalucía.

Fereres, E., Soriano, M.A., 2007. Deficit irrigation for reducing agricultural water use. Journal of Experimental Botany 58, 147–159.

Fonseca, D.P., Alves, C.M., Dellinghausen, B.C., Rosane, B.M.C., 2016. In: Avocado: Characteristics, Health Benefits and Uses, vol. 46. Ciência Rural, pp. 747–754.

Forzieri, G., Feyen, L., Rojas, R., Flörke, M., Wimmer, F., Bianchi, A., 2014. Ensemble projections of future streamflow droughts in Europe. Hydrology and Earth Systems Science 18, 85–108.

García, T.I.F., Hernández, A., Rodríguez, V.M., Ponce, J.R., Ramos, V., Muriel, J.L., Durán, Z.V.H., 2015. Estimating almond crop coefficients and physiological response to water stress in semiarid environments (SW Spain). Journal of Agricultural Science and Technology 17, 1255–1266.

Gardiazabal, F., Magdahl, C., Mena, F., Wilhelmy, C., 2003. Determinación del coeficiente de cultivo (Kc) para paltos cv Hass en Chile. In: Proceedings of the V World Avocado Congress, pp. 329–334.

Garner, L.C., Lovatt, C.J., 2008. The relationship between flower and fruit abscission and alternate bearing of 'Hass' avocado. Journal of the American Society for Horticultural Science 133, 3–10.

Gil, M.P., Ferreyra, R., Karlezi, D., Troncoso, C., Saavedra, J., Robledo, P., Defilippi, B., 2011. World Avocado Congress 2011. Available from: http://worldavocadocongress 2011.com/userfiles/file/Pilar%20Gil%201420-1440.pdf.

Gurovich, L., Saggé, O., 2005. Fine tuning irrigation scheduling with phytomonitoring technology in Chile. In: Proceedings of the EFITA/WCCA Joint Congress on IT in Agriculture. Vila Real, Portugal, pp. 727–732.

Hermoso, J.M., Soria, J.T., Cortés, S., Farré, J.M., 2003. Aplicación intermitente de agua sobre el follaje del aguacate Hass. In: Proceedings of the V World Avocado Congress. Granada, Spain, pp. 295–300.

Hermoso, J.M., Torres, M.D., Farré, J.M., 2007. Riego alterno del aguacate con potential matricial moderado en suelo. In: proceedings of the VI World Avocado Congress, Viña del Mar, Chile.

Hoffman, J.E., du Plessis, S.F., 1999. Seasonal water requirements of avocado trees grown under subtropical condition. In: Revista Chapingo Serie Horticultura, vol. 5, pp. 191–194.

Holzapfel, E., Alves de Souza, J., Jara, J., Carvallo, H.G., 2017. Responses of avocado production to variation in irrigation levels. Irrigation Science 35, 202–215.

Hueso, M.J.J., Cuevas, G.J., 2014. La fruticultura del siglo XXI en España. Serie agricultura. Grupo Cajamar, Spain.

IPCC: Climate Change, 2014. Impacts, adaptation, and vulnerability. In: Field, C.B., Barros, V.R., Dokken, D., Mach, K.J., Mastrandrea, M.D., Bilir, T.E., Chatterjee, M., Ebi, K.L., Estrada, Y.O., Genova, R.C., Girma, B., Kissel, E.S., Levy, A.N., MacCracken, S., Mastrandrea, P.R., White, L.L. (Eds.), Part A: Global and Sectorial Aspects, Contribution of Working Group II to the Fifth Assessment Report of the Intergovernmental Panel on Climate Change. Cambridge University Press, Cambridge, UK and New York, NY, USA.

Kaneko, T., 2016. Water Requirements for 'Hass' Avocado Flowering and Fruit Development in New Zealand (Ph.D. thesis). University of Waykato.

Kiggundu, N., Kati, W., Migliaccio, B.C., Li, Y., Crane, J.H., 2012. Water savings, nutrient leaching, and fruit yield in a young avocado orchard as affected by irrigation and nutrient management. Irrigation Science 30, 275–286.

Krishnajumar, J., Chan-Halbrendt, C., Zhang, Q., Sullivan, P., 2014. Effects of label on consumer preferences: focus on Hawaiian avocado industry. Journal of Food Products Marketing 20, 325–344.

Kurtz, C., Gull, I., Klein, I., 1992. Water rate effects on three avocado cultivars. In: Proceedings of the II World Avocado Congress, California, USA, pp. 357–364.

Lahav, E., Aycicegi-Lowengart, A., 2006. Re-evaluation of irrigation water amounts under changing growing conditions of avocados in Israel. In: proceedings of the seminar Manejo del Riego y Suelo en el Cultivo del Palto, La Cruz, Chile.

Lahav, A.W., Whiley, E., 2002. Irrigation and mineral nutrition. In: Whiley, A.W., Schaffer, B., Wolstenholme, B.N. (Eds.), The Avocado: Botany, Production and Uses. CBI Publishing, New York, pp. 259–297.

Lahav, E., Steinhardt, R., Kalmar, D., 1992. Water requirements and the effect of salinity in an avocado orchard on a clay soil. In: Proceedings of the II World Avocado Congress, California, USA, pp. 323–330.

Lionello, P., Malanotte-Rizzoli, P., Boscolo, R., Alpert, P., Artale, V., Li, L., Luterbacher, J., May, W., Trigo, R., Tsimplis, M., Ulbrich, U., Xoplaki, E., 2006. The mediterranean climate: an overview of the main characteristics and issues. In: Lionello, P., Malanotte-Rizzoli, P., Boscolo, R. (Eds.), Mediterranean Climate Variability, Developments in Earth and Environmental Sciences, vol. 4. Elsevier, Amsterdam.

Lorite, I.J., Santos, C., Testi, L., Fereres, E.C., 2012. Design and construction of a large weighing lysimeter in an almond orchard. Spanish Journal of Agricultural Research 10, 238–250.

MAPAMA, 2016. Anuario de Estadísticas Agrarias 2015. Available from: http://www.mapama.gob.es/estadistica/pags/anuario/2015/AE15.pdf.

Mbabazi, D., Migliaccio, K.W., Crane, J.H., Fraisse, C., Zotarelli, L., Morgan, K.T., Kiggundu, M., 2017. An irrigation schedule testing model for optimization of the Smartirrigation avocado app. Agricultural Water Management 179, 390–400.

Migliaccio, K.W., Schaffer, B., Crane, J.H., Davies, F.S., 2010. Plant response to evapotranspiration and soil water sensor irrigation scheduling methods for papaya production in south Florida. Agricultural Water Management 97, 1452–1460.

Naor, A., Cohen, S., 2003. Sensitivity and variability of maximum trunk shrinkage, midday stem water potential, and transpiration rate in response to withholding irrigation from field-grown apple trees. HortScience 38, 547–551.

Neuhaus, A., Turner, D.W., Colmer, T.D., Kuo, J., Eastham, J., 2007. Drying half the root-zone of potted avocado (*Persea americana* Mill., cv. Hass) trees avoids the symptoms of water deficit that occur under complete root-zone drying. The Journal of Horticultural Science and Biotechnology 82, 679–689.

Neuhaus, A., Turner, D.W., Colmer, T.D., Blight, A., 2009. Drying half of the root zone from mid fruit growth to maturity in 'Hass' avocado (*Persea americana* Mill.) trees for one season reduced fruit production in two years. Scientia Horticulturae 120, 437–442.

Neuhaus, A., 2003. Managing Irrigation for Yield and Fruit Quality in Avocado (Ph.D. thesis). University of Western Australia, Perth, Australia.

Observatorio de Precios y Mercados, 2015. El sector de los cultivos subtropicales en Andalucía. Agencia de Gestión Agraria y pesquera de Andalucía. Consejería de Agricultura, Pesca y Desarrollo Rural.

Olalla, L., Rodríguez, A.I., 1991. In: Some Environmental Factors and Yield Variabilities of Avocado Trees in a Spanish Commercial Orchard, vol. 75. Californian Avocado Society 1991 Yearbook, pp. 93–100.

Olalla, L., Salazar, A., Mira, A., Martin, A., Jurado, F., Lopez, C., 1992. The response of avocado trees in the south of Spain to different irrigation regimes and wetted areas. In: Proceedings of Second World Avocado Congress, California, USA, pp. 317–322.

Olalla, L., 1997. Informe final sobre el proyecto CA-9224', Economía del agua de riego, Junta de Andalucía, separatas I+D 1/97.

Olalla, L.M., 2012. Los frutales subtropicales en Málaga. Available from: https://www.google.es/url?sa=t&rct=j&q=&esrc=s&source=web&cd=1&ved=0ahUKEwjrg_rF2YLSAhVKGhQKHUKVCTcQFggjMAA&url=https://dialnet.unirioja.es/descarga/articulo/3938071.pdf&usg=AFQjCNHMv-T33uGxnzPufLr76UV4_7-FvQ&bvm=bv.146496531,d.d24&cad=rja.

Paul, M.J., Pellny, T.K., 2003. Carbon metabolite feedback regulation of leaf photosynthesis and development. Journal of Experimental Botany 54, 539–547.

Richards, S.J., Warneke, J.E., Bingham, F.T., 1962. In: Avocado Tree Growth Response to Irrigation, vol. 46. California Avocado Society Yearbook, pp. 83–87.

Rioux, D., Chamberland, H., Simard, M., Ouellette, G.B., 1995. Suberized tyloses in trees: an ultrastructural an cytochemical study. Planta 196, 125–140.

Rodríguez, P.C.R., Duran, Z.V.H., Martín, P.F., Franco, T.D., 2009. Impact of land-use change on soil degradation by establishment of terraces with subtropical orchards in sloping areas (Granada, SE Spain). In: proceedings of the congress avances en estudios sobre desertificación: aportaciones al Congreso Internacional sobre Desertificación en memoria del profesor John B. Thornes, Murcia, Spain, pp. 395–398.

Salgado, E., Cautín, R., 2008. Avocado root distribution in fine and coarse-textured soils under drip and microsprinkler irrigation. Agricultural Water Management 95, 817–824.

Santos, M.A.Z., Alicieo, T.V.R., Pereira, C.M.P., Ramos, R.G., Mendoça, C.R.B., 2014. Profile of bioactive compounds in avocado pulp oil: influence of dehydration temperature and extraction method. Journal of the American Oil Chemical Society 91, 19–27.

Schaffer, B., Whiley, A.W., 2002. Environmental physiology. In: Whiley, A.W., Schaffer, B., Wolstenholme, B.N. (Eds.), The Avocado: Botany, Production and Uses. CABI Publishing, Wallingford, UK, pp. 135–160.

Schaffer, B., Whiley, A.W., 2003. Environmental regulation of photosynthesis in avocado trees – a mini-review. In: Proceedings of the V World Avocado Congress, Spain, pp. 335–342.

Scholefield, P.B., Walcott, J.J., Kriedemann, P.E., Ramadasan, S., 1980. In: Some Environmental Effects on Photosynthesis and Water Relations of Avocado Leaves, vol. 64. California Avocado Society Yearbook, pp. 93–106.

Scora, R.W., Wolstenholme, B.N., Lavi, U., 2002. Taxonomy and botany. In: Whiley, A.W., Schaffer, B., Wolstenholme, B.N. (Eds.), The Avocado: Botany, Production and Uses. CABI Publishing, Wallingford, UK, pp. 25–37.

Sharon, Y., Bravdo, B., Bar, N., 2001. In: Aspects of the Water Economy of Avocado Trees (Persea Americana, cv. Hass), vol. 24. South African Avocado Growers' Association Yearbook, pp. 55–59.

Sharon, Y., 1999. In: Aspects of the Water Economy of Hass Avocado Trees (Persea Americana, cv. Hass) I. Plant Water Status and Gas Exchange, vol. 22. South African Avocado Growers' Association Yearbook, pp. 106–109.

Silber, A., Israeli, Y., Levi, M., Keinan, A., Shapira, O., Chudi, G., Golan, A., Noy, M., Levkovitch, I., Assouline, S., 2011. Response of 'Hass' avocado to temporal stresses: role of irrigation management and root constraint. In: Proceedings VII World Avocado Congress, Cairns, Australia.

Silber, A., Israeli, Y., Levi, M., Shapira, O., Ghudi, G., Golan, A., Noy, M., Levkovitch, I., Assouline, S., 2012. Response of 'Hass' avocado trees to irrigation management and root constraint. Agricultural Water Management 104, 95–103.

Silber, A., Israeli, Y., Levi, M., Keinan, A., Chudi, G., Golan, A., Noy, M., Levkovitch, I., Narkis, K., Maor, A., Assouiline, S., 2013. The roles of fruit sink in the regulation of gas exchange and water uptake: a case study for avocado. Agricultural Water Management 116, 21–28.

Simonne, E.H., Mills, H.A., Smittle, D.A., 1992. A simple method to measure on-farm pan evaporation for scheduling irrigation. In: Investigación Agraria, Produccion y Protección Vetegal, vol. 7, pp. 392–393.

Stoll, M., Loveys, B., Dry, P., 2000. Improving water use efficiency of irrigated horticultural crops. Journal of Experimental Botany 5, 1627–1634.

Tahi, H., Wahbi, S., Wakrim, R., Aganchich, B., Serraj, R., Centritto, M., 2007. Water relations, photosynthesis, growth and water-use efficiency in tomato plants subjected to partial rootzone drying and regulated deficit irrigation. Plant Biosystems 141, 265–274.

Turner, D.W., Neuhaus, A., Colmer, T., Blight, A., Whiley, A., 2001. Turning water into oil—physiology and efficiency. In: Talking Avocados, vol. 12, pp. 12–24.

Van Leeuwen, C., Chone, X., Tregoat, O., Gaudillere, J.P., 2001. In: The Use of Physiological Indicators to Assess Vine Water Uptake and to Manage Vineyard Irrigation, vol. 449. The Australian Grapegrower and Winemaker, pp. 18–24.

Whiley, A.W., Chapman, K.R., Saranah, J.B., 1988. Water loss by floral structures of avocado (*Persea Americana* cv. Fuerte) during flowering. Australian Journal of Agricultural Research 39, 457–467.

Whiley, A.W., Schaffer, B., 1994. 'Avocado' in Environmental Physiology of Fruit Crops Volume II: Subtropical and Tropical Crops. In: Schaffer, B., Andersen, P.C. (Eds.). CRC Press, Florida, pp. 3–35.

Wolstenholme, B.N., 1886. Energy costs of fruiting as a yield-limiting factor with special reference to avocado. Acta Horticulturae 175, 121–126.

Wolstenholme, B.N., Whiley, A.W., Saranah, J.B., 1990. Manipulating vegetative: reproductive growth in avocado (*Persea americana* Mill.) with paclobutrazol foliar sprays. Scientia Horticulturae 41, 315–327.

PART 3

Physiological and Molecular Responses to Drought

CHAPTER 15

Fruit Response to Water-Scarcity Scenarios. Water Relations and Biochemical Changes

Pedro Rodríguez[1], Alejandro Galindo[2], Jacinta Collado-González[3], Sonia Medina[4], Mireia Corell[5], Houssem Memmi[6,8], Ignacio F. Girón[7], Ana Centeno[6], María J. Martín-Palomo[5], Zulma N. Cruz[1], Angel Carbonell-Barrachina[3], Francisca Hernández[3], Arturo Torrecillas[9], Alfonso Moriana[5], David Pérez-López[6]

[1]Instituto Nacional de Ciencias Agrícolas (INCA), San José de Las Lajas, Cuba; [2]University of Twente, Enschede, The Netherlands; [3]Universidad Miguel Hernández de Elche, Alicante, Spain; [4]Corporación Universitaria Lasallista, Caldas, Colombia; [5]Universidad de Sevilla, Seville, Spain; [6]Technical University of Madrid, Madrid, Spain; [7]Instituto de Recursos Naturales y Agrobiología (CSIC), Seville, Spain; [8]IRIAF, Ciudad Real, Spain; [9]Centro de Edafología y Biología Aplicada del Segura (CEBAS-CSIC), Murcia, Spain

Contents

Water Scarcity and Sustainable Agriculture in Semiarid Environment
ISBN 978-0-12-813164-0
https://doi.org/10.1016/B978-0-12-813164-0.00015-6

1. INTRODUCTION

Maximum crop productivity is not possible without an adequate water supply because water is fundamental to the maintenance of normal physiological activity and membrane transport processes. It is the medium for long-distance transport of nutrients and growth regulatory compounds both in the phloem and in the xylem, the flow in these systems depending on their water relations (Jones and Tardieu, 1998).

As the population increases, water, especially in arid and semiarid regions, is increasingly a disputed commodity among different productive sectors because of the expansion of urban, touristic, and industrial activities that frequently lead to tensions, conflict between users, and extreme pressure on the environment. Compounding this situation, climate change will inevitably lead to very frequent and severe droughts in the near future, so that the pressure for a more sustainable use of water in agriculture will grow because agriculture wastes 60% or 1500 trillion liters of the 2500 trillion liters of water it uses each year, which represents 70% of the world's accessible water (Clay, 2004). As a consequence, sustainable agriculture in arid and semiarid agrosystems will be forced to cope with water scarcity by adopting deficit-irrigation strategies able to minimize the impact on crop yield and quality. In addition, there is increasing recognition on the part of consumers that fruits should not only have optimal physical characteristics but also an improved composition (phytonutrients and bioactive compounds). In this sense, increasing our knowledge of the water relations of developing fruit is essential for maintaining fruit quality because such relations regulate fruit growth and composition, and are related to the occurrence of fruit physiological disorders. Many fruits are sold by size or weight, as well as for their textural properties such as crispness. These characteristics, which are important contributors to quality, are the result of several factors, including cell wall properties, tissue structure, and turgor pressure. Deficit irrigation in fruit crops can improve fruit quality by raising the dry matter percentage and the levels of healthy bioactive compounds (e.g., Collado-González et al., 2014, 2015). Among these compounds, particular mention must be made of a huge variety of secondary metabolites, mainly phenolic compounds, which represent an important source of biological activities (Bourgaud et al., 2001) and have frequently been associated with beneficial effects for human health (Hooper and Cassidy, 2006). Nevertheless, studies on water relations have mainly focused on the vegetative part of the plant, while information on fruit water relations, especially under water deficit, is very scarce.

In light of all the above, our emphasis in this chapter was to review the state of knowledge on fruit water relations, water flow in developing fruits, and the incidence on some fruit physiological disorders. Moreover, the effect of water deficit on fruit quality characteristics, mainly those related to human health, and marketing strategies to promote fruits grown under deficit-irrigated conditions, have also been reviewed.

2. FRUIT WATER STATUS MEASUREMENT

2.1 Fruit Relative Water Content

Relative water content (RWC) is the water content of a fruit relative to its fully hydrated or fully turgid state; it is probably the most appropriate measure of fruit water status in terms of the physiological consequence of cellular water deficit. RWC is calculated as RWC = (FW-DW)/(TW-DW), being DW, FW and TW the dry, fresh and turgid fruit weight, respectively.

RWC is independent of anatomy and, because its changes are proportional to changes in turgor, it can be considered an indirect measure of the change in turgor under certain conditions such as a constant elastic modulus. Sources of error in the estimation of RWC could be summarized as: (1) change in dry weight (mainly due to respiratory losses), (2) increases in water content to above full turgidity levels, and (3) water accumulation in intercellular spaces.

RWC is not frequently used in fruit water relations studies mainly because it is difficult to rehydrate entire detached fruits to saturation and because of the frequent complexity in fruit structures that prevents sampling representative fruit portions. Nevertheless, in fruits without a complex structure, RWC can be measured using one section of a fruit or pericarp discs as Pomper and Breen (1997) showed in strawberry (*Fragaria ananassa* (Weston) Duchesne) and Aloni et al. (1999) in bell pepper (*Capsicum annum* L.) fruits.

2.2 Fruit Gas Exchange

Fruit transpiration and water vapor conductance from six species of platyopuntias (*Nopalea cochenillifera* L., *Opuntia ficus-indica* (L.) Miller, *Opuntia megacantha* Salm-Dyck, *Opuntia robusta* Wendland in Pfeiffer, *Opuntia streptacantha* Lemaire and *Opuntia undulata* Griffiths) were measured by Nobel and De la Barrera (2000) using a steady-state porometer. The acrylic top of the cuvette was removed so that it could be placed near the middle of the fruit. To achieve a good seal to the fruit, a foam rubber gasket was used.

Moreshet and Green (1980) measured orange (*Citrus sinensis* L. Osbeck) fruit photosynthesis using a $^{14}CO_2$ uptake meter with some improvements, and Nobel and De la Barrera (2000) measured fruit CO_2 uptake using a portable photosynthesis system replacing the cuvette with an acrylic plate, whose margin was covered with a foam rubber gasket to facilitate sealing to organ surface.

To measure apple (*Malus domestica* Borkh.) fruit transpiration rate, Yamada et al. (2004) used a plastic cylinder filled with calcium chloride connected to the upper end of a transparent chamber enclosing the fruit. The lower end of the chamber was also connected to an air pump. Another calcium chloride cylinder was connected to the air pump and ventilated at the same rate as a control. The fruit evapotranspiration rate was determined by subtracting the increased weight of calcium chloride cylinder in the control from that including water vapor from the surface of the treated fruit.

2.3 Fruit Water Potential and Components

Probably the most commonly used procedure for measuring fruit water potential (Ψ_{fruit}) is the isopiestic thermocouple psychometry technique, for which only very small pieces of tissue are used (Mills et al., 1996; Nobel and De la Barrera, 2000). Fruit disks are taken usually from the fruit equatorial portion, excluding the skin, and immediately sealed in the sample chamber to equilibrate before Ψ_{fruit} determination. Dehydration of the disk samples during the period elapsing between obtaining the sample to it being placed in the chamber is very important because measurements can be significantly altered. In addition, this methodology takes time for equilibrium to be reached and is prone to perturbations due to temperature gradients and the presence of minor contaminants (Jobling et al., 1997; Gelly et al., 2004).

To avoid these disadvantages, other authors such as Pomper and Breen (1995) proposed the use of temperature-corrected in situ psychrometers for measuring Ψ_{fruit}. Psychrometers are attached to a site on the fruit after abrading the fruit epidermis, thoroughly rinsing the abraded tissues with distilled water, and wiping the area dry. To clamp the instrument to the tissue, silicone stopcock grease is used to ensure a vapor seal. Minimum clamping pressure is used, and it is assumed that the elastic nature of the fruit will eliminate any turgor artifacts after equilibration (Johnson et al., 1992).

Ψ_{fruit} measurements with a pressure chamber have been performed in fruits with pedicel (Syvertsen and Albrigo, 1980; Dell'Amico et al., 2012; Galindo et al., 2014). However, measuring Ψ_{fruit} is difficult in fruits with scarce pedicels such as peach (*Prunus persica* [L.] Batsch) or a jujube (*Ziziphus jujuba* Mill.) (Gelly et al., 2004; Galindo et al., 2016). In these cases, McFadyen et al. (1996) and Morandi and Grappadelli (2009) suggested removing leaves from near a fruit borne toward the end of a shoot and sealing the cut surfaces with a quick-drying plastic paint or a hot-melt glue. When the paint is dried, the shoot can be cut above the fruit, and the fruit, with shoot attached, can be placed in the pressure chamber to carry out Ψ_{fruit} measurements in the usual way. Milad and Shackel (1992) estimated Ψ_{fruit} values measuring stem water potential (Ψ_{stem}) at predawn because they assumed that water potential gradients are minimal at this time.

Besides the fact that Ψ_{fruit} measurements using a pressure chamber demand considerable expertise and experience in order to determine the end point, the impact of pressurization on tissues is unclear. To overcome these weaknesses, Jobling et al. (1997) devised a nondestructive method to measure Ψ_{fruit}. The procedure is based on relative weight change of a series of salt solutions contained inside patches placed close to the fruit epidermis. The basis of the method is the assumption that the water potential of the salt solution in the patch that neither gains nor loses weight is equal to the water potential of the fruit. However, this simple and inexpensive method can accurately measure Ψ_{fruit} only within the range -1.0 to -1.5 MPa.

Hou et al. (2011) developed a novel centrifuge method that permits the water relation parameters of fleshy fruits to be determined. These authors obtained pressure—volume (PV) curves and Höfler diagrams of the cortex tissue of fresh apple (*M. domestica*) fruits and concluded that this methodology is a highly efficient and less time-consuming technique allowing for studying the roles that water plays in the growth and development of apple and other fruits in laboratories.

Gelly et al. (2004) found that Ψ_{fruit} measurements in peach (*P. persica*) trees with the pressure chamber showed considerable consistency and were more stable than other water potential assessments such as Ψ_{stem} and predawn leaf water potential. In this sense, Galindo et al. (2014) indicated that Ψ_{fruit} values in pomegranate (*Punica granatum* L.) showed significant first-order correlations with leaf water potential (Ψ_{leaf}) and Ψ_{stem}, the latter being the best predictor of Ψ_{fruit}.

One important aspect of water relations in fruits is that the levels of solutes, predominantly sugars, mainly at the end of fruit development, may be very high. Such fruit solutes or osmotic potential ($\Psi_{\pi\ \mathrm{fruit}}$) is determined in the same fruits as used for Ψ_{fruit} measurements. Fruits are covered with aluminium foil and immediately frozen in liquid nitrogen and stored at −80°C. $\Psi_{\pi\ \mathrm{fruit}}$ is measured after thawing the samples and expressing the sap using a psychrometer or an osmometer. Estimated fruit turgor potential ($\Psi_{\mathrm{p\ fruit}}$) is derived as the difference between $\Psi_{\pi\ \mathrm{fruit}}$ and Ψ_{fruit}.

3. GROWTH AND WATER FLOW IN DEVELOPING FRUITS

3.1 Fruit Growth

Fruit growth can exhibit a single- or a double-sigmoid pattern in which there are one or two periods of rapid growth, respectively. Pome and *Citrus* fruits show only a phase of rapid fruit growth (single-sigmoid pattern), which takes place after the initial period of cell division and minimal expansion, and is due mainly to a cell expansion process even though at the beginning some cell division can also take place. In stone fruits, which are characterized by a double-sigmoid pattern in fruit growth, the two periods of rapid growth are separated by a period of variable length during which little or no expansive growth occurs. So, in these fruits three different growth stages can be considered: stage I, which is the first growth period and is due to some cell division and cell expansion; stage II, at which the sclerification of the fruit endocarp takes place and fruit growth is extremely slow or does not occur; and stage III, which is the second period of rapid fruit growth and occurs due only to the expansion of existing cells and extends from the onset of this second growth period until maturity.

In general, fruit growth is considered a drought-sensitive phenological stage of fruit trees, particularly during the second rapid fruit growth period (stage III), when water deficit induces a reduction in stone fruit yield due to the smaller fruits at harvest (Torrecillas et al., 2000). In pome and *Citrus* fruits, the rapid fruit growth phase can be considered as a common critical period because water deficit causes a delay in attaining a marketable fruit size (Domingo et al., 1996).

It is near axiomatic that expansive cell growth requires the presence of cell turgor, because in many cases a direct relation between turgor and the rate of fruit growth has been found (e.g., Matthews and Shackel, 2005; Galindo et al., 2014). However, in some instances no such relation has been found. For example, Mills et al. (1997) showed that apple (*M. domestica*) fruit size was reduced in early deficit trees during the stress period while $\Psi_{\mathrm{p\ fruit}}$ was maintained. Turgor maintenance therefore does not necessarily result in the maintenance of growth in apple fruit under deficit irrigation conditions. Also, Galindo et al. (2016) studied the response of jujube (*Z. jujuba*) fruits to severe water stress (Ψ_{fruit} decrease from −0.76 to −3.70 MPa) and found that fruit size decreased despite the maintenance of fruit turgor ($\Psi_{p\ fruit} > 0$). There are several explanations for this behavior. Okello et al. (2015) showed that fruit size is strongly related to cell size rather than cell number. Moreover, if the elasticity of fruit cell walls increases (elastic adjustment) the fruit turgor can be maintained through a reduction in fruit size. This passive mechanism allows a decrease in cell volume with dehydration, slowing the rate of turgor loss by decreasing Ψ_{fruit}. Another hypothesis is that cell enlargement depends on the growth-effective turgor (difference between turgor pressure and wall yield threshold) rather than just turgor pressure (Van Volkenburg and Cleland, 1986).

3.2 Water Flow in Developing Fruits

In fact, fruits and leaves can be considered as important sinks for water and, from an energetic point of view, the gradient ($\Delta\Psi$) between Ψ_{fruit} and Ψ_{leaf} should determine the direction of water interchange between both plant organs. Nevertheless, taking into consideration that fruits and leaves share the same stem portion of water transport pathways, Matthews and Shackel (2005) suggested that the $\Delta\Psi$ between Ψ_{fruit} and Ψ_{stem} would be more appropriate for this analysis.

Fruits require a supply of water, which generally enters developing organs via the xylem. In this sense, Pomper and Breen (1997) showed that Ψ_{leaf} in control strawberry (*F. ananassa*) plants was always higher than Ψ_{fruit} in green–white fruit. However, in stressed plants Ψ_{leaf} was higher than Ψ_{fruit} in green–white fruit only in the dark, when these fruits expanded. When developing organs are not photosynthetically self-sufficient, water also enters via the phloem together with the solutes. In this way, substantial amounts of water enter the fruits of cotton (*Gossypium hirsutum* L.) (van Iersel et al., 1994). The literature describes substantially higher Ψ_{fruit} values than Ψ_{leaf} in some crops, such as apple (*M. domestica*) (Lang, 1990; Mills et al., 1997), Asian pear (*Pyrus serotina* L.) (Behboudian et al., 1994), avocado (*Persea americana* Mill.) (Blanke and Whiley, 1995), orange (*C. sinensis*) (Syvertsen and Albrigo, 1980), cotton (*G. hirsutum*) (Trolinder et al., 1993), jujube (*Z. jujuba*) (Galindo et al., 2016), pomegranate (*P. granatum*) (Galindo et al., 2014), and six species of platyopuntias (Nobel and de la Barrera, 2000), which could be partially due to a high resistance to water movement from fruit to the rest of the plant (Mills et al., 1997). Furthermore, according to Nobel and de la Barrera (2000), water cannot flow from leaves to the fruits since the xylem does not provide water to the fruits (Nobel et al., 1994). Thus, in these cases, the phloem rather than the xylem could be the preferential route for water to enter the fruits. In this sense, Johnson et al. (1992) suggested that in tomato (*Lycopersicon esculentum*) plants the bulk (90%) of the sap transported into the fruit occurs through the phloem, and, therefore, growth is a function of phloem transport. The fruit-stem water potential gradient would therefore be indirectly correlated with growth. However, other authors do not fully agree with these ideas. For example, Matthews and Shackel (2005) showed that in fully irrigated prunes (*Prunus domestica* L.) the relative importance of xylem and phloem in the water flowing to the fruit may be reversible. Lang (1990) found that apple (*M. domestica*) xylem contribution to overall fruit water balance progressively decreased as fruit developed. Also, Dell'Amico et al. (2012) showed that water flow in the olive (*Olea europaea* L.) fruit during pit hardening in fully irrigated conditions occurs via both xylem and phloem. In contrast, Greenspan et al. (1996) suggested that the bulk vascular water flow changes from xylem in pre-veraison to phloem in post-veraison in fully irrigated grape (*Vitis vinifera* L.) berry. Actually, at veraison xylem flow is reduced but not eliminated, contributing to the post-veraison recovery of berry volume during the night.

4. FRUIT WATER RELATIONS UNDER DROUGHT

The effect of water deficit on fruit water relations and composition clearly depend on the timing. In this sense, Mills et al. (1996) showed that water deficit early in the season may cause modifications in apple (*M. domestica*) fruit composition and water relations but has a minimal influence later in the season. Both water deficits led to a lower photosynthetic rate, Ψ_{leaf} and leaf conductance (g_{leaf}) values. Water deficit early in the season lowered Ψ_{fruit}, accompanied by a decrease in $\Psi_{\pi\ fruit}$ and the maintenance of $\Psi_{p\ fruit}$, despite which fruit weight was reduced. Moreover, titratable acidity, K^+, fructose, sorbitol, and total soluble sugar levels were higher in early season—stressed fruits than in control and late season—stressed fruits. However, the weight, sugar concentration and water relations of late season—stressed fruits were not affected by deficit irrigation, last mentioned perhaps due to the relative strength of the near-mature fruit as a sink for water and carbohydrates. Because $\Psi_{\pi\ fruit}$ of all fruit generally falls as fruit ripen (Berüter, 1989), fruits become more effective competitors for available water. Thus, fruit water relations are often unaffected despite a reduction in Ψ_{leaf} of the tree. Additionally, in some fruits of complex structure such as *Citrus* the presence of the albedo, the white part of the peel, confers peculiar characteristics to the fruit water relations from early in fruit development. In this sense, according to Schröder et al. (2004), in times of drought, the albedo tissue acts as a water reservoir for juice sacs, seeds, and leaves.

Cell wall elastic and active osmotic adjustment are two drought-tolerance mechanisms able to maintain symplastic volume and positive cell turgor in plant organs diurnally and during soil drying. Mills et al. (1997) showed that early deficit-irrigated apple (*M. domestica*) fruits maintained cell turgor but reduced fruit size, hypothesizing that factors such as an enhancement of cell wall elasticity (elastic adjustment) could explain such behavior. Galindo et al. (2016) also showed that jujube (*Z. jujuba*) fruits decreased in size under severe water stress even though fruit turgor was maintained. The active accumulation of solutes within cells is known as osmotic adjustment and may allow positive cell turgor to be maintained in fruits under water stress. Hexoses, amino acids, and inorganic salts usually accumulate in plant tissues that osmotically adjust (e.g., Wang et al., 1995). Jones et al. (1985) suggested osmotic adjustment may occur in fruit tissues, although to our knowledge, evidence of this mechanism is very scarce. Behboudian et al. (1994) mentioned the development of osmotic adjustment in Asian pear (*P. serotina*) fruit but not in leaves. Mills et al. (1996, 1997) suggested the possibility of osmotic adjustment in apple (*M. domestica*) fruit in response to water deficit in the early season. Pomper and Breen (1997) concluded that strawberry (*F. ananassa*) fruit expansion was very sensitive to water stress and only occurred when Ψ_{leaf} was higher than Ψ_{fruit}. However, green–white fruit underwent osmotic adjustment in response to water stress (fruit osmotic potential at full turgor 0.28 MPa lower than in control fruit), which was not sufficient to maintain fruit expansion during daylight hours when stress was greatest. Moreover, water stress did not have a significant effect on the osmotic potential at full turgor of red fruit. In Satsuma mandarin (*Citrus unshiu* L.) fruits, Yakushiji et al. (1996) suggested that sugar accumulation by water stress effect was not caused by dehydration but rather that sugars were accumulated by active osmoregulation, being glucose and fructose monosaccharides, arising from the sucrose degradation in vesicle cells, largely responsible for active osmoregulation in fruit under water stress conditions.

5. SOME FRUIT PHYSIOLOGICAL DISORDERS RELATED TO WATER STRESS

5.1 Fruit Cracking and Splitting

Fruit cracking and fruit splitting are major pre-harvest physiopathies that have been observed in many plant species including pome fruits (Kasai et al., 2008), stone fruits (Sekse, 1995), grapes (*V. vinifera*) (Clarke et al., 2010), figs (*Ficus carica* L.) (Kong et al., 2013), litchi (*Litchi chinensis* Sonn.) (Huang et al., 2008), *Citrus* (Almela et al., 1994), etc. However, the term cracking is frequently used inappropriately, because some studies use the term cracking to describe superficial cracks that expose the flesh (e.g., Galindo et al., 2014; Mesejo et al., 2016). For this reason, it is very important to underline that the meaning of both terms are clearly different (Kong et al., 2013). Cracking has been described as the physical failure of the fruit skin (Milad and Shackel, 1992) in the form of fractures in the cuticle or skin, which typically do not penetrate into the flesh (Fig. 15.1). In contrast, splitting is an extreme form of cracking that penetrates into the flesh (Opara et al., 1997) (Fig. 15.1). In any case, both physiological disorders are a result of changes in fruit water relations (Milad and Shackel, 1992) and fruit skin properties (Hankinson, 1979).

Albrigo (1975) indicated that *Citrus* fruit splitting occurs primarily when fruits have a high solute content and the peel has thinned. Fruits are under internal turgor pressure and this pressure increases as soil moisture increases after irrigation or heavy rainfall. Aloni et al. (1999) suggested that cuticular cracking in bell pepper (*C. annum*) fruit can be induced by weakening of the cuticle as a consequence of long-term alternating fruit shrinkage and swelling, possibly due to the diurnal changes in fruit turgor that occur under normal growth conditions. The same authors suggested that cuticle

(A) (B) (C) (D)

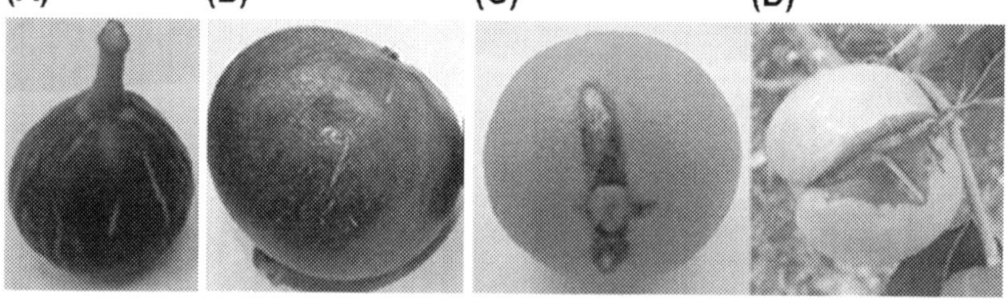

Figure 15.1 Fruit physiological disorders: Fig (*F. carica*) (A) and pomegranate (*P. granatum*) (B) fruit skin side cracking, and orange (*C. sinensis*) (C) and jujube (*Z. jujuba*) (D) fruit splitting. *(Photos courtesy Dr. Francisca Hernández.)*

weakening is possibly enhanced after the fruit reaches its full size and cuticle extensibility is reduced.

In *Citrus*, Mesejo et al. (2016) showed that irregularities in the tree water status due to interactions among soil moisture, rootstock, and climatic conditions leads to a number of substantial changes in the fruit growth rate, increasing the incidence of fruit splitting. In this sense, a close correlation between the degree of maximum daily trunk shrinkage and the incidence of fruit splitting during the rapid fruit growth period, but not later, was found. This is because xylem flow positively correlates with the stem-to-fruit pressure potential gradient during the initial, but not the final, part of the developing period, when xylem conductivity appears to be reduced. Also, fruit splitting was positively correlated with soil sand levels and inversely correlated with clay and silt levels. So, by reducing the frequency of irrigation, fruit splitting increased but fewer and smaller microcracks were found in the stylar end of nonsplit fruit from regularly irrigated trees. Regardless of soil conditions or species, an increase in the size of xylem vessels in the fruit pedicel enhanced splitting incidence because the larger the xylem vessels, the greater the instability in the tree water status, giving rise to more pronounced daily fruit shrinkage-expansion.

Measham (2011) showed that, regardless of the plant material, crop load, and skin properties, side cracking in sweet cherry (*Prunus avium* L.) fruit is associated with the uptake of water through the vascular system of the tree, while fruit cuticular cracks, which occur only in specific *loci* of the fruit, are associated with direct water uptake across the fruit skin. However, from the results it could not be concluded that the driving forces for water movement involved in crack development were purely osmotic in nature.

The shelf life of fresh figs (*F. carica.*) is very short and limited by decay and juice leaking from the ostiole end. Moreover, such postharvest decay increases as the degree of skin damage increases (Kong et al., 2013). The causes of fig skin side cracking (Fig. 15.1) and ostiole end splitting, which occur during fruit growth and development, have not been established but may be decreased in some cultivars by changes in irrigation management. In this sense, the above authors showed that deficit irrigation (55% ETc) applied when 25% of fruit were ripening reduced fruit skin damage and increased the percentage of sound fruit, without negatively impacting other quality attributes.

Fruit end cracking in French prune (*P. domestica*) was studied by Milad and Shackel (1992), who showed that a gradient in $\Psi_{\pi\ \mathrm{fruit}}$ existed along the vertical axis of fruit from all irrigation treatments, $\Psi_{\pi\ \mathrm{fruit}}$ always being lower at the stylar end than at the stem end. When previously water-stressed trees were irrigated, the $\Psi_{\pi\ \mathrm{fruit}}$ gradients exceeded those of all other treatments. Consequently, $\Psi_{\mathrm{p\ fruit}}$ at the stylar end of the fruit increased twofold within 24 h after irrigation, inducing the onset of fruit end cracking. In contrast, neither the well-watered controls nor the continuously drought-exposed fruit exhibited such changes. In addition to the overall recovery of Ψ_{fruit} after irrigation, there was also evidence of solute movement within the fruit, where cells near the stem end may be the first to be influenced by the recovery in plant Ψ_{fruit}. A subsequent movement of solutes to the distal part of the fruit (stylar end), together with overall higher fruit Ψ_{fruit}, may cause excessive turgor to build up, causing the observed end cracks.

Galindo et al. (2014) indicated that during the phenological period between the end of pomegranate (*P. granatum*) fruit growth and the ripening phases, water could enter the fruits via the phloem rather than via the xylem, and that the fruit is clearly sensitive to water deficit. In this sense, although leaf turgor was maintained during very severe water stress, fruit turgor was lost, inducing the reduced expansion of fruits. When rainfall affected previously water-stressed pomegranate plants, a very similar mechanism to that described in prune (*P. domestica*) by Milad and Shackel (1992) took place. In fact, when previously stressed trees were rewatered, there was an asymmetric increase in turgor pressure in this fruit, which has a very complex structure, because aril turgor increased to a much greater extent than peel turgor. This increase in aril pressure would put pressure on the peel and make it susceptible to splitting, especially if it is considered that under water stress peel mechanical properties change, peel extensibility tending to decline (Aloni et al., 1999), and the peel becomes thicker and stiffer (Lawlor and Leach, 1985).

5.2 Fruit Creasing

Creasing is a preharvest physiological disorder commonly found in many *Citrus* orchards, the occurrence of which varies in various species (Fig. 15.2). This disorder is characterized by separation of the tissue in the albedo to leave depressed lines in the peel. Chen et al. (2002) indicated that most creasing of orange (*C. sinensis*) fruit occurs in the shaded half of the fruit and mostly in very foggy and rainy days during fruit ripening. Jones et al. (1967) found that fruit on the south side of the tree developed creasing earlier and more severely, and the inside of each fruit was more likely to be

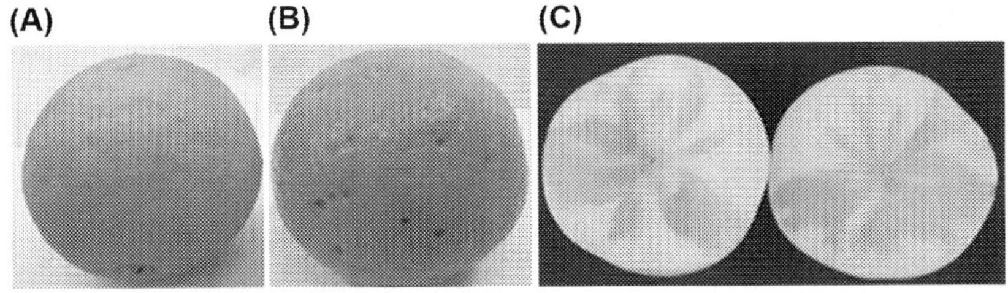

Figure 15.2 Fruit physiological disorders: orange (*C. sinensis*) fruit creasing (A), orange (*C. sinensis*) fruit pitting (B), and apple (*M. domestica*) watercore (C). *(Photos courtesy Dr. Francisca Hernández (A and B) and Dr. Ángela Melado and Dr. Pilar Barreiro (C).)*

damaged and suggested that radial temperature gradients might create a differential water stress across the fruit, leading to albedo stress and subsequent creasing. Also, Giles et al. (1974) showed that high water stress can lead to the breakdown of loosely packed mesophyll-type cell structures.

5.3 Fruit Pitting

Postharvest peel pitting at nonchilling temperatures is a severe disorder that affects fruit from several *Citrus* cultivars worldwide and results in important economic losses (Fig. 15.2). The disorder is characterized by the appearance of sunken areas of flavedo (pits) that eventually spread and affect oil glands. In severe cases, pits may turn bronze in color and become necrotic. These symptoms diminish the commercial value of fruit for the fresh market. In addition, the disorder is unpredictable, and its incidence and severity varies season to season or even during the same season (Alferez et al., 2004).

Previous reports related peel water status with peel pitting, regardless of the geographical origin of the samples, suggesting that variation in water relations within the fruit peel may contribute to this disorder (e.g., Lafuente and Sala, 2002). For these reasons, in grapefruit (*Citrus paradise* Mac.), Alferez et al. (2010) studied the effects of a postharvest commercial packing line or manual processing and subsequent storage under low and high relative humidity (RH) regimes on water, osmotic and turgor potentials in three layers of peel (flavedo and internal and external albedo), and the incidence of peel pitting. The results showed that at the end of the storage period at 45% RH, peel water potential was lower in fruit processed on a commercial packing line than in those processed manually, because commercial processing seemed to remove natural waxes on the fruit surface, inducing a higher rate of water loss. Similarly, turgor pressure decreased in all three peel tissues during fruit storage at 45% RH, especially in packing line—processed fruits as a consequence of the behavior of osmotic potential in peel tissues, which did not change as significantly as the water potential. When dehydrated fruit were transferred to 95% RH the water potential of flavedo only increased in fruits processed manually, an increase that was at the expense of the albedo cell layers, which underwent subsequent water loss. Likewise, turgor pressure increased only in manually wax-coated fruits. Thus, in the packing line—processed fruits severe peel pitting was probably due to cell collapse that avoided water adjustment.

5.4 Fruit Watercore

Watercore is a fruit physiological disorder characterized by water-soaked tissue around the vascular bundles or core area due to the spaces between cells becoming filled with fluid instead of air (Fig. 15.2). Generally, damage is not visible on the skin, and it is only when fruit is cut that the damage is seen. In severe cases, flesh can be affected right up to the skin, which then darkens in affected areas.

Inomata et al. (1999) established the importance of fruit transpiration in the occurrence of watercore in Japanese pear (*Pyrus pyrifolia* [Burm. F.] Nakai): ethephon (2-chloroethylphosphonic acid) inhibited fruit transpiration and increased watercore, whereas a spray of calcium carbonate increased fruit transpiration rate with a corresponding reduction in watercore symptom. Yamada et al. (2004) compared the water relations in apple (*M. domestica*) with and without watercore induced by preharvest fruit temperature (25 and 10 or 7°C) treatments on the tree. Fruits at 25°C presented the highest evapotranspiration rate and almost no watercore incidence, while during treatment at 10 and 7°C the degree of watercore increased. At 25°C, Ψ_{fruit} gradually decreased with a concomitant reduction of $\Psi_{\pi\ fruit}$ and $\Psi_{p\ fruit}$. In contrast, at the lower temperatures, Ψ_{fruit} and $\Psi_{\pi\ fruit}$ showed very small changes and $\Psi_{p\ fruit}$ tended to increase slightly, suggesting the sufficient or oversupply of water to the fruit more than the water loss by relatively low evapotranspiration. This situation can easily lead to water-soaked conditions around the core line tissue of the fruit.

6. FRUIT COMPOSITION UNDER DROUGHT

6.1 Effect of Water Stress on Primary Plant Metabolites

From the chemical point of view, plants produce an amazing diversity of reactions that generate low-molecular-weight biocompounds or metabolites (Fig. 15.3). Of this wide array of biocompounds, about 50,000 have

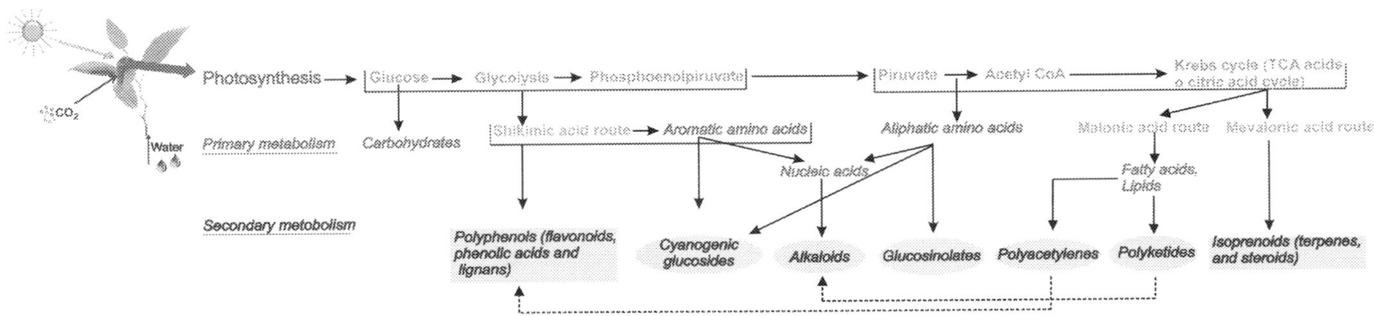

Figure 15.3 Basic scheme of primary and secondary metabolism in fruits.

already been elucidated. Primary metabolites are responsible for the maintenance of the vital state of a plant, allowing proper plant growth, development, and reproduction. In this sense, amino acids, lipids, nucleotides, and carbohydrates are considered primary metabolites (Yordanov et al., 2003) (Fig. 15.3).

Water deficit directly affects the primary metabolism of plants. In such a situation, plants experience decreased CO_2 availability, caused by diffusion limitations through the stomata and the mesophyll, growth inhibition, and reduced transpiration rates (Yordanov et al., 2003). Moreover, plants accumulate abscisic acid, proline, mannitol, and sorbitol, and increase the formation of radical scavenging compounds (ascorbate, glutathione, α-tocopherol, etc.) and the synthesis of new proteins and mRJNAs. Water deficit can also generate secondary effects, namely oxidative stress (Yordanov et al., 2003).

Water deficit activates the generation of reactive oxygen species (ROS) that affect the plant metabolism in different ways and cause cellular damage. A common response to water deficit in fruits and other plant tissues is activation of the protein breakdown pathway in order to synthesize specific amino acids (Krasensky and Jonak, 2012) that have a beneficial effect, allowing osmotic adjustment, antioxidant activity, and signaling (Collado-González et al., 2014).

Under drought conditions, the most important proteins are those acting as antioxidants and those protecting other proteins and plant membranes; these proteins are designated as late embryogenesis abundant (LEA) (Danon et al., 2004). The LEA proteins are accumulated during the mature stage of embryogenesis, just before the beginning of seed desiccation. Some of them even accumulate in vegetative tissues in response to the osmotic stress generated by various environmental conditions, such as dehydration, salinity, and low or very low temperatures (Colmenero-Flores et al., 1997).

Lipids play a key role in the resistance of plant cells to abiotic stresses because they can activate plant adaptation processes (Hou et al., 2016). Thus, drought affects both lipid content and composition. However, studies evaluating the contents of total lipids, phospholipids, and glycerolipids in pepper fruit (*Capsicum annuum*) and grapevine (*V. vinifera*) have provided contradictory results (Yordanov et al., 2003; Toumi et al., 2008), probably due to the absence of sufficient information on the duration and level of the water deficit achieved, the phenological period at which water deficit takes place, etc.

The biochemical responses to water stress include changes in gene expression, such as those encoding protective proteins, key enzymes in the osmolyte synthesis pathway, antioxidant enzymes, transcription factors, and other proteins involved in responses to water stress (Chaves et al. 2009). Early reports asserted that the number of genes or proteins capable of being altered by water deficit is related to stress intensity and to tissue type (Chaves et al., 2009). However, the expression of a gene during stress does not guarantee plant resistance to drought, and the expression of some genes may result from the injury occurring during stress. Whatever the case, the overexpression of many of these genes in plants can confer tolerance to water stress, as in the case of genes encoding LEA proteins (Yordanov et al., 2003).

6.2 Effect of Water Stress on Secondary Plant Metabolites

All plants produce a wide variety of secondary metabolites, but not all of these substances are found in all groups of plants (Fig. 15.3). They are synthesized from primary metabolites and their role in the maintenance of life processes in plants is unclear, even though they are known to be involved in active and potent defense mechanisms against herbivores, microbes, and viruses (Akula and Ravishankar, 2011) and play a major role in the adaptation of plants to a wide range of environmental stresses (high and low temperature, drought, alkalinity, salinity, UV stress, and pathogen infection) and in overcoming stress conditions (Akula and Ravishankar, 2011). Moreover, these compounds contribute to plant fitness by interacting with the ecosystems, and they also contribute to the specific odors, tastes, and colors of plants (Bourgaud et al., 2001), making them useful in the daily life of humans because they represent a unique potential source for food additives, flavorings, relaxing drugs, and especially essential oils (Akula and Ravishankar, 2011). Depending on their biosynthetic pathway, the main classification of secondary metabolites extracted from plants includes three major classes: terpenoids (terpenes and steroids), alkaloids, and phenolics (Fig. 15.3) (Bourgaud et al., 2001).

Studies on the effect of water deficit on terpenoid levels in fruits are very scarce. Nevertheless, some studies have shown that water deficit stimulates in a marked way the accumulation of terpenes in olive (*O. europaea*) fruit and apple (*M. domestica*) fruit (Bianco et al., 2008; Martinelli et al., 2013). Ou et al. (2010) reported that wines produced from water-stressed vines (*V. vinifera*), cv. Merlot, had higher concentrations of terpene alcohols, but the trend depended on the individual compound and vintage year. The effect of drought stress conditions in different herbs on terpenoid levels has been also studied, even though the results are not uniform. For example,

Ormeño et al. (2007) and Nowak et al. (2010) showed a great increase in monoterpene concentration in some herbs, while in other herbs Ormeño et al. (2007) indicated that the terpenoid content remains unchanged under drought conditions, probably due to differences in terpene storage structures, water stress conditions, or because sesquiterpenes could be replaced by monoterpenes when drought is prolonged.

Alkaloids are harmful as animal toxins and certainly serve as defense chemicals against predators (herbivores and carnivores) and, to a lesser degree, against bacteria, fungi, and viruses. Water deficit induces drought stress-related metabolic responses—the uptake of CO_2 decreases significantly and the CO_2 fixed by the plant increases. As a consequence, the synthesis of reduced compounds, such as alkaloids, is enhanced (Amirjani, 2013; Selmar and Kleinwächter, 2013), the changes in alkaloid content depending on the duration of the drought stress, stress level, and canopy structure because not all alkaloid compounds are able to withstand a high degree of water deficit (Amirjani, 2013).

Phenolic compounds are widespread in the plant kingdom and play a variety of functional roles in plant growth, reproduction, pigmentation, resistance to pathogens, and resistance to biotic and abiotic stresses (Akula and Ravishankar, 2011). The relation between phenolic compound content and osmotic stress is not linear because the phenolic response to water deficit depends on the water stress level achieved and on the phenological period in which it occurs. In this sense, short-term effects seemed to lead to increased production, whereas long-term effects generated opposite results (Horner, 1990; Gobbo-Neto and Lopes, 2007). This behavior may explain why there are conflicting reports in the literature concerning the effects of water deficit on the phenolic contents of fruits. Studies in olive (*O. europaea*) fruits have shown that plants exposed to drought stress accumulated higher concentrations of these secondary metabolites (Tovar et al., 2002). However, Collado-González et al. (2013) showed that jujube (*Z. jujuba*) fruits from trees suffering severe water stress lost some of their procyanidin content. Also, these authors showed that the fruit procyanidin content from full irrigated jujube trees increased during domestic cold storage. The increase in some phenolic compounds during a drought situation may also be related to the fact that water deficit can lead to an increase in the levels of their precursor (L-phenylalanine ammonia lyase) or in its synthesis (Tovar et al., 2002).

Smirnoff (1993) reported that drought stress induced an accumulation of ROS, which may damage many cellular components including lipids, proteins, carbohydrates, and nucleic acids. Membrane lipid peroxidation and protein oxidation constitute the simplest criteria for assessing the extent of oxidative damage. Lipid peroxidation may occur enzymatically or nonenzymatically. Enzymatic peroxidation was seen to produce oxylipins, such as jasmonic acid and its immediate precursor 12-oxo-phytodienoic acid, whereas the nonenzymatic route produced a variety of reactive aldehydes, including malondialdehyde or a large class of compounds known as phytoprostanes (PhytoPs) (Sattler et al., 2006; Hou et al., 2016).

PhytoPs are nonenzymatic lipid peroxidation products derived from α-linolenic acid (C18:3 w3, ALA), which is one of the most abundant polyunsaturated acids (PUFAs) in terrestrial higher plant membranes. Although it has been reported that PhytoPs have a wide range of biological activities in different plant species, their exact role and function in plant physiology remain to be elucidated in detail. Recently, interest in these compounds has grown due to the fact that several authors have hypothesized that PhytoPs can act in plants not only as defense signals synthesized in response to external stimuli but also as endogenous mediators capable of preventing cellular damage (Sattler et al., 2006). It has been concluded that the PhytoP content increases with the application of water stress. So, in deficit-irrigated olive (*O. europaea*) trees during pit hardening, the free PhytoPs (9-F_{1t}-PhytoP, 9-*epi*-9-F_{1t}-PhytoP, 16-B1-PhytoP, and 9-L_1-PhytyoP) content in the green table olive fruit flesh was increased (Collado-González et al., 2016). Also, water deficit during pit hardening or for a further period of 2 weeks is clearly critical for the exact composition of Cornicabra extra virgin olive oil because of the accompanying enhancement of free PhytoPs (9-F_{1t}-PhytoP, 9-*epi*-9-F_{1t}-PhytoP, 9-*epi*-9-D_{1t}-PhytoP, 9-D_{1t}-PhytoP, 16-B_1-PhytoP, and 9-L_1-PhytoP) (Collado-González et al., 2015) and hence their potential beneficial effects on human health.

7. MARKETING STRATEGIES TO PROMOTE FRUITS GROWN UNDER DEFICIT-IRRIGATED CONDITIONS

To date, very few research groups have focused their efforts on evaluating what consumers *think* about deficit irrigated fruits (Noguera-Artiaga et al., 2016) or their satisfaction or degree of acceptance of, for example, deficit-irrigated peaches (*P. persica*) (Lopez et al., 2016), pistachios (*P. vera*) (Carbonell-Barrachina et al., 2015), table olives (*O. europaea*) (Cano-Lamadrid et al., 2017), and olive (*O. europaea*) oils (Fernandes-Silva et al., 2013). There is a clear need to understand what consumers think about this type of fruit.

Even though there is much scientific evidence that deficit irrigation can have beneficial effects on the accumulation of bioactive compounds and on fruit quality, the consumer is still not aware of this fact, and fruits grown under deficit-irrigation conditions are to be found on supermarket shelves among all the other products on display. Any effort made by farmers to properly control cultivation conditions in order to achieve the sustainable use of irrigation water is not reflected in a higher price or improved reputation of the products. This is why Noguera-Artiaga et al. (2016) decided to study the opinion of Spanish consumers concerning products called *hydroSOStainable* or, in abbreviated form, *hydroSOS*. According to these authors, hydroSOS products have a solid identity based on two main factors: (1) water stress increases the plant secondary metabolite content and, thus, the functionality of the edible products (Ripoll et al., 2014), and (2) they are environmentally friendly because of the sustainable use of a very scarce resource—water. The authors argued that there is a need to differentiate hydroSOS products from others on the market. The first strategy to reach this goal was to design and use a logo (Fig. 15.4), whose original longer version has been improved and shortened, making it easier to remember and promote: hydroSOS.

Noguera-Artiaga et al. (2016) found that consumers are willing to pay a reasonably higher price for hydroSOS products, if they are properly labeled and identified. However, further research is needed to check whether this higher willingness to pay is the same for all fruits. Besides, there is a need to develop a hydroSOS index to certify that the products using the hydroSOS logo have been evaluated for their sustainable use of irrigation water. This index is under construction and will be based, among other factors, on farmers being able to demonstrate: (1) knowledge of the crop cultural practices, including water management during the

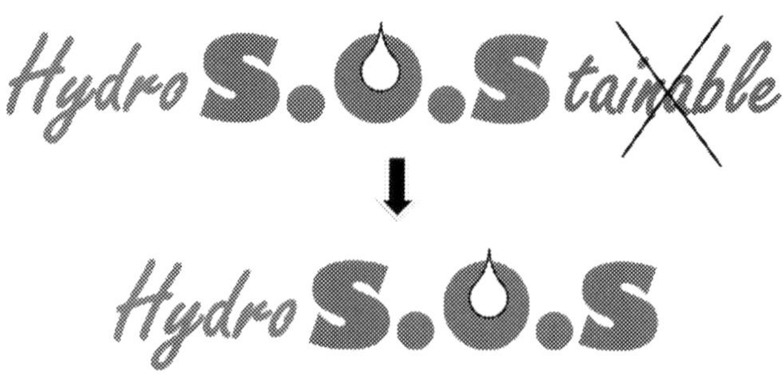

Figure 15.4 Logos designed for the differentiation of *hydrosustainable* fruits.

least-sensitive growing stages, (2) the timing, strength, and duration of the water stress, and (3) suitable monitoring and control of the stress applied, by measuring, for example, the water potential, etc.

8. CONCLUSIONS

Taking into consideration the aforementioned research, we can conclude the unquestionable importance of fruit water relations studies. In fact, fruit water status during the season affects fruit yield, some key fruit characteristics that are important contributors to quality, and the incidence of some fruit physiological disorders. Nevertheless, research on fruit water relations is very scarce in comparison to that developed on leaf water relations. This situation could be attributed to several reasons, among which are: (1) the interest of researchers has traditionally been focused mainly on the mechanisms developed by crop species to confront drought and the effect of deficit irrigation on yield, without paying special attention to the causes that modify fruit quality; (2) the complexity of some fruit water relations parameters measurements, mainly in fruits with a complex structure; (3) the degree of uncertainty in the information provided by some measurements, such as pressurization on fruit tissues, when Ψ_{fruit} is measured with a pressure chamber; and (4) the fact that some instrumental devices for specific fruit water relations study, as those for fruit gas exchange, are not commercially available. Research on the effect of water stress on fruit metabolites, especially secondary metabolites, is scarce. Also, detailed data on fruit or plant water status are frequently not provided. In order to deepen our knowledge of these aspects and to clarify conflicting reports in the literature, research must be redoubled and focused on establishing the relation between fruit metabolite content and the level of water stress applied, paying special attention to the rate at which water deficit is reached and the phenological period in which it occurs. In addition, greater effort to understand what consumers think about fruits or foodstuffs from deficit-irrigated trees is necessary. For example, it should be possible to raise public awareness of the health benefits to be gained from fruits obtained from deficit-irrigated trees due to their increased functionality, and of the environmental advantages that these fruits offer, being the result of a more sustainable use of the irrigation water. It is therefore necessary to differentiate hydroSOS products from others on the market, perhaps by developing an index to certify that the products using the hydroSOS logo have been evaluated regarding their sustainable use of irrigation water.

ACKNOWLEDGMENTS

This research was supported by Ministerio de Economía y Competitividad (MINECO) (CICYT/FEDER AGL2013-45922-C2-1-R and AGL2013-45922-C2-2-R) grants to the authors. AG, JC-G, and SM are funded by a postdoctoral Ramón Areces, Juan de la Cierva, and guest professor grants, respectively. Also, this work is a result of PR's internship (19925/IV/15) funded by the Fundación Séneca - Agencia de Ciencia y Tecnología de la Región de Murcia (Seneca Foundation - Agency for Science and Technology in the Region of Murcia) under the Jiménez de la Espada Program for Mobility, Cooperation and Internationalization.

REFERENCES

Akula, R., Ravishankar, G.A., 2011. Influence of abiotic stress signals on secondary metabolites in plants. Plant Signal Behaviour 6, 1720–1731.

Albrigo, L.G., 1975. Water relations and citrus fruit quality. In: Sauls, J.W., Jackson, L.K. (Eds.), Water Relations. University of Florida, Fruit Crops Department, Gainiville, pp. 41–48.

Alférez, F., Alquezar, B., Burns, J.K., Zacarias, L., 2010. Variation in water, osmotic and turgor potential in peel of 'Marsh' grapefruit during development of postharvest peel pitting. Postharvest Biology and Technology 56, 44–49.

Alférez, F., Burns, J., Zacarías, L., 2004. Postharvest peel pitting in Citrus is induced by changes in relative humidity. Proceedings of the Florida State Horticultural Society 117, 355–358.

Almela, V., Zaragoza, S., Primo Millo, E., Agusti, M., 1994. Hormonal control of splitting in nova mandarine fruit. The Journal of Horticultural Science and Biotechnology 69, 969–973.

Aloni, B., Karni, L., Moreshet, S., Yao, C., Stanghellini, C., 1999. Cuticular cracking in bell pepper fruit: II. Effects of fruit water relations and fruit expansion. Journal of Horticultural Science and Biotechnology 74, 1–5.

Amirjani, M.R., 2013. Effects of drought stress on the alkaloid contents and growth parameters of *Catharanthus roseus*. Journal of Agricultural and Biological Science 8, 745–750.

Behboudian, M.H., Lawes, G.S., Griffths, K.M., 1994. The influence of water deficit on water relations, photosynthesis and fruit growth in Asian pear (*Pyrus serotina* Rehd.). Scientia Horticulturae 60, 89–99.

Berüter, J., 1989. Carbohydrate partitioning and changes in water relations of growing apple fruit. Journal of Plant Physiology 135, 583–587.

Bianco, R.L., Farina, V., Avellone, G., Filizzola, F., Agozzino, P., 2008. Fruit quality and volatile fraction of 'Pink Lady'apple trees in response to rootstock vigor and partial rootzone drying. Journal of the Science of Food and Agriculture 88, 1325–1334.

Blanke, M.M., Whiley, A.W., 1995. Bioenergetics, respiration cost and water relations of developing avocado fruit. Journal of Plant Physiology 145, 87–92.

Bourgaud, F., Gravot, A., Milesi, S., Gontier, E., 2001. Production of plant secondary metabolites: a historical perspective. Plant Science vol 161, 839–851.

Cano-Lamadrid, M., Hernández, F., Corell, M., Burló, F., Moriana, A., Carbonell-Barrachina, A.A., 2017. Antioxidant capacity, fatty acids profile, and descriptive sensory analysis of table olives as affected by deficit irrigation. Journal of the Science of Food and Agriculture 97, 444–451.

Carbonell-Barrachina, A.A., Memmi, H., Noguera-Artiaga, L., Gijón-López, M.C., Ciapa, R., Pérez-López, D., 2015. Quality attributes of pistachio nuts as affected by rootstock and deficit irrigation. Journal of the Science of Food and Agriculture 95, 2866–2873.

Chaves, M.M., Flexas, J., Pinheiro, C., 2009. Photosynthesis under drought and salt stress: regulation mechanisms from whole plant to cell. Annals of Botany 103, 551–560.

Chen, J.Z., Lv, X.J., Ye, Z.X., Yao, Q., Wu, L., 2002. Study on the relation between mineral nutrition levels and creasing peel in mature orange. Plant Nutrition and Fertilizer Science 8, 367–371.

Clarke, S.J., Hardie, W.J., Rogiers, S.Y., 2010. Changes in susceptibility of grape berries to splitting are related to impaired osmotic water uptake associated with losses in cell vitality. Australian Journal of Grape and Wine Research 16, 469–476.

Clay, J., 2004. World Agriculture and the Environment: A Commodity-by-Commodity Guide to Impacts and Practices. Island Press, Washington.

Collado-González, J., Cruz, Z.N., Medina, S., Mellisho, C.D., Rodríguez, P., Galindo, A., Egea, I., Romojaro, F., Ferreres, F., Torrecillas, A., Gil-Izquierdo, A., 2014. Effects of water deficit during maturation on amino acids and jujube fruit eating quality. Macedonian Journal of Chemistry and Chemical Engineering 33, 105–119.

Collado-González, J., Cruz, Z.N., Rodríguez, P., Galindo, A., Díaz-Baños, F.G., García de la Torre, J., Ferreres, F., Medina, S., Torrecillas, A., Gil-Izquierdo, A., 2013. Effect of water deficit and domestic storage on the procyanidin profile, size, and aggregation process in pear-jujube (*Z. jujuba*) fruits. Journal of Agricultural and Food Chemistry 61, 6187–6197.

Collado-González, J., Moriana, A., Girón, I.F., Corell, M., Medina, S., Durand, T., Guy, A., Galano, J.M., Valero, E., Garrigues, T., Ferreres, F., Moreno, F., Torrecillas, A., Gil-Izquierdo, A., 2015. The phytoprostane content in green table olives is influenced by Spanish-style processing and regulated deficit irrigation. LWT – Food Science and Technology 64, 997–1003.

Collado-González, J., Pérez-López, D., Memmi, H., Gijón, M.C., Medina, S., Durand, T., Guy, A., Galano, J.M., Fernández, D.J., Carro, F., Ferreres, F., Torrecillas, A., Gil-Izquierdo, A., 2016. Effect of the season on the free phytoprostane content in Cornicabra extra virgin olive oil from deficit irrigated olive trees. Journal of the Science of Food and Agriculture 96, 1585–1592.

Colmenero-Flores, J.M., Campos, F., Garciarrubio, A., Covarrubias, A.A., 1997. Characterization of cDNA clones responsive to water deficit from *Phaseolus vulgaris*: a new late embryogenesis abundant protein. Plant Molecular Biology 35, 393–405.

Danon, A., Apel, K., Laloi, C., 2004. Reactive oxygen signalling: the latest news. Current Opinion in Plant Biology 7, 323–328.

Dell'Amico, J., Moriana, A., Corell, M., Girón, I.F., Morales, D., Torrecillas, A., Moreno, F., 2012. Low water stress conditions in table olive trees (*Olea europaea* L.) during pit hardening produced a different response of fruit and leaf water relations. Agricultural Water Management 114, 11–17.

Domingo, R., Ruiz-Sánchez, M.C., Sánchez-Blanco, M.J., Torrecillas, A., 1996. Water relations, growth and yield of Fino lemon trees under regulated deficit irrigation. Irrigation Science 16, 115–123.

Fernandes-Silva, A.A., Falco, V., Correia, C.M., Villalobos, F.J., 2013. Sensory analysis and volatile compounds of olive oil (cv. Cobrançosa) from different irrigation regimes. Grasas y Aceites 64, 59–67.

Galindo, A., Cruz, Z.N., Rodríguez, P., Collado-González, J., Corell, M., Memmi, H., Moreno, F., Moriana, A., Torrecillas, A., Pérez-López, D., 2016. Jujube fruit water relations at fruit maturation in response to water deficits. Agricultural Water Management 164, 110–117.

Galindo, A., Rodríguez, P., Collado-González, J., Cruz, Z.N., Torrecillas, E., Ondoño, S., Corell, M., Moriana, A., Torrecillas, A., 2014. Rainfall intensifies fruit peel cracking in water stressed pomegranate trees. Agricultural and Forest Meteorology 194, 29–35.

Gelly, M., Marsal, J., Mata, M., Arbonès, J., Rufat, J., Girona, J., 2004. Fruit water potential: a comparative study with other water potential parameters. Acta Horticulturae 646, 35–40.

Giles, K.L., Beardsell, M.F., Cohen, D., 1974. Cellular and ultrastructural changes in mesophyll and bundle sheath cells of maize in response to water stress. Plant Physiology 54, 208–212.

Gobbo-Neto, L., Lopes, N.P., 2007. Medicinal plants: factors of influence on the content of secondary metabolites. Química Nova 30, 374–381.

Greenspan, M.D., Schultz, H.R., Matthews, M.A., 1996. Field evaluation of water transport in grape during water deficits. Physiologia Plantarum 97, 55–62.

Hankinson, B., 1979. Histological and physical behaviour of tomato skins susceptible to cracking. Journal of the American Society for Horticultural Science 104, 577–581.

Hooper, L., Cassidy, A., 2006. A review of the health care potential of bioactive compounds. Journal of the Science of Food and Agriculture 86, 1805–1813.

Horner, J.D., 1990. Nonlinear effects of water deficits on foliar tannin concentration. Biochemical Systematic and Ecology 18, 211–213.

Hou, Q., Ufer, G., Bartels, D., 2016. Lipid signalling in plant responses to abiotic stress. Plant Cell and Environment 39, 1029–1048.

Hou, X.M., Wang, Z.H., Deng, X.M., Li, G.H., 2011. Water relation parameters of the cortex tissue of fresh 'Fuji' apple fruit determined by centrifugation. HortScience 46, 30–34.

Huang, X.M., Wang, H.C., Zhong, W.L., Yuan, W.Q., Lu, J.M., Li, J.G., 2008. Spraying calcium is not an effective way to increase structural calcium in litchipericarp. Scientia Horticulturae 117, 39–44.

Inomata, Y., Yaegaki, H., Suzuki, K., 1999. The effects of polyethylene bagging, calcium carbonate treatment and difference in fruit–air temperatures on the occurrence of watercore in Japanese pear 'Housui'. Journal of the Japanese Society for Horticultural Science 68, 336–342.

Jobling, J.J., Patterson, B.D., Moradi, S., Joyce, D., 1997. A non-destructive method for measuring the water potential of fruit and vegetables. Postharvest Biology and Technology 10, 1–8.

Johnson, R.W., Dixon, M.A., Lee, D.R., 1992. Water relations of the tomato during fruit growth. Plant Cell and Environment 15, 947–953.

Jones, H.G., Tardieu, F., 1998. Modelling water relations of horticultural crops: a review. Scientia Horticulturae 74, 21–46.

Jones, H.G., Lakso, A.N., Syvertsen, J.P., 1985. Physiological control of water status in temperate and subtropical fruit trees. Horticultural Reviews 7, 301–344.

Jones, W.W., Embleton, T.W., Garber, M.J., Cree, C.B., 1967. Creasing of orange fruit". Hilgardia 38, 231–244.

Kasai, S., Hayama, H., Kashimura, Y., Kudo, S., Osanai, Y., 2008. Relationship between fruit cracking and expression of the expansion gene MdEXPA3 in 'Fuji'apples (*Malus domestica* Borkh.). Scientia Horticulturae 116, 194–198.

Kong, M., Lampinen, B., Shackel, K., Crisosto, C.H., 2013. Fruit skin side cracking and ostiole-end splitting shorten postharvest life in fresh figs (*Ficus carica* L.), but are reduced by deficit irrigation. Postharvest Biology and Technology 85, 54–161.

Krasensky, J., Jonak, C., 2012. Drought, salt, and temperature stress-induced metabolic rearrangements and regulatory networks. Journal of Experimental Botany 63, 1593–1608.

Lafuente, M.T., Sala, J.M., 2002. Abscisic acid levels and the influence of ethylene, humidity and storage temperature on the incidence of postharvest rindstaning of 'Navelina' orange (*Citrus* sinensis L. Osbeck) fruit. Postharvest Biology and Technology 25, 49–57.

Lang, A., 1990. Xylem, phloem and transpiration flows in developing apple fruits. Journal of Experimental Botany 41, 645–651.

Lawlor, D.W., Leach, J.E., 1985. Leaf growth and water deficits: biochemistry in relation to biophysics". In: Baker, N.R., Davies, W.J., Ong, C.K. (Eds.), Control of Leaf Growth. Soc. Expt. Biol. Seminar Ser. 27. Cambridge Univ. Press, Cambridge, pp. 267–294.

Lopez, G., Echeverria, G., Bellvert, J., Mata, M., Behboudian, M.H., Girona, J., Marsal, J., 2016. Water stress for a short period before harvest in nectarine: yield, fruit composition, sensory quality, and consumer acceptance of fruit. Scientia Horticulturae 211, 1–7.

Martinelli, F., Remorini, D., Saia, S., Massai, R., Tonutti, P., 2013. Metabolic profiling of ripe olive fruit in response to moderate water stress. Scientia Horticulturae 159, 52–58.

Matthews, M.A., Shackel, K.A., 2005. Growth and water transport in fleshy fruit. In: Holbrook, N.M., Zwieniecki, M.A. (Eds.), Vascular Transport in Plants. Elsevier, Boston, pp. 181–197.

McFadyen, L.M., Hutton, R.J., Barlow, E.W.R., 1996. Effects of crop load on fruit water relations and fruit growth in peach. Journal of Horticultural Science 71, 469–480.

Measham, P., 2011. Rain-induced fruit cracking in sweet cherry (*Prunus avium* L.). Degree of Doctor of Philosophy. School of Agricultural Science, University of Tasmania, 170 p.

Mesejo, C., Reiga, C., Martínez-Fuentesa, A., Gambetta, G., Gravina, A., Agustí, M., 2016. Tree water status influences fruit splitting in Citrus. Scientia Horticulturae 209, 96–104.

Milad, R.E., Shackel, K.A., 1992. Water relations of fruit end cracking in French prune (*Prunus domestica* L. cv: French). Journal of the American Society for Horticultural Science 117, 824–828.

Mills, M.T., Behboudian, M.H., Clothier, B.E., 1997. The diurnal and seasonal water relations, and composition, of 'Braeburn' apple fruit under reduced plant water status. Plant Science 126, 145–154.

Mills, T.M., Behboudian, M.H., Clothier, B.E., 1996. Water relations, growth, and the composition of 'Braeburn' apple fruit under deficit irrigation. Journal of the American Society for Horticultural Science 121, 286–291.

Morandi, B., Grappadelli, L.C., 2009. Source and sink limitations in vascular flows in peach fruit. Journal of Horticultural Science and Biotechnology SI, 150–156.

Moreshet, S., Green, G.C., 1980. Photosynthesis and diffusion conductance of the Valencia orange fruit under field conditions. Journal of Experimental Botany 31, 15–27.

Nobel, P.S., De la Barrera, E., 2000. Carbon and water balances for young fruits of platyopuntias. Physiologia Plantarum 109, 160–166.

Nobel, P.S., Andrade, J.L., Wang, N., North, G.B., 1994. Water potential for developing cladodes and fruits of a succulent plant, including xylem-versus-floem implications for water movement. Journal of Experimental Botany 45, 1801–1807.

Noguera-Artiaga, L., Lipan, L., Vázquez-Araújo, L., Barber, X., Pérez-López, D., Carbonell-Barrachina, A.A., 2016. Opinion of Spanish consumers on hydrosustainable pistachios. Journal of Food Science 81, S2559–S2565.

Nowak, M., Manderscheid, R., Weigel, H.J., Kleinwächter, M., Selmar, D., 2010. Drought stress increases the accumulation of monoterpenes in sage (*Salvia officinalis*), an effect that is compensated by elevated carbon dioxide concentration. Journal of Applied Botany and Food Quality 83, 133–136.

Okello, R.C.O., Heuvelink, E., Visser, P.H.B., Strunk, P.C., Marcelis, L.F.M., 2015. What drives fruit growth? Functional Plant Biology 42, 817–827.

Opara, L.U., Studman, C.J., Banks, N.H., 1997. Fruit skin splitting and cracking. Horticultural Reviews 19, 217–262.

Ormeño, E., Mevy, J.P., Vila, B., Bousquet-Melou, A., Greff, S., Bonin, G., Fernandez, C., 2007. Water deficit stress induces different monoterpene and sesquiterpene emission changes in Mediterranean species. Relationship between terpene emissions and plant water potential. Chemosphere 67, 276–284.

Ou, C., Du, X., Shellie, K., Ross, C., Qian, M.C., 2010. Volatile compounds and sensory attributes of wine from cv. Merlot (Vitis vinifera L.) grown under differential levels of water deficit with or without a kaolin-based, foliar reflectant particle film. Journal of Agricultural and Food Chemistry 58, 12890–12898.

Pomper, K.W., Breen, P.J., 1995. Levels of apoplastic solutes in developing strawberry fruit. Journal of Experimental Botany 46, 743–752.

Pomper, K.W., Breen, P.J., 1997. Expansion and osmotic adjustment of strawberry fruit during water stress. Journal of the American Society for Horticultural Science 122, 183–189.

Ripoll, J., Urban, L., Staudt, M., Lopez-Lauri, F., Bidel, L.P.R., Bertin, N., 2014. Water shortage and quality of fleshy fruits-making the most of the unavoidable. Journal of Experimental Botany 65, 4097–4117.

Sattler, S.E., Mène-Saffrané, L., Farmer, E.E., Krischke, M., Mueller, M.J., DellaPenna, D., 2006. Nonenzymatic lipid peroxidation reprograms gene expression and activates defense markers in *Arabidopsis* tocopherol-deficient mutants. The Plant Cell 18, 3706–3720.

Schröder, R., Clark, C.J., Sharrock, K., Hallett, I.C., MacRae, E.A., 2004. Pectins from the albedo of immature lemon fruitlets have high water binding capacity. Journal of Plant Physiology 161, 371–379.

Sekse, L., 1995. Fruit cracking in sweet cherries (*Prunus avium* L.). Some physiological aspects-a mini review. Scientia Horticulturae 63, 135–141.

Selmar, D., Kleinwächter, M., 2013. Stress enhances the synthesis of secondary plant products: the impact of the stress-related over-reduction on the accumulation of natural products. Plant and Cell Physiology 54, 817–826.

Smirnoff, N., 1993. The role of active oxygen in the response of plants to water deficit and desiccation. New Phytologist 125, 27–58.

Syvertsen, J.P., Albrigo, L.G., 1980. Seasonal and diurnal citrus leaf and fruit water relations. Botanical Gazette 141, 440–446.

Torrecillas, A., Domingo, R., Galego, R., Ruiz-Sánchez, M.C., 2000. Apricot tree response to irrigation withholding at different phenological periods. Scientia Horticulturae 85, 201–215.

Toumi, I., Gargouri, M., Nouairi, I., Moschou, P.N., Salem-Fnayou, A.B., Mliki, A., Zarrouk, M., Ghorbel, A., 2008. Water stress induced changes in the leaf lipid composition of four grapevine genotypes with different drought tolerance. Biologia Plantarum 52, 161–164.

Tovar, M.J., Romero, M.P., Girona, J., Motilva, M.J., 2002. L-phenylalanine ammonia-lyase activity and concentration of phenolics in developing olive (*Olea europaea* L cv Arbequina) fruit grown under different irrigation regimes. Journal of the Science of Food and Agriculture 82, 892–898.

Trolinder, N.L., McMichael, B.L., Upchurch, D.R., 1993. Water relations of cotton flower petals and fruit. Plant, Cell and Environment 16, 755–760.

van Iersel, M.W., Oosterhuis, D.M., Harris, W.M., 1994. Apoplastic water flow to cotton leaves and fruits during development. Journal of Experimental Botany 45, 163–167.

Van Volkenburg, E., Cleland, R.E., 1986. Wall yield threshold and effective turgor in growing bean leaves. Planta 167, 37–43.

Wang, Z., Quebedeaux, B., Stutte, G.W., 1995. Osmotic adjustment – effect of water-stress on carbohydrates in leaves, stems and roots of apple. Australian Journal of Plant Physiology 22, 747–754.

Yakushiji, H., Nonami, H., Fukuyama, T., Ono, S., Takagi, N., Hashimoto, Y., 1996. Sugar accumulation enhanced by osmoregulation in Satsuma mandarin fruit. Journal of the American Society for Horticultural Science 121, 466–472.

Yamada, H., Takechi, K., Hoshi, A., Amano, S., 2004. Comparison of water relations in watercored and non-watercored apples induced by fruit temperature treatment. Scientia Horticulturae 99, 309–318.

Yordanov, I., Velikova, V., Tsonev, T., 2003. Plant responses to drought and stress tolerance. Bulgarian Journal of Plant Physiology 29, 187–206.

CHAPTER 16

Genetic Improvement of Grapevine (*Vitis vinifera* L.) Water Use Efficiency: Variability Among Varieties and Clones

Hipólito Medrano, Ignacio Tortosa, Esther Montes, Alicia Pou, Pedro Balda, Josefina Bota, José M. Escalona
Universitat de les Illes Balears, Palma de Mallorca, Spain

Contents

1. GENERAL INTRODUCTION

Water use is an important issue for the environmental sustainability of viticulture in semiarid areas. Around 60% of this crop is located under such conditions and therefore water scarcity is the main limitation for grapevine crops (Medrano et al., 2015a,b; Flexas et al., 2010). High water requirements

Water Scarcity and Sustainable Agriculture in Semiarid Environment
ISBN 978-0-12-813164-0
https://doi.org/10.1016/B978-0-12-813164-0.00016-8

are necessary for a reasonable yield in grapevines because the growth season coincides with the driest months, making irrigation necessary (Rozak and Senge, 2016; Chaves et al., 2010; Williams and Ayars, 2005). In dry areas, water use for irrigation can result in a compromise between environmental sustainability for this crop and other critical human uses (Chaves et al., 2007). Moreover, climate change forecasts a clear increase in ambient temperatures, as well as an increment in the frequency and intensity of climatic anomalies for the near future, such as droughts and heat waves (Hannah et al., 2013; Jones and Vaughan, 2010; IPCC, 2013). All these circumstances are becoming a prevalent condition for most of the Mediterranean and semiarid regions used for grapevine production, as well as most of the "new world" viticulture due to the high irrigation volumes required to obtain a reasonable harvest. Consequently, the optimization of water use for vineyards, by improving water use efficiency (WUE), is a subject of main interest to ensure the sustainability of viticulture. This is reflected in the important volumes of applied and fundamental research focused on optimizing grapevine water use both by WUE-oriented agronomical practices (Medrano et al., 2015a,b; Costa et al., 2016) and by the identification of higher WUE cultivars (Tomás et al., 2014a,b; Bota et al., 2016).

In line with this general trend to enhance crop sustainability, genetics emerges as a promising field. The genetic improvement of crops has been an important basis for the general increase in productivity in the last few decades because of different selection criteria and existing genetic variability (Alleweldt and Possingham, 1988; Galet, 2000). For grapevines, the great variety of commercial cultivars, considered to be one of the biggest among actual crops, speaks in favor of a promising way for genetic improvement of WUE (Tortosa et al., 2016).

In fact, thousands of grapevine cultivars have been described around the world (This et al., 2006; OIV, 2009), showing an impressive genetic variability and plasticity of the grapevine genome. However, until now there has been no reference to the use of WUE as selection criteria, even though there are cultivars reputed to be more adapted to drought-prone conditions, which presumably should also present high WUE. Genetic variability in WUE was demonstrated by Bota et al. (2001) and Gaudillère et al. (2002), measuring leaf intrinsic WUE (WUE_i) as A_N/g_s and the surrogate character carbon isotope discrimination ($\delta^{13}C$), respectively, in different collections of grapevine cultivars, and more recently by comparison of leaf WUE in different cultivars (Tomás et al., 2012, 2014b; Zsófi et al., 2009) as in large collections of local varieties (Bota et al., 2016).

However, differences in many physiological parameters such as A_N and g_s, and consequently in WUE_i, are widely affected by environmental conditions such as soil water content, climatic conditions, and year-to-year variability in field assays (Costa et al., 2012; Prieto et al., 2010; Tomás et al., 2012), thus making it difficult to achieve a proper assessment and evaluation of genotype performance. Moreover, most of the measurements of WUE in grapevines are obtained from studies performed at the leaf level, while at the whole-plant level information is scarce and does not always correspond to the expected information on the basis of single-leaf taken data (Medrano et al., 2012). However, the large variability of leaf A_N/g_s among cultivars observed under field conditions offers an opportunity to explore genetic selection using this character as representative of plant WUE.

A way to estimate WUE in a wider scale is to use isotopic discrimination against $\delta^{13}C$, which is dependent mainly on the stomatal aperture. This parameter reflects the average water economy (thus the average A_N/g_s) during dry mass accumulation in a certain organ such as leaf, fruit, shoot, etc., providing an interesting way to reduce the particular measuring conditions for a certain A_N/g_s measurement. There are different reports on the genetic variability of this character among different cultivars in grapevines (Costa et al., 2012; Gaudillère et al., 2002; Flexas et al., 2010; Tomás et al., 2014a; Bota et al., 2016), even though with this parameter the relationship between $\delta^{13}C$ and whole-plant WUE is not always significant (Gibberd et al., 2001; Tomás et al., 2012).

On the other hand, the importance of rootstocks for the total water use and WUE of the crop should be considered. Rootstock variability on plant adaptation to water scarcity has been shown usually to be linked to their differential capacity to extract water from the soil to adjust water supply to plant transpiration (Alsina et al., 2011; Marguerit et al., 2012; Tramontini et al., 2013). However, the variability of WUE in rootstocks has largely been unexplored, in addition to obtaining contradictory results (Satisha et al., 2007; Jacobs 2010). However, the main role of rootstocks in plant water economy leads to the consideration of the genetic variability of WUE as a main target for future research.

Knowing these limitations for the evaluation of genotype performance in WUE, but also realizing the clear improvements achieved by genetic selection in other crop characteristics, the present revision shows the magnitude of variation of this character both among grapevine cultivars and inside a specific grapevine cultivar. The initial results encourage further work to identify cultivars and clones with enhanced WUE.

2. GRAPEVINE GENETIC VARIABILITY AND OPPORTUNITIES FOR GENETIC SELECTION. LESSONS FROM THE RECENT SELECTION PROGRAMS

2.1 The History

The history and cultural practices of grapevine since the beginning of viticulture have largely determined the genetic diversity of this crop. This diversity gives a high potential for genetic improvement to reach the main factors influencing high productivity and water efficient varieties, but also high-quality grapes. Evaluation and conservation of *Vitis* genetic diversity in old landraces and wild species is therefore of great interest for future breeding programs and for preservation.

Wild grapes were among the first domesticated plants (Mc Govern, 2003) and are thought to have begun during the Neolithic era (6000−6500 BP) in the Caucasus region with later expansion along trade routes to adjacent areas with suitable climatic conditions. Therefore through diverse civilizations, grape varieties of eastern origin reached the Mediterranean and Western Europe (5500−5000 BP). The encounter of different grape populations from Eastern and Western Europe with local *Vitis vinifera* L. ssp. *sylvestris* gave rise to a multitude of new grape varieties. Although many studies of divers domestication processes have already been published, considerable disagreement persists regarding the existence of diverse domestication pools or a single major domestication event followed by a selection of diverse genotypes. Several authors propose a secondary domestication event in the Iberian Peninsula (Arroyo-García et al., 2006; This et al., 2006) and confirm the idea of a mixed origin of Iberian cultivars from imported oriental germplasm, together with the use of local secondarily domesticated genotypes. Moreover, over the course of history it is important to consider a few cycles of spontaneous pollination of newly domesticated plants with pollen from imported cultivars and selection of hybrids with oriental phenotypes, which would have resulted in new cultivars showing genotypes more closely related to oriental characteristics. In conclusion, the results of the analysis of grapevine genotypes support these introgressions of genetic material from western *sylvestris* in the current western cultivars (Myles et al., 2011). After so many crosses and such diverse origins, it is almost impossible to determine nowadays if the characteristics associated with some varieties, such as phenological cycle, pest resistance productivity, abiotic stress tolerance, or WUE, correspond to a certain geographical origin or any genetic introgression of wild forms, etc.

Actually, the Vitis International Variety Catalog (http://www.vivc.de/) lists 137 collections in the world with a total of 48,826 accessions, most of them held in Asia, North America, and Europe under subtropical, Mediterranean, and continental-temperate climatic conditions. However, in spite of the contrast between cultivars mentioned in the literature (Viala and Vermorel, 1902; Alleweldt and Dettweiler, 1994) and those actually existing in situ or in collections, there is a general consensus that nowadays there are more than 5000 cultivars (Galet, 2000), even though there has been the loss of a multitude of genotypes. This reported decline in the number of grapevine cultivars is related to the existence of multiple synonyms and homonyms in different cultivar collections. Moreover, the large number of diseases (such as the phylloxera during the 20th century) and other human activities that occurred during the recent history of viticulture, as well as the generalized use of certain global cultivars around the world in the last few decades, severely contributed to reducing the genetic background of the *Vitis* genus. From all these varieties that are believed to exist in the world, only 200–300 are cultivated on a large scale (Galet, 2000). So, there is a pressing need to preserve the accumulated diversity of wild and cultivated grapevines that have played and still play a key role as genetic diversity sources. To do this, it is necessary to go back in history and visit our old vineyards in the less-developed wine regions, where traditional viticulture still exists and where silent wild grapevines keep climbing the natural vegetation.

2.2 Lessons From the Recent Selection Programs

It is well known that *V. vinifera* L. was one of the first domesticated and widely cultivated species along the Mediterranean basin. Since the beginning of modern viticulture, some few centuries ago, vegetative propagation has been the only technique used for grapevine multiplication. The obtained vine material for any new plantation of vines was usually collected from the old vineyards showing interesting agronomical characters. This way of propagation has permitted the generation and preservation of a large part of the genetic material, which is at the present time preserved in the grape-growing industry.

Nowadays, a large diversity of cultivars exists, not only because of the high number of different varieties, but also because of the great diversity found inside each variety. Most of the current varieties resulted from natural hybridizations that took place hundreds of years ago. However, even now their respective clones are identical to the former ones, although some of them may show divergent genotypes and, to some extent, different phenotypes (Pelsy, 2010). Intravariety diversity begins at the origin of any variety, mainly because of the somatic mutations carried out during cell division. The progressive accumulation of genetic mutations explains the great genetic variability of many grape varieties (Rives, 1961; Ulanovsky et al., 2002; Sefc et al., 2000). Since grapevine domestication and considering hybridization and somatic mutations, we have inherited between 5000 and 10,000 different cultivars of *V. vinifera* L. (Alleweldt and Possingham, 1988; Galet, 2000). In some of them, intravariety diversity reaches around 1000 biotypes, as it is in the case of Tempranillo, which includes more than 1200 biotypes in several Spanish collections. The case of the Tempranillo cultivar contrasts with that observed in *Citrus* sp., which has a low intraspecific diversity in cultivated species such as sweet orange and lemon (Herrero et al., 1996), and also in *Prunus* sp., which includes only 400–430 species, 98 of them considered to be important and listed in the Genetic Resource Information System (Biswajit et al., 2011). In the case of *Olea europaea* we also have a similar trend. Although its importance throughout history along the entire Mediterranean basin is noted, there are only around 200 cultivars described worldwide (http://www.bioinfo-cbs.org/ogdd/).

Thus, contrary to other perennial crops, the grapevine-growing industry has always been focused on the preservation of varietal heritage. Its objective is to produce wines with personality and thus, in this case, the varietal factor is underlined as very positive. Moreover, every wine region has its own traditional varieties and they are a symbol of distinction and differentiation, which has been carefully preserved. Indeed, many research projects in many regions of Spain (Rioja, Galicia, Aragón, Castilla y León, Balearic Islands, etc.) (Martínez de Toda and Sancha, 1997; Yuste et al., 2006; Buhner-Zaharieva et al., 2010; Escalona et al., 2017) and Europe have been developed to recover, preserve, and study the endangered grapevine germplasms. This philosophy of the grapevine industry is not shared with other woody cultivars, such as *Citrus* or *Prunus* (peach, cherry, apricot, plum, almond), whose products are not linked to their origin. This is an important consideration to understand the differences in the genetic diversity between *Vitis* and other woody plant cultivars.

Since the beginning of viticulture, the main strategy to improve grapevine material has been the clonal selection programs, more than breeding new varieties. Moreover, the restriction of EU legislation, local governments, and the rules of the appellations have impeded or restricted the development of genetic engineering and breeding programs. Several clonal selection programs have been developed worldwide. In Europe, this process started in Germany in 1876 because of the low yield they obtained with virus-infected plants. Later, France and Italy also introduced their own grapevine clonal selection programs, and finally Spain, which began in the 1970s. The only objective of these programs was to increase productivity by selecting virus-free plant material. During this process, the inherited diversity resulted in a few individuals with a very specific agronomic characteristic: high productivity and good phytosanitary standards. However, during the last few decades, the clonal selection programs carried out by private enterprises or public research centers have changed the selection criteria, trying to preserve the major range of variability within a small number of clones, and not only the selection of the best clones for the selection criteria. Some characteristics such as grape quality, high temperature, and water scarcity tolerance are nowadays the main objectives to acquire plant material adapted to climatic exchange and restrictive environmental growing conditions.

3. WATER USE EFFICIENCY CONCEPT AND WAYS TO MEASURE IT. SURROGATE CHARACTERS FOR WATER USE EFFICIENCY

The idea of WUE has been used to evaluate water management for many years. The original efficiency term was called "duty of water" and was expressed as "the ratio between a given quantity of water and the growing area of mature crop" (Wilson, 1912). Nowadays, water use (i.e., the water consumed) and WUE (i.e., the efficiency of this consumed water to assimilate carbon, produce biomass, or grain yield) are crucial parameters, especially in areas in which water is scarce, e.g., south Europe, West Asia, Western Australia, Chile, North Africa, and parts of South Africa. In viticulture, wine-growing regions are located widely around the world; however, the climates of most wine regions are categorized (Köppen and Wladimir, 1884) as being part of a Mediterranean, maritime, or continental climate, which are found between the latitudes of 30 and 50 degrees in both Northern and Southern Hemispheres. In the case of the Mediterranean Basin, the climate is characterized by relatively scarce and erratic precipitation and wet winters and dry and hot summers (Acevedo et al., 1999). Thus in these areas, progressive soil water deficits and high leaf-to-air vapor pressure gradients, together with high irradiance and temperatures, exert large constraints on crop yield and quality. In this sense, WUE has become a suitable tool for analyzing the strategies that allow the best use of water in viticulture to be attained through the management of physiological water-saving measures.

The term WUE reflects the balance between carbon gain and the associated cost in water, and can be measured and/or estimated at different spatial and temporal scales, from leaf scale to crop field (Medrano et al., 2015a,b). At the leaf level (leaf water use efficiency, WUE_{leaf}), WUE is defined either as the ratio between CO_2 assimilation rate (A_N) and stomatal conductance (g_s) (A_N/g_s, WUE_i) or between A_N and leaf transpiration rate (E) (A_N/E, instantaneous leaf water use efficiency, WUE_{inst}) per leaf unit area. At the plant level, plant water use efficiency (WUE_{plant}) can be defined as the amount of total biomass gained (ground dry matter) or net carbon gain per unit of water applied or transpired; it is also called biomass water productivity (Steduto et al., 2007). Finally, in the agronomic approach, at the crop level (water productivity, WUE_{yield}), the concept of crop WUE refers to final yield and total water consumed (i.e., evapotranspiration). In many crops, the concept WUE_{yield} has been summarized as "more crop per drop,"; however, in grapevines the main concern is not to reach a high productivity and water saving, but rather an efficient deficit irrigation scheduling, allowing the highest quality of grapes for premium wine production. This also suggests that select drought-tolerant genotypes taking into account only the highest WUE_i as a selection target, regardless other criteria, may be appropriate in terms of drought tolerance, but not in terms of quality. Thus to cope with market forces, WUE_i has to be complemented with the analysis of other parameters, such as plant production, grape quality, etc.

3.1 Ways to Measure Water Use Efficiency and Other Surrogate Characters

At the leaf level, portable gas exchange measurements allow instantaneous measurements of A_N/g_s and A_N/E to be obtained, which are mostly used to characterize genetic and environmental effects, respectively. While soil moisture and g_s are important determinants of A_N/g_s, vapor pressure deficit (VPD) is the main driver for A_N/E (Rogiers et al., 2011). Although instantaneous measurements are widely used to understand the global results obtained from the agronomical approach based on the assumption that they are representative of whole plants, this comparison does not always reveal a clear relationship (Tomás et al., 2012; Medrano et al., 2012;

Poni et al., 2009; Tarara et al., 2011). The uncertainty of this measurement as representative of whole-plant WUE has been widely analyzed (Medrano et al., 2015a,b; Bchir et al., 2016; Tomás et al., 2014a,b) because good and bad correspondences are found. Nevertheless, the characteristic of this parameter as a direct measurement of instantaneous WUE_{leaf} is widely appreciated for different assessments in field and pot studies and provides a key parameter to qualify WUE of a certain genotype.

More integrative measurements are taking place at the plant and crop levels. $\delta^{13}C$ in dry mass has been developed as a method for estimating seasonal WUE_{plant} (Farquhar and Richards, 1984; Condon et al., 2004). Generally, positive relationships between $\delta^{13}C$ and grain yield are mostly found under moderate to well-watered conditions, whereas for severely stressed environments (low yields) such relationships are absent or negative (Voltas et al., 1999; Araus et al., 2003). In this stressful condition, lower $\delta^{13}C$ (i.e., higher WUE_{inst}) is associated with higher grain yield. Thus conflicting results on the association between $\delta^{13}C$ and yield in various crops under different growing conditions and the expense and feasibility of this technique in large segregating breeding populations hinder its use.

An attempt to ameliorate the divergence between scales is the use of two other WUE surrogate characters. On the one hand, specific leaf area (SLA), which was demonstrated to be closely and negatively correlated with WUE, although sometimes influenced by factors such as time of sampling and leaf age (Wright et al., 1988, Wright and Hammer, 1994; Nageswara Rao Wright, 1994). Significant and high correlations between SLA and specific leaf nitrogen (Nageswara Rao and Wright, 1994) and SLA and ribulose 1,5-bisphosphate carboxylase (Nageswara Rao et al., 1995) suggested that photosynthetic capacity per unit leaf area is one of the major contributors to variations in WUE. On the other hand, a handheld portable chlorophyll meter (SPAD), which measures chlorophyll content for rapidly assessing drought tolerance in plants, was observed to be significantly and positively correlated with SLA (Marenco et al., 2009), suggesting an achievement of a rapid and reliable method to identify genotypes with low SLA and hence high transpiration efficiency. Thus understanding the genetic nature of both surrogate characters and its possible association with yield parameters enables their deployment in large-scale screening programs for improved drought resistance.

The most appropriate ways to scale up in WUE and to better estimate WUE at different levels have been frequently discussed (Fig. 16.1) (Escalona et al., 2003; Tomás et al., 2014a,b; Medrano et al., 2015a,b); however, the

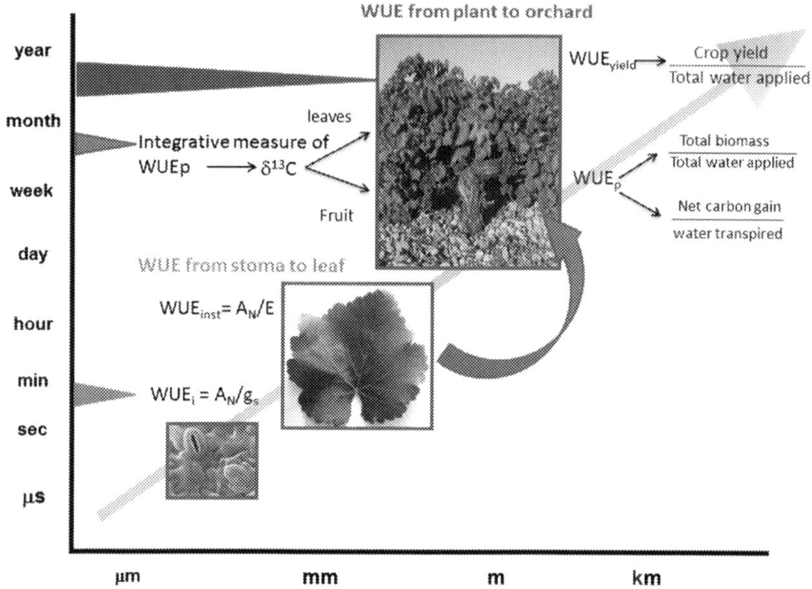

Figure 16.1 Water use efficiency (WUE) defined as the balance between carbon gain and the associated cost of water, measured and/or estimated at different spatial and temporal scales, from leaf scale to orchard.

decision mostly depends on the capacity, facilities, and scale of the specific study. The use of whole-plant chambers is taking place in many studies trying to investigate plant physiological processes on an intermediate scale between leaf-scale cuvette measurements and landscape-scale eddy covariance measurements. Thus two general approaches to whole-canopy gas exchange measurements are enclosure (i.e., whole-canopy chambers) (Poni et al., 1997) and meteorological methods (i.e., flux-gradient or eddy correlation) (Field and Davis, 1989; Garcia et al., 1990; Long and Hällgren, 1989). However, scaling-up from measuring single-leaf gas exchange to whole-canopy gas exchange is not straightforward because the latter includes leaves of different ages and degrees of light exposure, and integrates organs such as fruits, shoots, and trunks. Indeed, to date, a lack of good correlation between single-leaf and whole-vine measurements has been frequently reported because single-leaf measurements have been overestimated (Edson et al., 1993; Intrieri et al., 1997; Poni et al., 1997). Moreover, plant respiration is an important part of the carbon economy and final dry mass accumulation (Gifford, 1995; Escalona et al., 2012), so that whole-plant net carbon gain results are difficult to apply for the evaluation of large genotype collections.

4. VARIABILITY BETWEEN AND WITHIN CULTIVARS. THE CASE OF TEMPRANILLO CLONES

Grapevine crop is widely distributed around the world with thousands of cultivars, and thus it is well adapted to a large range of environmental conditions. Because of this variability, it is recognized as having important genotypic plasticity, so that grapevines are good candidates to test in breeding programs (Reynolds, 2015), which are usually focused on improved plant yield or quality characteristics. As a consequence, new cultivars are being registered around the world every year for this crop, even though most of them are related to grapevine as fresh fruit crops. Breeding usually consists in hybridization between two or more commercial varieties, allowing specific selection according to a certain selected criteria. For example, in southern Spain, Monastrell cv. is well adapted to dry and hot conditions and is being hybridized with Cabernet and Syrah to improve its quality under such semiarid conditions (Bayo-Canha et al., 2012). Genetic variability was reported in different studies showing important differences in WUE_i between grapevine cultivars (Bota et al., 2016; Souza et al., 2005; Soar et al., 2006; Pou et al., 2008; Prieto et al., 2010; Costa et al., 2012; Tomás et al., 2014a,b) under both controlled and field conditions, thus showing that selection programs for this character could be successful.

However, for most of the regions producing reputed wines, there are commercial protection figures such as in France with VQPRD (Wines of Quality Produced in Determined Region, in French), which avoid or limit the introduction of new cultivars in a given region. In consequence, the only way to improve genetically particular cultivars is to look at intra-cultivar variability. For this purpose, clonal variability can be a good alternative (Ibáñez et al., 2015).

4.1 Clonal Selection

Clonal selection was used in the past century with success, for instance, for improving crop production or for conferring on some clones resistance to certain biotic stresses, such as downy mildew and powdery mildew (Reynolds, 2015). More recently, variability of quality traits has also been a focus of interest for clonal selection programs, and specific characters related with quality parameters have been shown to express considerable genetic variation. For example, Sá Borges et al. (2013) found differences in grape juice anthocyanin and tannin composition in six Concord cv. clones.

Fidelibus et al. (2006) showed a difference in yield and grape quality traits in six clones of Merlot cv. Muñoz et al. (2014) showed the existence of clonal variation for the anthocyanin profiles among 131 Malbec cv. clones, and demonstrated the putative involvement of genes related to hydroxylation, methylation, and transport of anthocyanins.

In parallel, genetic viability inside one cultivar at the molecular level has been observed. For example, using simple sequence repeat markers, Hocquigny et al. (2003) found differences in 145 Pinot Noir clones and Cervera et al. (2002), using amplified fragment length polymorphism markers, found variability in 32 Tempranillo clones, among others.

As a consequence, there is clear evidence that there is wide variability among clones at both physiological and genetic levels. This strongly suggests that variability for WUE will also be present for different clones of a certain cultivar.

4.2 The Case of Tempranillo Clones

The cultivar Tempranillo is the second most cultivated variety in Spain and is gaining interest around the world and the one with older references about its use for premium wines in Spain (Valcárcel, 1791). Nowadays, Tempranillo distribution in Spain exceeds 200,000 ha (Ministerio de Agricultura, Alimentación y Medio Ambiente), being the second most planted cultivar in this country and the one with the highest cropping area. Tempranillo is a preferential or authorized cultivar in 56 out of the 72 Spanish Denominations of Origin and it is officially authorized not only in Spain, but also in Portugal, France, Italy, Greece, Malta, and Cyprus (Lacombe, 2011). This variety is thought to originate from two old varieties with minor distribution, Albillo Mayor and Benedicto de Aragón (Ibáñez et al., 2012) of northern Spain, and presents around 50 commercial clones, which show important differences in agronomical and quality traits.

The measurement of WUE at the leaf level showed considerable variability among a collection of 30 clones measured under field conditions (Tortosa et al., 2016). Under similar conditions, the intercultivar variability for this character was clearly higher, but, nevertheless, clonal variability achieved around 80% of that found between the 23 cultivars (Tortosa et al., 2016). This means an important basis for selection programs for this character.

According to Tortosa et al. (2016), several steps were needed to correctly compare WUE_i between genotypes. Because of the large collection to be measured under field conditions, and the diurnal time variation of environmental conditions (particularly VPD) during the hot and dry summer days, the effect of measuring time (from 10:30 a.m. to 13 p.m.) can result in a serious interference for the correct comparison of genotypes. Therefore the first important step was to ensure the absence of any effect of the measurement time. For this, we checked the changes in WUE_i and WUE_{inst} between 10:30 a.m and 13 p.m.). For A_N/E we observed an important time effect, which came mainly from variations in VPD. Instead, the A_N/g_s was to be fully independent from the measuring time variations (Rogiers et al., 2011).

Because of the presumed variation in soil water content of the experimental fields, a wide range of g_s and stem water potential was found and, as is widely reported, there is an important effect of soil humidity on g_s. In consequence, to compare the WUE_i data we arranged three categories based on their respective g_s values according to Medrano et al. (2002): plants under nonwater stress conditions ($g_s > 0.150$ mol H_2O m^{-2} s^{-1}), plants under moderate water stress conditions (g_s between 0.150 and 0.075 mol H_2O m^{-2} s^{-1}), and plants under severe water stress conditions ($g_s < 0.075$ mol H_2O m^{-2} s^{-1}). This classification allowed a comparison between WUE_i and each one of the genotypes in the same range of g_s (Fig. 16.2).

In consequence, to compare WUE_i under different soil water availability conditions we plotted WUE_i against g_s and obtained a regression line of very high r^2 coefficient (0.87), which expresses the known dependency of WUE_i from g_s. To evaluate the performance in WUE_i between genotypes at a given g_s, we calculated the residuals of this relationship associated with each genotype. This evaluation consists of comparing the measured WUE_i of a given genotype with the one predicted by the regression line for a given measured g_s. To compare all the genotypes between themselves we decided to calculate their respective deviations as a percentage. For example, a given genotype with a positive high deviation will have a larger WUE_i compared to the others. Considering the overall range of g_s measured, the total genotypic variation was between +18% and −16%.

Moreover, three genotypes were compared at different water stress levels (Table 16.1). Statistical differences in WUE_i at the same genotype (RJ26) among different water stress levels were found (*P* value <.001). These results showed the high influence of g_s to WUE_i. Furthermore, the RJ26 clone showed results clearly above the RJ43 clone in well-watered conditions (*P* value <.0001), but this difference was not significant under

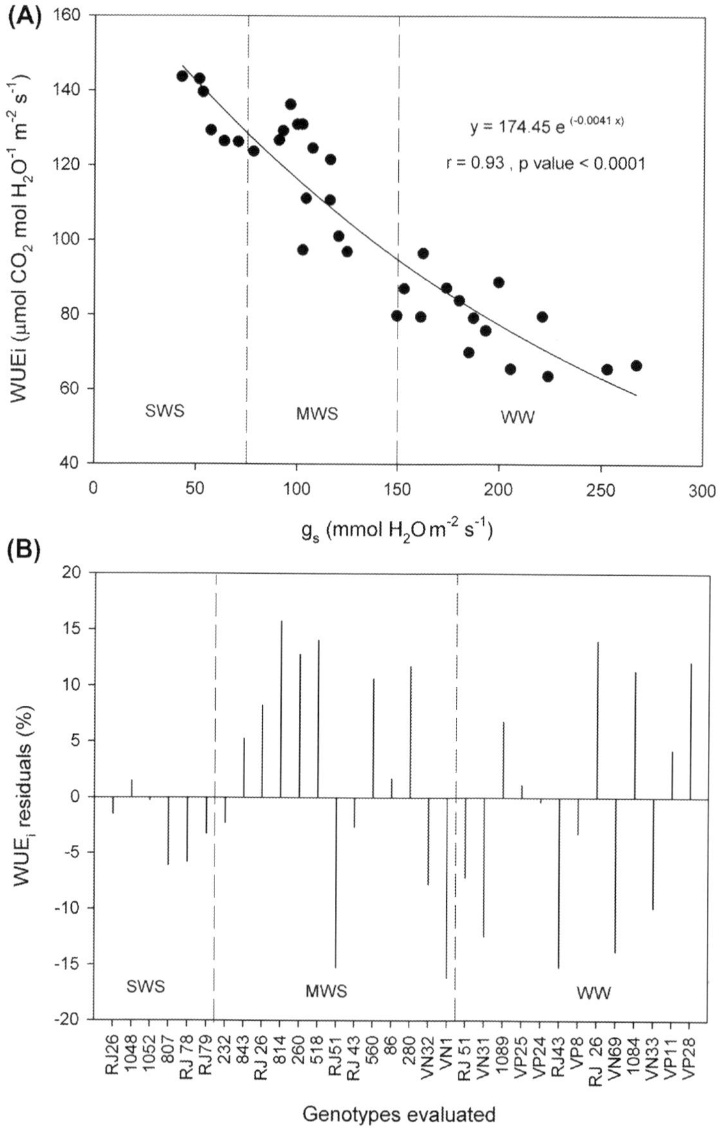

Figure 16.2 Correlations between intrinsic water use efficiency (WUE_i) and stomatal conductance (g_s) for the measured clones of Tempranillo variety (A) and residuals (observed−expected values) for each genotype expressed as a percentage (B). *Dashed lines* show the results grouped by water status. *MWS*, moderate water stress; *SWS*, severe water stress; *WW*, well watered.

Table 16.1 Genotypes measured at different stress levels

		RJ43	RJ26	RJ78
SWS	g_s (mmol H_2O $m^{-2}s^{-1}$)		42.9 ± 20.8	63.4 ± 17.8
	A_N (μmol CO_2 m^{-2} s^{-1})		6.0 ± 2.5	8.0 ± 2.2
	E (mmol H_2O m^{-2} s^{-1})		2.8 ± 1.1	3.3 ± 0.8
	A/E (g C L^{-1} H_2O)		2.1 ± 0.3	2.4 ± 0.2
	A/g_s (μmol CO_2 $mmol^{-1}$ H_2O)		143.6 ± 15.5^{A}	126.5 ± 14.5
MWS	g_s (mmol H_2O $m^{-2}s^{-1}$)	104.1 ± 20.8	92.6 ± 29.1	120.2 ± 37.6
	A_N (μmol CO_2 m^{-2} s^{-1})	11.4 ± 1.4	11.4 ± 1.5	11.8 ± 2.2
	E (mmol H_2O m^{-2} s^{-1})	3.8 ± 1.8	4.0 ± 0.4	3.2 ± 0.7
	A/E (g C L^{-1} H_2O)	3.5 ± 1.3	2.8 ± 0.2	3.7 ± 0.2
	A/g_s (μmol CO_2 $mmol^{-1}$ H_2O)	111.1 ± 11.5	129.3 ± 23.6^{B}	101.1 ± 18.6
WW	g_s (mmol H_2O $m^{-2}s^{-1}$)	185.1 ± 35.1	199.5 ± 22.9	
	A_N (μmol CO_2 m^{-2} s^{-1})	12.8 ± 1.4^{b}	17.7 ± 1.7^{a}	
	E (mmol H_2O m^{-2} s^{-1})	5.7 ± 1.9	4.9 ± 0.5	
	A/E (g C L^{-1} H_2O)	2.5 ± 0.8^{b}	3.6 ± 0.1^{a}	
	A/g_s (μmol CO_2 $mmol^{-1}$ H_2O)	70.0 ± 6.8^{b}	88.8 ± 2.3^{aC}	

A, B, and C indicate statistical significance (Tukey test at $P = .05$ level) within the same column, and a and b indicate statistical differences between genotypes.
MWS, moderate water stress; *SWS*, severe water stress; *WW*, well watered.

moderate water stress. These results support the importance of comparing WUE_i values from plants under similar water availability.

The previous results showed that is possible to identify genotypes by higher and lower WUE_i than the average of the cultivar. Nevertheless, it will be necessary to compare those field data in more controlled experiments with several replications for each genotype to confirm the genotype ranking recorded under field conditions.

5. NEW TOOLS FOR WIDE GENOTYPE NUMBER SELECTION PROGRAMS. PERSPECTIVES FOR NEAR-FUTURE PROGRAMS

5.1 Breeding for Future Viticulture; Water Use Efficiency as Selection Target

The described changes in viticulture worldwide regions due to climate change associated with plant growth, yield and quality, and pest and disease issues are likely to continue in the future (Bois et al., 2016). In this changing environment, the selection targets for breeding programs should be redefined. This is imperative to address the selection and breeding programs to identify genotypes more adapted to warm and dry climates to overcome those future climatic changes. Traditional clonal selection or cross-breeding can help to obtain improved abiotic resistance traits combined with high quality. In this sense, drought tolerance or WUE are good candidates for selection targets. Thousands of grapevine cultivars have been described around the world (OIV, 2009; This et al., 2006), showing an impressive genetic variability and plasticity of the grapevine genome. Moreover, new hybrid grape varieties obtained by crossing are also available. Among all this genetic variability, the old and minor cultivars can also be good candidates, as only with well-adapted new cultivars can the threats of the future be overcome.

This variability offers an invaluable genetic resource to cope with crop adaptation to the different environmental conditions and potentially to climate change. There is no reference to the use of WUE as selection criteria, but there are cultivars reputed as more adapted to drought-prone conditions, which presumably should also present high WUE. Water stress response can be related to root and leaf variables. Since grapes are normally grafted, both rootstock and scion can be objective to drought tolerance improvement. Then, breeding for scion varieties can only take into account the leaf-related variables. Many studies performed at the leaf level have shown a wide range of intraspecific variability of WUE_i among cultivars (Bota et al., 2001; Costa et al., 2012; Gaudillère et al., 2002; Koundouras et al., 2008; Pou et al., 2008; Prieto et al., 2010; Rogiers et al., 2011; Schultz, 2003; Souza et al., 2005; Tomás et al., 2012, 2014b; Zsófi et al., 2009). This variability is mostly associated with different g_s responses (Tomás et al., 2014a,b). However, attempts to correlate values of stomatal conductance or WUE_i with certain genetic markers is not an easy task, because these variables are simultaneously affected by a number of environmental variables. Advances in molecular and genomic studies have provided valuable information on response diversity to those stresses, response at the molecular level, and identification of key genetics involved.

The analysis of gene expression has provided a rich source of biological information, which allows breeders to understand the molecular basis of complex plant processes, leading to the identification of new targets for manipulating these processes (Perez de Castro et al., 2012). Some studies have used large-scale gene expression analysis to study the water stress response of grapevine to improve the knowledge of transcriptional regulation during drought (Cramer et al., 2007; Deluc et al., 2009; Rocheta et al., 2014; Berdeja et al., 2015; Corso et al., 2015). In addition, reports have demonstrated changes in the metabolome in response to drought and data assessing the effect of drought in grapevines are now available (Deluc et al., 2009; Conde et al., 2015). Table 16.2 shows the different publications concerning omics studies related to water stress and/or drought tolerance in grapevines.

Among the different physiological responses to drought, stomatal control is a key issue in drought tolerance, and WUE_i in grapevine is primarily mediated by changes in g_s (Chaves et al., 2010; Flexas et al., 2010). A study of genome-wide transcriptomic comparison of different stomatal responses under water stress (iso- or anisohydric) revealed relevant genotype response (Dal Santo et al., 2016). Advances in next-generation sequencing technology and the different omics approaches can contribute to increase this knowledge and are of prime interest for applying knowledge-based breeding as marker-assisted selection (MAS). MAS is an indirect process where selection is carried out based on a marker instead of the trait itself. Techniques such as RNA-Seq enable gene discovery to be combined with the identification of allelic variation, and sequence variants associated with differentially expressed genes in response to a treatment can be identified and used in advanced breeding programs (Perez de Castro et al., 2012; Hübner et al., 2015). On the other hand, the metabolic plasticity, synthesis,

Table 16.2 Omics studies related to water stress and/or drought tolerance in grapevine

Material	Type of omics	Technique used	Trait	References
Shoot tips, *Vitis vinifera* L.	Transcriptomics, metabolomics	Microarray, GC-MS	Water stress, salt stress	Cramer et al. (2007)
Berry, *V. vinifera* L.	Transcriptomics, metabolomics	Microarray, GC-MS	Water stress, berry metabolism	Deluc et al. (2009)
Leaf, *V. vinifera* L.	Transcriptomics	Microarray	Water stress	Perrone et al. (2012)
Leaf, *V. vinifera* L.	Transcriptomics	Microarray	Heat and water stress	Rocheta el al. (2014)
Berry, *V. vinifera* L. on different rootstock	Transcriptomics	Microarray	Water stress, primary and secondary metabolism	Berdeja et al. (2015)
Leaf, Berry, *V. vinifera* L.	Transcriptomics	Microarray	Water stress, stomatal behavior	Dal Santo et al. (2016)
Leaf, root, *V. vinifera* L.	Transcriptomics	RNA-seq	Water stress, secondary metabolite (polyphenolic compound)	Corso et al. (2015)
Leaf, berry, *V. vinifera* L.	Metabolomics	GC-MS	Water stress	Conde et al. (2015)
Leaf, *V. vinifera* L.	Metabolomics	GC-MS, LC-MS	Water stress, primary metabolite, secondary metabolite (polyphenolic compound, volatile compound)	Griesser et al. (2015)

GC-MS, gas chromatography-mass spectrometry; *LC-MS*, liquid chromatography coupled to mass spectrometry.

and accumulation of primary and secondary metabolites during water stress offer considerable promise as biomarkers used as selective traits for plant adaptation to future environmental constraints (Conde et al., 2015; Griesser et al., 2015).

6. CONCLUSIONS

The genus *Vitis* shows an impressive genetic variability for many agronomic characteristics, which is important for present and future selection programs.

This variability is also present when selection is being done inside each cultivar to identify a "clone" with determined characteristics such as productivity, fruit quality, or disease resistance.

Grapevine WUE is a subject of particular interest for a wide number of viticulture regions located in semiarid areas. This interest is expected to increase in the near future due to climatic change predictions.

Moreover, even though the precise determination of WUE is still difficult, recent work suggests that there is enough WUE variability to be exploited inside the Tempranillo cultivar.

These results and the increasing progress of omics enable advancement of efforts to improve WUE in commercial grapevine cultivars.

ACKNOWLEDGMENTS

This work was performed with financial support from the Spanish Ministry of Science and Technology (project AGL2014-30408-C04-01), from INIA (project RTA2013-00068-C03-03 and 652615 ERA-NET), and a predoctoral fellowship BES-2015-073331. We would like to thank Mr. Miquel Truyols and collaborators of the UIB Experimental Field and Greenhouses, who are supported by the UIBGrant 15/2015.

REFERENCES

Acevedo, E.H., Silva, P.C., Silva, H.R., Solar, B.R., 1999. Wheat production in mediterranean environments. In: Satorre, E.H., Slafer, G.A. (Eds.), Wheat: Ecology and Physiology of Yield Determination. Food Products Press, New York, pp. 295–323.

Alleweldt, G., Dettweiler, E., 1994. In: Geilweilerhof (Ed.), The Genetic Resources of Vitis: World List of Grapevine Collections, second ed.

Alleweldt, G., Possingham, J.V., 1988. Progress in grapevine breeding. Theoretical Applied Genetics 75, 669–673.

Alsina, M.M., Smart, D.R., Bauerle, T., De Herralde, F., Biel, C., Stockert, C., Save, R., 2011. Seasonal changes of whole root system conductance by a drought-tolerant grape root system. Environmental and Experimental Botany 62, 99–109.

Araus, J.L., Villegas, D., Aparicio, N., 2003. Environmental factors determining carbon isotope discrimination and yield in durum wheat under Mediterranean conditions. Crop Science 43, 170–180.

Arroyo-García, R., Ruiz-García, L., Bolling, L., Ocete, R., Lopez, M.A., Arnold, C., Aradhya, M.K., Ergul, E., Söylemezo, G., Uzun, H.I., Cabello, F., Ibáñez, J., Aradhya, M.K., Atanassov, A., Aranassov, I., Balint, S., Cenis, J.L., Costantini, L., Gorislavets, S., Grando, M.S., Klein, B.Y., McGovern, P.E., Merdinoglu, D., Pejic, I., Pelsy, F., Primikiros, N., Risovannya, V., Roubelakis-Angelakis, K.A., Snoussi, H., Sotiri, P., Tamhankar, S., This, P., Troshin, L., Malpica, J.M., Lefort, F., Martínez-Zapater, J.M., 2006. Multiple origins of cultivated grapevine (*Vitis vinifera* L. ssp. sativa) based on chloroplast DNA polymorphisms. Molecular Ecology 15, 3707–3714.

Bayo-Canha, A., Fernández-Fernández, J.I., Martínez-Cutillas, A., Ruiz-García, L., 2012. Phenotypic segregation and relationships of agronomic traits in Monastrell × Syrah wine grape progeny. Euphytica 186 (2), 393–407.

Bchir, A., Escalona, J.M., Gallé, A., Hernández-Montes, E., Tortosa, I., Braham, M., Medrano, H., 2016. Carbon isotope discrimination (δ 13 C) as an indicator of vine water status and water use efficiency (WUE): looking for the most representative sample and sampling time. Agricultural Water Management 167, 11–20.

Berdeja, M., Nicolas, P., Kappel, C., Dai, Z.W., Hilbert, G., Peccoux, A., Lafontaine, M., Ollat, N., Gomès, E., Delrot, S., 2015. Water limitation and rootstock genotype interact to alter grape berry metabolism through transcriptome reprogramming. Horticulture Research 2, 15012.

Biswajit, D., Ahmed, N., Pushkar, S., 2011. Prunus diversity-early and present development: a review. International Journal of Biodiversity and Conservation 14, 721–734.

Bois, B., Gavrilescu, C., Moriondo, M., Jones, G.V., July 2016. Wine growing regions global climate analysis. In: Van Jones, Gregory, Doran, Nicole (Eds.), Proceedings of the 11th International Terroir Congress, pp. 9–14.

Bota, J., Flexas, J., Medrano, H., 2001. Genetic variability of photosynthesis and water use in Balearic grapevine cultivars. Annals of Applied Biology 138 (3), 353–361.

Bota, J., Tomás, M., Flexas, J., Medrano, H., Escalona, J.M., 2016. Differences among grapevine cultivars in their stomatal behavior and water use efficiency under progressive water stress. Agricultural Water Management 164, 91–99.

Buhner-Zaharieva, T., Moussaoui, S., Lorente, M., Andreu, J., Núñez, R., Ortiz, J.M., Gogorcena, Y., 2010. Preservation and molecular characterization of ancient varieties in Spanish grapevine germplasm collections. American Journal of Enology and Viticulture 61, 557–562.

Cervera, M.T., Cabezas, J.A., Rodríguez-Torres, I., Chavez, J., Cabello, F., Martínez-Zapater, J.M., 2002. Varietal diversity within grapevine accessions of cv. Tempranillo. Vitis – Journal of Grapevine Research 41, 33–36.

Chaves, M.M., Santos, T.P., Souza, C.R., Ortuño, D., Rodrigues, M.L., Lopes, C.M., Maroco, J.P., Pereira, J.S., 2007. Deficit irrigation in grapevine improves water-use efficiency while controlling vigour and production quality. Annals of Applied Biology 150 (2), 237–252.

Chaves, M.M., Zarrouk, O., Francisco, R., Costa, J.M., Santos, T., Regalado, A.P., Rodríguez, M.L., Lopes, C.M., 2010. Grapevine under deficit irrigation: hints from physiological and molecular data. Annals of Botany 105 (5), 661–676.

Conde, A., Regalado, A., Rodrigues, D., Costa, J.M., Blumwald, E., Chaves, M.M., Gerós, H., 2015. Polyols in grape berry: transport and metabolic adjustments as a physiological strategy for water deficit stress tolerance in grapevine. Journal of Experimental Botany 66 (3), 889–906.

Condon, A.G., Richards, R.A., Rebetzke, G.J., Farquhar, G.D., 2004. Breeding for high water-use efficiency. Journal of Experimental Botany 55, 2447–2460.

Corso, M., Vannozzi, A., Maza, E., Vitulo, N., Meggio, F., Pitacco, A., Telatin, A., D'Angelo, M., Feltrin, E., Negri, A.S., Prinsi, B., Valle, G., Ramina, A., Bouzayen, M., Bonghi, C., Lucchin, M., 2015. Comprehensive transcript profiling of two grapevine rootstock genotypes contrasting in drought susceptibility links the phenylpropanoid pathway to enhanced tolerance. Journal of Experimental Botany 66, 5739–5752.

Costa, J.M., Ortuño, M.F., Lopes, C.M., Chaves, M.M., 2012. Grapevine varieties exhibiting differences in stomatal response to water deficit. Functional Plant Biology 39 (3), 179–189.

Costa, J.M., Vaz, M., Escalona, J.M., Egipto, R., Lopes, C., Medrano, H., Chaves, M.M., 2016. Modern viticulture in southern Europe: vulnerabilities and strategies for adaptation to water scarcity. Agricultural Water Management 164, 5–18.

Cramer, G.R., Ergul, A., Grimplet, J., Tillett, R.L., Tattersall, E.A.R., Bohlman, M.C., Vincent, D., Sonderegger, J., Evans, J., Osborne, C., 2007. Water and salinity stress in grapevines: early and late changes in transcript and metabolite profiles. Functional and Integrative Genomics 7 (2), 111–134.

Dal Santo, S., Palliotti, A., Zenoni, S., Tornielli, G.B., Fasoli, M., Paci, P., d'Onofrio, C., 2016. Distinct transcriptome responses to water limitation in isohydric and anisohydric grapevine cultivars. BMC Genomics 17 (1), 815.

Deluc, L.G., Quilici, D.R., Decendit, A., Grimplet, J., Wheatley, M.D., Schlauch, K.A., Merillon, J.M., Cushman, J.C., Cramer, G.R., 2009. Water deficit alters differentially metabolic pathways affecting important flavor and quality traits in grape berries of Cabernet Sauvignon and Chardonnay. BMC Genomics 10, 212.

Edson, C.E., Howell, G.S., Flore, J.A., 1993. Influence of crop load on photosynthesis and dry matter partitioning of Seyval grapevines. I. Single leaf and whole-vine response pre- and post-harvest. American Journal of Enology and Viticulture 44, 139–147.

Escalona, J.M., Flexas, J., Bota, J., Medrano, H., 2003. Distribution of leaf photosynthesis and transpiration within grapevine canopies under different drought conditions. Vitis – Journal of Grapevine Research 42 (2), 57–64.

Escalona, J.M., Tomás, M., Martorell, S., Medrano, H., Ribas-Carbo, M., Flexas, J., 2012. Carbon balance in grapevines under different soil water supply: importance of whole plant respiration. Australian Journal of Grape and Wine Research 18 (3), 308–318.

Escalona, J.M., Luna, J.M., Bota, J., Garau, C., Martorell, A., 2017. Las Variedades de vid de las Islas Baleares. In: Conselleria de Medi Ambient, Agricultura i Pesca CAIB.

Farquhar, G.D., Richards, R.A., 1984. Isotopic composition of plant carbon correlates with water-use efficiency of wheat genotypes. Australian Journal of Plant Physiology 11, 539–552.

Fidelibus, M.W., Christensen, L.P., Katayama, D.G., Verdenal, P.T., 2006. Yield components and fruit composition of six Chardonnay grapevine clones in the Central San Joaquin Valley, California. American Journal of Enology and Viticulture 57 (4), 503–506.

Field, C.B., Davis, S.D., 1989. Physiological ecology. The California chaparral: paradigms reexamined. Science Series 34, 154–164.

Flexas, J., Galmés, J., Gallé, A., Gulías, J., Pou, A., Ribas-Carbó, M., Tomás, M., Medrano, H., 2010. Improving water use efficiency in grapevines: potential physiological targets for biotechnological improvement. Australian Journal of Grape and Wine Research 16 (s1), 106–121.

Galet, P., 2000. In: Dictionnaire encyclopédique des cépages. Hachette Livre, p. 935.

Garcia, R., Norman, J.M., McDErmitt, D.K., 1990. Measurements of canopy gas exchange using an open chamber system. Remote Sensing Reviews 5, 141–162.

Gaudillère, J.P., Van Leeuwen, C., Ollat, N., 2002. Carbon isotope composition of sugars in grapevine, an integrated indicator of vineyard water status. Journal of Experimental Botany 53 (369), 757–763.

Gibberd, M.R., Walker, R.R., Blackmore, D.H., Condon, A.G., 2001. Transpiration efficiency and carbon-isotope discrimination of grapevines grown under well-watered conditions in either glasshouse or vineyard. Australian Journal of Grape and Wine Research 7 (3), 110–117.

Gifford, R.M., 1995. Whole plant respiration and photosynthesis of wheat under increased CO_2 concentration and temperature: long-term vs. short term distinctions for modelling. Global Change Biology 1, 385–396.

Griesser, M., Weingart, G., Schoedl-Hummel, K., Neumann, N., Becker, M., Varmuza, K., Liebner, F., Schuhmacher, R., Forneck, A., 2015. Severe drought stress is affecting selected primary metabolites, polyphenols, and volatile metabolites in grapevine leaves (*Vitis vinifera* cv. Pinot noir). Plant Physiology and Biochemistry 88, 17–26.

Hannah, L., Roehrdanz, P.R., Ikegami, M., Shepard, A.V., Shaw, M.R., Tabor, G., Zhi, L., Marquet, P.A., Hijmans, R.J., 2013. Climate change, wine, and conservation. Proceedings of the National Academy of Sciences 110 (17), 6907–6912.

Herrero, R., Asins, M.J., Carbonell, A.E., Navarro, L., 1996. Genetic diversity in the orange subfamily Aurantioideae. I. Intraspecies and intragenus genetic variability. Theoretical and Applied Genetics 92, 599–609.

Hocquigny, S., Merdinoglu, D., Heloir, M.C., Relsy, F., 2003. Chimerism and genetic diversity within the cultivar group of Pinots. Acta Horticulturae 603, 535–544.

Hübner, S., Korol, A.B., Schmid, K.J., 2015. RNA-Seq analysis identifies genes associated with differential reproductive success under drought-stress in accessions of wild barley Hordeum spontaneum. BMC Plant Biology 15, 134.

Ibáñez, J., Muñoz-Organero, G., Zinelabidine, L.H., de Andrés, M.T., Cabello, F., Martínez-Zapater, J.M., 2012. Genetic origin of the grapevine cultivar Tempranillo. American Journal of Enology and Viticulture 63 (4), 549–553.

Ibáñez, J., Carreño, J., Yuste, J., Martínez-Zapater, J.M., 2015. Grapevine breeding and clonal selection programmes in Spain. In: Grapevine Breeding Programs for the Wine Industry, first ed. Woodhead Publishing, Cambridge, pp. 183–209.

Intrieri, C., Poni, S., Rebucci, B., Magnanini, E., 1997. Effects of canopy manipulations on whole-vine photosynthesis: results from pot and field experiments. Vitis – Journal of Grapevine Research 36, 167–173.

Change, IPCC Climate, 2013. The Physical Science Basis. Contribution of Working Group I to the Fifth Assessment Report of the Intergovernmental Panel on Climate Change, p. 1535.

Jacobs, S.D., 2010. Effect of rootstock and water stress on gas exchange, water relations, and water-use efficiency in Petite Sirah grapevines (Doctoral dissertation). California State University.

Jones, H.G., Vaughan, R.A., 2010. Remote Sensing of Vegetation: Principles, Techniques, and Applications. Oxford University Press, Oxford.

Köppen, E., Wladimir, S., 1884. Die Wärmezonen der Erde, nach der Dauer der heissen, gemässigten und kalten Zeit und nach der Wirkung der Wärme auf die organische Welt betrachtet (The thermal zones of the earth according to the duration of hot, moderate and cold periods and to the impact of heat on the organic world). Translated by Volken, E, Brönnimann, S Meteorologische Zeitschrift 20 (3), 351–360.

Koundouras, S., Tsialtas, I.T., Zioziou, E., Nikolaou, N., 2008. Rootstock effects on the adaptive strategies of grapevine (*Vitis vinifera* L. cv. Cabernet–Sauvignon) under contrasting water status: leaf physiological and structural responses. Agriculture, Ecosystems and Environment 128 (1), 86–96.

Lacombe, T., 2011. Grapevine European catalogue: towards a comprehensive list. Vitis – Journal of Grapevine Research 50, 65–68.

Long, S.P., Hällgren, J.E., 1989. Measurement of CO_2 assimilation by plants in the field and the laboratory. In: Coombs, J., Hall, D.O. (Eds.), Techniques in Bioproductivity and Photosynthesis. Chapman and Hall Ltd, London, pp. 62–94.

Marenco, R.A., Antezana-Vera, S.A., Nascimento, H.C.S., 2009. Relationship between specific leaf area, leaf thickness, leaf water content and SPAD-502 readings in six Amazonian tree species. Photosynthetica 47, 184–190.
Marguerit, E., Brendel, O., Lebon, E., Van Leeuwen, C., Ollat, N., 2012. Rootstock control of scion transpiration and its acclimation to water deficit are controlled by different genes. New Phytologist 194 (2), 416–429.
Martínez de Toda, F., Sancha, J.C., 1997. Ampelographical characterization of red *Vitis vinifera* L. cultivars preserved in Rioja. International Organisation of Vine and Wine 70, 220–234.
Mc Govern, P.E., 2003. Ancient Wine: The Search for the Origins of Viniculture. Press, Princeton, p. 400.
Medrano, H., Pou, A., Tomás, M., Martorell, S., Gulias, J., Flexas, J., Escalona, J.M., 2012. Average daily light interception determines leaf water use efficiency among different canopy locations in grapevine. Agricultural Water Management 114, 4–10.
Medrano, H., Tomàs, M., Martorell, S., Escalona, J.M., Pou, A., Fuentes, S., Flexas, J., Bota, J., 2015a. Improving water use efficiency of vineyards in semi-arid regions. A review. Agronomy for Sustainable Development 35, 499–517.
Medrano, H., Tomás, M., Martorell, S., Flexas, J., Hernández, E., Rosselló, J., Pou, A., Escalona, J.M., Bota, J., 2015b. From leaf to whole plant water use efficiency (WUE) in complex canopies: limitations of leaf WUE as selection target. Crop Journal 3, 220–228.
Medrano, H., Escalona, J.M., Bota, J., Gulias, J., Flexas, J., 2002. Regulation of photosynthesis of C3 plants in response to progressive drought: stomatal conductance as a reference parameter. Annals of Botany 89 (7), 895–905.
Muñoz, C., Gomez-Talquenca, S., Chialva, C., Ibáñez, J., Martinez-Zapater, J.M., Peña-Neira, A., Lijavetzky, D., 2014. Relationships among gene expression and anthocyanin composition of Malbec grapevine clones. Journal of Agricultural and Food Chemistry 62 (28), 6716–6725.
Myles, S., Boyko, A.R., Owens, C.L., Brown, P.J., Grassi, F., Aradhya, M.K., Prins, B., Reynolds, A., Chia, J.M., Ware, D., Bustamante, C.D., Bucklera, E.S., 2011. Genetic structure and domestication history of the grape. Proceedings of the National Academy of Sciences 108, 3530–3535.
Nageswara Rao, R.C., Wright, G.C., 1994. Stability of the relationship between specific leaf area and carbon isotope discrimination across environments in peanut. Crop Science 34, 98–103.
Nageswara Rao, R.C., Udayakumar, M., Farquhar, G.D., Talwar, H.S., Prasad, T.G., 1995. Variation in carbon isotope discrimination and its relationship to specific leaf area and ribulose-1,5-bisphosphate carboxylase content in groundnut genotypes. Australian Journal of Plant Physiology 22, 545–551.
OIV, 2009. Descriptor List for Grape Varieties and Vitis Species. OIV Publications, Paris.
Pelsy, F., 2010. Molecular and cellular mechanisms of diversity within grapevine varieties. Heredity 104 (4), 331–340.
Perez-de-Castro, A.M., Vilanova, S., Canizares, J., Pascual, L., Blanca, J.M., Diez, M.J., 2012. Application of genomic tools in plant breeding. Current Genomics 13, 179–195.
Perrone, I., Pagliarani, C., Lovisolo, C., Chitarra, W., Roman, F., Schubert, A., 2012. Recovery from water stress affects grape leaf petiole transcriptome. Planta 235 (6), 1383–1396.
Poni, S., Magnanini, E., Rebucci, B., 1997. An automated chamber system for measurements of whole-vine gas exchange. Horticultural Science 32, 64–67.
Poni, S., Bernizzoni, F., Civardi, S., Gatti, M., Porro, D., Camin, F., 2009. Performance and water-use efficiency (single-leaf vs. whole-canopy) of well-watered and half-stressed

split-root Lambrusco grapevines grown in Po Valley (Italy). Agriculture, Ecosystems and Environment 129, 97–106.

Pou, A., Flexas, J., Alsina, M.D.M., Bota, J., Carambula, C., De Herralde, F., Galmés, J., Lovisolo, C., Jiménes, M., Ribás-Carbó, M., Rusjan, D., Secchi, F., Tomás, M., Zsófi, Z., Medrano, H., 2008. Adjustments of water use efficiency by stomatal regulation during drought and recovery in the drought-adapted Vitis hybrid Richter-110 (V. berlandieri × V. rupestris). Physiologia Plantarum 134 (2), 313–323.

Prieto, J.A., Lebon, É., Ojeda, H., 2010. Stomatal behavior of different grapevine cultivars in response to soil water status and air water vapor pressure deficit. OENO One 44 (1), 9–20.

Reynolds, A.G., 2015. In: Grapevine Breeding Programs for the Wine Industry. Elsevier.

Rives, M., 1961. Bases génétiques de la sélection clonale chez la vigne. Ann. Amélior. Plantes 11, 337–348.

Rocheta, M., Becker, J.D., Coito, J.L., Carvalho, L., Amâncio, S., 2014. Heat and water stress induce unique transcriptional signatures of heat-shock proteins and transcription factors in grapevine. Functional and Integrative Genomics 14 (1), 135–148.

Rogiers, S., Greer, D., Hutton, R., Clarke, S., 2011. Transpiration efficiency of the grapevine cv. Semillon is tied to VPD in warm climates. Annals of Applied Biology 158, 106–114.

Rozak, Z., Senge, M., 2016. Crucial limit points in deficit irrigation. In: Proceedings of International Conference on Climate Change no. 196 ICCC/856.

Sá Borges, R., da Silva, G.A., Roberto, S.R., de Assis, A.M., Yamamoto, L.Y., 2013. Phenolic compounds, favorable oxi-redox activity and juice color of 'Concord'grapevine clones. Scientia Horticulturae 161, 188–192.

Satisha, J., Ramteke, S.D., Karibasappa, G.S., 2007. Physiological and biochemical characterisation of grape rootstocks. South African Journal for Enology and Viticulture 28 (2), 163.

Schultz, H.R., 2003. Differences in hydraulic architecture account for near-isohydric and anisohydric behaviour of two field-grown *Vitis vinifera* L. cultivars during drought. Plant, Cell and Environment 26 (8), 1393–1405.

Sefc, K.M., Lopes, M.S., Lefort, F., Botta, R., Roubelakis-Angelakis, K.A., Ibanez, J., Steinkellner, H., 2000. Microsatellite variability in grapevine cultivars from different European regions and evaluation of assignment testing to assess the geographic origin of cultivars. Theoretical and Applied Genetics 100 (3–4), 498–505.

Soar, C.J., Dry, P.R., Loveys, B.R., 2006. Scion photosynthesis and leaf gas exchange in *Vitis vinifera* L. cv. Shiraz: mediation of rootstock effects via xylem sap ABA. Australian Journal of Grape and Wine Research 12 (2), 82–96.

Souza, C.R., Maroco, J.P., dos Santos, T.P., Rodrigues, M.L., Lopes, C., Pereira, J.S., Chaves, M.M., 2005. Control of stomatal aperture and carbon uptake by deficit irrigation in two grapevine cultivars. Agriculture, Ecosystems and Environment 106 (2), 261–274.

Steduto, P., Hsiao, T.C., Fereres, E., 2007. On the conservative behaviour of biomass water productivity. Irrigation Science 25, 189–207.

Tarara, J.M., Peña, J.E.P., Keller, M., Schreiner, R.P., Smithyman, R.P., 2011. Net carbon exchange in grapevine canopies responds rapidly to timing and extent of regulated deficit irrigation. Functional Plant Biology 38 (5), 386–400.

This, P., Lacombe, T., Thomas, M.R., 2006. Historical origins and genetic diversity of wine grapes. Trends in Genetics 22, 511–519.

Tomás, M., Medrano, H., Pou, A., Escalona, J.M., Martorell, S., Ribas-Carbó, M., Flexas, J., 2012. Water-use efficiency in grapevine cultivars grown under controlled conditions: effects of water stress at the leaf and whole-plant level. Australian Journal of Grape and Wine Research 18 (2), 164–172.

Tomás, M., Medrano, H., Brugnoli, E., Escalona, J.M., Martorell, S., Pou, A., Ribas-Carbó, M., Flexas, J., 2014a. Variability of mesophyll conductance in grapevines cultivars under water stress conditions in relation to leaf anatomy and water use efficiency. Australian Journal of Grape and Wine Research 20, 272–280.

Tomás, M., Medrano, H., Escalona, J.M., Martorell, S., Pou, A., Ribas-Carbó, M., Flexas, J., 2014b. Variability of water use efficiency in grapevines. Environmental and Experimental Botany 103, 148–157.

Tortosa, I., Escalona, J.M., Bota, J., Tomás, M., Hernández, E., Escudero, E.G., Medrano, H., 2016. Exploring the genetic variability in water use efficiency: evaluation of inter and intra cultivar genetic diversity in grapevines. Plant Science 251, 35–43.

Tramontini, S., van Leeuwen, C., Domec, J.C., Destrac-Irvine, A., Basteau, C., Vitali, M., Mosbach-Schulz, O., Lovisolo, C., 2013. Impact of soil texture and water availability on the hydraulic control of plant and grape-berry development. Plant and Soil 368 (1–2), 215–230.

Ulanovsky, S., Gogorcena, Y., de Toda, F.M., Ortiz, J.M., 2002. Use of molecular markers in detection of synonymies and homonymies in grapevines (Vitisvinifera L.). Scientia Horticulturae 92 (3), 241–254.

Valcárcel, J.A., 1971. Agricultura general y gobierno de la casa de campo (Tomo VIII).

Viala, P., Vermorel, V., 1902–1910. In: Traité Général D'Ampélographie (Masson, Paris).

Voltas, J., Romagosa, I., Lafarga, A., Armesto, A.P., Sombrero, A., Araus, J.L., 1999. Genotype by environment interaction for grain yield and carbon isotope discrimination of barley in Mediterranean Spain. Crop and Pasture Science 50 (7).

Williams, L.E., Ayars, J.E., 2005. Grapevine water use and the crop coefficient are linear functions of the shaded area measured beneath the canopy. Agricultural and Forest Meteorology 132 (3), 201–211.

Wilson, H.M., 1912. Irrigation Engineering, sixth ed. John Wiley and Sons, New York, pp. 47–57.

Wright, G.C., Hammer, G.L., 1994. Distribution of nitrogen and radiation use efficiency in peanut canopies. Australian Journal of Agricultural Research 45, 565–574.

Wright, G.C., Hubick, K.T., Farquhar, G.D., 1988. Discrimination in carbon isotopes of leaves correlates with water-use efficiency of field grown peanut cultivars. Australian Journal of Plant Physiology 15, 815–825.

Yuste, J., Martín, J.P., Rubio, J.A., Hidalgo, E., Recio, P., Santana, J.C., Arranz, C., Ortiz, J.M., 2006. Identification of autochthonous grapevine varieties in the germoplasm collection at the ITA of "Castilla y León" in Zamadueñas Station, Valladolid, Spain. Spanish Journal of Agricultural Research 4, 31–36.

Zsófi, Z.S., Gál, L., Szilágyi, Z., Szűcs, E., Marschall, M., Nagy, Z., Bálo, B., 2009. Use of stomatal conductance and pre-dawn water potential to classify terroir for the grape variety Kékfrankos. Australian Journal of Grape and Wine Research 15 (1), 36–47.

CHAPTER 17

Mesophyll Conductance to CO_2 Diffusion: Effects of Drought and Opportunities for Improvement

Miquel Nadal, Jaume Flexas
Universitat de les Illes Balears, Palma de Mallorca, Spain

Contents

Water Scarcity and Sustainable Agriculture in Semiarid Environment
ISBN 978-0-12-813164-0
https://doi.org/10.1016/B978-0-12-813164-0.00017-X

1. INTRODUCTION TO MESOPHYLL CONDUCTANCE

1.1 Photosynthesis: A Biochemical and Diffusive Process

Photosynthesis is the primary source of carbon input in plants; it can be described as a dual process, including photochemistry-light capture and its use for electron transport and generation of adenosine triphosphate and nicotinamide adenide dinucleotide phosphate and carbon fixation. Therefore the photosynthetic capacity of C_3 plants is determined by two main components: diffusional limitations to CO_2 across the air—leaf path and biochemical characteristics (mainly light-harvesting complexes and Calvin cycle enzymes) (Sharkey et al., 2007). For a complete depiction of carbon assimilation, see von Caemmerer et al. (2009).

Carbon diffusion from the atmosphere to the sites of carboxylation by the enzyme Ribulose-1,5-bisphosphate carboxylase/oxygenase (Rubisco) is one of the most important aspects of the photosynthetic process. Within this process, two separate steps take place: diffusion through stomatal pores, described mainly by the stomatal conductance (g_s) and the CO_2 concentration differential between the atmosphere and the substomatal cavity (C_a and C_i, respectively) (Gago et al., 2014; Warren, 2008b), and diffusion from the substomatal cavity to the carboxylation sites within the chloroplasts, described by the internal or mesophyll conductance (g_m) and the CO_2 concentration differential between the substomatal cavity and the carboxylation sites (C_c) (Niinemets et al., 2009; von Caemmerer et al., 2009). In Fig. 17.1A these parameters are schematically represented in a leaf section. The ratio between assimilation rate (A_n) and stomatal conductance to water vapor (g_{sw}) accounts for the intrinsic water use efficiency at the leaf level (WUE_i). In the last 10 years, an increasing amount of evidence has recognized CO_2 diffusion within the leaf as one of the key steps affecting the carbon assimilation of the leaf (Ethier and Livingston, 2004; Flexas et al., 2008, 2012; Niinemets et al., 2009).

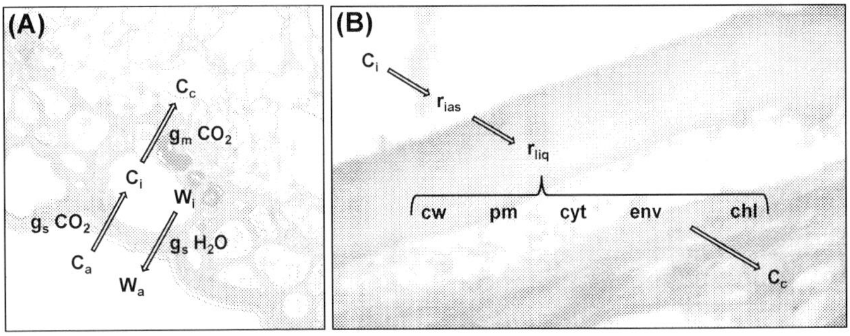

Figure 17.1 Schematic representation of CO_2 and H_2O diffusion within the leaf. (A) Micrograph of a *Cycas revoluta* abaxial leaf surface where the CO_2 path from the air surrounding the leaf to the site of carboxylation is represented. Note the shared path with H_2O through the stomata. C_a, [CO_2] in the atmosphere; C_c, [CO_2] in the chloroplast stroma; C_i, [CO_2] in the substomatal cavity; g_m, mesophyll conductance to CO_2 diffusion; g_s, stomatal conductance to either CO_2 or H_2O diffusion; W_a, [water vapor] in the atmosphere; W_i, [water vapor] in the substomatal cavity. (B) Electron micrograph of a mesophyll cell of the fern *Dryopteris tyrrhena* where the anatomical components determining g_m are indicated (in terms of resistances). *Within the liquid phase, diverse serial elements contribute to the CO_2 diffusion resistance (indicated within the braket). chl*, Chloroplast stroma; *cw*, cell wall; *cyt*, cytosol; *env*, chloroplast envelope; *pm*, plasmalemma; r_{ias}, gas phase resistance to CO_2 diffusion; r_{liq}, liquid phase resistance to CO_2 diffusion. *(*C. revoluta *photo by M. Nadal and* D. tyrrhena *by M. Carriquí.)*

There is a large amount of literature regarding photosynthesis response to water stress (Chaves et al., 2003, 2009; Flexas et al., 2004; Lawlor and Tezara, 2009; among others); here, the focus will be on mesophyll conductance as an important factor that determines photosynthetic performance under drought and its potential role for improving yield and water productivity.

1.2 History of Mesophyll Conductance to CO_2: From Early Assumptions to Current Knowledge

1.2.1 Early Descriptions

In one of the earliest depictions of photosynthesis, Gaastra (1959) suggested that, besides stomata, CO_2 diffusion through the mesophyll could impose a significant limitation to photosynthesis. In this pioneer study, mesophyll anatomy was already thought to be the main constraint to CO_2 internal diffusion. The very first method for estimating mesophyll conductance, proposed by Jones and Slatyer (1972), was based on Gaastra's (1959) assumption that C_c was near zero (i.e., at the point of CO_2 compensation). The development of the early methods allowed the first estimations of mesophyll conductance (Jones and Slatyer, 1972; Samsuddin and Impens, 1979), confirming its limiting role in assimilation. Moreover, mesophyll conductance appeared to decrease in response to water stress (Beadle et al., 1973; Hansen, 1971; Troughton, 1969). It is worth mentioning that most early studies working in CO_2 diffusion referred to what nowadays is known as mesophyll conductance with terms such as "mesophyll resistance" (r_m, the inverse of conductance), "leaf internal conductance" (g_i) or "CO_2 transfer conductance" (Flexas et al., 2008). Moreover, some of the first reports on mesophyll conductance used the relationship of net assimilation and substomatal CO_2 concentration (C_i) as a proxy of g_m (Dang and Guy, 1994; Wan et al., 1993; among others), but the slope of A_n/C_i curves should not be referred as mesophyll conductance but rather as "apparent carboxylation efficiency" or "apparent carboxylation conductance," as such slope includes both diffusion and biochemical effects (Flexas et al., 2008).

1.2.2 Development of the Current Methodology

The introduction of the vastly used FvCB model for C_3 photosynthesis (Farquhar et al., 1980) implied revision of the early assumption of C_c being near zero (Gaastra, 1959). Instead, in the 1980s and early 1990s it was assumed that C_c was equal to C_i (i.e., infinite and nonlimiting g_m). However, the development of new techniques based on gas exchange coupled with isotope discrimination (Evans et al., 1986) and fluorescence (Di Marco et al., 1990; Harley et al., 1992) allowed reliable estimations of C_c. These techniques showed that C_c was lower than C_i and that g_m could not be neglected in photosynthesis studies. The accurate estimation of g_m is one of the main topics of its study since its importance as a central photosynthesis component was established in the 2000s (Flexas et al., 2008; Warren, 2008b). Besides isotope discrimination and fluorescence, A_n/C_i curve fitting using gas exchange measurements has also been proposed for simultaneous estimation of g_m and biochemical parameters, mainly maximum carboxylation velocity (V_{cmax}) and electron transport rate (J_{max}) (Bellasio et al., 2016; Ethier and Livingston, 2004; Long and Bernacchi, 2003; Sharkey et al., 2007). For a complete depiction of the most common techniques used for g_m estimation, see the reviews by Pons et al. (2009) and Warren (2006). However, it is worth noting that mesophyll conductance is a rather tricky parameter to estimate, and care must be taken in experiments willing to measure g_m. It is important for reliable g_m studies to present at least two independent g_m estimations (e.g., in Flexas et al., 2007; Tosens et al., 2016; Vrábl et al., 2009), especially under stress conditions where added technical difficulties in the measures may affect sensitive parameters and introduce artifactual responses (Galmés et al., 2006).

The methods mentioned earlier rely on two-dimensional models of CO_2 diffusion through the leaf, whereas it may be more accurate to describe it as a three-dimensional process (Flexas et al., 2008 and references within). However, three-dimensional models are not as straightforward as the mentioned above and far from widespread use due to its complexity. Keeping in mind their limitations, two-dimensional methods are, to date, the best option for photosynthesis studies willing to consider g_m, and the source of all data and knowledge used in this chapter.

1.2.3 The Variable and Dynamic Nature of Mesophyll Conductance

The development of techniques for accurate measurement of mesophyll conductance allowed its study in a wide range of species and environmental studies. Early work showed that g_m was variable among species (Loreto et al., 1992; von Caemmerer and Evans, 1991) and leaf age (Loreto et al., 1994), and environmental factors such as drought (Renou et al., 1990) or light regime (Lloyd et al., 1992). These early studies illustrated the importance of mesophyll conductance as an important factor determining the rate of photosynthesis (Grassi and Magnani, 2005; Warren, 2008b). The increasing amount of data indicates that g_m varies across plant forms, from the highest in fast-growing crops such as cotton or birch to the lowest in sclerophyllous trees and ferns (Flexas et al., 2012; Tosens et al., 2016), and depends on growing and short-term environmental conditions, such as water availability, light, temperature, CO_2 atmospheric concentration, or salt stress (see Flexas et al., 2008 and Niinemets et al., 2009, for a compilation and quantification of factors affecting g_m).

The reported variability of mesophyll conductance upon different factors stimulated studies trying to elucidate its underlying mechanistic basis (see next section). Current knowledge of the determinants of g_m allows for the first attempts to manipulate photosynthesis through the modification of g_m, as it has been shown that mesophyll conductance constitutes a good target for improving yield and "water production" (Flexas et al., 2016).

2. FACTORS DETERMINING MESOPHYLL CONDUCTANCE

2.1 The Path of Carbon Dioxide Within the Leaf

Mesophyll conductance to CO_2 integrates several diffusional steps for a CO_2 molecule to reach the chloroplast stroma (Evans et al., 2009; Flexas et al., 2012): CO_2 diffuses from the intercellular air spaces to the mesophyll cells, then overcomes the resistance of the polymer matrix that constitutes the cell wall, passes through the lipid bilayers (both plasmalemma and chloroplast envelopes) and, finally, reaches the sites of carboxylation of the enzyme Rubisco inside the chloroplast stroma. Using a two-dimensional approach, CO_2 diffusion can be described as the sum of the diffusion resistances along the leaf mesophyll (Evans et al., 2009); see Fig. 17.1B for the complete picture of the CO_2 path and the resistances involved. Mesophyll resistance (r_m, the inverse of g_m) can be disclosed as follows:

$$r_m = r_{ias} + r_{liq} = r_{ias} + r_{cw} + r_{pm} + r_{cyt} + r_{env} + r_{chl} \qquad (17.1)$$

where r_{ias} and r_{liq} refer to the resistances in the gas and liquid phases, respectively. Within the liquid phase can be distinguished resistances imposed by the cell wall (r_{cw}), the plasmalemma (r_{pm}), the cytosol (r_{cyt}), the chloroplast envelope (r_{env}), and the chloroplast inner membrane (r_{chl}) (Flexas et al., 2008). The relative contributions of each diffusive step to the total g_m depend on its relative length and its "effective" path, i.e., implications of the tortuosity and porosity of the medium (Flexas et al., 2012). In the next sections, each of these steps will be described; see the review by Evans et al. (2009) for an exhaustive depiction of CO_2 diffusion within the leaf.

2.2 Gas Phase

Once CO_2 has entered the leaf via stomata, it has to diffuse through the intercellular air space fraction of the leaf or porosity (f_{ias}), which, together with the leaf thickness, determines the gas-phase conductance (g_{ias}) (Tomàs et al., 2013). The relative contribution of the gaseous phase is usually thought to be low compared to the liquid phase, as air imposes much less resistance to CO_2 movement (Evans et al., 2009; Flexas et al., 2008). However, some studies using air with modified diffusive properties (*hedox*, nitrogen replaced by helium) have found an increase in CO_2 diffusivity and enhanced photosynthesis rates (Parkhurst and Mott, 1990), although others did not observe such effect (Genty et al., 1998).

2.3 Liquid Phase

2.3.1 Entering the Medium: Carbonic Anhydrases

CO_2 has to enter the liquid phase to reach the chloroplast stroma. Liquid-phase conductance (g_{ias}) involves different steps: from the cell wall pores, the plasmalemma, cytosol, and chloroplast envelopes to the stroma (Flexas et al., 2012). A role for carbonic anhydrases (CA), which catalyze the conversion of CO_2 to $^{-}HCO_3$, has been proposed as an enhancer of carbon diffusion through the liquid phase (Flexas et al., 2008), although no strong effects have been detected so far in g_m when modifying CA activity (Han et al., 2016; Price et al., 1994; Williams et al., 1996). Nonetheless, a study in poplar found a positive relationship of the activity of CA with mesophyll conductance in the field (Momayyezi and Guy, 2017). Hence the role of CA is still unclear and may only be determining under specific circumstances.

2.3.2 Cell Wall

Once CO_2 is in the liquid phase, the cell wall constitutes one of the main obstacles for its path. The detailed physical characteristics of the cell wall (porosity and tortuosity) have been difficult to measure so far; however, cell wall thickness (usually assessed with transmission electron microscopy) is thought to be a major driver of mesophyll resistance (Evans et al., 2009; Terashima et al., 2011; Tomàs et al., 2013). In fact, low g_m values have been reported in angiosperm species with thick mesophyll cell walls (Tomàs et al., 2013), and cell wall thickness is key to explain the lower photosynthesis capacity of ferns compared to higher plants (Carriquí et al., 2015; Tosens et al., 2016). It has been suggested that cell wall composition may also play a role in CO_2 diffusion, as the relative contribution of each polymer (cellulose, hemicellulose, pectin, among others) may determine the whole matrix characteristics (Gago et al., 2016), but this has not been demonstrated yet.

2.3.3 Diffusion Through Membranes: The Role of Aquaporins

Cell membrane or plasmalemma is the next step of the liquid phase. CO_2 is able to pass through the bulk of the lipid bilayer; however, in the early 2000s a possible role of aquaporins as facilitators of CO_2 transport across membranes was suggested (Terashima and Ono, 2002). The function of aquaporins (or "cooporins," to account for their transport of both CO_2 and

H_2O) in enhancing CO_2 diffusion was later confirmed in vivo in plants with modified aquaporin expression, where both photosynthesis and g_m were altered (Flexas et al., 2006b; Hanba et al., 2004). Aquaporins have been proposed as the nexus of the g_m and hydraulic conductance relationship (Flexas et al., 2013b); in fact, the proportion of tetramers that constitute the whole protein may determine its relative contribution to H_2O and CO_2 transport (Uehlein et al., 2008).

2.3.4 Cytosol

The spacing between the cell wall and chloroplasts is difficult to determine, as the uncoupling of chloroplasts from the cell's edge observed in some cases may be due to fixation artifacts (Evans et al., 2009). Tobacco plants with altered chloroplast arrangement, resulting in increased cytosol path, showed reduced assimilation and g_m (Sharkey et al., 1991), indicating an important added resistance due to the increased diffusion distance through the cytosol. However, the characteristics of this medium regarding CO_2 diffusion are poorly understood (Evans et al., 2009).

2.3.5 Chloroplast

The last major step of CO_2 diffusion is its entrance within the chloroplasts. In this regard, the surface of chloroplasts exposed to the intercellular air spaces (S_c/S) has been revealed as a strong determinant of mesophyll conductance (Tomàs et al., 2013) as it reduces the path length of CO_2 diffusion, thus enhancing its arrival at the sites of carboxylation in the stroma (Terashima et al., 2011). Chloroplast number, size, and arrangement determine S_c/S (Oguchi et al., 2005); and even chloroplast movements can induce short-term variations of g_m (Tholen et al., 2008).

Chloroplast envelope and inner membrane constitute the last barriers that CO_2 has to overcome to reach the stroma; in this regard, aquaporins are also thought to facilitate its diffusion through these membranes (Uehlein et al., 2008). However, the permeability of these last steps and their relative contribution to total mesophyll resistance are not clear.

2.4 The Importance of Anatomy Constraining Photosynthesis

Several data compilation studies have shown the relationship between structural characteristics of the leaf and its carbon fixation capacity, resulting in what has been established as the "leaf economics spectrum" (Wright et al., 2004). Bulk leaf anatomy parameters such as leaf mass per area (LMA), related mainly to the leaf structure (thickness, density; see the review by Poorter et al., 2009) and the mesophyll cell wall and vein characteristics (Flexas et al., 2013b; Hikosaka and Shigeno, 2009), can determine leaf assimilation through CO_2 diffusion (Flexas et al., 2008; Niinemets et al., 2009). In fact, when chloroplast arrangement is taken into account, cell wall thickness determines most of the variation observed in mesophyll resistance (Evans et al., 2009; Terashima et al., 2011). The asymptotic relationship between LMA and g_m across species (Niinemets et al., 2009), where LMA imposes a maximum value of g_m, indicates the importance of leaf structure in constraining maximum CO_2 diffusion and photosynthesis. This limitation imposed by leaf structure is stronger in evergreen and sclerophyllous species, which usually present higher investment in leaf structural components such as cell walls (Niinemets et al., 2011; Warren and Adams, 2004). Mediterranean species, despite presenting high LMA values, constitute an intriguing exception to the assimilation—leaf structure relationship: although g_m values are low, assimilation rates are higher than temperate species for a given LMA value due to enhanced biochemistry (V_{cmax}) (Flexas et al., 2014). A study in sclerophyllous oaks (*Quercus* sp.) showed another mechanism of Mediterranean species to overcome the diffusive limitation imposed by high LMA: in the case of oaks, high chloroplast exposure (S_c/S) correlated with higher g_m and assimilation rates, similar to congeneric deciduous species (Peguero-Pina et al., 2017).

Thus, the increasing amount of data suggests that maximum assimilation is constrained by leaf anatomy characteristics through the imposed mesophyll resistance to CO_2 diffusion. However, g_m is much more variable in the short term than leaf anatomy; biochemical components (CA and aquaporins) may be more related to its short-term dynamics (Pérez-Martín et al., 2014).

3. MESOPHYLL CONDUCTANCE UNDER WATER STRESS

3.1 Mesophyll Conductance Under Water Stress Conditions: An Important Factor Limiting Photosynthesis

Photosynthesis performance under water stress is one of the central topics of plant ecophysiology, as water stress is the main limiting factor for plant production (Loreto and Centritto, 2008). A decrease in photosynthesis under reduced water supply can be due to diffusive and biochemical limitations (Chaves et al., 2003). Diffusional limitations include the whole CO_2 path from the atmosphere to the site of carboxylation inside the chloroplast. Stomatal closure in response to water stress (thus decreasing g_s) is a well-described process involving sensitivity to both abscisic acid (ABA) and leaf turgor (Chaves et al., 2003; Rodríguez-Domínguez et al., 2016).

Stomatal closure is the first clear response to a decrease in water availability and it can be used as a reference parameter for monitoring the intensity of water stress endured by leaves. Flexas et al. (2004) showed that a threshold of g_s of about 0.05 mol H_2O m^{-2} s^{-1} is indicative of a shift from diffusion-limited photosynthesis to metabolic impairment, i.e., biochemistry limitations, and the occurrence of more severe damage such as increased generation of reactive oxygen species or impairment of the electron transport components (see Lawlor and Tezara, 2009, for an extensive review of biochemical responses to water stress). Nonetheless, this depiction of drought stress by Flexas et al. (2004) did not include an exhaustive description of g_m responses to water availability; nowadays, thanks to the increasing amount of data, it is possible to provide a more complete picture of the role of g_m in limiting photosynthesis upon drought.

In the first studies on photosynthesis response to water stress it was usually assumed that stomatal and—to a lower extent—biochemical limitations solely accounted for the decrease in assimilation (reviewed in Chaves et al., 2009). CO_2 diffusion through the mesophyll was thought to be unaffected and C_i/C_c variations were attributed to artifacts; only a few studies eventually argued in favor of a modified mesophyll conductance in response to stress (Renou et al., 1990; Roupsard et al., 1996; Tournex and Peltier, 1995). Later studies confirmed the decrease in g_m in response to water stress in different species (Flexas et al., 2002; Warren et al., 2004), and pointed out that mesophyll conductance may affect ^{13}C isotope discrimination in leaf sugars under drought (Scartazza et al., 1998). With the accurate analysis of photosynthesis limitations in tree species during a seasonal field study, Grassi and Magnani (2005) reported a significant contribution of g_m in constraining photosynthesis (high mesophyll limitation) during the driest year. These early studies demonstrated the importance of including g_m as an important limiting factor for photosynthesis under stress (Grassi and Magnani, 2005) and for accurately predicting assimilation responses to drought (Keenan et al., 2010b).

3.2 Response of Mesophyll Conductance: Effect on Photosynthesis

Photosynthesis is usually thought to be limited by a combination of diffusive and biochemical factors, the former including both stomatal and mesophyll limitations and the latter carboxylation and photochemistry (Chaves et al., 2009; Flexas et al., 2004). When a reference value (non-stressed) is taken, it is possible to estimate the relative contributions of the stomatal, mesophyll, and biochemical limitations; their sum accounts for the reduction of photosynthesis compared to the reference value (see Grassi and Magnani, 2005, for the calculation). These authors described a limitation pattern of water stress in ash (*Fraxinus oxyphylla*) and oak (*Quercus robur*) trees that has been supported by later studies in different species, including a variety of growth forms (Flexas et al., 2009; Galle et al., 2009; Galmés et al., 2007); this pattern is schematically represented in Fig. 17.2.

The very first response to water withholding is the decrease of g_s as a result of stomatal closure, resulting in photosynthesis being mostly limited by stomata (mild stress conditions). In the first stages of water depletion, g_m usually remains constant or only slightly affected (Delfine et al., 2001; Flexas et al., 2009; Warren, 2008a); however, as drought progresses, g_m decreases and its relative contribution to photosynthesis limitation increases (moderate stress conditions; Galmés et al., 2007). This pattern appears to be more related to drought duration than severity (Cano et al., 2014; Flexas et al., 2009). Only under severe stress conditions do biochemical limitations and metabolic impairment seem to occur (Flexas et al., 2004; Grassi and Magnani, 2005). The different responses of diffusive and biochemical

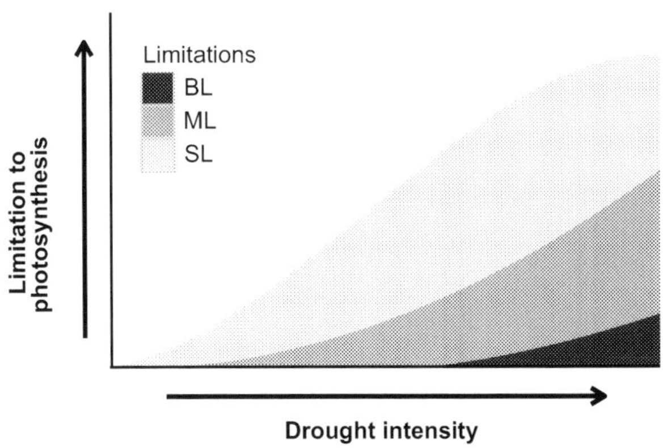

Figure 17.2 Scheme of the general pattern of the contributions of stomata, mesophyll, and biochemistry (Calvin cycle enzymes and photochemistry) to photosynthesis decrease upon drought stress. Stomatal limitation (SL) is the first sign of drought but its relative contribution to photosynthesis limitation decreases with greater drought intensity, when mesophyll limitation (ML) becomes increasingly important. Biochemical limitation (BL) appears only under severe drought conditions.

limitations to decreasing water potential support the described pattern (Niinemets and Keenan, 2014; Zhou et al., 2016). Diffusive and biochemical limitations do not have to be taken as "watertight compartments" but as connected processes that influence one another: for example, the decrease in mesophyll conductance reduces the CO_2 concentration at the site of carboxylation, which in turn may lead to the deactivation of Rubisco, therefore increasing biochemical limitations (Flexas et al., 2006a; Galmés et al., 2011b).

Usually, stomatal limitations are the most important in limiting photosynthesis under mild and moderate stress, with mesophyll limitations accounting for approximately 25% of photosynthesis decrease (Cano et al., 2014; Grassi and Magnani, 2005). However, in some species, g_m is the most important of the diffusive limitations: this has been reported in several Mediterranean species (Galmés et al., 2007), the alpine plant *Rhododendron delavayi* (Cai et al., 2015), and wheat (Olsovska et al., 2016). Usually, evergreen species are also more limited by mesophyll conductance than stomatal conductance, especially as drought progresses (Niinemets and Keenan, 2014).

Currently, there are a limited amount of studies that exhaustively describe the photosynthesis limitations in response to drought (or recovery); more data are needed to confirm the dynamic role of g_m in constraining photosynthesis during drought, but also to assess the generality of its role across species. In any case, the available work demonstrates the importance of mesophyll conductance in limiting photosynthesis under water stress. In addition, the inclusion of g_m is necessary for proper estimation of biochemical parameters, especially V_{cmax} (Ethier and Livingston, 2004; Limousin et al., 2010).

3.2.1 Coordination of Diffusive Limitations

The first studies monitoring the response of photosynthetic parameters generally showed that mesophyll conductance presented a somewhat delayed response to the decrease in g_s, e.g., in pepper (*Capsicum annuum*) and grapevine (*Vitis vinifera*) (Delfine et al., 2002; Flexas et al., 2002). This response of g_m in regards to g_s was later found for more species (Bunce, 2009; Cano et al., 2014; Flexas et al., 2009; Warren, 2008a). This relationship is in accordance with the photosynthesis limitation dynamics explained earlier, as stomatal limitation increases earlier than mesophyll limitation in most studied species. The delayed response of g_m may explain the poor correlation found when pooling data from multiple studies (Fig. 17.3A).

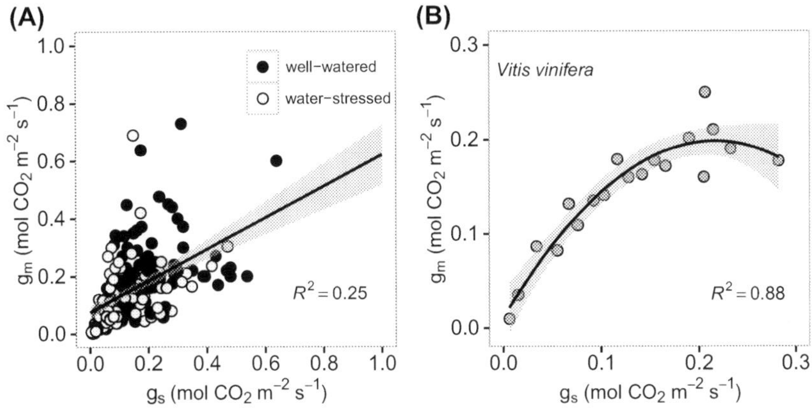

Figure 17.3 Coordination of the CO_2 diffusion conductances, stomatal (g_s) and mesophyll (g_m) conductance. (A) Multispecies g_s–g_m relationship with different irrigation treatments. (B) g_s–g_m relationship in *Vitis vinifera* cv. Tempranillo grown in the field. In both graphs, *black line* represents a quadratic polynomial fit ($P < .0001$) and shaded area is the 95% confidence interval. *(A) Dataset from Flexas et al. (2013a), and complemented with data from Aganchich et al. (2009), Bunce (2009), Levi et al. (2009), Fleck et al. (2010), Galmés et al. (2011a), Li et al. (2012), Gu et al. (2012), Carmo-Silva et al. (2012), Aranda et al. (2012), Rho et al. (2012), Brilli et al. (2013), Aranjuelo et al. (2013), Rzigui et al. (2013), Cano et al. (2013), Cai et al. (2015) and Olsovska et al. (2016). (B) Data from Flexas et al. (2002).)*

Despite the poor correlation of g_s and g_m shown in Fig. 17.3A, the two conductances usually present a tight intraspecific relationship in response to water stress (Flexas et al., 2013a; Sun et al., 2014). For example, in grapevine cultivars differing in their anisohydric and isohydric behavior, g_m presented a highly coordinated response with g_s during water withholding (Pou et al., 2012). As an example of within-species coordination, Fig. 17.3B shows the relationship between g_s and g_m in grapevine; note that g_m remains unaffected over a wide range of g_s (0.1–0.3 mol CO_2 m^{-2} s^{-1}). Due to their intraspecific coordination, the ratio g_m/g_s has been proposed as a good parameter for improving WUE_i and genotype selection (see Section 4; Flexas et al., 2010; Galmés et al., 2011a; Han et al., 2016; Théroux-Rancourt et al., 2015).

An environmental factor usually coupled with drought under field conditions is high vapor pressure deficit (VPD). High VPD induces stomatal closure (McAdam et al., 2016); however, few reports exist describing its effects on g_m. Warren (2008a) observed a decoupling of g_s and g_m under variable VPD (short term) in three different species: g_s decreased in response to higher VPD (from 1 to 2 kPa), whereas g_m remained constant.

Pérez-Martín et al. (2009) also found little effects of VPD in olive (*Olea europaea*) and grapevine. The lack of response of g_m to VPD despite its substantial effects on g_s suggests that g_m is not directly linked with stomatal conductance but rather with leaf water status.

The coordinated response of stomatal and mesophyll conductances to soil drought suggests a common regulation mechanism. One of the most important regulators of stomatal conductance is ABA, which is produced in response to water stress and induces stomatal closure (Rodríguez-Domínguez et al., 2016; Wilkinson and Davies, 2002). Although some studies did not find a response of g_m to ABA treatment (Vrábl et al., 2009), Duan et al. (2011) reported a negative relationship between ABA leaf content and g_m, and the application of ABA reduced both g_s and g_m in different species (Sorrentino et al., 2016). Furthermore, the role of ABA in the coordinated response of both conductances was confirmed by Mizokami et al. (2015) using an ABA-deficient tobacco mutant, which showed no g_s or g_m response upon drought. Thus, these results support a role of ABA in regulating not only g_s but also g_m, although the underlying mechanisms of g_m response to ABA signaling remain unclear. The rapid nature of the g_m response suggests that aquaporins and/or CA may be involved (Sorrentino et al., 2016).

3.2.2 Response of Mesophyll Conductance: Causes of Decrease Under Water Depletion

Rapid (short-term) g_m decrease upon water stress may be related to modified activity and/or expression of aquaporins and CA, possibly related to ABA signaling (as explained earlier). Aquaporin activity has been reported to be inhibited in response to prolonged water stress in tobacco (Miyazawa et al., 2008). Few studies have shown the role of aquaporins and/or CA during short-term drought; in this regard, Pérez-Martín et al. (2014) showed that the reduced expression of aquaporin genes *OePIP1.1* and especially *OePIP2.1* and CA upon water stress was related to the decrease in g_m in olive. In contrast, other studies have reported that anatomical factors may be an important cause of g_m decrease upon short-term water stress.

As explained earlier, chloroplast size and distribution is one of the most important anatomical features that influence maximum g_m (Terashima et al., 2011). Li et al. (2012), in a short-term polyethylene glycol (PEG)-induced study in rice, suggested that chloroplast downsizing (in response to low water potential) may result in reduced g_m, therefore constraining assimilation. Although no exhaustive anatomical data are provided, chloroplast downsizing upon water stress has been reported to be a common response in different species (Li et al., 2012; and references within); the reduction in chloroplast size may imply a decreased chloroplast surface exposed to the mesophyll spaces (S_c/S), which would result in lower g_m (Evans et al., 2009; Terashima et al., 2011).

Besides chloroplast size, water stress affects other leaf morphological parameters such as leaf density and cell wall thickness (Niinemets et al., 2009; Poorter et al., 2009). Regarding more long-term drought responses, Galmés et al. (2013) showed that different morphological parameters influenced assimilation in tomato (*Solanum lycopersicum*) genotypes after 2 weeks of reduced irrigation in the field. In this case, reduction of the intercellular air space fraction and a possible increase in cell wall thickness may be responsible for increased CO_2 diffusion resistance.

3.3 Role of g_m During Acclimation to Drought

Most g_m studies in relation to drought are focused on its quick response to water stress (usually reached by stopping irrigation). Fewer studies take into account the possible acclimation of g_m to maintained drought conditions, repeated drought cycles, or even drought-developed leaves. The term "acclimation" is used in many studies where leaves are subjected to long-term drought (weeks to seasons) by maintaining water irrigation at a desired level (usually assessed by evapotranspiration, soil water content, or stomatal conductance).

Some studies have monitored the response of g_m and other photosynthetic parameters under an extended stress period after reaching a mild and/or severe stress level (usually several days or weeks). Mesophyll conductance values are usually constant during the acclimation period, following a pattern similar to g_s (Galle et al., 2010; Pou et al., 2012), although some discrepancies have been found. Flexas et al. (2009) reported a relative increase in mesophyll limitations to photosynthesis at the latter stages of a short-term (1-week) acclimation in a drought-resistant *Vitis* hybrid. On the other hand, Galle et al. (2009) found in tobacco that during an acclimation period of 1 week g_m experienced an increase, even reaching similar values to the well-watered plants; nonetheless, this increase was influenced by the growing conditions and did not produce an enhanced assimilation during drought.

Acclimation to drought may also be referred to successive drought events: Galle et al. (2011) performed three withholding–rewatering cycles in holm oak (*Quercus ilex*) and rockrose (*Cistus albidus*), two Mediterranean species with contrasting growth forms (evergreen and semideciduous, respectively). In this study, different g_m and WUE_i responses were found between the two species. Interestingly, rockrose showed an increased g_m in the second and third water stress events, which resulted in higher assimilation rates during drought. Similar results have been reported in cotton (Han et al., 2016), where drought pretreatment resulted in reduced photosynthetic decline and higher WUE_i during a later drought. This response was attributed to g_m acclimation (lower g_m decrease in second drought) through an enhanced expression of aquaporin *GhAQP1*.

As g_m is partly determined by anatomical features (Flexas et al., 2012), and drought can modify leaf structure (Poorter et al., 2009), it may be expected some CO_2 diffusion constrain as prolonged drought exposure hardens leaves, especially in evergreen species (Limousin et al., 2010; Niinemets et al., 2009). Tosens et al. (2012) described in poplar that new leaves developed under moderate water stress (PEG-induced to avoid any major nutrient shortage) presented higher leaf density and also thicker cell walls, resulting in an increased mesophyll resistance. However, a similar experiment performed in tobacco found that aquaporin deactivation, rather than anatomy, was the main factor limiting g_m in drought-developed leaves (Miyazawa et al., 2008). These results could suggest that the limiting mechanics during g_m acclimation may be species or growth form dependent (Galle et al., 2011); additional studies are required to assess the underlying factors of decreased CO_2 diffusion under long-term stress in different growth forms.

Acclimation studies are especially important in describing plant performance in the field, as prolonged drought episodes and/or mild water scarcity are recurrent limiting factors in many areas of the world (Loreto and Centritto, 2008). Field and seasonal studies are usually affected by many uncontrolled factors that have to be taken into account, as exemplified in Galle et al. (2009); however, they allow to assess the performance of plant processes under "real" conditions. For example, light regime is a very strong determinant of g_m variation within the canopy (Díaz-Espejo et al., 2007; Egea et al., 2012). When taking into account light and drought effects, Cano et al. (2013) found that drought-induced increases in mesophyll limitations were largest in the shaded leaves from the lowest canopy in sessile oak (*Quercus petraea*) and beech (*Fagus sylvatica*). Age is also an important factor to consider in field and seasonal studies: 1-year-old leaves of holm oak subjected to prolonged water scarcity suffered morphological changes that led to increased mesophyll limitations to photosynthesis in comparison to younger leaves (Limousin et al., 2010).

Accurately describing g_m responses to drought and the possible role of its acclimation in the field is necessary for a better understanding of the production dynamics under stress at the ecosystem level, as several model-based studies (Keenan et al., 2010a,b; Niinemets et al., 2009; Zhou et al., 2013) have shown that including g_m significantly improves carbon assimilation predictions.

3.4 Role of g_m During Recovery

Recovery after a water stress event is also a major part of how drought affects plant photosynthesis and productivity, yet little attention has been paid to photosynthesis—including mesophyll conductance—recovery since early water stress studies (Chaves et al., 2009). The first study that extensively described the photosynthesis limitations upon recovery after severe stress was performed in several Mediterranean species by Galmés et al. (2007). In these species, it was found that mesophyll conductance was the main restriction in photosynthesis during recovery, although the rate of recovery was dependent on growth form. On the other hand, some discrepancies exist regarding whether other factors are more limiting for assimilation recovery. Resco et al. (2008) found that stomatal limitations—a consequence of impaired vascular water transport capacity—were more determinant than mesophyll or biochemical limitations upon rewatering.

In contrast to these first results, later studies found, in general, fast g_m recovery in different species. A drought-adapted *Vitis* hybrid showed rapid (1-day) g_m recovery after drought acclimation, even under severe stress conditions (Flexas et al., 2009). Similar results have been found in holm oak, olive, and poplar (Oliver et al., 2012; Sun et al., 2014; Tsonev et al., 2014). During drought in *Eucalyptus dumosa* and poplar hybrids, fast g_m increase after rewatering (in contrast to slower g_s increase) was the main responsible factor for assimilation return to predrought values and enhanced WUE_i after drought (Cano et al., 2014; Théroux-Rancourt et al., 2015). The fast g_m (and assimilation) recovery found in these studies has been attributed to the drought-adapted nature of most of the plants used (Cano et al., 2014; Oliver et al., 2012).

It has been found that the rate of recovery can be variable depending not only on the species but also on the prevailing environmental conditions. Galle et al. (2009) compared the water stress dynamics of *Nicotiana sylvestris* grown in both the field and in a growth chamber, and found a delayed recovery in the field, attributed to high temperature and irradiance conditions.

For a complete understanding of mesophyll conductance recovery upon rewatering and its role limiting assimilation, more studies are needed to unravel the mechanistic basis of g_m recovery—fast or slow—and the observed differences between species and environmental conditions.

The dynamics of mesophyll conductance have been studied in a wide range of species, mostly in model plants such as tobacco (*Nicotiana* sp.) and poplar (*Populus* sp.). Among woody species, oak (*Quercus* sp.) is the most commonly measured species. Regarding herbaceous crops, species such as rice (*Oryza sativa*) and cotton (*Gossypium* sp.) have been studied. On the other hand, grapevine (*Vitis* sp.) and olive (*Olea europaea*) are the woody crops with the highest number of studies. Table 17.1 presents studies where mesophyll conductance response to drought (along other parameters) has been reported for woody crops.

4. IMPROVING MESOPHYLL CONDUCTANCE IN A WATER-LIMITED SCENARIO

4.1 Need for Increased "Water Productivity"

The increase in world population and the effects of global climate change are two capital problems that must be faced in the 21st century. Besides the atmospheric CO_2 concentration and temperature increment, water regime changes are one of the main factors to take into account for ecosystem and crop production (Loreto and Centritto, 2008). Local reduction of water availability is expected in the upcoming years, especially in the Mediterranean Basin (IPCC, 2014); in this context, the need for higher water production ("more crop per drop") and efficient water management is of extreme importance for agriculture (Morison et al., 2007).

An improved balance between yield and water loss can be achieved by either enhancing carbon gain or minimizing transpiration (Flexas et al., 2013a). At the leaf level, WUE_i is defined as the ratio between carbon assimilation and stomatal conductance to water vapor (A_n/g_s). This ratio can vary among species (Gago et al., 2014) and under some conditions (e.g., moderate drought; Galmés et al., 2011a). However, intraspecific enhancement of WUE_i usually results in lower yield (Flexas et al., 2016). The best-case scenario would be the simultaneous improvement of both water use efficiency and plant production; in this case, mesophyll conductance to CO_2 may be a suitable target for achieving this goal.

Table 17.1 List of studies of woody crops subjected to drought or water stress treatment(s) where mesophyll conductance is measured (with any of the currently "accepted" methods described in Pons et al., 2009)

Species	Genotypes/ cultivars	Location	Duration	WS type	A_n limitations analysis	"Complements" to g_m	References
Castanea sativa	No[a]	Outdoors	Summer	Rain fed	No	LMA	Lauteri et al. (1997)
Vitis vinifera	Tempranillo, Mantonegro, 22 cv.	Field, outdoors[b]	Summer	Reduced irrigation	No	No	Flexas et al. (2002)
Olea europaea	Not indicated	Field	April—August	Rain fed	Yes	LMA	Díaz-Espejo et al. (2007)
Vitis berlandieri × *Vitis rupestris*	Hybrid Richter-110	Outdoors	1 month	Withholding, reduced irrigation[c]	Yes	No	Flexas et al. (2009)
V. vinifera, *O. europaea*	Tempranillo (*V*), Manzanilla (*O*)	Outdoors	1—2 months	Reduced irrigation, VPD[d]	Yes	LMA	Pérez-Martín et al. (2009)
O. europaea	Picholine marocaine	Greenhouse	9 months	Partial root zone drying	No	No	Aganchich et al. (2009)
O. europaea	Manzanilla	Outdoors	2 months	Withholding[e]	No	Aquaporins, carbonic anhydrase	Pérez-Martín et al. (2014)
Prunus dulcis	Marta, grafted on Mayor	Field	3 years[f]	Reduced irrigation	No	LMA	Egea et al. (2012)

Continued

Table 17.1 List of studies of woody crops subjected to drought or water stress treatment(s) where mesophyll conductance is measured (with any of the currently "accepted" methods described in Pons et al., 2009)—cont'd

Species	Genotypes/ cultivars	Location	Duration	WS type	A_n limitations analysis	"Complements" to g_m	References
V. vinifera	Chardonnay, Grenache, Syrah	Outdoors	25 days	Withholding, reduced irrigation[c]	No	No	Pou et al. (2012)
Vaccinium corymbosum	Bluecrop	Greenhouse	3 weeks	Withholding	No	No	Rho et al. (2012)
O. europaea	Leccino	Greenhouse	30 days	Withholding[e]	No	No	Sun et al. (2014)
Jatropha curcas	FT1	Greenhouse	20 days	Withholding[e]	No	No	Silva et al. (2015)
V. vinifera	Malvasia, Giro-Ros	Outdoors	4 weeks	Reduced irrigation	Yes	Leaf anatomy	El Aou-ouad et al. (2016)

"*Complements*" *to* g_m, measurement of its possible mechanistic basis; *Duration*, refers to the drought and/or time of measurements; *Location*: "field" refers to sown in the field, "outdoors" refers to plants sown in pots.
[a]Chestnut (*Castanea sativa*) from seven different provenances in Turkey.
[b]Tempranillo and Mantonegro grown in the field; 22 cultivars (see Bota et al., 2001) in pots.
[c]Including both acclimation and recovery.
[d]Including acclimation.
[e]Including recovery.
[f]Measurements performed only in the last year.

4.2 Mesophyll Conductance: Link With Water Use Efficiency

4.2.1 Theoretical Evidence

As stated earlier, carbon assimilation is dependent on both diffusional and biochemical factors. It scales positively with both stomatal and mesophyll conductances, according to Fick's first law of diffusion (Flexas et al., 2013a):

$$A_n = g_s(C_a - C_i) \tag{17.2}$$

$$A_n = g_m(C_c - C_i) \tag{17.3}$$

Combining both equations shows that A_n/g_s depends on the ratio g_m/g_s, following a hyperbolic relationship (Flexas et al., 2016):

$$A_n/g_s = \frac{g_m/g_s}{1 + g_m/g_s}(C_a - C_c) \tag{17.4}$$

This equation indicates that WUE_i is more related to g_m/g_s than g_m per se. This relationship is expected: whereas g_m refers to CO_2, g_s implies not only carbon gain but also water loss; conductances to CO_2 (g_{sc}) and H_2O (g_{sw}) are linked as both molecules share the same path (stomata). These conductances are easily interchangeable as $g_{sc} = g_{sw}/1.6$, where 1.6 is the ratio of binary diffusion for H_2O and CO_2 (Flexas et al., 2016). However, not only conductances determine WUE_i; because the CO_2 drawdown from the atmosphere to the site of carboxylation ($C_a - C_c$) is also included in the equation, the photosynthetic capacity (a multitrait component involving V_{cmax}, J_{max}, and Rubisco kinetics) also influences WUE_i. Nonetheless, and keeping in mind the complexity of the factors determining WUE_i, the positive scaling of A_n with g_m and A_n/g_s with g_m/g_s shows that the improvement of mesophyll conductance may result, theoretically, in yield and water productivity enhancement. Fig. 17.4 shows a simulated response of A_n (Fig. 17.4A) and WUE_i (Fig. 17.4B) in olive when any major parameter determining photosynthesis is modified.

There is a slight increment in A_n and WUE_i when increasing g_m (about 10% when g_m is increased by 50%); however, this increase is significantly higher when parameters related to photosynthetic capacity ("biochemistry") are also improved (about 25%). In any case, modification of stomatal conductance alone results in improved yield but at the expense of reduced water use efficiency (decreased by 25% when g_s is enhanced by 50%), showing the importance of the g_m/g_s ratio in determining WUE_i. Other modeling approaches, considering daily gas exchange variations and parameters such as stomatal speed (in response to

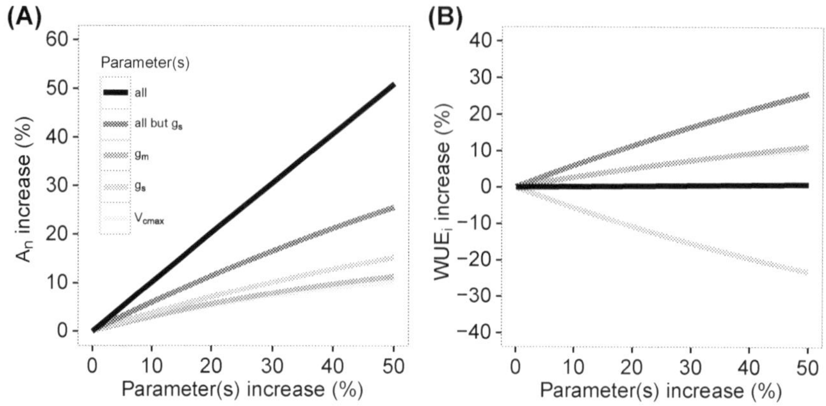

Figure 17.4 Modeling approach as in Flexas (2016). Simulated response of photosynthesis (A_n; A) and water use efficiency (WUE_i; B) if one or more of the following components of photosynthesis were improved to a certain level: stomatal (g_s) and mesophyll (g_m) conductances to CO_2 diffusion and maximum velocity of carboxylation by the enzyme Rubisco (V_{cmax}). The simulations were carried out using the FvCB model, as described in von Caemmerer et al. (2009), where $A_n = 12.2\ \mu mol\ CO_2\ m^{-2}\ s^{-1}$, $g_s = 0.095\ mol\ CO_2\ m^{-2}\ s^{-1}$, $g_m = 0.126\ mol\ CO_2\ m^{-2}\ s^{-1}$, and $V_{cmax} = 258.3\ \mu mol\ CO_2\ m^{-2}\ s^{-1}$ at a CO_2 ambient concentration of 360 ppm, a temperature of 30°C, and saturating light. *(Initial data were taken from olive (*Olea europaea *cv. Manzanilla) from Pérez-Martín et al. (2009).)*

environmental variables), have also shown that increased g_m results in higher WUE_i (Moualeu-Ngangue et al., 2016).

4.2.2 Empirical Evidence

The theoretical relationship of g_m with A_n and WUE_i has been depicted; now, does this relationship really occur among plants? Data compilation for a large number of species and publications has shown a positive relationship of A_n with g_m, as expected (Flexas et al., 2013a,b; Niinemets et al., 2009) (Fig. 17.5A). This relationship is also found within studies when subjecting plants to different water regimes (Galle et al., 2009; Galmés et al., 2011a; among others), in accordance with the limitation imposed by g_m to photosynthesis during drought. Furthermore, higher values of g_m may translate into increased yields in crops with g_m-limited photosynthesis, such as rice (Lauteri et al., 2014).

The direct relationship of g_m with WUE_i is, however, more elusive. Although few studies have found a positive relationship of g_m with WUE_i (Hommel et al., 2014), most of them found no relationship or even negative correlation (Aranda et al., 2012; Barbour and Kaiser, 2016; Bunce, 2016).

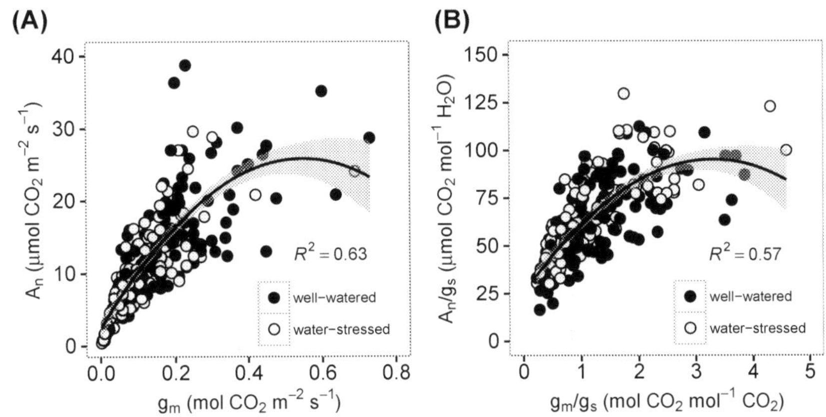

Figure 17.5 Empirical evidence of the influence of mesophyll conductance (g_m) on photosynthesis rate (A_n) and water use efficiency (A_n/g_s or WUE_i). (A) Multispecies A_n–g_m relationship with different irrigation treatments. (B) Multispecies A_n/g_s–g_m/g_s relationship with different irrigation treatments. In both graphs, *black line* represents a quadratic polynomial fit for both treatments ($P < .0001$) and *shaded area* is the 95% confidence interval. *(Data as in Fig. 17.3.)*

On the other hand, there is empirical evidence of the positive relationship of WUE_i with g_m/g_s when analyzing a large number of studies (Gago et al., 2014), especially under water stress (Flexas et al., 2013a) (Fig. 17.5B). As mentioned in earlier sections, this ratio may be useful for WUE_i selection within drought-tolerant genotypes (Galmés et al., 2011a; Théroux-Rancourt et al., 2015), although the complexity of measuring g_m difficults its use for high-throughput screening in the field (Barbour and Kaiser, 2016).

The empirical evidence shows that an improvement of both A_n and WUE_i through an increase in g_m is possible, if g_s is not modified. However, both conductances are tightly coordinated (as explained earlier) implying an added difficulty for water use improvement. In the next section, attempts to improve water productivity and yield with g_m modifications will be explained, as well as the success achieved so far.

4.3 Manipulating Mesophyll Conductance

4.3.1 Case Studies

Mesophyll conductance is a complex multitrait parameter, as is determined by both anatomy (chloroplast size and distribution, cell wall properties) and biochemistry (aquaporins, CA). As a result, precise modification of any of its components requires a complete understanding of its genetic and mechanistic basis. In the case of leaf anatomy, the complexity of the gene networks involved makes it extremely difficult to modify (Tholen et al., 2012). However, it has been possible to manipulate chloroplast arrangement (Tholen et al., 2008), which resulted in modified g_m through changes in S_c/S.

Biochemical factors influencing g_m may be a better target for g_m improvement. Several studies have observed g_m modifications when altering the levels of CA and, especially, aquaporins. Although CA activity has been related to g_m, their modification through genetic engineering has not resulted in significant improvements of either g_m or WUE_i (Flexas et al., 2016). Aquaporins, on the other hand, offer more promising results. Overexpression of aquaporin *NtAQP1* in tobacco resulted in enhanced g_m, whereas antisense mutants showed lower g_m (Flexas et al., 2006b). Although g_m alteration enhanced assimilation rates, simultaneous modification in g_s resulted in lower WUE_i, exemplifying the capital problem of single-parameter modification: the tight coregulation of all photosynthesis limitations (Flexas, 2016). Other studies have modified the levels of aquaporins; however, most of them yielded similar results (for a complete list of these studies, see Flexas et al., 2016). The concomitant increase in g_s may be the result of H_2O and CO_2 sharing a common path through the "aquaporin" (Sade et al., 2010).

Few studies have reported an improvement in either yield or WUE_i due to aquaporins. Kawase et al. (2013) showed that the overexpression of *McMIPB* (aquaporin from the crassulacean acid metabolism plant *Mesembryanthemum crystallinum*) enhanced A_n and g_m by 45% and 53%, respectively, with no modification of WUE_i. On the other hand, Sade et al. (2010) reported that overexpression of *NtAQP1* in tomato resulted in increased A_n (by 41%) and improved whole-plant water use efficiency, as *NtAQP1* was also expressed in roots. In this case, however, g_m was not measured and it is not clear if other pleiotropic effects were involved in the enhancement of A_n. These results show that, although not straightforward, g_m improvement using aquaporins may be possible. In addition, aquaporin levels may be modified by water management techniques such as drought pretreatment (Han et al., 2016). Despite increased WUE_i after drought pretreatment, treated plants presented lower A_n and, potentially, reduced yields due to a water-limited cycle.

In general, modifications of the potential mechanistic regulators of g_m resulted not only in variation of g_m but also in pleiotropic effects on other photosynthetic parameters, including V_{cmax} and J_{max} (Heckwolf et al., 2011) and, most importantly, g_s (Flexas et al., 2016). The decoupling of diffusion conductances (being able to increase the g_m/g_s ratio) remains one of the main goals for yield and water use efficiency improvement.

4.3.2 Potential New Targets

Attempts to modify g_m through any of its mechanistic determinants have obtained little success so far, as explained earlier. In addition, it has been shown that in order to accomplish a significant improvement of both photosynthesis and WUE_i, it is best to induce the simultaneous improvement of mesophyll and biochemical limitations while keeping g_s as constant as possible (Fig. 17.4). In this regard, some studies have reported general photosynthesis modifications that show promising results. Masle et al. (2005) identified the *ERECTA* gene, which is implied in several traits such as stomatal density, epidermis cell development, and mesophyll anatomical characteristics. The overexpression of this gene in *Arabidopsis thaliana* mutants resulted in enhanced whole-plant water use efficiency and photosynthetic capacity and, although no g_m values were reported, mesophyll modifications indicate that g_m may play a significant role (Flexas et al., 2016). This study shows that leaf anatomy could be modified to improve water productivity. Modifications of cell wall properties would also be an interesting target for g_m improvement, not only thickness but also its composition. Several of its components have been related to g_m, mainly structural sugars (arabinose, galactose, raffinose), possibly due to changes in cell wall porosity and/or tortuosity (Gago et al., 2016). However, to the best of our knowledge, no studies regarding cell wall engineering for photosynthesis improvement have been performed so far.

Although not directly linked with g_m, cytokinins (proteins involved in plant responses to abiotic stress) may be a potential target for WUE_i and, moreover, plant drought tolerance. The overexpression of cytokinins in transgenic tobacco plants expressing P_{SARK}::*IPT* (for senescence-associated receptor kinase::isopentenyltransferase) resulted in greater WUE_i under well-irrigated conditions and maintenance of photosynthesis under drought (Rivero et al., 2009). Although g_m was not measured (which may result in misleading conclusions regarding photosynthetic parameters), the improvement of WUE_i in the transgenic plants may be best explained by increased mesophyll conductance (Flexas, 2016). More research is needed to confirm the possible role of g_m in the assimilation and water use improvement reported in these studies.

One study has shown increased yield when cyanobacteria *Ictb* and *FBP/Sbpase* genes were expressed in transgenic rice (Gong et al., 2015). These genes are thought to be involved in CO_2 concentration and Ribulose bisphosphate regeneration. Although g_m and A_n showed higher values, g_s was also higher, resulting in lowered WUE_i. Nonetheless, this novel approach using genes from disparate phototrophic organisms may prove useful in the future.

The decoupling of CO_2 diffusion conductances has proven to be one of the main problems for improving WUE_i. A new target has been described that may overcome this limitation. Yang et al. (2016) reported that overexpression of ABA receptors in the leaf (the study included 14 different ABA receptors [regulatory component of ABA receptors]) resulted in higher ABA sensitivity and stomatal closure, yet mesophyll conductance was not affected. Some *Arabidopsis* mutant lines showed greater water use efficiency (at the leaf and whole-plant levels) without any yield compromise. One of the possible explanations of the decoupling of both conductances is related to the photosynthesis response pattern described in previous sections: upon first signs of stomatal closure, mesophyll conductance is still unaffected. This kind of modification may be possible in species with saturated assimilation operating over a wide range of g_s, such as grapevine (see Flexas et al., 2010 for graphical depiction).

5. CONCLUSIONS

CO_2 diffusion through the mesophyll is a complex process involving different biochemical and anatomical factors, mainly aquaporins, chloroplast distribution, and cell wall characteristics. Early photosynthesis studies neglected its role in determining photosynthesis; nowadays, there is an increasing amount of evidence that establishes g_m as one of the major factors limiting photosynthesis, including both maximum assimilation values (anatomy constraints) and photosynthetic performance under stress. It has been shown that, under drought, mesophyll conductance decreases and limits photosynthesis to a significant extent, usually less than or similar to stomatal conductance but more than biochemistry. Moreover, its inclusion in exhaustive photosynthesis studies is necessary to accurately estimate biochemical parameters, and also for prediction of assimilation rates under stress. The role of g_m under drought acclimation and recovery has been less studied; nonetheless, it is a major player in determining plant performance under long-term drought and photosynthesis recovery may be limited in some cases by mesophyll conductance.

Theoretically, mesophyll conductance and the ratio g_m/g_s are presented as potential targets for improving both yield and water use efficiency, one of two main goals for plant productivity in the upcoming years. However, most attempts reported thus far have accomplished little success, mostly due to the tight coordination of stomatal and mesophyll conductances. Despite these limitations, there are promising targets for improving mesophyll conductance and/or the g_m/g_s ratio, such as cytokinins and ABA receptors. On the other hand, leaf anatomy remains the "holy grail" for mesophyll conductance improvement. Although difficult, understanding of the precise mechanisms that regulate g_m and the underlying gene network controlling leaf development would provide a highly useful tool for photosynthesis and water productivity enhancement.

ACKNOWLEDGMENTS

MN was supported by a predoctoral fellowship BES-2015-072578 from the Ministerio de Economía y Competitividad (MINECO, Spain) cofinanced by the ESF. Research by MN and JF on mesophyll conductance is supported by the project CTM2014-53902-C2-1-P from the Ministerio de Economía y Competitividad (MINECO, Spain) and the ERDF (FEDER).

REFERENCES

Aganchich, B., Wahbi, S., Loreto, F., Centritto, M., 2009. Partial root zone drying: regulation of photosynthetic limitations and antioxidant enzymatic activities in young olive (*Olea europaea*) saplings. Tree Physiology 29, 685–696.

Aranda, I., Rodríguez-Calcerrada, J., Robson, T.M., Cano, F.J., Alte, L., Sánchez-Gómez, D., 2012. Stomatal and non-stomatal limitations on leaf carbon assimilation in beech (*Fagus sylvatica* L.) seedlings under natural conditions. Forest Systems 21 (3), 405–417.

Aranjuelo, I., Tcherkez, G., Molero, G., Gilard, F., Avice, J.-C., Nogués, S., 2013. Concerted changes in N and C primary metabolism in alfalfa (*Medicago sativa*) under water restriction. Journal of Experimental Botany 64 (4), 885–897.

Barbour, M.M., Kaiser, B.N., 2016. The response of mesophyll conductance to nitrogen and water availability differs between wheat genotypes. Plant Science 251, 119–127.

Beadle, C.L., Stevenson, K.R., Neumann, H.H., Thurtell, G.W., King, K.M., 1973. Diffusive resistance, transpiration, and photosynthesis in single leaves of corn and sorghum in relation to leaf water potential. Canadian Journal of Plant Science 53 (3), 537–544.

Bellasio, C., Beerling, D.J., Griffiths, H., 2016. An Excel tool for deriving key photosynthetic parameters from combined gas exchange and chlorophyll fluorescence: theory and practice. Plant, Cell and Environment 39 (6), 1180–1197.

Bota, J., Flexas, J., Medrano, H., 2001. Genetic variability of photosynthesis and water use in Balearic grapevine cultivars, Annals of Applied Biology, 138, 353–361.

Brilli, F., Tsonev, T., Mahmood, T., Velikova, V., Loreto, F., Centritto, M., 2013. Ultradian variation of isoprene emission, photosynthesis, mesophyll conductance, and optimum temperature sensitivity for isoprene emission in water-stressed *Eucalyptus citriodora* saplings. Journal of Experimental Botany 64 (2), 519–528.

Bunce, J.A., 2009. Use of the response of photosynthesis to oxygen to estimate mesophyll conductance to carbon dioxide in water-stressed soybean leaves. Plant, Cell and Environment 32, 875–881.

Bunce, J., 2016. Variation among soybean cultivars in mesophyll conductance and leaf water use efficiency. Plants 5 (44).

Cai, Y., Wang, J., Li, S., Zhang, L., Peng, L., Xie, W., Liu, F., 2015. Photosynthetic response of an alpine plant, *Rhododendron delavayi* Franch, to water stress and recovery: the role of mesophyll conductance. Frontiers in Plant Science 6 (1089).

Cano, F., Sánchez-Gómez, D., Rodríguez-Calcerrada, J., Warren, C.R., Gil, L., Aranda, I., 2013. Effects of drought on mesophyll conductance and photosynthetic limitations at different tree canopy layers. Plant, Cell and Environment 36, 1961–1980.

Cano, F.J., López, R., Warren, C., 2014. Implications of the mesophyll conductance to CO_2 for photosynthesis and water-use efficiency during long-term water stress and recovery in two contrasting *Eucalyptus* species. Plant, Cell and Environment 37 (11), 2470–2490.

Carmo-Silva, A.E., Gore, M.A., Andrade-Sanchez, P., French, A.N., Hunsaker, D.J., Salvucci, M.E., 2012. Decreased CO_2 availability and inactivation of Rubisco limit photosynthesis in cotton plants under heat and drought stress in the field. Environmental and Experimental Botany 83, 1–11.

Carriquí, M., Cabrera, H.M., Conesa, M.À., Coopman, R.E., Douthe, C., Gago, J., Gallé, A., Galmés, J., Ribas-Carbó, M., Tomàs, M., Flexas, J., 2015. Diffusional limitations explain the lower photosynthetic capacity of ferns as compared with angiosperms in a common garden study. Plant, Cell and Environment 38 (3), 448–460.

Chaves, M.M., Maroco, J.P., Pereira, J.S., 2003. Understanding plant responses to drought – from genes to the whole plant. Functional Plant Biology 30, 239–264.

Chaves, M.M., Flexas, J., Pinheiro, C., 2009. Photosynthesis under drought and salt stress: regulation mechanisms from whole plant to cell. Annals of Botany 103, 551–560.

Dang, Q.L., Guy, R.D., 1994. Genetic variation of ecophysiological trait in red alder (*Alnus rubra* Bong.). Canadian Journal of Forest Research 24, 2150–2156.

Delfine, S., Loreto, F., Alvino, A., 2001. Drought-stress effects on physiology, growth and biomass production of rainfed and irrigated bell pepper plants in the Mediterranean region. Journal of the American Society for Horticultural Science 126 (3), 297–304.

Delfine, S., Tognetti, R., Loreto, F., Alvino, A., 2002. Physiological and growth responses to water stress in field-grown bell pepper (*Capsicum annuum* L.). Journal of Horticultural Science and Biotechnology 77 (6), 697–704.

Di Marco, G., Manes, F., Tricoli, D., Vitale, E., 1990. Fluorescence parameters measured concurrently with net photosynthesis to investigate chloroplastic CO_2 concentration in leaves of *Quercus ilex* L. Journal of Plant Physiology 136, 538–543.

Díaz-Espejo, A., Nicolás, E., Fernández, J.E., 2007. Seasonal evolution of diffusional limitations and photosynthetic capacity in olive under drought. Plant, Cell and Environment 30, 922–933.

Duan, B., Ran, F., Zhang, X., Zhang, Y., Korpelainen, H., Li, C., 2011. Long-term acclimation of mesophyll conductance, carbon isotope discrimination and growth in two contrasting *Picea asperata* population exposed to drought and enhanced UV-B radiation for three years. Agricultural and Forest Meteorology 151 (1), 116–126.

Egea, G., González-Real, M.M., Baille, A., Nortes, A., Conesa, M.R., Ruiz-Salleres, I., 2012. Effects of water stress on irradiance acclimation of leaf traits in almond trees. Tree Physiology 32 (4), 450–463.

El Aou-ouad, H., Montero, R., Medrano, H., Bota, J., 2016. Interactive effects of grapevine leafroll-associated virus 3 (GLRaV-3) and water stress on the physiology of *Vitis vinifera* L. cv. Malvasia de Banyalbufar and Giro-Ros, Journal of Plant Physiology 196–197, 106–115.

Ethier, G.J., Livingston, N.J., 2004. On the need to incorporate sensitivity to CO_2 transfer conductance into the Farquhar-von Caemmerer-Berry leaf photosynthesis model. Plant, Cell and Environment 27, 137–153.

Evans, J.R., Sharkey, T.D., Berry, J.A., Farquhar, G.D., 1986. Carbon isotope discrimination measured concurrently with gas exchange to investigate CO_2 diffusion in leaves of higher plants. Australian Journal of Plant Physiology 13, 281–292.

Evans, J.R., Kaldenhoff, R., Genty, B., Terashima, I., 2009. Resistances along the CO_2 diffusion pathway inside leaves. Journal of Experimental Botany 60 (8), 2235–2248.

Farquhar, G.D., von Caemmerer, S., Berry, J.A., 1980. A biochemical model of photosynthetic CO_2 assimilation in leaves of C_3 species. Planta 149, 78–90.

Fleck, I., Peña-Rojas, K., Aranda, X., 2010. Mesophyll conductance to CO_2 and leaf morphological characteristics under drought stress during *Quercus ilex* L. resprouting. Annals of Forest Science 67 (3), 308–308.

Flexas, J., Bota, J., Escalona, J.M., Sampol, B., Medrano, H., 2002. Effects of drought on photosynthesis in grapevines under field conditions: an evaluation of stomatal and mesophyll limitations. Functional Plant Biology 29, 461–471.

Flexas, J., Bota, J., Loreto, F., Cornic, G., Sharkey, T.D., 2004. Diffusive and metabolic limitations to photosynthesis under drought and salinity in C_3 plants. Plant Biology 6, 269–279.

Flexas, J., Ribas-Carbó, M., Bota, J., Galmés, J., Henkle, M., Martínez-Cañellas, S., Medrano, H., 2006a. Decreased Rubisco activity during water stress is not induced by decreased relative water content but related to conditions of low stomatal conductance and chloroplast CO_2 concentration. New Phytologist 172, 73–82.

Flexas, J., Ribas-Carbó, M., Hanson, D.T., Bota, J., Otto, B., Cifre, J., McDowell, N., Medrano, H., Kaldenhoff, R., 2006b. Tobacco aquaporin NtAQP1 is involved in involved in mesophyll conductance to CO_2 *in vivo*. The Plant Journal 48, 427–439.

Flexas, J., Díaz-Espejo, A., Galmés, J., Kaldenhoff, R., Medrano, H., Ribas-Carbó, M., 2007. Rapid variation of mesophyll conductance in response to changes in CO_2 concentration around leaves. Plant, Cell and Environment 30, 1284–1298.

Flexas, J., Ribas-Carbó, M., Díaz-Espejo, A., Galmés, J., Medrano, H., 2008. Mesophyll conductance to CO_2: current knowledge and future prospects. Plant, Cell and Environment 31, 602–621.

Flexas, J., Barón, M., Bota, J., Ducruet, J.-M., Gallé, A., Galmés, J., Jiménez, M., Pou, A., Ribas-Carbó, M., Sajnani, C., Tomàs, M., Medrano, H., 2009. Photosynthesis limitations during water stress acclimation and recovery in the drought-adapted *Vitis* hybrid Richter-110 (*V. berlandieri* × *V. rupestris*). Journal of Experimental Botany 60 (8), 2361–2377.

Flexas, J., Galmés, J., Gallé, A., Gulías, J., Pou, A., Ribas-Carbó, M., Tomàs, M., Medrano, H., 2010. Improving water use efficiency in grapevines: potential physiological targets for biotechnological improvement. Australian Journal of Grape and Wine Research 16, 106–121.

Flexas, J., Barbour, M.M., Brendel, O., Cabrera, H.M., Carriquí, M., Díaz-Espejo, A., Douthe, C., Dreyer, E., Ferrio, J.P., Gago, J., Gallé, A., Galmés, J., Kodama, N., Medrano, H., Niinemets, Ü., Peguero-Pina, J.J., Pou, A., Ribas-Carbó, M., Tomàs, M., Tosens, T., Warren, C., 2012. Mesophyll conductance to CO_2: an unappreciated central player in photosynthesis. Plant Science 193–194, 70–84.

Flexas, J., Niinemets, Ü., Gallé, A., Barbour, M.M., Centritto, M., Díaz-Espejo, A., Douthe, C., Galmés, J., Ribas-Carbó, M., Rodríguez, P.L., Rosselló, F., Soolanayakanahally, R., Tomàs, M., Wright, I.J., Farquhar, G.D., Medrano, H., 2013a. Diffusional conductances to CO_2 as a target for increasing photosynthesis and photosynthetic water-use efficiency. Photosynthesis Research 117, 45–59.

Flexas, J., Scoffoni, C., Gago, J., Sack, L., 2013b. Leaf mesophyll conductance and leaf hydraulic conductance: an introduction to their measurement and coordination. Journal of Experimental Botany 64 (13), 3965–3981.

Flexas, J., Díaz-Espejo, A., Gago, J., Gallé, A., Galmés, J., Gulías, J., Medrano, H., 2014. Photosynthetic limitations in Mediterranean plants: a review. Environmental and Experimental Botany 103, 12–23.

Flexas, J., 2016. Genetic improvement of leaf photosynthesis and intrinsic water use efficiency in C_3 plants: why so much little success? Plant Science 251, 155–161.

Flexas, J., Díaz-Espejo, A., Conesa, M.A., Coopman, R.E., Douthe, C., Gago, J., Gallé, A., Galmés, J., Medrano, H., Ribas-Carbó, M., Tomàs, M., Niinemets, Ü., 2016. Mesophyll conductance to CO_2 and Rubisco as targets for improving intrinsic water use efficiency in C_3 plants. Plant, Cell and Environment 39, 965–982.

Gaastra, P., 1959. Photosynthesis of crops plants as influenced by light, carbon dioxide, temperature, and stomatal diffusion resistance. In: Mededelingen van de Landbouwhogeschool te Wageningen, Nederland, vol. 59, pp. 1–68 no. 13.

Gago, J., Douthe, C., Florez-Sarasa, I., Escalona, J., Galmés, J., Fernie, A.R., Flexas, J., Medrano, H., 2014. Opportunities for improving leaf water use efficiency under climate change conditions. Plant Science 226, 108–119.

Gago, J., Daloso, D.M., Figueroa, C.M., Flexas, J., Fernie, A.R., Nikoloski, Z., 2016. Relationships of leaf net photosynthesis, stomatal conductance, and mesophyll conductance to primary metabolism: a multispecies meta-analysis approach. Plant Physiology 171, 1–15.

Galle, A., Florez-Sarasa, I., Tomàs, M., Pou, A., Medrano, H., Ribas-Carbó, M., Flexas, J., 2009. The role of mesophyll conductance during water stress and recovery in tobacco (*Nicotiana sylvestris*): acclimation or limitation? Journal of Experimental Botany 60 (8), 2379–2390.

Galle, A., Florez-Sarasa, I., Thameur, A., de Paepe, R., Flexas, J., Ribas-Carbó, M., 2010. Effects of drought stress and subsequent rewatering on photosynthetic and respiratory pathways in *Nicotiana sylvestris* wild type and the mitochondrial complex I-deficient CMSII mutant. Journal of Experimental Botany 61 (3), 765–775.

Galle, A., Florez-Sarasa, I., El Aououad, H., Flexas, J., 2011. The Mediterranean evergreen *Quercus ilex* and the semi-deciduous *Cistus albidus* differ in their lead gas exchange regulation and acclimation to repeated drought and re-watering cycles. Journal of Experimental Botany 62 (14), 5207–5216.

Galmés, J., Medrano, H., Flexas, J., 2006. Acclimation of Rubisco specificity factor to drought in tobacco: discrepancies between *in vitro* and *in vivo* estimations. Journal of Experimental Botany 57 (14), 3659–3667.

Galmés, J., Medrano, H., Flexas, J., 2007. Photosynthetic limitations in response to water stress and recovery in Mediterranean plants with different growth forms. New Phytologist 75, 81–93.

Galmés, J., Conesa, M.A., Ochogavía, J.M., Perdomo, J.A., Francis, D.M., Ribas-Carbó, M., Savé, R., Flexas, J., Medrano, H., Cifre, J., 2011a. Physiological and morphological adaptations in relation to water use efficiency in Mediterranean accessions of *Solanum lycopersicum*. Plant, Cell and Environment 34, 245–260.

Galmés, J., Ribas-Carbó, M., Medrano, H., Flexas, J., 2011b. Rubisco activity in Mediterranean species is regulated by the chloroplastic CO_2 concentration under water stress. Journal of Experimental Botany 62 (2), 653–665.

Galmés, J., Ochogavía, J.M., Gago, J., Roldán, E.J., Cifre, J., Conesa, M.A., 2013. Leaf responses to drought stress in Mediterranean accessions of *Solanum lycopersicum*: anatomical adaptations in relation to gas exchange parameters. Plant, Cell and Environment 36, 920–935.

Genty, B., Meyer, S., Piel, C., Badeck, F., Liozon, R., 1998. CO_2 diffusion inside leaf mesophyll of ligneous plants. In: Garab, G. (Ed.), Photosynthesis: Mechanisms and Effects, vol. 5. Kluwer Academic Publishers, Dordrecht, pp. 3961–3966.

Gong, H.Y., Li, Y., Fang, G., Hu, D.H., Jin, W.B., Wang, Z.H., Li, Y.S., 2015. Transgenic rice expressing *Ictb* and *FBP/Sbpase* derived from cyanobacteria exhibits enhanced photosynthesis and mesophyll conductance to CO_2. PLoS One 10 (10), e0140928.

Grassi, G., Magnani, F., 2005. Stomatal, mesophyll conductance and biochemical limitations to photosynthesis as affected by drought and leaf ontogeny in ash and oak trees. Plant, Cell and Environment 28, 834–849.

Gu, J., Yin, X., Stomph, T.-J., Wang, H., Struik, P.C., 2012. Physiological basis of genetic variation in leaf photosynthesis among rice (*Oryza sativa* L.) introgression lines under drought and well-watered conditions. Journal of Experimental Botany 63 (14), 5137–5153.

Han, J., Meng, H., Wang, S., Jiang, C., Liu, F., Zhang, W., Zhang, Y., 2016. Variability of mesophyll conductance and its relationship with water use efficiency in cotton leaves under drought pretreatment. Plant Physiology 194, 61–71.

Hanba, Y.T., Shibasaka, M., Hayashi, Y., Hayakawa, T., Kasamo, K., Terashima, I., Katsuhara, M., 2004. Overexpression of the barley aquaporin HvPIP2;1 increases internal CO_2 conductance and CO_2 assimilation in the leaves of transgenic rice plants. Plant and Cell Physiology 45, 521–529.

Hansen, G.K., 1971. Photosynthesis, transpiration and diffusion resistance in relation to water potential in leaves during water stress. Acta Agriculturae Scandinavica 21 (3), 163–171.

Harley, P.C., Loreto, F., Di Marco, G., Sharkey, T.D., 1992. Theoretical considerations when estimating the mesophyll conductance to CO_2 flux by analysis of the response of photosynthesis to CO_2. Plant Physiology 98, 1429–1436.

Heckwolf, M., Pater, D., Hanson, D.T., Kaldenhoff, R., 2011. The *Arabidopsis thaliana* aquaporin AtPIP1;2 is a physiologically relevant CO_2 transport facilitator. The Plant Journal 67, 795–804.

Hikosaka, K., Shigeno, A., 2009. The role of Rubisco and cell walls in the interspecific variation in photosynthetic capacity. Oecologia 160, 443–451.

Hommel, R., Siegwolf, R., Saurer, M., Farquhar, G.D., Kayler, Z., Ferrio, J.P., Gessler, A., 2014. Drought response of mesophyll conductance in forest understory species – impacts on water-use efficiency and interactions with leaf water movement. Physiologia Plantarum 152 (1), 98–114.

IPCC, 2014. In: Core Writing Team, Pachauri, R.K., Meyer, L.A. (Eds.), Climate Change 2014: Synthesis Report. Contribution of Working Groups I, II and III to the Fifth Assessment Report of the Intergovernmental Panel on Climate Change, IPCC, Geneva, Switzerland.

Jones, H.G., Slatyer, R.O., 1972. Estimation of the transport and carboxylation components of the intracellular limitation to leaf photosynthesis. Plant Physiology 50, 283–288.

Kawase, M., Hanba, Y.T., Katsuhara, M., 2013. The photosynthetic response of tobacco plants overexpressing ice plant aquaporin McMIPB to a soil water deficit and high vapor pressure deficit. Journal of Plant Research 126, 517–527.

Keenan, T., Sabate, S., Gracia, C., 2010a. Soil water stress and coupled photosynthesis-conductance models: bridging the gap between conflicting reports on the relative roles of stomatal, mesophyll conductance and biochemical limitations to photosynthesis. Agricultural and Forest Meteorology 150 (3), 443–453.

Keenan, T., Sabate, S., Gracia, C., 2010b. The importance of mesophyll conductance in regulating forest ecosystem productivity during drought periods. Global Change Biology 16 (3), 1019–1034.

Lauteri, M., Haworth, M., Serraj, R., Monteverdi, M.C., Centritto, M., 2014. Photosynthetic diffusional constraints affect yield in drought stressed rice cultivars during flowering. PLoS One 9 (10), e109054.

Lauteri, M., Scartazza, A., Guido, M.C., Brugnoli, E., 1997. Genetic variation in photosynthetic capacity, carbon isotope discrimination and mesophyll conductance in provenances of *Castanea sativa* adapted to different environments. Functional Ecology 11, 675–683.

Lawlor, D., Tezara, W., 2009. Causes of decreased photosynthetic rate and metabolic capacity in water-deficient leaf cells: a critical evaluation of mechanisms and integration of processes. Annals of Botany 103 (4), 561–579.

Levi, A., Ovnat, L., Paterson, A.H., Saranga, Y., 2009. Photosynthesis of cotton near-isogenic lines introgressed with QTLs for productivity and drought related traits. Plant Science 177, 88–96.

Li, Y., Ren, B., Yang, X., Xu, G., Shen, Q., Guo, S., 2012. Chloroplast downsizing under nitrate nutrition restrained mesophyll conductance and photosynthesis in rice (*Oryza sativa* L.) under drought conditions. Plant and Cell Physiology 53 (5), 892–900.

Limousin, J.-M., Misson, L., Lavoir, A.-V., Martin, N.K., Rambal, S., 2010. Do photosynthetic limitations of evergreen *Quercus ilex* leaves change with long-term increased drought severity? Plant, Cell and Environment 33 (5), 863–875.

Lloyd, J., Syvertsen, J.P., Kriedemann, P.E., Farquhar, G.D., 1992. Low conductances for CO_2 diffusion from stomata to the sites of carboxylation in leaves of woody species. Plant, Cell and Environment 15, 873–899.

Long, S.P., Bernacchi, C.J., 2003. Gas exchange measurements, what can they tell us about the underlying limitations to photosynthesis? Procedures and sources of error. Journal of Experimental Botany 54 (392), 2393–2401.

Loreto, F., Centritto, M., 2008. Leaf carbon assimilation in a water-limited world. Plant Biosystems 142 (1), 154–161.

Loreto, F., Harley, P.C., Di Marco, G., Sharkey, T.D., 1992. Estimation of mesophyll conductance to CO_2 flux by three different methods. Plant Physiology 98, 1437–1443.

Loreto, F., Di Marco, G., Tricoli, D., Sharkey, T.D., 1994. Measurements of mesophyll conductance, photosynthetic electron transport and alternative electron sinks of field grown wheat leaves. Photosynthesis Research 41 (3), 397–403.

Masle, J., Gilmore, S.R., Farquhar, G.D., 2005. The *ERECTA* gene regulates plant transpiration efficiency in *Arabidopsis*. Nature 436, 866–870.

McAdam, S.A., Sussmilch, F.C., Brodribb, T.J., 2016. Stomatal responses to vapour pressure deficit are regulated by high speed gene expression in angiosperms. Plant, Cell and Environment 39, 485–491.

Miyazawa, S.-I., Yoshimura, S., Shinzaki, Y., Maeshima, M., Miyake, C., 2008. Deactivation of aquaporins decreases internal conductance to CO_2 diffusion in tobacco leaves grown under long-term drought. Functional Plant Biology 35, 553–564.

Mizokami, Y., Noguchi, K., Kojima, M., Sakakibara, H., Terashima, I., 2015. Mesophyll conductance decreases in the wild type but not in an ABA-deficient mutant (*aba1*) of *Nicotiana plumbaginifolia* under drought conditions. Plant, Cell and Environment 38, 388–398.

Momayyezi, M., Guy, R.D., 2017. Substantial role for carbonic anhydrase in latitudinal variation in mesophyll conductance of *Populus trichocarpa* Torr. & Gray. Plant, Cell and Environment 40 (1), 138–149.

Morison, J.I.L., Baker, N.R., Mullineaux, P.M., Davies, W.J., 2007. Improving water use in crop production. Philosophical Transactions of the Royal Society B 363, 639–658.

Moualeu-Ngangue, D.P., Chen, T.-W., Stützel, H., 2016. A modelling approach to quantify the effects of stomatal behavior and mesophyll conductance on leaf water use efficiency. Frontiers in Plant Science 7 (875).

Niinemets, Ü., Keenan, T., 2014. Photosynthetic responses to stress in Mediterranean evergreens: mechanisms and models. Environmental and Experimental Botany 103, 24–41.

Niinemets, Ü., Díaz-Espejo, A., Flexas, J., Galmés, J., Warren, C.R., 2009. Role of mesophyll conductance in constraining potential photosynthetic productivity in the field. Journal of Experimental Botany 60 (8), 2249–2270.

Niinemets, Ü., Flexas, J., Peñuelas, J., 2011. Evergreens favored by higher responsiveness to increased CO_2. Trends in Ecology and Evolution 26 (3), 136–142.

Oguchi, R., Hikosaka, K., Hirose, T., 2005. Leaf anatomy as a constraint for photosynthetic acclimation: differential responses in leaf anatomy to increasing growth irradiance among three deciduous trees. Plant, Cell and Environment 28, 916–927.

Oliver, R.J., Taylor, G., Finch, J.W., 2012. Assessing the impact of internal conductance to CO_2 in a land-surface scheme: measurement and modelling of photosynthesis in *Populus nigra*. Agricultural and Forest Meteorology 152, 240–251.

Olsovska, K., Kovar, M., Brestic, M., Zivcak, M., Slamka, P., Shao, H.B., 2016. Genotypically identifying wheat mesophyll conductance regulation under progressive drought stress. Frontiers in Plant Science 7 (1111).

Parkhurst, D.F., Mott, K., 1990. Intercellular diffusion limits CO_2 uptake in leaves. Plant Physiology 94, 1024–1032.

Peguero-Pina, J.J., Sisó, S., Flexas, J., Galmés, J., García-Nogales, A., Niinemets, Ü., Sancho-Knapik, D., Saz, M.A., Gil-Pelegrín, E., 2017. Cell-level anatomical characteristics explain high mesophyll conductance and photosynthetic capacity in sclerophyllous Mediterranean oaks. New Phytologist. https://doi.org/10.1111/nph.14406.

Pérez-Martín, A., Flexas, J., Ribas-Carbó, M., Bota, J., Tomás, M., Infante, J.M., Díaz-Espejo, A., 2009. Interactive effects of soil water deficit and air vapour pressure deficit on mesophyll conductance to CO_2 in *Vitis vinifera* and *Olea europaea*. Journal of Experimental Botany 60 (8), 2391–2405.

Pérez-Martín, A., Michelazzo, C., Torres-Ruiz, J.M., Flexas, J., Fernández, J.E., Sebastiani, L., Díaz-Espejo, A., 2014. Regulation of photosynthesis and stomatal and mesophyll conductance under water stress and recovery in olive trees: correlation with gene expression of carbonic anhydrase and aquaporins. Journal of Experimental Botany 65 (12), 3143–3156.

Pons, T.L., Flexas, J., von Caemmerer, S., Evans, J.R., Genty, B., Ribas-Carbó, M., Brugnoli, E., 2009. Estimating mesophyll conductance to CO_2: methodology, potential errors, and recommendations. Journal of Experimental Botany 60 (8), 2217–2234.

Poorter, H., Niinemets, Ü., Poorter, L., Wright, I.J., Villar, R., 2009. Causes and consequences of variation in leaf mass per area (LMA): a meta-analysis. New Phytologist 182, 565–588.

Pou, A., Medrano, H., Tomàs, M., Martorell, S., Ribas-Carbó, M., Flexas, J., 2012. Anisohydric behaviour in grapevines results in better performance under moderate water stress and recovery than isohydric behaviour. Plant and Soil 359, 335–349.

Price, G.D., von Caemmerer, S., Evans, J.R., Yu, J.-W., Lloyd, J., Oja, V., Kell, P., Harrison, K., Gallagher, A., Badger, M.R., 1994. Specific reduction of chloroplast carbonic anhydrase activity by antisense RNA in transgenic tobacco plants has a minor effect on photosynthetic CO_2 assimilation. Planta 193 (3), 331–340.

Renou, J.-L., Gerbaud, A., Just, D., André, M., 1990. Differing substomatal and chloroplastic CO_2 concentrations in water-stressed wheat. Planta 182, 415.

Resco, V., Ewers, B.E., Sun, W., Huxman, T.E., Weltzin, J.F., Williams, D.G., 2008. Drought-induced hydraulic limitations constrain leaf gas exchange recovery after precipitation pulses in the C_3 woody legume, *Prosopis velutina*. New Phytologist 181, 672–682.

Rho, H., Yu, D.J., Kim, S.J., Lee, H.J., 2012. Limitation factors for photosynthesis in 'Bluecrop' highbush blueberry (*Vaccinium corymbosum*) leaves in response to moderate water stress. Journal of Plant Biology 55, 450–457.

Rivero, R.M., Shulaev, V., Blumwald, E., 2009. Cytokinin-dependent photorespiration and the protection of photosynthesis during water deficit. Plant Physiology 150 (3), 1530–1540.

Rodríguez-Domínguez, C.M., Buckley, T.N., Egea, G., de Cires, A., Hernández-Santana, V., Martorell, S., Díaz-Espejo, A., 2016. Most stomatal closure in woody species under moderate drought can be explained by stomatal responses to leaf turgor. Plant, Cell and Environment 39, 2014–2026.

Roupsard, O., Gross, P., Dreyer, E., 1996. Limitation of photosynthetic activity by CO_2 availability in the chloroplasts of oak leaves from different species and during drought. Annals of Forest Science 53, 243–254.

Rzigui, T., De Paepe, R., Cornic, G., Streb, P., 2013. In the mitochondrial CMSII mutant of *Nicotiana sylvestris* photosynthetic activity remains higher than in the WT under persisting mild water stress. Plant Science 205–206, 20–28.

Sade, N., Gebretsadik, M., Seligmann, R., Schwartz, A., Wallach, R., Moshelion, M., 2010. The role of tobacco aquaporin1 in improving water use efficiency, hydraulic conductivity, and yield production under salt stress. Plant Physiology 152 (1), 245–254.

Samsuddin, Z., Impens, I., 1979. Photosynthesis and diffusion resistances to carbon dioxide in *Hevea brasiliensis* muel. agr. clones. Oecologia 37 (3), 361–363.

Scartazza, A., Lauteri, M., Guido, M.C., Brugnoli, E., 1998. Carbon isotope discrimination in leaf and stem sugars, water-use efficiency and mesophyll conductance during different developmental stages in rice subjected to drought. Australian Journal of Plant Physiology 25, 489–498.

Sharkey, T.D., Vassey, T.L., Vanderveer, P.J., Vierstra, R.D., 1991. Carbon metabolism enzymes and photosynthesis in transgenic tobacco (*Nicotiana tabacum* L.) having excess phytochrome. Planta 185, 287–296.

Sharkey, T.D., Bernacchi, C.J., Farquhar, G.D., Singsaas, E.L., 2007. Fitting photosynthetic carbon dioxide response curves for C_3 leaves. Plant, Cell and Environment 30, 1035–1040.

Silva, E.N., Silvera, J.A.G., Ribeiro, R.V., Vieira, S.A., 2015. Photoprotective function of energy dissipation by thermal processes and photorespiratory mechanisms in *Jatropha curcas* plants during different intensities of drought and after recovery. Environmental and Experimental Botany 110, 36–45.

Sorrentino, G., Haworth, M., Wahbi, S., Mahmood, T., Zuomin, S., Centritto, M., 2016. Abscisic acid induces rapid reductions in mesophyll conductance to carbon dioxide. PLoS One 11 (2), e018554.

Sun, P., Wahbi, S., Tsonev, T., Haworth, M., Liu, S., Centritto, M., 2014. On the use of land spectral indices to assess water status and photosynthetic limitations in *Olea europaea* L. during water-stress and recovery. PLoS One 9 (8), e105165.

Terashima, I., Ono, K., 2002. Effects of $HgCl_2$ on CO_2 dependence of leaf photosynthesis: evidence indicating involvement of aquaporins in CO_2 diffusion across the plasma membrane. Plant and Cell Physiology 43 (1), 70–78.

Terashima, I., Hanba, Y.T., Tholen, D., Niinemets, Ü., 2011. Leaf functional anatomy in relation to photosynthesis. Plant Physiology 155, 108–116.

Théroux-Rancourt, G., Éthier, G., Pepin, S., 2015. Greater efficiency of water use in poplar clones having a delayed response of mesophyll conductance to drought. Tree Physiology 35, 172–184.

Tholen, D., Boom, C., Noguchi, K., Ueda, S., Katase, T., Terashima, I., 2008. The chloroplast avoidance response decreases internal conductance to CO_2 diffusion in *Arabidopsis thaliana*. Plant, Cell and Environment 31, 1688–1700.

Tholen, D., Boom, C., Zhu, X.-G., 2012. Opinion: prospects for improving photosynthesis by altering leaf anatomy. Plant Science 197, 92–101.

Tomàs, M., Flexas, J., Copolovici, L., Galmés, J., Hallik, L., Medrano, H., Ribas-Carbó, M., Tosens, T., Vislap, V., Niinements, Ü., 2013. Importance of leaf anatomy in determining mesophyll diffusion conductance to CO_2 across species: quantitative limitations and scaling up by models. Journal of Experimental Botany 64 (8), 2269–2281.

Tosens, T., Niinemets, Ü., Vislap, V., Eichelmann, H., Castro Díez, P., 2012. Developmental changes in mesophyll diffusion conductance and photosynthetic capacity under different light and water availabilities in *Populus tremula*: how structure constrains function. Plant, Cell and Environment 35, 839–856.

Tosens, T., Nishida, K., Gago, J., Coopman, R.E., Cabrera, H.M., Carriquí, M., Laanisto, L., Morales, L., Nadal, M., Rojas, R., Talts, E., Tomàs, M., Hanba, Y., Niinemets, Ü., Flexas, J., 2016. The photosynthetic capacity in 35 ferns and fern allies: mesophyll CO_2 diffusion as a key trait. New Phytologist 209 (4), 1576–1590.

Tournex, C., Peltier, G., 1995. Effect of water deficit on photosynthetic oxygen exchange measured using $^{18}O_2$ and mass spectrometry in *Solanum tuberosum* L. leaf discs. Planta 195 (4), 570–577.

Troughton, J.H., 1969. Plant water status and carbon dioxide exchange of cotton leaves. Australian Journal of Biological Sciences 22 (2), 289–302.

Tsonev, T., Wahbi, S., Sun, P., Sorrentino, G., Centritto, M., 2014. Gas exchange, water relations and their relationships with photochemical reflectance index in *Quercus ilex* plants during water stress and recovery. International Journal of Agriculture and Biology 16, 335–341.

Uehlein, N., Otto, B., Hanson, D.T., Fischer, M., McDowell, N., Kaldenhoff, R., 2008. Function of *Nicotiana tabacum* aquaporins as chloroplast gas pores challenges the concept of membrane CO_2 permeability. The Plant Cell 20, 648–657.

von Caemmerer, S., Evans, J.R., 1991. Determination of the average partial pressure of CO_2 in chloroplasts from leaves of several C_3 plants. Australian Journal of Plant Physiology 18, 287–305.

von Caemmerer, S., Farquhar, G., Berry, J., 2009. Biochemical model of C_3 photosynthesis. In: Laisk, A., Nedbal, L., Govindjee (Eds.), Photosynthesis in Silico: Understanding Complexity from Molecules to Ecosystems. Springer Science + Business Media B.V., pp. 209–230

Vrábl, D., Vaškova, M., Hronková, M., Flexas, J., Šantrůček, J., 2009. Mesophyll conductance to CO_2 transport estimated by two independent methods: effect of variable CO_2 concentration and abscisic acid. Journal of Experimental Botany 60 (8), 2315–2323.

Wan, C., Sosebee, R.E., McMichael, B.L., 1993. Broom snakeweed responses to drought: I. Photosynthesis, conductance and water-use efficiency. Journal of Range Management 46, 355–359.

Warren, C., 2006. Estimating the internal conductance to CO_2 movement. Functional Plant Biology 33, 431–442.

Warren, C.R., 2008a. Soil water deficits decrease the internal conductance to CO_2 transfer but atmospheric water deficits do not. Journal of Experimental Botany 59 (2), 327–334.

Warren, C.R., 2008b. Stand aside stomata, another actor deserves centre stage: the forgotten role of the internal conductance to CO_2 transfer. Journal of Experimental Botany 59 (7), 1475–1487.

Warren, C.R., Adams, M.A., 2004. Evergreen trees do not maximize instantaneous photosynthesis. Trends in Plant Science 9 (6), 270–274.

Warren, C.R., Livingston, N.J., Turpin, D.H., 2004. Water stress decreases the transfer conductance of Douglas-fir (*Pseudotsuga menziesii*) seedlings. Tree Physiology 24, 971–979.

Wilkinson, S., Davies, W.J., 2002. ABA-based chemical signaling: the co-ordination of responses to stress in plants. Plant, Cell and Environment 25, 195–210.

Williams, T.G., Flanagan, L.B., Coleman, J.R., 1996. Photosynthetic gas exchange and discrimination against $^{13}CO_2$ and $C^{18}O^{16}O$ in tobacco plants modified by an antisense construct to have low chloroplastic carbonic anhydrase. Plant Physiology 112, 319–326.

Wright, I.J., Reich, P.B., Westoby, M., Ackerly, D.D., Baruch, Z., Bongers, F., Cavender-Bares, J., Chapin, T., Cornelissen, J.H.C., Diemer, M., Flexas, J., Garnier, E., Groom, P.K., Gulías, J., Hikosaka, K., Lamont, B.B., Lee, T., Lee, W., Lusk, C., Midgley, J.J., Navas, M.-L., Niinemets, Ü., Oleksyn, J., Osada, N., Poorter, H., Poot, P., Prior, L., Pyankov, V.I., Roumet, C., Thomas, S.C., Tjoelker, M.G., Veneklaas, E.J., Villar, R., 2004. The worldwide leaf economics spectrum. Nature 428, 821–827.

Yang, Z., Liu, J., Tischer, S.V., Christmann, A., Windisch, W., Schnyder, H., Grill, E., 2016. Leveraging abscisic acid receptors for efficient water use in *Arabidopsis*. Proceedings of the National Academy of Sciences of the United States of America 113 (24), 6791–6796.

Zhou, S., Duursma, R.A., Medlyn, B.E., Kelly, J.W.G., Prentice, I.C., 2013. How should we model plant responses to drought? An analysis of stomatal and non-stomatal responses to water stress. Agricultural and Forest Meteorology 182–183, 204–214.

Zhou, S.-X., Medlyn, B.E., Prentice, I.C., 2016. Long-term water stress leads to acclimation of drought sensitivity of photosynthetic capacity in xeric but not riparian *Eucalyptus* species. Annals of Botany 177, 133–144.

CHAPTER 18

The Olive Tree Under Water Stress: Fitting the Pieces of Response Mechanisms in the Crop Performance Puzzle

Antonio Diaz-Espejo[1], José E. Fernández[1], Jose M. Torres-Ruiz[2], Celia M. Rodriguez-Dominguez[1,3], Alfonso Perez-Martin[1], Virginia Hernandez-Santana[1]
[1]Instituto de Recursos Naturales y Agrobiología de Sevilla (IRNAS, CSIC), Seville, Spain; [2]BIOGECO, INRA, University Bordeaux, Pessac, France; [3]University of Tasmania, Hobart, TAS, Australia

Contents

If God had consulted me before embarking on the Creation, I would have suggested something simpler.
Alfonso X the wise, king of Castile (13th century).

1. THE OLIVE TREE IN THE SPOTLIGHT OF PLANT RESPONSE TO WATER STRESS

Olive is a major woody crop in arid and semiarid regions, where it has great economic importance (Carr, 2013). The olive tree is very resistant to drought (Fernández, 2014), able to keep a positive carbon balance

Water Scarcity and Sustainable Agriculture in Semiarid Environment
ISBN 978-0-12-813164-0
https://doi.org/10.1016/B978-0-12-813164-0.00018-1

(Angelopoulos et al., 1996; Giorio et al., 1999; Moriana et al., 2002; Hernandez-Santana et al., 2017), and to achieve reasonable yields, in such harsh environmental conditions (Moriana et al., 2003; Tognetti et al., 2006; Gucci et al., 2007; Fernández et al., 2013; Hernandez-Santana et al., 2017). These physiological traits made olive a woody plant model for ecophysiological studies on the response to water stress. Indeed, many research works and excellent reviews on its biology and physiology have been published in recent years (Connor and Fereres, 2005; Carr, 2013; Fernández, 2014). A simple search in the database of ISI Web of Science, filtered by "Agronomy AND Plant Science" reported a total of 415 works on "olive AND irrigation," 181 works on "olive AND water stress," 149 on "olive AND leaf water potential," 132 on "olive AND stomatal conductance," 156 on "olive AND (photosynthesis OR net CO_2 assimilation rate)," 49 on "olive AND hydraulic conduct*," and 38 on "olive AND (abscisic acid OR ABA)." Although these data are far below those of other fruit tree species, they represent a substantial amount of research done in the last decades in olive. Remarkably, some of these studies made on olive represent a breakthrough in plant science. They include the use of cutting-edge techniques, like X-ray microtomography in intact olive plants, to confirm the long-vessel artifactual effect incurred with classical measurements on loss of hydraulic conductivity, which has enhanced even more the idea of high embolism resistance in olive (Torres-Ruiz et al., 2015a, 2017); the quantification of the roles of leaf turgor-mediated and leaf turgor-independent mechanisms in stomatal decline in woody angiosperms during drought, by a process-based model (Rodriguez-Dominguez et al., 2016); and the confirmation that the rapid de novo biosynthesis of abscisic acid (ABA) in response to changes in vapor pressure deficit is triggered by reductions in leaf turgor (McAdam and Brodribb, 2016). These studies highlight olive as one of the most interesting woody species to be studied in response to water stress, and have depicted a clearer picture of the physiological mechanisms involved in this response to the lack of water. Taking a step forward, integrating all these mechanisms will allow us to understand the crop response to water stress and to improve sustainable irrigation strategies such as regulated deficit irrigation. When applying regulated deficit irrigation it is of outstanding importance to know at which stages the plant is most sensitive to water stress and when we could save water without penalizing yield. The correct use of deficit-irrigation strategies requires both a good understanding of physiological mechanisms related to the response of the plant to the lack of water and the use of reliable and sensitive indicators of water stress (Fernández, 2014).

Despite all the knowledge on the response of olive to water stress, we are still far from interpreting correctly the role of all the involved physiological variables. This curtails our ability to develop better irrigation strategies. The use of plant-based sensors for irrigation management, like dendrometers or sap flow and leaf patch clamp pressure probes, demands a clearer picture of the actual meaning of their outputs in the context of plant water stress. Our intention in this chapter is to place the meaning of physiological variables frequently measured in irrigation trials into the context of water use by the olive tree, and then integrate main physiological mechanisms of olive in its response to water stress, to unravel which are the consequences, trade-offs, and limits for implementing those mechanisms in agronomy.

2. FROM LEAF TO PLANT SCALE: IMPLICATIONS OF STOMATAL REGULATION ON TRANSPIRATION

For olive, as for most plant species, functioning requires an equilibrium between the amount of water demanded by the canopy and that taken up by the roots. In the soil-plant-atmosphere continuum the amount of water supplied to the canopy depends on both the hydraulic conductance of the plant (K) and the water potential gradient between the soil (Ψ_s) and the leaves (Ψ_{leaf}), while the water demanded by the leaves is directly proportional to both air water vapor pressure deficit (VPD) and stomatal conductance (g_s). This can be expressed as Eq. (18.1), which represents the equilibrium between liquid and vapor fluxes.

$$K(\psi_s - \psi_{leaf}) = g_s \, \text{VPD} \tag{18.1}$$

Obviously, this estimation of water demand from the tree crown is a simplification of the Penman–Monteith equation (Penman, 1958; Allen et al., 1998), since transpiration is driven not only by VPD but also by radiation. However, the olive tree has some characteristics that allows for applying Eq. (18.1) for water demand based on VPD only. Thus, olive trees have open canopies and small leaves, two features for high aerodynamic conductance even under low wind conditions. Under conditions of high aerodynamic conductance, the canopy is well coupled to the atmosphere (Jarvis and McNaughton, 1986). Several studies have reported on this high coupling estimating that over 90% of the olive transpiration is driven by VPD (Moreno et al., 1996; Tognetti et al., 2009; López-Bernal et al., 2015). Recently this high coupling of olive leaves with the surrounding atmosphere has been used to estimate g_s from sap flow—related measurements in commercial olive orchards (Hernandez-Santana et al., 2016a), which opens the door to new possibilities for managing irrigation.

Full open stomata allow maximizing CO_2 assimilation by leaves, which directly impacts on yield and growth (Hernandez-Santana et al., 2017). Under conditions of low VPD and full water availability in the soil (i.e., $\Psi_{soil} = 0$) both K and g_s achieve maximum values, resulting in a high Ψ_{leaf}. As VPD increases, the equilibrium between liquid and vapor water fluxes in the tree can be maintained by the increase of the water potential gradient between soil and the leaves, i.e., by decreasing Ψ_{leaf} values. However, these compensation mechanisms have limits, because the decrease in Ψ_{leaf} below a certain threshold value would induce embolism formation in the xylem vessels, with the subsequent loss in K and, therefore, a reduction in water supply to the leaves. Olive is a pretty resistant species to embolism formation in xylem, being able to keep 50% of its hydraulic functioning at water potential values as low as −4.0 and −5.5 MPa, depending on the variety (Torres-Ruiz et al., 2013, 2015a), with first xylem embolisms appearing at water potential below −2.5 MPa (Torres-Ruiz et al., 2017). So, under high atmospheric demand, it is challenging for the plant to achieve an equilibrium between transpiration and water taken up by roots, and the only way to avoid reaching water potentials in the xylem low enough for causing embolism is to reduce g_s through stomatal closure.

But how efficient is the stomatal closure on controlling plant transpiration? Fig. 18.1 shows solar radiation, VPD, and plant transpiration estimated from sap flow measurements in 35-year-old olive trees, in two different days and under two contrasting water treatments. Details on how transpiration was estimated for these trees can be found in Fernández et al. (2001). VPD was lower in July than in September, and hardly increased during the day in July, while in the sampled day of September it increased threefold in the afternoon. However, this increase in VPD in September did not enhance tree transpiration. Following Eq. (18.1), the increase in VPD in September should have augmented water use by the plant. However, the olive trees were able to control transpiration in the central hours of the day (red lines in Fig. 18.1). The most plausible explanation for this behavior is the occurrence of a marked stomatal closure, which was able to counterbalance the increase in VPD. Fig. 18.2 supports this hypothesis,

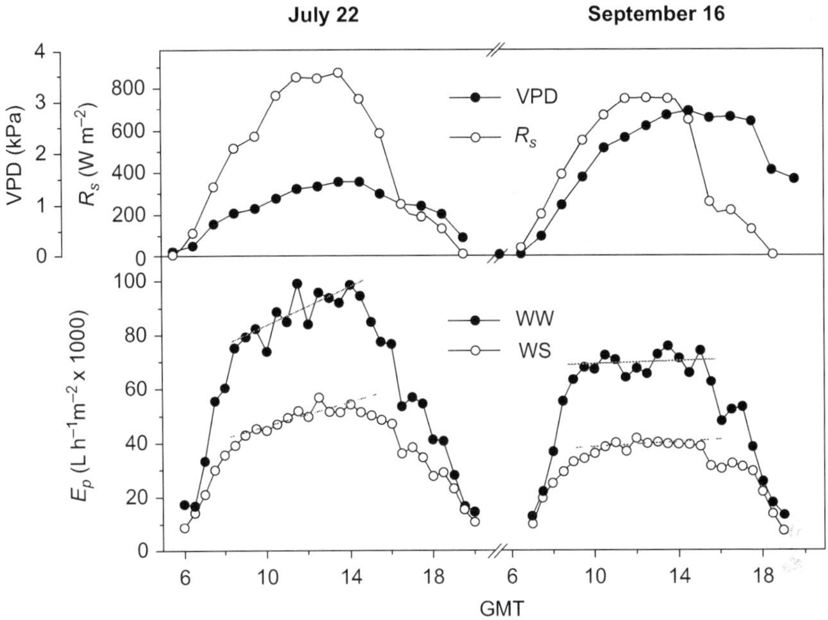

Figure 18.1 Diurnal evolution of tree transpiration (E_p) in 35-year-old olive trees at the beginning (July) and end (September) of summer for well-watered (WW) and water-stressed (WS) trees. Also represented are the air vapor pressure deficit (VPD) and solar radiation (R_s). *Red (dark gray in print version) lines* show the trend of transpiration during the central hours of the day, for both the WW (*black circles*) and WS (*white circles*) trees. Trees were able to maintain E_p during the central hours of the day in September despite the larger VPD on that date.

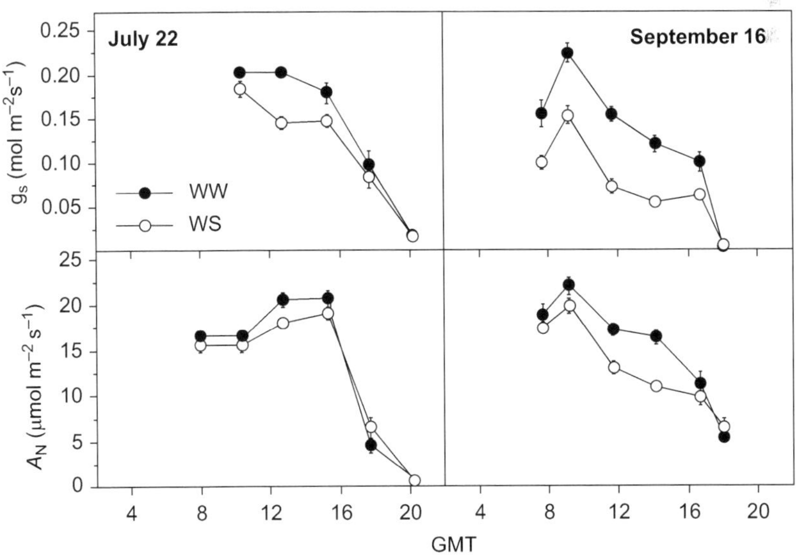

Figure 18.2 Diurnal evolution of stomatal conductance (g_s) and photosynthesis (A_N) in the same trees and dates shown in Fig. 18.1. Each point represents the average of six leaves. In September a strong stomatal closure in both water treatments is seen. See text for details.

and, while in July (low VPD) g_s was kept high during the central hours of the day (see red lines in Fig. 18.1), g_s decreased in September from early in the morning. In September, VPD doubled its value between 10:00 GMT and 14:00 GMT, meanwhile g_s halved it. This is the first example reported in the literature for olive tree of such a tight correspondence between stomatal closure, performed at the leaf level, and water use at the tree level. These findings illustrate the highly effective control of transpiration in olive when atmospheric demand is too high to be satisfied by water uptake. Stomatal closure, however, has a cost in the CO_2 assimilated by the leaf (A_N). As stomata close to reduce water loss by transpiration, the diffusion of CO_2 into the mesophyll is also reduced. In fact, a considerable reduction on A_N was observed in September in well-watered (WW) trees during the central hours of the day, increasing this limitation to 45% in water-stressed (WS) trees with respect to their maximum (Fig. 18.2), similarly to WS treatment. Photosynthesis limitation in olive, in response to water stress, has been attributed not only to stomatal conductance but also to mesophyll conductance to CO_2 and biochemical capacity (Diaz-Espejo et al., 2007; Perez-Martin et al., 2009, 2014). It is clear, however, that short-term responses under moderate water stress, like the commented reaction to VPD increase, are mainly a consequence of a diffusional restriction to CO_2 influx trough stomata. Why this stomatal response seems to be so pronounced in olive tree is a consequence of how water relation parameters are set in olive and the characteristics of its hydraulic system.

3. IMPORTANCE OF FINE-TUNING OF WATER RELATION PARAMETERS IN OLIVE LEAVES

Plant water status can be expressed in different ways. The choice of the variable depends on what physiological issue we are considering. If we are dealing with water transport in the plant, and the gradient in water potential between soil and plant driving transpiration is important, we could refer to either leaf or stem water potential. When studying the metabolic or biochemical reactions in the cells matrix, probably it is a better choice to measure leaf relative water content (RWC). However, if working with growth or mechanisms related to guard cells functioning, we could better consider turgor pressure. The most informative parameters related to leaf water relations that we use in the context of plant ecophysiology are osmotic pressure, at both full turgor (π_{FT}) and turgor loss point (π_{TLP}), Ψ_{leaf} at turgor loss point (Ψ_{TLP}), relative water content at turgor loss point

(RWC_{TLP}), and bulk elastic modulus (ε). All these variables are related and can be obtained from a pressure–volume (P–V) curve. A P–V curve is a plot of Ψ_{leaf} against water volume in drying leaves. There are several examples in the literature of P–V curves in olive, especially on how the parameters change under water stress conditions (Bacelar et al., 2009; Chartzoulakis et al., 1999; Dichio et al., 2003, 2006, 2009; Sofo et al., 2008). Among them, π_{TLP} has the highest impact on the capacity of the olive tree to cope with water stress. Actually, π_{TLP} is recognized as a major physiological determinant of plant water stress response (Bartlett et al., 2012). Bartlett et al. (2012) demonstrated that π_{TLP} has a primary role in determining physiological drought tolerance, and they established the importance of π_{TLP} as a functional trait at species and biome scale. π_{TLP} is mainly determined by π_{FT}, which means that it depends on the accumulation of solutes in the cells (Bartlett et al., 2012). Dichio et al. (2003) reported values of π_{TLP} in olive of −3.07 MPa and −3.85 MPa in well-irrigated and stressed plants, respectively. These π_{TLP} corresponded to values of π_{FT} of −2.06 MPa and −2.81, respectively. These values are similar to π_{FT} measured seasonally in a hedgerow "Arbequina" olive orchard (Fig. 18.3, see Fernández et al., 2011 for details of experimental conditions). Olive maintained their Ψ_{leaf} between −1.1 MPa and −1.5 MPa under well-watered conditions at the same time that their π_{FT} fluctuated

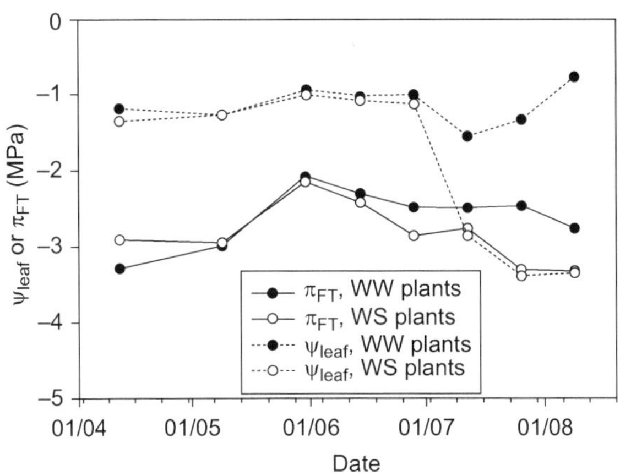

Figure 18.3 Seasonal evolution of leaf water potential (Ψ_{leaf}, *dashed lines*) and osmotic potential at full turgor (π_{FT}, *solid lines*) in well-watered (WW, *black circles*) and water-stressed (WS, *white circles*) trees. Ψ_{leaf} was measured with a Scholander chamber and π_{FT} with a Wescor psychrometer. Each point represents the average of four to six leaves.

between −2.0 MPa and −3.0 MPa. This implies a large safety margin with respect to the threshold for turgor loss point. No differences in either Ψ_{leaf} or π_{FT} were found until Ψs drop below −1.0 MPa and soil water became limited. Under moderate water stress conditions, water stressed plants reduced their π_{FT} and both Ψ_{leaf} and π_{FT} became very similar, which means that olive leaves were much closer to π_{TLP}. But, which are the benefits of lowering π_{TLP} values for the olive leaf? The reduction of π_{TLP} implies that cell turgor is lost at lower Ψ_{leaf}. Therefore, olive trees are capable of taking up water from drying soils at lower water potentials while maintaining turgor. Nice examples of the maintenance of turgor despite Ψ_{leaf} values are provided by the use of turgor-related probes (ZIM probes) in olive trees growing in drying soils (Fernández et al., 2011). In Fig. 7 of that cited work, it was observed that the slope of the linear correlation between Ψ_{leaf} and probe output was maintained along the season, but the offset shifted to lower values of Ψ_{leaf}. This change was related to osmotic adjustment. This is empirical evidence of the maintenance of turgor at lower Ψ_{leaf}. In summary, the low π_{TLP} values in olive are indicative of three main aspects related to its tolerance to drought. First, olive has a high capacity of adjusting its functional range of Ψ_{leaf}, achieving very low Ψ_{leaf} values before reaching permanent loss of turgor, i.e., permanent wilting point. Astonishing examples can be found in the literature in which olive plants recovered with irrigation after reaching Ψ_{leaf} values as low as −6 MPa (Perez-Martin et al., 2014) or −8.0 MPa (Moriana et al., 2003). Second, olive trees under severe water stress ($\Psi_{leaf} < -5.0$ MPa) are able to maintain residual, but still useful, values of g_s (around 0.035 mol m^{-2} s^{-1}) since guard cells need some turgor to keep stomata pores open. That marginal stomatal conductance might be critical to keep a positive carbon balance under drought conditions (Perez-Martin et al., 2014). And third, if olive is able to reduce its Ψ_{leaf} to such low values and still afford recovery, it means that its hydraulic system is very resistant to cavitation.

Additional adjustments, however, seems to be necessary for the leaves to be fully functional under severe water stress conditions. As most species, olive shows a close negative relationship between π_{FT} and ε (Bartlett et al., 2012; Dichio et al., 2003). Fig. 18.4 shows the seasonal evolution of leaf RWC in the same olive trees studied by Fernández et al. (2011) (see that work for details of experimental conditions). The figure illustrates that minimum values of RWC are constant along season, despite a larger decrease in Ψ_{leaf} along the dry season. The drop in minimum Ψ_{leaf} was

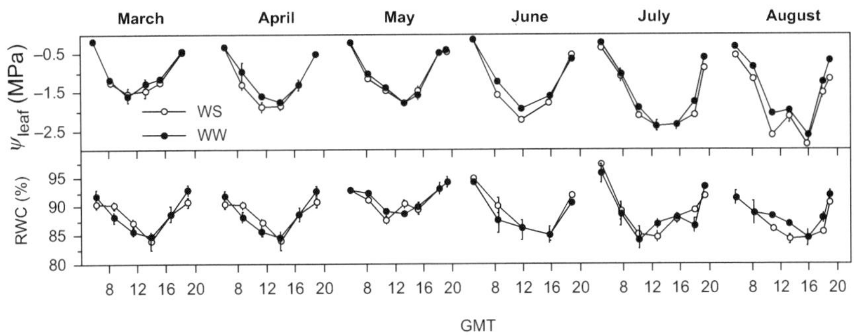

Figure 18.4 Diurnal evolution along the season of leaf water potential (Ψ_{leaf}) and leaf relative water content (RWC) for well-watered (WW) and water-stressed (WS) trees. Each point represents the average of six leaves. These trees correspond to the experimental trial of trees in Figs. 18.1 and 18.2.

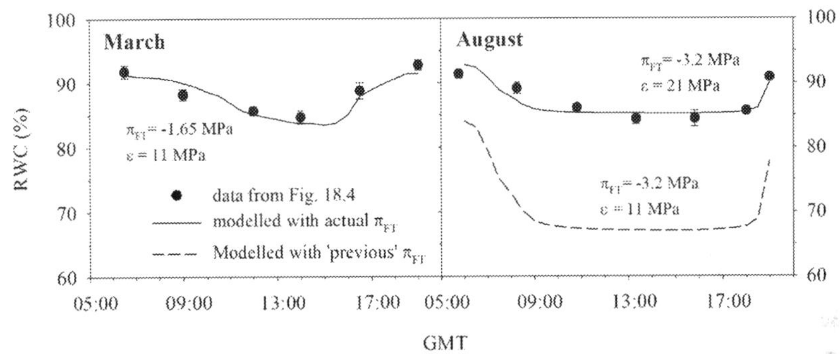

Figure 18.5 Simulation of the diurnal evolution of leaf relative water content by using the model BMF is explained in more detailed in Section 7. A simple water relation model was coupled to BMF model to simulate the effect of bulk elastic modulus in RWC after osmotic adjustment in August with respect to March.

mirrored by a decrease in π_{FT} and ε. As shown in Fig. 18.5, changes in ε were just enough to keep minimum RWC within a narrow range. π_{FT} decreased from −1.65 MPa in March to −3.2 MPa in August, at the same time that ε increased from 11 to 21 MPa. In that figure, the dashed line indicates the diurnal RWC dynamics in August for ε values similar to those in March. Minimum RWC under those conditions would have dropped to 63%, impairing the metabolism of the leaf (Lawlor and Cornic, 2002).

In summary, olive can adjust its π_{FT} to keep turgor despite low Ψ_{leaf}. Although in Fig. 18.3 we have seen evidence for the maintenance of a constant minimum Ψ_{leaf}, suggesting isohydric behavior, the drop in Ψ_s might be so large that a further decrease in Ψ_{leaf} is the only option to keep a water potential gradient that satisfies the water demand when g_s is already at its minimum values (Eq. 18.1). Both Figs. 18.3 and 18.4 show that, at some point, Ψ_{leaf} cannot be kept constant anymore and falls to values close to turgor loss point. To what extent the isohydric behavior is typical of the olive tree is discussed in the next section.

4. LOCATING THE OLIVE TREE IN THE "HYDROSCAPE" ALONG THE ISO- TO ANISOHYDRIC CONTINUUM

In olive, the effective control of tree transpiration by stomata closure in response to VPD keeps the balance between water supply and water demand. When a plant is able to maintain high and constant Ψ_{leaf} values at the central hours of the day by regulating stomatal opening, we say that it shows isohydric behavior. Is that at all the case of the olive tree? Some insights on the control of Ψ_{leaf} by stomata have been reported in the literature for olive (Fernández et al., 1997; Torres-Ruiz et al., 2013; Fernández, 2014). In their results, Fernández (2014) described what they labeled as a near-isohydric behavior in olive trees. However, recently Rodriguez-Dominguez et al. (2016), when studying the regulation of stomatal conductance in response to water stress, used olive as an example of anisohydric behavior. In Fig. 18.6 we can understand the reason for this

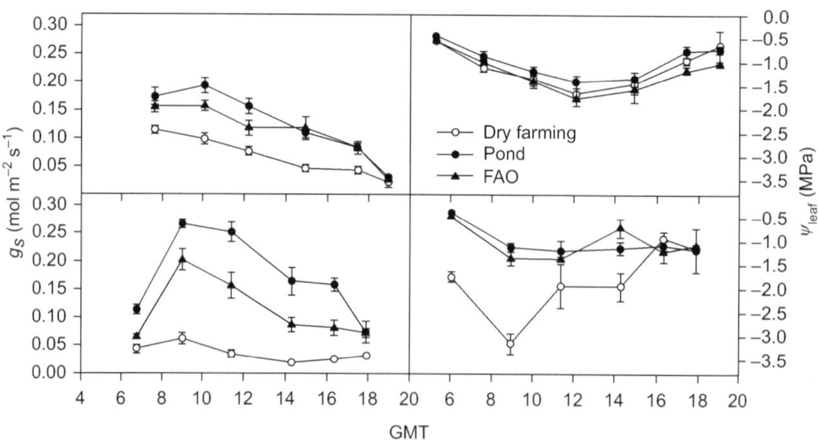

Figure 18.6 Diurnal evolution of stomatal conductance (g_s) and leaf water potential (Ψ_{leaf}) in 3-year-old olive trees growing in 50-L pots. *Black circles* represent a pond irrigation treatment (Pond) in which water was applied to maintain soil at field capacity, *black triangles* water was added to satisfy daily water used by the plant estimated by the FAO method (FAO), *white circles* corresponding to trees to which irrigation water was withhold (Dry farming).

discrepancy. Olive trees were able to show nearly perfect isohydric behavior in June, when water was still relatively available in the soil. Under those conditions, the reduction in g_s was enough to achieve this goal successfully. As soil dried out, however, stomatal closure was not sufficient to avoid Ψ_{leaf} from decreasing. These results reinforce the most recent opinion in the scientific community that plants are not strictly isohydric or anisohydric (Klein, 2014; Martínez-Vilalta et al., 2014; Garcia-Forner et al., 2016), but they move in a continuum that is determined not only by the physiological characteristics of the species but also by the environment in which they grow and the conditions they are currently experiencing (Schultz, 2003; Domec and Johnson, 2012; Rogiers et al., 2012). In fact, results from a recent meta-analysis carried out compiling information from more than 100 species from different climates (Martínez-Vilalta and Garcia-Forner, 2017) have shown how the regulation of Ψ_{leaf} is not necessarily related to a great capacity for controlling the stomatal behavior. Therefore, it is difficult to classify and quantify the control of Ψ_{leaf} in different species and to infer different strategies in response to water stress from such an analysis.

Martínez-Vilalta et al. (2014) turned up with a theoretical framework to quantify the degree of isohydry. Following Eq. (18.1) we can obtain

$$\psi_{leaf} = \psi_s - \frac{\text{VPD}\, g_s}{K} \tag{18.2}$$

$$\psi_{leaf} = \psi_s + \Lambda \frac{f_g(\psi_s)}{f_k(\psi_s)} \tag{18.3}$$

$$\psi_{leaf} = \psi_s + \Lambda \cdot \Theta(\psi_s) \approx \Lambda + \sigma \cdot \psi_s \tag{18.4}$$

where $f_g(\Psi_s)$ and $f_k(\Psi_s)$ are functions describing the reduction of g_s and K, respectively, with declining Ψ_s; Λ gives an estimate of the maximum transpiration when $\Psi_s = 0$, being indicative of the hydraulic capacity of the plant; Θ is the ratio between $f_g(\Psi_s)$ and $f_k(\Psi_s)$. As explained by Martínez-Vilalta et al. (2014), Θ can be transformed into σ, which determines the reduction in Ψ_{leaf} as Ψ_s declines. Therefore, σ characterizes the relative sensitivity of transpiration rate and plant hydraulic conductance to declining Ψ_s, and represents the level of isohydry. If $\sigma = 0$, our species is strictly isohydric, showing constant Ψ_{leaf} as Ψ_s declines. If $\sigma = 1$, it is a strict anisohydric species, i.e., the difference between Ψ_{leaf} and Ψ_s remains constant as Ψ_s decreases. If $0 < \sigma < 1$, then there is a partial isohydry, i.e., the difference between Ψ_{leaf} and Ψ_s is reduced as Ψ_s declines. Fig. 18.7A represents the relationship between midday leaf water potential (Ψ_{md})

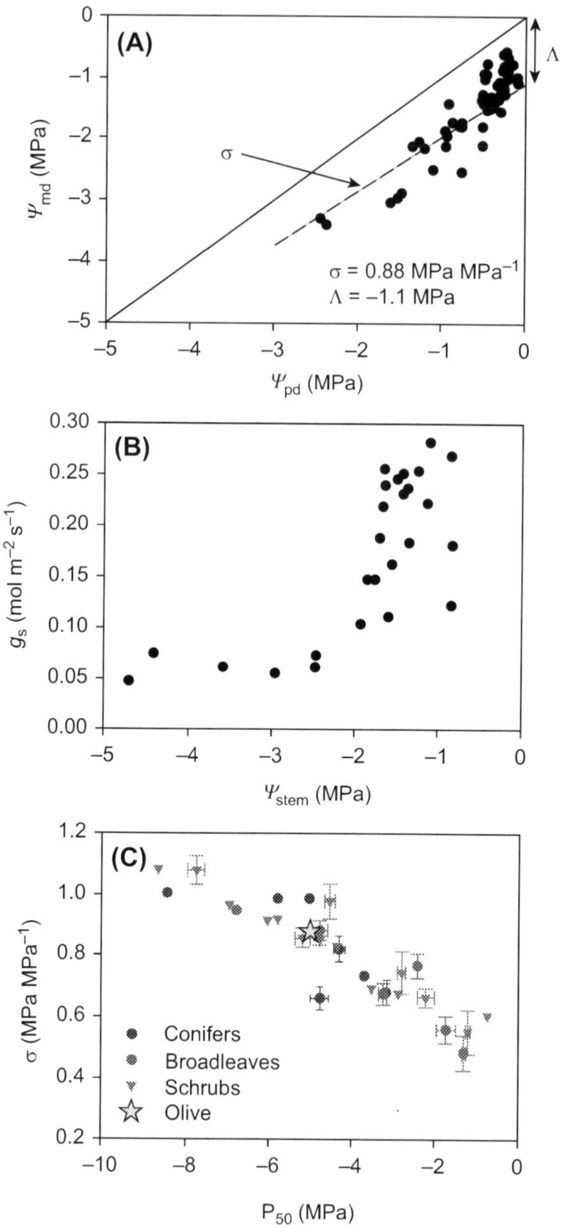

Figure 18.7 (A) Midday leaf water potential (Ψ_{md}) versus predawn leaf water potential (Ψ_{pd}, used as surrogate of soil water potential, Ψ_s). Line 1:1 is plotted. The relative sensitivity of transpiration rate and plant hydraulic conductance to declining Ψ_s (σ) and an estimate of the maximum transpiration when $\Psi_s = 0$, surrogate of the hydraulic capacity of the plant (Λ) are estimated (Martínez-Vilalta et al., 2014; Meinzer et al., 2016). (B) Stomatal conductance (g_s) versus stem water potential (Ψ_{stem}) corresponding to trees of panel (A) are shown. Each point represents the average of six and four leaves for g_s and Ψ_{stem}, respectively. (C) The relationship between σ and Ψ_{stem} at which 50% of loss of hydraulic conductivity is represented. Olive tree is represented by a *yellow (light gray in print version) star. (Data on conifers, broadleaves, and shrubs modified from Martínez-Vilalta, J., Poyatos, R., Aguade, D., Retana, J., Mencuccini, M., 2014. A new look at water transport regulation in plants. New Phytologist 204, 105–115.)*

and predawn leaf water potential (Ψ_{pd}, a surrogate of Ψ_s), for "Arbequina" olive trees during 2012 and 2013 (see Fernández et al., 2011 for details of experimental conditions). Each point is the average of a minimum of four leaves taken from April to October. Plot in Fig. 18.7A was analyzed after estimating Λ according to Meinzer et al. (2016). Ψ_{md} points close to $\Psi_{pd} = 0$ were restricted to the minimum observed value since at high Ψ_{pd} values Ψ_{md} is affected by nonsaturating light conditions or very low values of VPD (Meinzer et al., 2016) and do not represent the control by stomata. The slope of the regression, σ, is an estimate of the stringency of stomatal control of plant water status in drying soil. In olive, σ is equal to 0.88 MPa MPa^{-1}, which suggests that olive tree behaved more anisohydrically than isohydrically. This pattern cannot be considered as representative of what Franks et al. (2007) considered as isohydrodynamic behavior, characterized by a constant gradient between soil and leaf ($\Delta\Psi$) as Ψ_s declines, which is not the case in olive. Fig. 18.7A and results in Perez-Martin et al. (2014) reported marked $\Delta\Psi$ decrease as Ψ_s became lower. The value of 0.88 MPa MPa^{-1} is also very close to the average value reported by Martínez-Vilalta et al. (2014) for 102 species covering all biomes, 0.86 MPa MPa^{-1}. The main implication of a value of σ lower than 1 is that the sensitivity of stomatal conductance to declining Ψ_s is larger than that of hydraulic conductance. This is clearly depicted in Fig. 18.7B, where g_s shows 80% closure at relatively high values of stem water potential (Ψ_{stem}). In this sense, the correlation between σ and vulnerability to cavitation fits well in the meta-analysis carried out by Martínez-Vilalta et al. (2014) (Fig. 18.7C, olive tree represented by a yellow star). Species with σ closer to 1 tend to show lower percentage of cavitation at lower values of Ψ_{stem}. Although the value of σ for olive fits well for species living in environments with high atmospheric demand, as those associated to their area of distribution, Λ is relatively high (close to 0) for a species in such dry environments (Martínez-Vilalta et al., 2014). But there is another hypothesis that could explain the high Λ, which disagrees somehow with our idea of olive being a species very typical of arid environments. Terral et al. (2004) analyzed subfossil records in charcoal fragments and studied their wood vessels anatomy. Then they applied a hydraulic model to reinterpret the history and ecology of *Olea* evolution since the Neogene. They concluded that *Olea* had to survive in well-protected areas linked to riparian habitats during the Last Glacial Maximum. Riverbanks played a major role as refuge areas for *Olea*, and, afterwards, during the Holocene olive trees retained their underlying riparian characteristics, i.e., no need for large drop in Ψ_{leaf}

since Ψ_s was always high. That could explain why the sensitivity of stomata to Ψ_s, reflected in a value of σ close to 1, co-occurred in olive with a high Λ.

Meinzer et al. (2016) gave another step forward in our understanding of the complex regulation of Ψ_{leaf} by g_s. They selected eight woody species representative of a range of vegetation types and made them grow in a common environment before bringing them under water stress conditions by withholding irrigation. Their initial hypothesis aimed to prove that, since stomatal behavior is determined largely by leaf turgor, P—V traits must have a determinant role in explaining the iso- versus anisohydric behavior displayed by different species. They started their analysis similarly to Martínez-Vilalta et al. (2014), but they realized that similar values of σ could be estimated in either drought-resistant or drought-avoidance riparian species. The reason for this result is because the slope of Ψ_{md} versus Ψ_{pd} is affected by both the range of minimum Ψ_{leaf} in moist soil and the extent to which Ψ_{leaf} drops beyond the point at which stomata closure no longer affects Ψ_{leaf}. To avoid this incertitude, Meinzer et al. (2016) proposed to analyze the hydroscape area of the species. The hydroscape area of a species is defined as the area of a triangle bounded by the regression line between Ψ_{md} and Ψ_{pd} and the 1:1 line. They named this area "hydroscape" because it represents the water potential "landscape" over which stomata is effective in controlling Ψ_{leaf} as soil dries out. Our analysis shows that for olive this area was 5.54 MPa^2. When the hydroscape area of olive is plotted against its π_{FT} (Fig. 18.8), olive again fits well in the trend

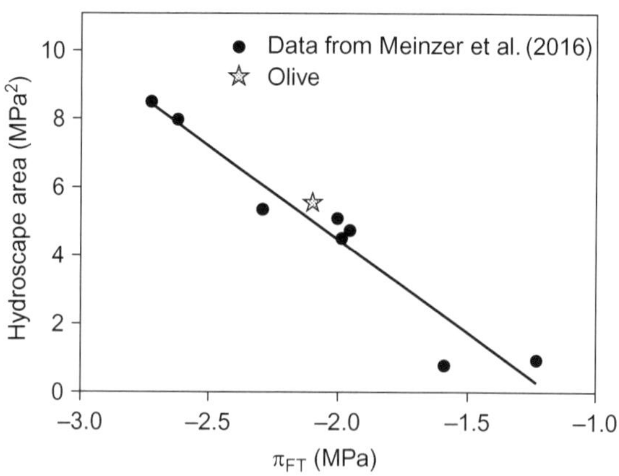

Figure 18.8 Hydroscape area (see text for details) versus osmotic potential at full turgor (π_{FT}) for olive *(yellow [light gray in print version] star)* in the context of the analysis performed by Meinzer et al. (2016) with eight woody species.

reported by Meinzer et al. (2016). We have plotted a π_{FT} value for well-watered plants (see Section 3), like Meinzer et al. (2016) did. Our analysis confirms that olive tree is located toward the anisohydric area. Species like olive, with a high capacity for osmotic adjustment, and therefore for lowering its π_{TLP}, are more prone to display an anisohydric-like behavior. This plasticity in water relation properties confers a high capacity to occupy a wider operating range of water potentials, resulting in the observed transition between apparent isohydric behavior at high Ψ_s and nearly anisohydric behavior as soil dries out (Fig. 18.6). Still, our perception of olive tree as a species usually showing isohydric behavior is unavoidable when patterns like the ones shown in Fig. 18.6 are observed. The transient Ψ_{leaf} homeostasis in olive can be explained on the basis of two main facts. First, olive is a species with a low hydraulic conductance growing, in general, in environments with high atmospheric demands for several months along the year. The high VPD values push the hydraulic system to its capacity limit, setting the minimum Ψ_{leaf} under well-watered conditions, which in olive is around -1.5 MPa. Olive could reduce further Ψ_{leaf} and increase $\Delta\Psi$ to fulfill VPD. However, there is a reason for not working under large $\Delta\Psi$. Franks (2006) observed in several species that the operational $\Delta\Psi$ is generally small, and he speculated that there were two main causes for this constraint. One is related to the control of cavitation and air embolisms, and the second one is to the influence on plasmodesmata function. Apparently, the resistance to flow through plasmodesmata increases if the pressure difference between two adjacent cells exceeds a certain threshold. A reduced $\Delta\Psi$ helps to keep this pressure differential. The second fact that favors our perception of olive-like isohydric depends on soil and root characteristics. Olive trees living in deep and well-structured soils will have larger root systems, such that soil water depletion in the root zone will occur later than for similar trees growing in shallow, sandy soils. This will allow for a slow water potential drop in the plant. That explains why in the surroundings of Seville (southern Spain), where the data shown in the figures come from, under the same climate some olive trees tend to show isohydric behavior (Fernández et al., 1997, 2006) and others do not (Cuevas et al., 2013; Fernández et al., 2013). Diaz-Espejo et al. (2012) analyzed this discrepancy with a hydraulic model and confirmed the limitation imposed by the roots and soil in olive orchards with a low ratio of root:leaves growing in sandy soil. Similar arguments were used by Martínez-Vilalta and Garcia-Forner (2017) to explain why the water potential regulation and stomatal behavior is decoupled across species living in the same environment.

Therefore stringency of stomatal regulation of plant water potential and the classification of olive tree as iso- or anisohydric species depend mainly on π_{FT} and K. Both are key physiological traits affecting the response of stomata to environmental changes. But we know that g_s is mainly regulated by hydraulic and chemical factors (ABA), or, in agreement with Rodriguez-Dominguez et al. (2016), by turgor-related and nonturgor-related factors. In the next two sections we review the hydraulic characteristics of olive tree and what we know about its response to ABA.

5. BEING SLOW BUT RESISTANT

Hydraulic efficiency of olive, defined as the capacity to supply water to downstream plant organs depending on the conducting elements available and the area and length of organs to satisfy, is among the lowest in fruit tree species. We cannot confound hydraulic efficiency (capacity for water transport) with hydraulic safety, which is related to the capacity of a species to maintain some water transport at low water potentials. An example of the low hydraulic efficiency of olive was given by Rodriguez-Dominguez et al. (2016), who compared K values in olive with those of almond and grapevine, also two species of high agronomical interest. Its low K determines largely the response of olive to a changing environment. High VPD increases the demand of water and, since the hydraulic efficiency is low, the drop in Ψ_{leaf} must be larger to equilibrate water supply with demand (see Section 2). Similarly, as Ψ_s declines, Ψ_{leaf} must decrease as well to maintain the $\Delta\Psi$. As we saw in Fig. 18.7A, olive reduces $\Delta\Psi$ progressively as soil drought progresses. Both large increases in VPD and reductions in $\Delta\Psi$ force stomata to close (see Fig. 18.6), although olive can be considered as a pretty resistant species to embolism, with 50% of their xylem vessels still functional at water potentials between −4.0 and −5.5 MPa depending on the variety (Torres-Ruiz et al., 2015a, 2017). Indeed, the K of stem (K_{stem}) only showed symptoms of hydraulic conductance loss below −2.5 MPa, as reported previously in different vulnerability curves to embolism for this species (Diaz-Espejo et al., 2012; Torres-Ruiz et al., 2013). A more detailed analysis of the K_{stem} response curve to Ψ_{stem} in olive allows us to estimate the air entry Ψ_{stem}, which represents the threshold pressure at which loss of conductivity begins to increase rapidly (Meinzer et al., 2009). For "Manzanilla" olive trees, this value is around −1.25 MPa (Torres-Ruiz et al., 2013). Recently, however, Torres-Ruiz et al. (2017) have rebuilt vulnerability curves in intact olive plants "var. Arbequina" by the novel method of X-ray microtomography, confirming the high embolism resistance for olive, with almost no embolism formation until the water potential reaches a threshold value of −4.0 MPa. This means a higher embolism resistance in olive stems than previously suggested, and it was confirmed that artifactual issues were involved as a result of the combination of centrifuge methods with the long-vessel nature of olive trees (Torres-Ruiz et al., 2014). Therefore, the strong g_s regulation observed at relatively high water potentials (Fig. 18.7B) cannot be related directly to the maintenance of the integrity of the hydraulic system at the stem level (Cochard and Delzon, 2013; Hochberg et al., 2017). Thus, K_{stem} cannot be a good candidate to regulate g_s. With hardly any change in the values of K_{stem} at Ψ_{stem} above −2.0 MPa (Torres-Ruiz et al., 2014, 2017), this canceled any chance for a hydraulic signal originating in the stem to trigger stomatal closure. However, when K is estimated at the plant level from sap flow measurements (Fig. 18.9), we observe a large decline in K at Ψ_{leaf} at noon between −0.5 and −2.0 MPa, a range in which g_s exerts 80% of its regulation. Similar results were observed in olive by Hernandez-Santana et al. (2016b) estimating K by Ohm's law and Tognetti et al. (2009). Obviously other components of the hydraulic system of olives must be the responsible for this drop of K at the

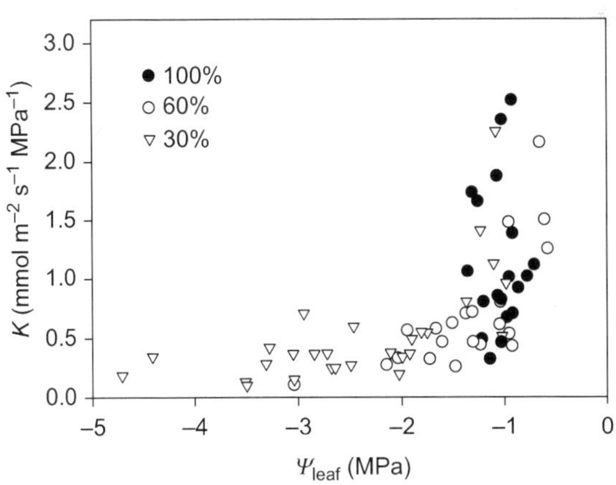

Figure 18.9 Relationship between plant hydraulic conductance (K) and leaf water potential at midday (Ψ_{md}). K was estimated from sap flow measurements and predawn water potential (Ψ_{pd}, surrogate of soil water potential) and Ψ_{md} by using Ohm's law. *Black circles* represent well-watered plants irrigated to satisfy 100% of their daily water consumption, *white circles* trees irrigated at 60% of their water needs, and *white triangles* irrigated at 30%. Each point represents the average of three and four replicates for sap flow and Ψ_{pd} and Ψ_{md}, respectively.

plant level, as roots and leaves are good candidates to drive such a change. The hydraulic of roots in olive have not drawn much attention in the literature except a few works studying the effect of soil temperature (Pavel and Fereres, 1998), salinity (Rewald et al., 2011), and drought (Torres-Ruiz et al., 2015b) in the hydraulic conductivity of the roots. Although Torres-Ruiz et al. (2015b) pointed out roots, in a similar way as leaves, as a quite vulnerable organ to embolisms, with around 80% of their hydraulic functionality lost at Ψ_{stem} of −2 MPa, a very recent study (Rodriguez-Dominguez et al., unpublished) have found a much greater hydraulic resistance at this level. In this study, embolism events were monitored simultaneously in the suberized roots, stems, and leaf midribs of intact 2-year-old olive seedlings using the recently developed optical method (Brodribb et al., 2016a,b, 2017). Interestingly, these authors did not observe any embolism in the root xylem until the plant reached at least a Ψ_{stem} of −4 MPa, which questions the hydraulic segmentation previously found in olive species. This might be likely explained by the necessity of the plant to preserve the functionality of the roots as long as possible to be able to resprout during more favorable conditions. Accordingly, that finding definitely leaves the hydraulic regulation of g_s to the leaf level, which, thinking about the proximity with leaf stomata, would make more sense.

On the other hand, K_{leaf} has been reported to be highly dynamic at high water potentials above −1.5 MPa. Hernandez-Santana et al. (2016b) estimated for olive that 50% of K_{leaf} is already lost at Ψ_{leaf} over −1.5 MPa. Although it might appear counterintuitive, this large loss of hydraulic conductance at high water potentials daily overcome by olive in its dry environment, this is a common result found in the literature (Scoffoni et al., 2012; Sack et al., 2016), and this is now a focus of attention for scientists of the plant hydraulic community. Although not specifically in olive but in other species, it has been reported that the xylematic component of K_{leaf} is as resistant to embolism as it is in the stem (e.g., *Pinus pinaster*, Bouche et al., 2016, Brodribb et al., 2016a; Trifiló et al., 2016), although that was not the case in others (e.g., *Vitis vinifera*, Charrier et al., 2016). So, for those species showing similar resistance to embolism of the xylematic component at leaf and stem level, the extraxylematic component seems to be a plausible approach to account for the changes observed in response to Ψ_{leaf} (Scoffoni et al., 2017). Although this is currently a controversial issue, it is very appealing option to think of a dynamic K_{leaf} that could modulate the response of g_s to changes in the environment. As the extraxylematic component depends mainly on mesophyll anatomy and/or changes in aquaporin abundance and activity (Buckley, 2015; Kim and Steudle, 2007), it is not subjected to irreversible losses of hydraulic functionality as those occurring in the xylem (Charrier et al., 2016). Nevertheless, more specific experiments focused on that question (the huge decline in K_{leaf} at high Ψ_{leaf}) are needed to isolate possible artefactual results emerging from the techniques used. Moreover, within a context of incorporating the role of hormonal factors, Pantin et al. (2012) suggested that K_{leaf} could be regulated by ABA produced under water stress conditions.

6. DISENTANGLING THE ROLE OF ABSCISIC ACID IN THE RESPONSE TO WATER STRESS IN OLIVE

Abscisic acid is undoubtedly a promoter of stomatal closure in angiosperms (Brodribb and McAdam, 2011). When applied to detached leaves it triggers, in minutes, a response of guard cells independently of plant water status (Jia et al., 1996; Pantin et al., 2012). This response, and the fact that increasing concentrations of ABA are usually measured in plants under water stress conditions, are behind ABA being considered to play a major role in g_s regulation. Several authors have related ABA to the response of olive to water stress (Guerfel et al., 2009; Kitsaki and Drossopoulos, 2005; Torres-Ruiz et al., 2014; Hernandez-Santana et al., 2016b), as well as to flower induction and development (Ulger et al., 2004) and fruit development (Kitsaki et al., 1995). For many fruit tree species, ABA has been linked to partial root drying (PRD). This irrigation strategy implies alternating irrigation from one-half of the root zone to the other every 2–3 weeks, which is supposed to induce ABA synthesis in the drying half. This ABA generated in the roots is assumed to travel to the leaf and induce stomatal closure. This long-distance root-to-shoot signaling mechanism has been stated to increase water use efficiency and water productivity (Dry et al., 2001; Dodd et al., 2006). Other authors, however, reported no benefits of PRD as compared to other deficit-irrigation strategies, implying uniform wetting of the whole root zone for a variety of species (Sadras, 2009; Egea et al., 2010; Perez-Perez et al., 2012), including olive (Fernández et al., 2006; Dbara et al., 2016). The reason for this disappointing effectivity of PRD must be sought in the conceptual role attributed to ABA in the stomata response to water stress. Early evidence on the role of root-sourced ABA (Zhang and Davies, 1989; Tardieu and Davies, 1993) has been challenged by recent empirical data. ABA was believed to be mostly synthesized and accumulated in the roots as soil dries out, presumably as a function of Ψ_s (Tardieu and Simonneau, 1998). However, experiments with tomato and Arabidopsis using reciprocal grafting between wild-type (WT) and ABA-deficient mutants demonstrated that ABA promoted in leaves is the main contributor to the total shoot hormone, whereas root-sourced ABA has a negligible contribution (Holbrook et al., 2002; Christmann et al., 2007; Dodd et al., 2009; Albacete et al., 2015). Manzi et al. (2015, 2016, 2017) have further demonstrated that the appearance of ABA in detached roots in response to dehydration (Simonneau et al., 1998; Zhang and Tardieu, 1996) is, in fact, mediated by previous accumulation of ABA precursors from the shoot. Therefore, the presence of ABA in roots might be a consequence of shoot-to-root, instead of root-to-shoot, signaling. Phloem is very effective in transporting ABA basipetally, i.e., toward the roots (Boursiac et al., 2013; Jeschke et al., 1997; Zhong et al., 1996), even under water stress conditions (Vernieri et al., 2001; Ikegami et al., 2009). Girdling experiments in drought in *Pinus radiata* plants have demonstrated that ABA accumulates largely in the top millimeters of the bark over the girdling cut, at the same time that root ABA is gradually reduced (Mitchell et al., 2016). But why would plants be interested in this basipetal signaling? Do leaves require information on the soil water depletion to adjust stomatal opening accordingly? Likely not, since leaves can feel instantaneously the soil water status by hydraulic signals. Classical experiments by Passioura and Tanner (1985) and Comstock and Mencuccini (1998) showed unequivocally how stomatal conductance was regulated in seconds in response to a sudden perturbation of soil pressure. A chemical shoot-to-root signal is, however, very useful for the plant. Leaves are the ultimate component of the soil-plant-atmosphere continuum and have the "duty" of integrating both water and CO_2 fluxes while sensing soil and atmospheric conditions. They are just in the right location in the plant for sensing better whether a given Ψ_s is enough for satisfying the atmospheric demand and maintaining a positive carbon balance. If soil water becomes insufficient to replace the plant needs, then ABA might play an important role inducing root growing and ramification (De Smet et al., 2003; Deak and Malamy, 2005; McAdam et al., 2016b; Spollen et al., 2000) for increasing the root hydraulic conductivity (Kudoyarova et al., 2011; Mahdieh and Mostajeran, 2009). It is well known that the root:leaf ratio increases in plants under water stress (Poorter et al., 2012). Overall, these actions aim to increase the supply of water to the canopy.

But to what extent does ABA play a role in the regulation of g_s in olive? It is difficult to answer to this question unequivocally, since all processes are

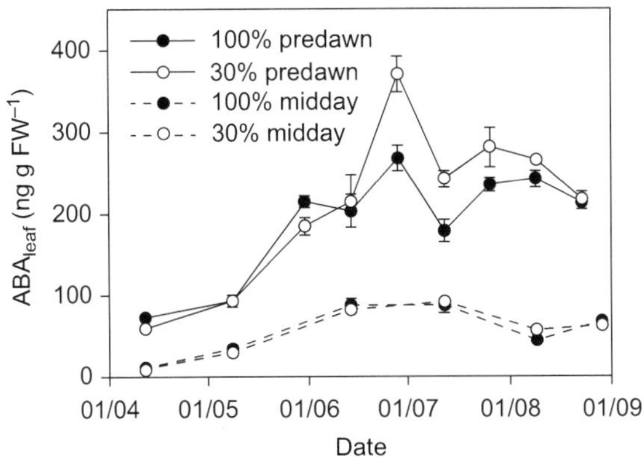

Figure 18.10 Seasonal evolution of leaf abscisic acid (ABA) in well-watered plants (100%, *black circles*) and water-stressed plants (30%, *white circles*) at predawn (*solid line*) and noon (*dashed line*). Samples were collected in field-grown olive trees. ABA was determined by HPLC-MS-MS. Each point is the average of six leaves.

very much interrelated. Up to fourfold higher concentrations of leaf ABA are found at predawn than at midday in olive, independently of irrigation treatment (Fig. 18.10, unpublished results). The higher concentration of ABA before the onset of the day and during the night has been explained by Tallman (2004) as a means to regulate the initial phase of stomatal opening in the morning and to ensure that stomata remain closed in darkness. At dawn, most of this ABA is catabolized and stomata are allowed to open again. During the day new ABA is synthetized depending on atmospheric and drought conditions. A correlation between g_s and ABA does not solve the question of the role played by ABA since they do not mean cause and effect. Fig. 18.10 showed that the differences in ABA concentration found between treatments at dawn vanished at midday, hindering the role of ABA as a major driver of g_s regulation. Fig. 18.11 shows the relationship between g_s and ABA in sap and leaves of olive trees (see Fernández et al., 2011 for details). No significant changes in sap ABA between well-watered and stressed trees were observed. Only when g_s was at its minimum did ABA increase significantly. But in the case of leaf ABA, which is supposed to induce the response of guard cells, no clear pattern was observed. This lack of correlation between g_s and ABA is often reported in the literature. It is usually found that no relationship appears and, even when present, it varies with the time of the day, level of stress, or individual (Holbrook et al., 2002). This perception of an unnecessary role of ABA in the response of g_s to soil drought agrees well with findings reported by

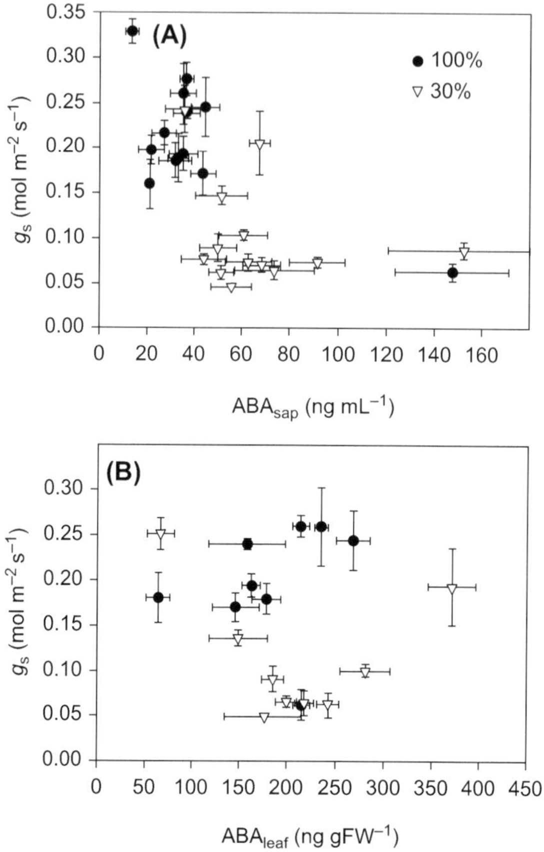

Figure 18.11 Relationship between stomatal conductance (g_s) and (A) sap abscisic acid (ABA) and (B) leaf ABA at noon. Samples were collected in field-grown olive trees of well-watered plants (100%, *black circles*) and water-stressed plants (30%, *white triangles*). ABA was determined as in Fig. 18.10. Each point represents the average of six measurements.

Rodriguez-Dominguez et al. (2016). They used a mechanistic model to assess the role of turgor-mediated and nonturgor-mediated (ABA related) mechanisms in g_s regulation in response to soil drying for three woody species of high agronomical value (grapevine, almond, and olive). They concluded that the putative role of ABA in the response of stomata to drought only accounted for 13% of the observed response in almond and grapevine, and 0% in olive.

The total lack of nonturgor-related mechanisms in the response of stomata to drought in olive does not mean that ABA has an unessential function in relation to g_s. Recently, McAdam et al. (2016a) demonstrated that VPD response is mediated by rapid de novo synthesis of ABA in the leaf. This finding solved one of the "black holes" in our understanding of stomatal response to air humidity, and validates the most plausible explanation given until now, which is that the guard cell osmotic pressure is actively regulated in response to leaf water status. This is the so-called hydroactive feedback hypothesis, and McAdam et al. (2016a) filled this gap and provided the evidence to explain how leaves overcome the wrong-way response due to the mechanical advantage of epidermal cells over guard cells as discussed by Buckley (2016). Furthermore, McAdam and Brodribb (2016) linked this rapid de novo synthesis of ABA with turgor pressure for several species including olive. Their results with detached shoots show that ABA synthesis can be promoted in olive leaves artificially by pressurizing at pressures over 1.2 MPa. ABA synthesis in leaves is, therefore, triggered before turgor loss point, and it is expected to occur even in well-watered plants exposed to moderate increase in VPD. Our ABA data from well-watered olive trees growing under field conditions confirm this result (Fig. 18.12). During our measurements, VPD increased from 1.2 to 4.5 kPa. Despite a minimum Ψ_{leaf} value of only −1.0 MPa, ABA increased at midday by threefold with respect to early morning values (Fig. 18.12A). This result does not contradict those shown in Fig. 18.10 since, as explained by Tallman (2004), most of the ABA is catabolized at dawn under low fluences of blue light during the process of stomatal opening. Therefore, at the relatively high Ψ_{leaf} value of −0.5 MPa, ABA had already reached maximum values (Fig. 18.12B). Interestingly, the increase in ABA was related to the loss of K_{leaf} measured in opposite leaves to those sampled for ABA in the shoot, which were experiencing very similar conditions. These findings sustain the results by Pantin et al. (2012) suggesting that ABA downregulates K_{leaf}. And these results likewise support the step response of K_{leaf} to dehydration in olive leaves at high Ψ_{leaf} (Hernandez-Santana et al., 2016b).

Comprehension of the role of ABA in the regulation of stomatal conductance in olive requires additional studies. Since transpiration of the whole tree is so much determined by g_s, as we demonstrated in Section 1, we must comprehend thoroughly what mechanisms related to the response to water stress have to be considered to model g_s satisfactorily. Although ABA is putatively associated with chemical signaling, or nonturgor-related component of regulation of g_s in response to drought, its close relationship with leaf turgor in its synthesis makes its role much more complex to understand.

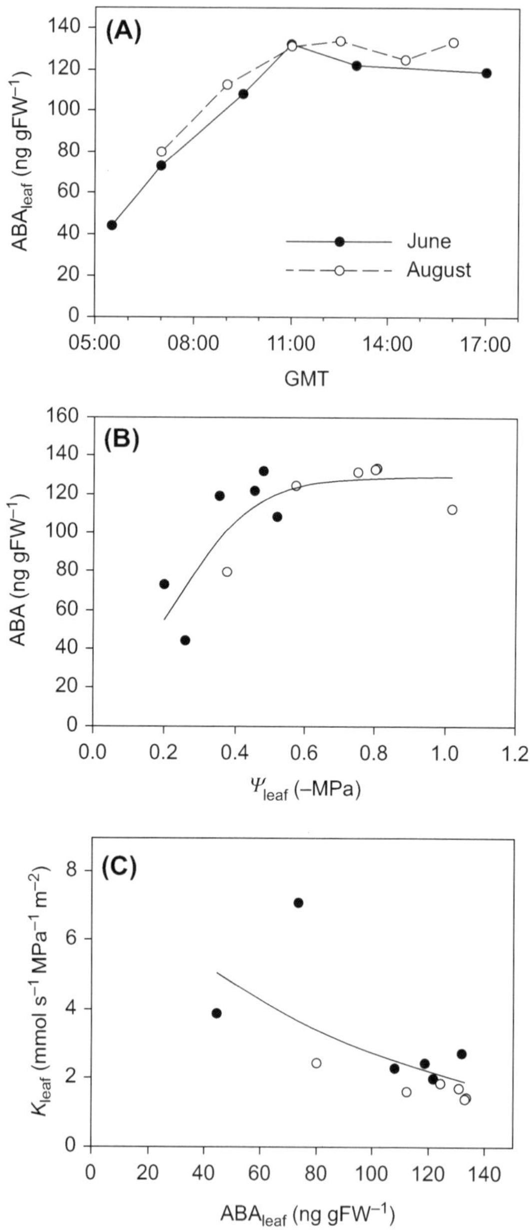

Figure 18.12 (A) Diurnal evolution of leaf abscisic acid (ABA) in well-watered olive trees under field conditions in June and August. ABA was determined as in Fig. 18.10. (B) Relationship between ABA in leaves of (A) and leaf water potential (Ψ_{leaf}) of its opposite leaf in the shoot. (C) Relationship between leaf hydraulic conductance (K_{leaf}) and leaf ABA. Measurements were taken in the same leaf. K_{leaf} was measured in situ in the field by the dynamic rehydration method (Blackman and Brodribb, 2011) immediately after leaf sampling.

7. FITTING TOGETHER THE PIECES OF THE PUZZLE

In previous sections we have reviewed main topics related to the water use of olive trees, focusing on the main determinants of g_s regulation under water stress conditions. However, it is complex to understand how this plethora of mechanisms assembles under any given environmental conditions to produce the observed response. For example, we have seen that olive is able to reduce its π_{TLP} to keep turgor at low Ψ_{leaf}. However, that does not result in maintaining stomata wide open, since at low Ψ_{leaf} there is an important loss of K_{leaf}. Process-based models (i.e., mechanistic) are effective tools for bringing all these physiological traits and responses to environment variables together. These models incorporate all these key physiological features. They must consider the adjustment of osmotic potential, the loss of hydraulic conductance with decreasing water potential, and how chemical signals, like ABA, interact with each other under given environmental conditions to produce the right observed response. The hydromechanical model of stomatal conductance (hereafter BMF) developed by Buckley et al. (2003) captures all current knowledge about stomatal hydromechanics, being able, therefore, to predict both steady-state and transient responses of stomata to environmental perturbations (Buckley, 2005). This model predicts g_s from the balance of opposing hydromechanical and biochemical influences in and around guard cells. The model has been simplified from its original form for a friendly application in ecophysiological studies (Buckley et al., 2012; Diaz-Espejo et al., 2012; Rodriguez-Dominguez et al., 2016). The model is based on biophysical process descriptions for water relations, gas exchange, and stomatal hydromechanics, and on the hydroactive feedback hypothesis, which states that guard cell osmotic pressure is actively regulated in proportion to epidermal turgor pressure (which acts as a sensor for changes in leaf water status). In its simplified form, the model is expressed as

$$g_s = \frac{naK(\psi_s + \pi)}{K + na\Delta w} \tag{18.5}$$

where K is leaf-specific hydraulic conductance, ψ_s is soil water potential, π is bulk leaf osmotic pressure, and Δw is leaf to air water vapor mole fraction gradient (equivalent to VPD); n and a capture nonhydraulic effects: a is mesophyll ATP concentration expressed relative to its maximum value and n is a lumped parameter representing other nonhydraulic factors.

Rodriguez-Dominguez et al. (2016) further developed a procedure to quantify how much of the stomatal closure in response to drought is due to leaf turgor—mediated or leaf turgor—independent mechanisms.

The use of the terms turgor-related and turgor-independent, instead of the classical hydraulic versus chemical, was included to accommodate the role of osmotic pressure on turgor. Therefore, turgor-related parameters comprise Ψ_s, K, and π, meanwhile turgor-independent accounts for n and a.

The BMF model can be used as a tool that integrates most of the knowledge concerning the response of olive tree to water stress, reviewed previously. We used it, in fact, to understand what is limiting g_s and A_N under different conditions of VPD and Ψ_s. We focused on the hydraulic limitation by K, which is the most important in olive (Rodriguez-Dominguez et al., 2016), and then we analyzed the limitation of photosynthesis by applying the method proposed by Buckley and Diaz-Espejo (2015). Furthermore, to understand better the implications of these physiological traits (K, π, and n) for olive, we performed the analysis with olive and grapevine, two species that, despite coexisting in the same environment, show different resistance to drought and different hydraulic characteristics. Table 18.1 shows the values of the parameters used, as well as their reference source for each species.

The first clear difference we observed between olive and grapevine is the hydraulic limitation for stomatal conductance (Fig. 18.13). At high Ψ_s (=close to zero) hydraulic limitation in olive is expressed even at very low VPD. This is a consequence of its low K in comparison with grapevine. If water is available in the soil, grapevine can sustain larger demand of water induced by high VPD, for a wider range of Ψ_s than olive. However, olive can maintain its activity for lowest values of Ψ_s, as a consequence of its more resistant hydraulic system. For clarity, we merged both analyses into a single plot, where the differences between both species were depicted (Fig. 18.13). There is a region of Ψ_s over -1.5 MPa where conditions favor grapevine with respect to olive. But at the lowest range of Ψ_s olive is more favored than grapevine.

We can extend the analysis to estimate the implication of this stomatal closure in terms of photosynthesis limitation (Fig. 18.14). Grapevine photosynthesis was mainly affected by low water availability in the soil, and when Ψ_s is high it can maintain large rates of A_N. That is not the case of olive, which again shows restricted A_N more by high VPD than by low Ψ_s. Its more resistant hydraulic system, and its larger capacity for osmotic adjustment with respect to grapevine, makes possible this difference. Two pristine areas outlined at both sides of $\Psi_s = -1.5$ MPa segregate the conditions under which each species is more favored with respect to the

Table 18.1 Physiological parameters used to run the simulations of BMF model for olive (*Olea*) and grapevine (*Vitis*) in Figs. 18.13 and 18.14

Trait	Species	Values	References
Leaf hydraulic vulnerability	*Olea*	Curve slope = −0.8 MPa P_{50} = −1.3 MPa	Hernandez-Santana et al. (2016b)
	Vitis	Curve slope = −4 MPa P_{50} = −1.5 MPa	Martorell et al. (2015)
Maximum leaf hydraulic conductance	*Olea*	4 mmol m^{-2} s^{-1} MPa^{-1}	Hernandez-Santana et al. (2016b)
	Vitis	6 mmol m^{-2} s^{-1} MPa^{-1}	Martorell et al. (2015)
Turgor loss point	*Olea*	Maximum: −1.5 MPa Minimum: −3.5 MPa	Dichio et al. (2003)
	Vitis	Maximum: −1.2 MPa Minimum: −2 MPa	Martorell et al. (2015)
Maximum rate of carboxylation	*Olea*	153.1 μmol m^{-2} s^{-1}	Rodriguez-Dominguez et al. (2016)
	Vitis	120.9 μmol m^{-2} s^{-1}	Rodriguez-Dominguez et al. (2016)
Maximum rate of electron transport	*Olea*	92.6 μmol m^{-2} s^{-1}	Rodriguez-Dominguez et al. (2016)
	Vitis	134.7 μmol m^{-2} s^{-1}	Rodriguez-Dominguez et al. (2016)
Mesophyll conductance	*Olea*	0.043 mol m^{-2} s^{-1}	Rodriguez-Dominguez et al. (2016)
	Vitis	0.249 mol m^{-2} s^{-1}	Rodriguez-Dominguez et al. (2016)

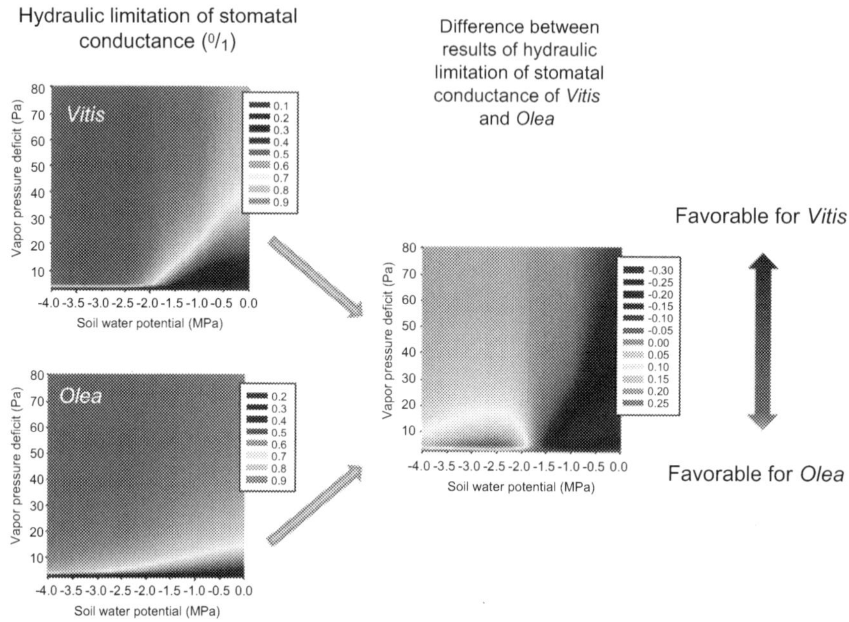

Figure 18.13 Estimation of hydraulic limitation of stomatal conductance in grapevine (*Vitis*) and olive (*Olea*) under changing soil water potential (Ψ_s) and air vapor pressure deficit (VPD). Hydraulic limitation was estimated following methodology proposed by Rodriguez-Dominguez et al. (2016) based on the BMF model. Results from both species were subtracted to visualize under which conditions of VPD and Ψ_s one species is favored over the other one.

other. A threshold value of $\Psi_s = -1.5$ MPa sets the difference for grapevine, which loses most of its K_{leaf} above that value (Martorell et al., 2015).

This sort of analysis, although simplistic, is very useful to understand better how a crop species will face the foreseen conditions of climate change in the future. As we complete our knowledge on the response of key physiological variables to water stress for each species, we should be able to evaluate not only under which conditions one species is using all its potential but also to quantify the level of performance for those conditions or for different management practices (e.g., irrigation). Crop distribution areas might not always be the most optimal ones for our species, and for sure they will vary with climate change. Process-based models integrating all the accumulated physiological knowledge are perceived as effective tools to generate new hypotheses and infer physiological mechanisms of regulation, as well as to manage smart irrigation in the future.

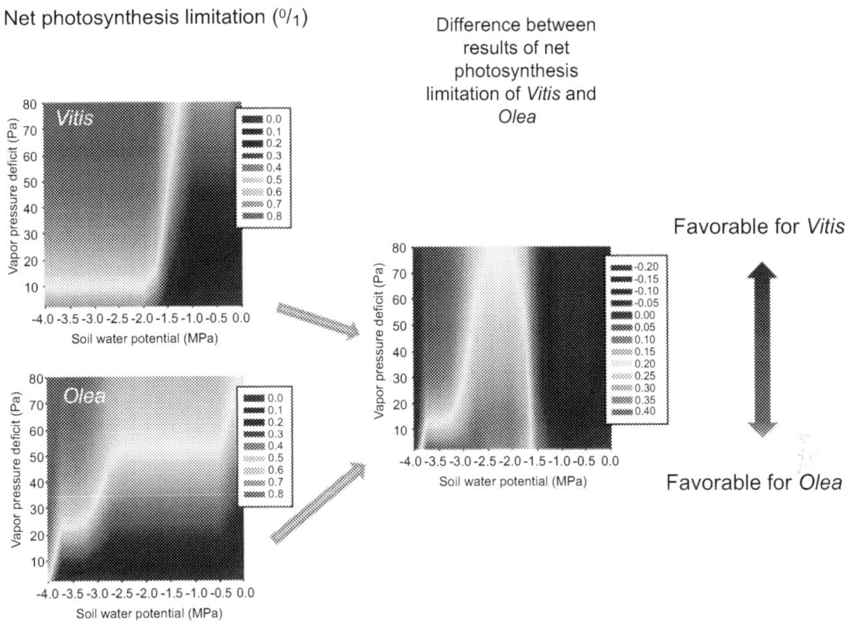

Figure 18.14 Estimation of photosynthesis limitation in grapevine (*Vitis*) and olive (*Olea*) under changing soil water potential (Ψ_s) and air vapor pressure deficit (VPD). Hydraulic limitation was estimated following methodology proposed by Rodriguez-Dominguez et al. (2016) based on the BMF model. Results from both species were subtracted to visualize under which conditions of VPD and Ψ_s one species is favored over the other one.

8. SETTING THE TARGET: WHAT PHYSIOLOGICAL VARIABLE SHOULD WE TRACE BACK FOR IRRIGATION SCHEDULING?

After reviewing in the previous sections how olive trees respond to environmental conditions and how they adjust their physiological parameters accordingly, we can better understand the actual implication of variables like water potential, transpiration, or stomatal conductance in crop performance. Still, the key question is not so easy to answer: Which variable should we measure for precise information on the plant water stress and to take decisions on irrigation scheduling? We must bear in mind that olive is a tree crop and yield is what really matters in the last term. Thus, we should focus on the main physiological determinants of fruit growth: turgor and photosynthesis. Fruits require turgor for growing, so new cells can elongate and reproduce. It is believed that there is a threshold in turgor pressure that determines fruit growth (Green and Cummins, 1974; Green et al., 1971). This concept of a threshold value has been widely used in growth models (Génard et al., 2001; Steppe et al., 2006; Lechaudel et al., 2007; Martre et al., 2011). In addition to turgor, fruit needs photoassimilates to increase their biomass, to accumulate sugars, and to produce new cells. Therefore, photosynthesis of adjacent leaves has to provide, via phloem, this carbon. Although a relationship between photosynthesis and yield is intuitive, it has not been described until recently, when it has been demonstrated empirically for olive (Hernandez-Santana et al., 2017) for several water treatments. Different irrigation treatments reached different maximum photosynthesis rates along the season, explaining differences in vegetative growth and yield. Then it appears that A_N might be a good choice, along with turgor, to control the degree of water stress in olive trees.

The leaf patch clamp pressure (LPCP) is a closely turgor-related variable that can be easily measured in olive trees under field conditions (Fernández et al., 2011; Rodriguez-Dominguez et al., 2012; Padilla-Díaz et al., 2016; Marino et al., 2016). It has been shown to be related to cell turgor pressure (Ehrenberger et al., 2012) and Ψ_{leaf} (Fernández et al., 2011). Might any of these variables, turgor or Ψ_{leaf}, be a good choice as targets for a fine assessment of the level of stress of the plant and irrigation scheduling? Leaf water potential has been one of the most recommended variables to be seasonally followed up to assess the water status of our plants (Fereres and Goldhamer, 2003; Jones, 2004). A threshold of Ψ_{stem} larger than -1.2 MPa, corresponding to Ψ_{leaf} of -1.5 MPa, has been proposed as good indicator in olive for well-watered plants (Padilla-Díaz et al., 2016; Fernández et al., 2011). However, those high water potentials occur in the region where most of the stomatal regulation is taking place in olive (up to 80% stomatal closure) and, therefore, a significant limitation of photosynthesis. Consequently, as seen in Fig. 18.6, Ψ_{leaf} might not be a very sensitive indicator of how much reduction in photosynthesis our trees are experiencing (Hernandez-Santana et al., 2017). Turgor pressure, as well, might be difficult to interpret for precise irrigation scheduling. As described in Section 3, olive has a large capacity for osmotic adjustment, which modifies π_{TLP}. This means, as reported in Fernández et al. (2011), that olive leaves can maintain turgor at Ψ_{leaf} around -3.0 MPa, which corresponds to low g_s (Fig. 18.7B) and A_N. However, the LPCP output has been proposed as a good indicator of water stress, suitable for scheduling irrigation (Fernández et al., 2011; Rodriguez-Dominguez et al., 2012; Padilla-Díaz et al., 2016; Marino et al., 2016). Its use with irrigation purposes is based on the identification of three states determined by the shape of the daily outputs (Fernández et al., 2011; Ehrenberger et al., 2012). The approach showed to be effective for the application of a regulated deficit irrigation (Padilla-Díaz et al., 2016). This irrigation strategy relies on supplying enough water to replace the crop water needs, or nearly, in the periods in which the plant is most sensitive to water stress, and reducing, or even withholding irrigation for the rest of the cycle. Fernández et al. (2013) reviewed the literature and described three periods during the growing season when olive tree is especially sensitive to water stress: blooming, active pit hardening, and the recovery period at the end of summer. As explained in Fernández et al. (2011), Padilla-Díaz et al. (2016), and Chapter 9, State I in the LPCP daily curve suggests low water stress, typical of fully irrigated periods during the most sensitive phenological stages. States II and III suggest greater water stress and can be allowed to occur at nonsensitive stages. But Fig. 18.15 shows how scattered this relation is. We plotted together values from trees of four water treatments (control, RDI 60%, RDI 40%, and RDI 30%)

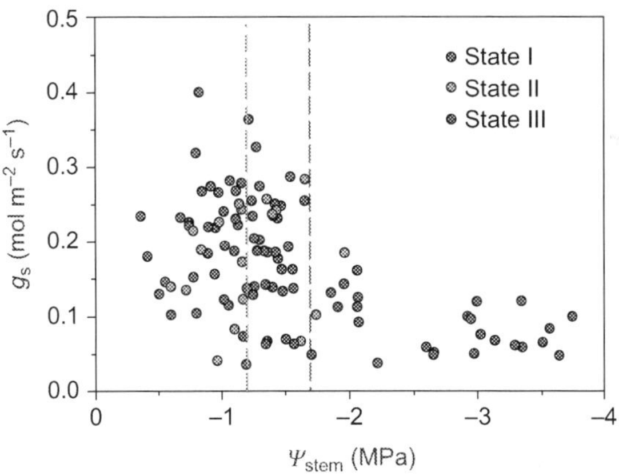

Figure 18.15 Relationship between stomatal conductance (g_s) and midday stem water potential (Ψ_{stem}) measured in olive trees from a hedgerow orchard during four field campaigns (2010–13). Each point represents a tree where these variables were measured. Leaf patch clamp pressure (LPCP) output was monitored on the same trees and the three different states (State I - green [gray in print version], State II - orange [light gray in print version], State III - red [dark gray in print version]) were identified from the shape of the LPCP output curves. The *vertical dashed lines* represent the Ψ_{stem} thresholds at which transitions between State I and II (−1.2 MPa, *orange [light gray in print version] dashed line*) and between State II and III (−1.7 MPa, *red [dark gray in print version] dashed line*) likely occur according to Fernández et al. (2011) and Padilla-Díaz et al. (2016).

corresponding to four field campaigns (2010–13). At the same time that g_s and Ψ_{stem} were measured, the state shown by the LPCP curve was determined in the same trees (shown in Fig. 18.15 with green, orange, and red color for State I, II, and III, respectively). Both states I and II extended over a wide range of g_s and Ψ_{leaf}, from 0.400 mol m^{-2} s^{-1} to 0.040 mol m^{-2} s^{-1} and −0.36 MPa to −2.07 MPa, respectively. This was expected since, as discussed in Section 4, the sensitivities of plant water potential and stomatal conductance to drying soil are different, and olive trees have a high capacity to acclimate and adjust their water-relation parameters to water stress. These findings suggest that if we are interested in maintaining a certain A_N, we must monitor a variable that holds a tighter correlation with it. Fig. 18.1 shows g_s as a suitable option.

Diffusional limitations, including g_s and mesophyll conductance to CO_2, are the main constraints of A_N in olive tree response to water stress (Diaz-Espejo et al., 2007; Perez-Martin et al., 2009, 2014). However, we cannot measure g_s values automatically and continuously in trees growing in the field. A new approach, however, has been recently proposed to overcome this limitation. Sap flow methods have highly improved, increasing the accuracy both for low (Burgess et al., 2001) and high flows (Buckley et al., 2017). As a result, sap flow–related measurements can be made continuously and automatically for reliable information on plant water stress and plant water use. In addition to the inversion of the Penman–Monteith equation, new and simpler methods have been developed to estimate g_s (Hernandez-Santana et al., 2016a; Kučera et al., 2017) and A_N (López-Bernal et al., 2015). Hernandez-Santana et al. (2016a) derived a relationship between sap flux density values normalized by VPD and g_s in olive. This approach seems promising for irrigation scheduling in olive orchards. Stomatal conductance not only responds very fast to water stress, with stomatal opening reflecting leaf turgor, but it also controls fruit growth through its effect on A_N (Flexas et al., 2014). Both the close relationship between g_s and A_N and the estimation of g_s from continuous and automatic sap flow measurements open the possibility of using g_s as a reliable water stress indicator. In addition, the derived g_s values can be input in photosynthesis models such as Farquhar's (Farquhar et al., 1980; von Caemmerer and Farquhar, 1981; Farquhar and von Caemmerer, 1982), which when applied to fruit tree orchards helps to predict yield. This allows for more precise irrigation scheduling.

ACKNOWLEDGMENTS

Results from our group reported in this work are from experiments carried out in the frame of research projects AGL2009-11310/AGR (Spanish Ministry of Science and innovation), AGL2012-34544 and AGL2015-71585-R (Spanish Ministry of Economy and Competitiveness), AGR-6456-2010 (Junta de Andalucía), and the Operational Programme of the European Social Fund (ESF) 2007–2013 "Andalusia is moving with Europe." Other funds were from the FEDER program. Thanks to the owners of Internacional Olivarera, S.A. (Interoliva), for allowing us to conduct the experiments at the Sanabria farm.

REFERENCES

Albacete, A., Martinez-Andujar, C., Martinez-Perez, A., Thompson, J., Dodd, I.C., Perez-Alfocea, F., 2015. Unravelling rootstock-scion interactions to improve food security. Journal of Experimental Botany 66, 2211–2226.

Allen, R., Pereira, L.S., Raes, D., Smith, M., 1998. Crop Evapotranspiration – Guidelines for Computing Crop Water Requirements. FAO Irrigation and Drainage Paper, No. 56 (Rome, Italy).

Angelopoulos, K., Dicchio, B., Xiloyannis, C., 1996. Inhibition of photosynthesis in olive trees (*Olea europaea* L.) during water stress and rewatering. Journal of Experimental Botany 47, 1093–1100.

Bacelar, E.A., Moutinho-Pereira, J.M., Gonçalves, B.C., Lopes, J.I., Correia, C.M., 2009. Physiological responses of different olive genotypes to drought conditions. Acta Physiologiae Plantarum 31, 611–621.

Bartlett, M.K., Scoffoni, C., Sack, L., 2012. The determinants of leaf turgor loss point and prediction of drought tolerance of species and biomes: a global meta-analysis. Ecology Letters 15, 393–405.

Blackman, C.J., Brodribb, T.J., 2011. Two measures of leaf capacitance: insights into the water transport pathway and hydraulic conductance in leaves. Functional Plant Biology 38, 118–126.

Bouche, P.S., Delzon, S., Choat, B., Badel, E., Brodribb, T., Burlett, R., Cochard, H., Charra-Vaskou, K., Lavigne, B., Li, S., Mayr, S., Morris, H., Torres-Ruiz, J., Zufferey, V., Jansen, S., 2016. Are needles of *Pinus pinaster* more vulnerable to xylem embolism than branches? New insights from X-ray computed tomography. Plant, Cell and Environment 39, 860–870.

Boursiac, Y., Léran, S., Corratge-Faillie, C., Gojon, A., Krouk, G., Lacombe, B., 2013. ABA transport and transporters. Trends in Plant Science 18, 325–333.

Brodribb, T.J., McAdam, S.A.M., 2011. Passive origins of stomatal control in vascular plants. Science 331, 582–585.

Brodribb, T.J., Skelton, R.P., McAdam, S.A.M., Bienaimé, D., Lucani, C.J., Marmottant, P., 2016a. Visual quantification of embolism reveals leaf vulnerability to hydraulic failure. New Phytologist 209, 1403–1409.

Brodribb, T.J., Bienaimé, D., Marmottant, P., 2016b. Revealing catastrophic failure of leaf networks under stress. Proceedings of the National Academy of Sciences 113, 4865–4869.

Brodribb, T.J., Carriqui, M., Delzon, S., Lucani, C., 2017. Optical measurement of stem xylem vulnerability. Plant Physiology. https://doi.org/10.1104/pp.17.00552.

Buckley, T.N., Diaz-Espejo, A., 2015. Partitioning changes in photosynthetic rate into contributions from different variables. Plant, Cell and Environment 38, 1200–1211.

Buckley, T.N., Mott, K.A., Farquhar, G.D., 2003. A hydromechanical and biochemical model of stomatal conductance. Plant, Cell and Environment 26, 1767–1785.

Buckley, T.N., Turnbull, T.L., Adams, M.A., 2012. Simple models for stomatal conductance derived from a process model: cross-validation against sap flux data. Plant, Cell and Environment 35, 1647–1662.

Buckley, T., Deng, Z.R., Salter, W., Adams, M., 2017. DRM (double-ratio method): a simple method for measuring high, low and reverse sap flow rates. In: 10th International Workshop on Sap Flow, 2017.

Buckley, T.N., 2005. The control of stomata by water balance. New Phytologist 168, 275–292.

Buckley, T.N., 2015. The contributions of apoplastic, symplastic and gas phase pathways for water transport outside the bundle sheath in leaves. Plant, Cell and Environment 38, 7–22.

Buckley, T.N., 2016. Stomatal responses to humidity: has the "black box" finally been opened? Plant, Cell and Environment 39, 482–484.

Burgess, S.S., Adams, M.A., Turner, N.C., Beverly, C.R., Ong, C.K., Khan, A.A., Bleby, T.M., 2001. An improved heat pulse method to measure low and reverse rates of sap flow in woody plants. Tree Physiology 21, 589–598.

Carr, M.K.V., 2013. Water relations and irrigation requirements of olive (*Olea europaea* L.): a review. Experimental Agriculture 49, 597–639.

Charrier, G., Torres-Ruiz, J.M., Badel, E., Burlett, R., Choat, B., Cochard, H., Delmas, C.E.L., Domec, J.C., Jansen, S., King, A., et al., 2016. Evidence for hydraulic vulnerability segmentation and lack of xylem refilling under tension. Plant Physiology 172, 1657–1668.

Chartzoulakis, K., Patakas, A., Bosabalidis, A.M., 1999. Changes in water relations, photosynthesis and leaf anatomy induced by intermit- tent drought in two olive cultivars. Environmental and Experimental Botany 42, 113–120.

Christmann, A., Weiler, E.W., Steudle, E., Grill, E., 2007. A hydraulic signal in root-to-shoot signalling of water shortage. The Plant Journal 52, 167–174.

Cochard, H., Delzon, S., 2013. Hydraulic failure and repair are not routine in trees. Annals of Forest Science 70, 659–661.

Comstock, J.P., Mencuccini, M., 1998. Control of stomatal conductance by leaf water potential in *Hymenoclea salsola* (T. & G.), a desert subshrub. Plant, Cell and Environment 21, 1029–1038.

Connor, D.J., Fereres, E., 2005. The physiology of adaptation and yield expression in olive. Horticultural Reviews 34, 155–229.

Cuevas, M.V., Martín-Palomo, M.J., Diaz-Espejo, A., Torres-Ruiz, J.M., Rodríguez-Domínguez, C.M., Pérez-Martín, A., Pino-Mejías, R., Fernández, J.E., 2013. Assessing water stress in a hedgerow olive orchard from sap flow and trunk diameter measurements. Irrigation Science 31, 729–746.

Dbara, S., Haworth, M., Emiliani, G., Mimoun, M.B., Gomez-Cadenas, A., Centritto, M., 2016. Partial root-zone drying of olive (*Olea europaea* var. Chetoui) induces reduced yield under field conditions. PLoS One 11, 1–20.

De Smet, I., Signora, L., Beeckman, T., Inzé, D., Foyer, C.H., Zhang, H., 2003. An abscisic acid-sensitive checkpoint in lateral root development of Arabidopsis. The Plant Journal 33, 543–555.

Deak, K.I., Malamy, J., 2005. Osmotic regulation of root system architecture. The Plant Journal 43, 17–28.

Diaz-Espejo, A., Nicolás, E., Fernández, J.E., 2007. Seasonal evolution of diffusional limitations and photosynthetic capacity in olive under drought. Plant, Cell and Environment 30, 922–933.

Diaz-Espejo, A., Buckley, T.N., Sperry, J.S., Cuevas, M.V., de Cires, A., Elsayed-Farag, S., Martin-Palomo, M.J., Muriel, J.L., Perez-Martin, A., Rodriguez-Dominguez, C.M., Rubio-Casal, A.E., Torres-Ruiz, J.M., Fernández, J.E., 2012. Steps toward an improvement in process-based models of water use by fruit trees: a case studyin olive. Agricultural Water Management 114, 37–49.

Dichio, B., Xiloyannis, C., Angelopoulos, K., Nuzzo, V., Bufo, S.A., Celano, G., 2003. Drought-induced variations of water relations parameters in *Olea europaea*. Plant and Soil 257, 381–389.

Dichio, B., Xiloyannis, C., Sofo, A., Montanaro, G., 2006. Osmotic regulation in leaves and roots of olive trees during a water deficit and rewatering. Tree Physiology 6, 179–185.

Dichio, B., Margiotta, G., Xiloyannis, C., Bufo, S.A., Sofo, A., Cataldi, T.R.I., 2009. Changes in water status and osmolyte contents in leaves and roots of olive plants (*Olea europaea* L.) subjected to water deficit. Trees 23, 247–256.

Dodd, I.C., Theobald, J.C., Bacon, M.A., Davies, W.J., 2006. Alternation of wet and dry sides during partial rootzone drying irrigation alters root-to-shoot signalling of abscisic acid. Functional Plant Biology 33, 1081–1089.

Dodd, I.C., Theobald, J.C., Richer, S.K., Davies, W.J., 2009. Partial phenotypic reversion of ABA-deficient flacca tomato (*Solanum lycopersicum*) scions by a wild-type rootstock: normalizing shoot ethylene relations promotes leaf area but does not diminish whole plant transpiration rate. Journal of Experimental Botany 60, 4029–4039.

Domec, J.-C., Johnson, D.M., 2012. Does homeostasis or disturbance of homeostasis in minimum leaf water potential explain the isohydric versus anisohydric behavior of *Vitis vinifera* L. cultivars? Tree Physiology 32, 245–248.

Dry, P.R., Loveys, B.R., McCarthy, M.G., Stoll, M., 2001. Strategic irrigation management in Australian vineyards. Journal International des Sciences de la Vigne et du Vin 35, 129–139.

Egea, G., Nortes, P.A., González-Real, M.M., Baille, A., Domingo, R., 2010. Agronomic response and water productivity of almond trees under contrasted deficit irrigation regimes. Agricultural Water Management 97, 171–181.

Ehrenberger, W., Rüger, S., Rodríguez-Domínguez, C.M., Díaz-Espejo, A., Fernández, J.E., Moreno, J., Zimmermann, D., Sukhorukov, V.L., Zimmermann, U., 2012. Leaf patch clamp pressure probe measurements on olive leaves in a nearly turgorless State. Plant Biology 14, 666–674.

Farquhar, G.D., von Caemmerer, S., 1982. Modelling of photosynthetic response to environmental conditions. Physiological Plant Ecology II 549–587 (Springer Berlin Heidelberg).

Farquhar, G.V., von Caemmerer, S.V., Berry, J.A., 1980. A biochemical model of photosynthetic CO2 assimilation in leaves of C3 species. Planta 149, 78–90.

Fereres, E., Goldhamer, D.A., 2003. Suitability of stem diameter variations and water potential as indicators for irrigation scheduling of almond trees. The Journal of Horticultural Science and Biotechnology 78, 139–144.

Fernández, J.E., Moreno, F., Girón, I.F., Blázquez, O.M., 1997. Stomatal control of water use in olive tree leaves. Plant Soil 190, 179–192.

Fernández, J.E., Palomo, M.J., Diaz-Espejo, A., Clothier, B.E., Green, S.R., Girón, I.F., 2001. Heat-pulse measurement of sap flow in olives: tests, root flow and diagnostics of water stress. Agricultural Water Management 51, 99–123.

Fernández, J.E., Diaz-Espejo, A., Infante, J.M., Duran, P., Palomo, M.J., Chamorro, V., Giron, I.F., Villagarcıa, L., 2006. Water relations and gas exchange in olive trees under regulated deficit irrigation and partial rootzone drying. Plant Soil 284, 271–287.

Fernández, J.E., Rodriguez-Dominguez, C.M., Perez-Martin, A., Zimmermann, U., Rüger, S., Martín-Palomo, M.J., Torres-Ruiz, J.M., Cuevas, M.V., Sann, C., Ehrenberger, W., Diaz-Espejo, A., 2011. Online-monitoring of tree water stress in a hedgerow olive orchard using the leaf patch clamp pressure probe. Agricultural Water Management 100, 25–35.

Fernández, J.E., Perez-Martin, A., Torres-Ruiz, J.M., Cuevas, M.V., Rodriguez-Dominguez, C.M., Elsayed-Farag, S., Morales-Sillero, A., García, J.M., Hernandez-Santana, V., Diaz-Espejo, A., 2013. A regulated deficit irrigation strategy for hedgerow olive orchards with high plant density. Plant Soil 372, 279–295.

Fernández, J.E., 2014. Understanding olive adaptation to abiotic stresses as a tool to increase crop performance. Environmental and Experimental Botany 103, 158–179.

Flexas, J., Diaz-Espejo, A., Gago, J., Gallé, A., Galmés, J., Gulías, J., Medrano, H., 2014. Photosynthetic limitations in Mediterranean plants: a review. Environmental and Experimental Botany 103, 12–23.

Franks, P.J., Drake, P.L., Froend, R.H., 2007. Anisohydric but isohydrodynamic: seasonally constant plant water potential gradient explained by a stomatal control mechanism incorporating variable plant hydraulic conductance. Plant, Cell and Environment 30, 19–30.

Franks, P.J., 2006. Higher rates of leaf gas exchange are associated with higher leaf hydrodynamic pressure gradients. Plant, Cell and Environment 29, 584–592.

Garcia-Forner, N., Adams, H.D., Sevanto, S., Collins, A.D., Dickman, L.T., Hudson, P.J., Zeppel, M.J.B., Jenkins, M.W., Powers, H., Martínez-Vilalta, J., McDowell, N.G., 2016. Responses of two semiarid conifer tree species to reduced precipitation and warming reveal new perspectives for stomatal regulation. Plant, Cell and Environment 39, 38–49.

Génard, M., Fishman, S., Vercambre, G., Huguet, J.G., Bussi, C., Besset, J., Habib, R., 2001. A biophysical analysis of stem and root diameter variations in woody plants. Plant Physiology 126, 188–202.

Giorio, P., Sorrentino, G., d'Andria, R., 1999. Stomatal behaviour, leaf water status and photosynthetic response in field-grown olive trees under water deficit. Environmental and Experimental Botany 42, 95–104.

Green, P.B., Cummins, W.R., 1974. Growth rate and turgor pressure: auxin effect studied with an automated apparatus for single coleoptiles. Plant Physiology 54, 863–869.

Green, P.B., Erickson, R.O., Buggy, J., 1971. Metabolic and physical control of cell elongation rate: in vivo studies in Nitella. Plant Physiology 47, 423–430.

Gucci, R., Lodolini, E., Rapoport, H.F., 2007. Productivity of olive trees with different water status and crop load. The Journal of Horticultural Science and Biotechnology 82, 648–656.

Guerfel, M., Beis, A., Zotos, T., Boujnah, D., Zarrouk, M., Patakas, A., 2009. Differences in abscisic acid concentration in roots and leaves of two young Olive (*Olea europaea* L.) cultivars in response to water deficit. Acta Physiologiae Plantarum 31, 825–831.

Hernandez-Santana, V., Fernández, J.E., Rodriguez-Dominguez, C., Romero, R., Diaz-Espejo, A., 2016a. The dynamics of radial sap flux density reflects changes in stomatal conductance in response to soil and air water deficit. Agricultural and Forest Meteorology 218, 92–101.

Hernandez-Santana, V., Rodriguez-Dominguez, C.M., Fernández, J.E., Diaz-Espejo, A., 2016b. Role of leaf hydraulic conductance in the regulation of stomatal conductance in almond and olive in response to water stress. Tree Physiology 36, 725–735.

Hernandez-Santana, V., Fernández, J.E., Cuevas, M.V., Perez-Martin, A., Diaz-Espejo, A., 2017. Photosynthetic limitations by water deficit: effect on fruit and olive oil yield, leaf area and trunk diameter and its potential use to control vegetative growth of super-high density olive orchards. Agricultural Water Management 184, 9–18.

Hochberg, U., Windt, C.W., Ponomarenko, A., Zhang, Y.-J., Gersony, J., Rockwell, F.E., Holbrook, N.M., 2017. Stomatal closure, basal leaf embolism, and shedding protect the hydraulic integrity of grape stems. Plant Physiology 174, 764–775.

Holbrook, N.M., Shashidhar, V.R., James, R.A., Munns, R., 2002. Stomatal control in tomato with ABA-deficient roots: response of grafted plants to soil drying. Journal of Experimental Botany 53, 1503–1514.

Ikegami, K., Okamoto, M., Seo, M., Koshiba, T., 2009. Activation of abscisic acid biosynthesis in the leaves of *Arabidopsis thaliana* in response to water deficit. Journal of Plant Research 122, 235–243.

Jarvis, P.G., McNaughton, K.G., 1986. Stomatal control of transpiration: scaling up from leaf to region. Advances in Ecological Research 15, 1–49.

Jeschke, W.D., Holobradá, M., Hartung, W., 1997. Growth of Zeamays L. plants with their seminal roots only. Effects on plant development, xylem transport, mineral nutrition and the flow and distribution of abscisic acid (ABA) as a possible shoot to root signal. Journal of Experimental Botany 48, 1229–1239.

Jia, W.S., Zhang, J.H., Zhang, D.P., 1996. Metabolism of xylem-delivered ABA in relation to ABA flux and concentration in leaves of maize and Commelina communis. Journal of Experimental Botany 47, 1085–1091.

Jones, H.G., 2004. Irrigation scheduling: advantages and pitfalls of plant-based methods. Journal of Experimental Botany 407, 2427–2436.

Kim, Y.X., Steudle, E., 2007. Light and turgor affect the water permeability (aquaporins) of parenchyma cells in the midrib of leaves of Zea mays. Journal of Experimental Botany 58, 4119–4129.

Kitsaki, C.K., Drossopoulos, J.B., 2005. Environmental effect on ABA concentration and water potential in olive leaves (*Olea europaea* L. cv " Koroneiki ") under non-irrigated field conditions. Environmental and Experimental Botany 54, 77–89. http://doi.org/10.1016/j.envexpbot.2004.06.002.

Kitsaki, C.K., Drossopoulos, J.B., Terzis, S., 1995. Endogenous free abscisic acid in floral, bark and leaf tissues of olive during anthesis and early fruit development. Scientia Horticulturae 64, 95–102.

Klein, T., 2014. The variability of stomatal sensitivity to leaf water potential across tree species indicates a continuum between isohydric and anisohydric behaviours. Functional Ecology 28, 1313–1320.

Kučera, J., Brito, P., Jiménez, M.S., Urban, J., 2017. Direct Penman–Monteith parameterization for estimating stomatal conductance and modeling sap flow. Trees 31, 873–885.

Kudoyarova, G., Veselova, S., Hartung, W., Farhutdinov, R., 2011. Involvement of root ABA and hydraulic conductivity in the control of water relations in wheat plants exposed to increased evaporative demand. Planta 233, 87–94.

Lawlor, D.W., Cornic, G., 2002. Photosynthetic carbon assimilation and associated metabolism in relation to water deficits in higher plants. Plant, Cell and Environment 25, 275–294.

Lechaudel, M., Vercambre, G., Lescourret, F., Normand, F., Génard, M., 2007. An analysis of elastic and plastic fruit growth of mango in response to various assimilate supplies. Tree Physiology 27, 219–230.

López-Bernal, Á., García-Tejera, O., Vega, V.A., Hidalgo, J.C., Testi, L., Orgaz, F., Villalobos, F.J., 2015. Using sap flow measurements to estimate net assimilation in olive trees under different irrigation regimes. Irrigation Science 33, 357–366.

Mahdieh, M., Mostajeran, A., 2009. Abscisic acid regulates root hydraulic conductance via aquaporin expression modulation in *Nicotiana tabacum*. Journal of Plant Physiology 166, 1993–2003.

Manzi, M., Lado, J., Rodrigo, M.J., Zacariás, L., Arbona, V., Gómez-Cadenas, A., 2015. Root ABA accumulation in long-term water-stressed plants is sustained by hormone transport from aerial organs. Plant and Cell Physiology 56, 2457–2466.

Manzi, M., Lado, J., Jesús, M., Arbona, V., Gómez-Cadenas, A., 2016. ABA accumulation in water-stressed *Citrus* roots does not rely on carotenoid content in this organ. Plant Science 252, 151–161.

Manzi, M., Pitarch-Bielsa, M., Arbona, V., Gómez-Cadenas, A., 2017. Leaf dehydration is needed to induce abscisic acid accumulation in roots of citrus plants. Environmental and Experimental Botany 139, 116–126.

Marino, G., Pernice, F., Marra, F.P., Caruso, T., 2016. Validation of an online system for the continuous monitoring of tree water status for sustainable irrigation managements in olive (*Olea europaea* L.). Agricultural Water Management 177, 298–307.

Martínez-Vilalta, J., Garcia-Forner, N., 2017. Water potential regulation, stomatal behaviour and hydraulic transport under drought: deconstructing the iso/anisohydric concept. Plant, Cell and Environment 40, 962–976.

Martínez-Vilalta, J., Poyatos, R., Aguade, D., Retana, J., Mencuccini, M., 2014. A new look at water transport regulation in plants. New Phytologist 204, 105–115.

Martorell, S., Medrano, H., Tomàs, M., Escalona, J.M., Flexas, J., Diaz-Espejo, A., 2015. Plasticity of vulnerability to leaf hydraulic dysfunction during acclimation to drought in grapevines: an osmotic-mediated process. Physiologia Plantarum 153, 381–391.

Martre, P., Bertin, N., Salon, C., Genard, M., 2011. Modelling the size and composition of fruit, grain and seed by process-based simulation models. New Phytologist 191, 601–618.

McAdam, S.A.M., Brodribb, T.J., 2016. Linking turgor with ABA biosynthesis : implications for stomatal responses to vapor pressure deficit across land plants. Plant Physiology 171, 2008–2016.

McAdam, S.A.M., Sussmilch, F.C., Brodribb, T.J., 2016a. Stomatal responses to vapour pressure deficit are regulated by high speed gene expression in angiosperms. Plant, Cell and Environment 39, 485–491.

McAdam, S.A.M., Brodribb, T.J., Ross, J.J., 2016b. Shoot-derived abscisic acid promotes root growth. Plant, Cell and Environment 39, 652–659.

Meinzer, F.C., Johnson, D.M., Lachenbruch, B., McCulloh, K.A., Woodruff, D.R., 2009. Xylem hydraulic safety margins in woody plants: coordination of stomatal control of xylem tension with hydraulic capacitance. Functional Ecology 23, 922–930.

Meinzer, F.C., Woodruff, D.R., Marias, D.E., Smith, D.D., McCulloh, K.A., Howard, A.R., Magedman, A.L., 2016. Mapping "hydroscapes" along the iso- to anisohydric continuum of stomatal regulation of plant water status. Ecology Letters 19, 1343–1352.

Mitchell, P.J., Mcadam, S.A.M., Pinkard, E.A., Brodribb, T.J., 2016. Significant contribution from foliage-derived ABA in regulating gas exchange in *Pinus radiata*. Tree Physiology 37, 236–245.

Moreno, F., Fernandez, J.E., Clothier, B.E., Green, S.R., 1996. Transpiration and root water uptake by olive trees. Plant and Soil 184, 85–96.

Moriana, A., Villalobos, F.J., Fereres, E., 2002. Stomatal and photosynthetic response of olive *(Olea europaea* L.) leaves to water deficits. Plant, Cell and Environment 25, 395–405.

Moriana, A., Orgaz, F., Pastor, M., Fereres, E., 2003. Yield responses of a mature olive orchard to water deficits. Journal of the American Society for Horticultural Science 128, 425–431.

Padilla-Díaz, C.M., Rodriguez-Dominguez, C.M., Hernandez-Santana, V., Perez-Martin, A., Fernández, J.E., 2016. Scheduling regulated deficit irrigation in a hedgerow olive orchard from leaf turgor pressure related measurements. Agricultural Water Management 164, 28–37.

Pantin, F., Monnet, F., Jannaud, D., Costa, J.M., Renaud, J., Muller, B., Genty, B., 2012. The dual effect of abscisic acid on stomata. New Phytologist 197, 65–72.

Passioura, J.B., Tanner, C.B., 1985. Oscillations in apparent hydraulic conductance of cotton plants. Australian Journal of Plant Physiology 12, 455–461.

Pavel, E.W., Fereres, E., 1998. Low soil temperatures induce water deficits in olive (*Olea europaea*) trees. Physiologia Plantarum 104, 525–532.

Penman, H.L., 1958. Natural evaporation from open water, bare soil and grass. Proceedings of the Royal Society of London Series A 193, 120–145.

Perez-Martin, A., Flexas, J., Ribas-Carbó, M., Bota, J., Tomás, M., Infante, J.M., Diaz-Espejo, A., 2009. Interactive effects of soil water deficit and air vapour pressure deficit on mesophyll conductance to CO_2 in *Vitis vinifera* and *Olea europea*. Journal of Experimental Botany 60, 2391–2405.

Perez-Martin, A., Michelazzo, C., Torres-Ruiz, J.M., Flexas, J., Fernández, J.E., Sebastiani, L., Diaz-Espejo, A., 2014. Physiological and genetic response of olive leaves to water stress and recovery: implications of mesophyll conductance and genetic expression of aquaporins and carbonic anhidrase. Acta Horticulturae 922, 99–106.

Perez-Perez, J.G., Dodd, I.C., Botia, P., 2012. Partial rootzone drying increases water-use efficiency of lemon Fino 49 trees independently of root-to-shoot ABA signaling. Functional Plant Biology 39, 366–378.

Poorter, H., Niklas, K.J., Reich, P.B., Oleksyn, J., Poot, P., Mommer, L., 2012. Tansley review Biomass allocation to leaves, stems and roots: meta-analyses of interspecific variation and environmental control. New Phytologist 193, 30–50.

Rewald, B., Leuschner, C., Wiesman, Z., Ephrath, J.E., 2011. Influence of salinity on root hydraulic properties of three olive varieties. Plant Biosystems 145, 12–22.

Rodriguez-Dominguez, C.M., Ehrenberger, W., Sann, C., Rüger, S., Sukhorukov, V., Martín-Palomo, M.J., Diaz-Espejo, A., Cuevas, M.V., Torres-Ruiz, J.M., Perez-Martin, A., Zimmermann, U., Fernández, J.E., 2012. Concomitant measurements of stem sap flow and leaf turgor pressure in olive trees using the leaf patch clamp pressure probe. Agricultural Water Management 114, 50–58.

Rodriguez-Dominguez, C.M., Buckley, T.N., Egea, G., de Cires, A., Hernandez-Santana, V., Martorell, S., Diaz-Espejo, A., 2016. Most stomatal closure in woody species under moderate drought can be explained by stomatal responses to leaf turgor. Plant, Cell and Environment 39, 2014–2026.

Rogiers, S.Y., Greer, D.H., Hatfield, J.M., Hutton, R.J., Clarke, S.J., Hutchinson, P.A., et al., 2012. Stomatal response of an anisohydric grapevine cultivar to evaporative demand, available soil moisture and abscisic acid. Tree Physiology 32, 249–261.

Sack, L., Buckley, T.N., Scoffoni, C., 2016. Why are leaves hydraulically vulnerable? Journal of Experimental Botany 67, 4917–4919.

Sadras, V.O., 2009. Does partial root-zone drying improve irrigation water productivityin the field? A meta-analysis. Irrigation Science 27, 183–190.

Schultz, H.R., 2003. Differences in hydraulic architecture account for nearisohydric and anisohydric behaviour of two field-grown *Vitis vinifera* L. cultivars during drought. Plant, Cell and Environment 26, 1393–1405.

Scoffoni, C., McKown, A.D., Rawls, M., Sack, L., 2012. Dynamics of leaf hydraulic conductance with water status: quantification and analysis of species differences under steady-state. Journal of Experimental Botany 63, 643–658.

Scoffoni, C., Albuquerque, C., Brodersen, C.R., Townes, S.V., Grace, P.J., Bartlett, M.K., Buckley, T.N., McElrone, A.J., Sack, L., 2017. Outside-xylem vulnerability, not xylem embolism, controls leaf hydraulic decline during dehydration. Plant Physiology 173, 1197–1210.

Simonneau, T., Barrieu, P., Tardieu, F., 1998. Accumulation rate of ABA in detached maize roots correlates with root water potential regardless of age and branching order. Plant, Cell and Environment 21, 1113–1122.

Sofo, A., Manfreda, S., Fiorentino, M., Dichio, B., Xiloyannis, C., 2008. The olive tree: a paradigm for drought tolerance in Mediterranean climates. Hydrology and Earth System Sciences 12, 293–301.

Spollen, W.G., LeNoble, M.E., Samuels, T.D., Bernstein, N., Sharp, R.E., 2000. Abscisic acid accumulation maintains maize primary root elongations at low water potentials by restricting ethylene production. Plant Physiology 122, 967–976.

Steppe, K., De Pauw, D.J.W., Lemeur, R., Vanrolleghem, P.A., 2006. A mathematical model linking tree sap flow dynamics to daily stem diameter fluctuations and radial stem growth. Tree Physiology 26, 257–273.

Tallman, G., 2004. Are diurnal patterns of stomatal movement the result of alternating metabolism of endogenous guard cell ABA and accumulation of ABA delivered to the apoplast around guard cells by transpiration? Journal of Experimental Botany 55, 1963–1976.

Tardieu, F., Davies, W.J., 1993. Integration of hydraulic and chemical signalling in the control of stomatal conductance and water status of drought plants. Plant, Cell and Environment 16, 341–349.

Tardieu, F., Simonneau, T., 1998. Variability among species of stomatal control under fluctuating soil water status and evaporative demand: modelling isohydric and anisohydric behaviours. Journal of Experimental Botany 49, 419–432.

Terral, J.F., Badal, E., Heinz, C., Roiron, P., Thiebault, S., Figueiral, I., 2004. A hydraulic conductivity model points to post-Neogene survival of the Mediterranean olive. Ecology 85, 3158–3165.

Tognetti, R., d'Andria, R., Lavini, A., Morelli, G., 2006. The effect of deficit irrigation on crop yield and vegetative development of *Olea europaea* L. (cvs Frantoio and Leccino). European Journal of Agronomy 25, 356–364.

Tognetti, R., Giovannelli, A., Lavini, A., Morelli, G., Fragnito, F., d'Andria, R., 2009. Assessing environmental controls over conductances through the soil–plant–atmosphere continuum in an experimental olive tree plantation of southern Italy. Agricultural and Forest Meteorology 149, 1229–1243.

Torres-Ruiz, J.M., Diaz-Espejo, A., Morales-Sillero, A., Martin-Palomo, M.J., Mayr, S., Beikircher, B., Fernandez, J.E., 2013. Shoot hydraulic characteristics, plant water status and stomatal response in olive trees under different soil water conditions. Plant and Soil 373, 77–87.

Torres-Ruiz, J.M., Cochard, H., Mayr, S., Beikircher, B., Diaz-Espejo, A., Rodriguez-Dominguez, C.M., Badel, E., Fernández, J.E., 2014. Vulnerability to cavitation in *Olea europaea* current-year shoots: further evidence of an open-vessel artifact associated with centrifuge and air-injection techniques. Physiologia Plantarum 152, 465–474.

Torres-Ruiz, J.M., Jansen, S., Choat, B., McElrone, A.J., Cochard, H., Brodribb, T.J., Badel, E., Burlett, R., Bouche, P.S., Brodersen, C.R., Li, S., Morris, H., Delzon, S., 2015a. Direct X-ray microtomography observation confirms the induction of embolism upon xylem cutting under tension. Plant Physiology 167, 40–43.

Torres-Ruiz, J.M., Diaz-Espejo, A., Perez-Martin, A., Hernandez-Santana, V., 2015b. Role of hydraulic and chemical signals in leaves, stems and roots in the stomatal behaviour of olive trees under water stress and recovery conditions. Tree Physiology 35, 415–424.

Torres-Ruiz, J.M., Cochard, H., Choat, B., Jansen, S., Rosana, L., Tom, I., Padilla-Diaz, C., Badel, E., Burlett, R., King, A., Lenoir, N., Martin-StPaul, N.K., Delzon, S., 2017. Xylem resistance to embolism : presenting a simple diagnostic test for the open vessel artefact. The New Phytologist 215, 489–499.

Trifilò, P., Raimondo, F., Savi, T., Lo Gullo, M.A., Nardini, A., 2016. The contribution of vascular and extra-vascular water pathways to drought induced decline of leaf hydraulic conductance. Journal of Experimental Botany 67, 5029–5039.

Ulger, S., Sonmez, S., Karkacier, M., Ertoy, N., Akdesir, O., 2004. Determination of endogenous hormones, sugars and mineral nutrition levels during the induction, initiation and differentiation stage and their effects on flower formation in olive. Plant Growth Regulation 42, 89–95.

Vernieri, P., Lenzi, A., Figaro, M., Tognoni, F., Pardossi, A., 2001. How the roots contribute to the ability of Phaseolus vulgaris L. to cope with chilling-induced water stress. Journal of Experimental Botany 52, 2199–2206.

von Caemmerer, S., Farquhar, G.D., 1981. Some relationships between the biochemistry of photosynthesis and the gas exchange of leaves. Planta 153, 376–387.

Zhang, J., Davies, W.J., 1989. Sequential responses of whole plant water relations towards prolonged soil drying and the mediation by xylem sap ABA concentrations in the regulation of stomatal behaviour of sunflower plants. New Phytologist 113, 167–174.

Zhang, J., Tardieu, F., 1996. Relative contribution of Apices and mature tissues to ABA synthesis in droughted maize root systems. Plant and Cell Physiology 37, 598–605.

Zhong, W., Hartung, W., Komor, E., Schobert, C., 1996. Phloem transport of abscisic acid in *Ricinus communis* L. seedlings. Plant, Cell and Environment 19, 471–477.

CHAPTER 19

Environmental Factors Controlling Carbon Assimilation, Growth, and Yield of Papaya (*Carica papaya* L.) Under Water-Scarcity Scenarios

Eliemar Campostrini[1], Bruce Schaffer[2], José D.C. Ramalho[3,4], Julián C. González[5], Weverton P. Rodrigues[1], Jefferson R. da Silva[1], Roberta S.N. Lima[6]

[1]Universidade Estadual do Norte Fluminense, Rio de Janeiro, Brazil; [2]University of Florida, Homestead, FL, United States; [3]Universidade de Lisboa, Oeiras, Portugal; [4]Universidade NOVA de Lisboa, Caparica, Portugal; [5]Universidad de Almeria, Almeria, Spain; [6]Universidade Federal de Alagoas, Arapiraca, Brazil

Contents

1. INTRODUCTION

Papaya (*Carica papaya* L.) is a fast-growing plant, probably native to Central America (southern Mexico and Nicaragua), that rarely branches, has a short juvenile phase (3–8 months), and is considered a semiwoody plant (Ming et al., 2007). The number of genes involved in lignin synthesis is intermediate between that of herbaceous *Arabidopsis thaliana* and that of woody

Water Scarcity and Sustainable Agriculture in Semiarid Environment
ISBN 978-0-12-813164-0
https://doi.org/10.1016/B978-0-12-813164-0.00019-3

poplar trees (*Populus* sp.). Thus, the authors of this chapter have concluded that papaya is evolving from an herbaceous to a woody plant.

The main commercial papaya production areas in the world are located between latitudes 23°N and S. However, the crop can be found at latitudes up to 36°N and S (Chan, 2009). For example, in Almeria (36°50′N) in southern Spain and in the Canary Islands (28°32′N) there are high yielding commercial papaya plantations protected with screen structures or transparent plastic covers (greenhouse) up to 5 m in height (Fig. 19.1).

Papaya exhibits spiraled phyllotaxis in a 3:8 configuration. There are three leaves positioned clockwise or counterclockwise within each 360 degrees turn around the trunk (Fig. 19.2). The number eight means that eight leaves are needed so that the leaf located on the bottom part is perfectly aligned with the leaf located on the upper part of the canopy. This spiral leaf insertion on the trunk, associated with a long petiole, allows efficient light distribution in the papaya canopy (Ferraz et al., 2016).

The papaya plant has a hollow, straight cylindrical green stem, with a diameter at the base of up to 40 cm, which is thinner at the top (about 5–7.5 cm) and can reach up to 10 m in height. Depending on the climatic conditions and water availability, the longevity of a papaya leaf varies from 2.5 to 8 months, and new leaves are produced at the rate of one and a half to four leaves per week (Australian Government, 2008). Under greenhouse conditions, at an average air temperature of 27°C, maximum light intensity (photosynthetic photon flux) of 1000 $\mu mol\ m^{-2}\ s^{-1}$, and mean vapor pressure deficit (VPD) of 1 kPa, "Golden" papaya plants with adequate available soil water produce about three leaves per week (Lima, 2014).

Figure 19.1 Commercial-scale protected papaya cultivation in the Canary Archipelago.

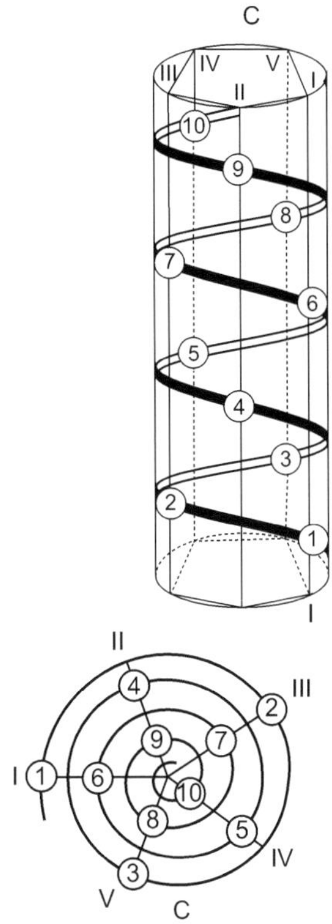

Figure 19.2 3:8 phyllotaxis of the papaya plant.

In the tropics, papaya can be harvested 8–9 months after sowing, and in the bred varieties, yield can be more than 100 t ha^{-1} $year^{-1}$ (Chan, 2009). In the commercial papaya areas in Brazil, the harvest is two to four fruits per plant each week. In field plantations in the north of the state of Espírito Santo, Brazil, a papaya plant of the Solo group, considered high yielding, produces about 95 t fruit ha^{-1}, and each plant produces on average 95 fruits and has about 24 leaves. It is important to point out that members of the Solo papaya group usually have less leaf area than members of the Formosa papaya group. Formosa papaya plantations can reach up to 150 t ha^{-1} production, which corresponds to an average of 73 fruits per plant, and each plant of the Formosa group has about 25 leaves (Geraldo Ferreguetti, personal communication). These reports show that one leaf of the papaya plant of the Solo group provides photoassimilates for about four fruits, and one leaf of the Formosa papaya group provides photoassimilates for about three fruits. It is important to emphasize that, although the literature reports that papaya can live up to 20 years (Marler et al., 1994), commercial plantations are renewed every 2–3 years, plants older than 3 years become tall and harvesting them is not economically viable.

Papaya is one of the most important fruit trees cultivated in tropical regions, and the ripe fruit of this plant, known as *pawpaw* in England, *lechosa* in Spain, *mugua* in China, *melonenbaum* in Germany, *malakor, loko*, or *ma kuai thet* in Thailand, *papayer* in France, *du du* in Vietnam, and *mamão* in Brazil, can be consumed fresh, whereas unripe fruit can be used in salads or cooked to produce sweets (Dhekney, 2004; Crane, 2005). Papaya is ranked first on a list of 38 nutritionally important fruits, based on the percentage of recommended daily human nutritional requirements in the United States for vitamin A, C, potassium, folic acid, niacin, thiamine, riboflavin, iron, calcium, and fiber (Ming et al., 2008). The yellow pigment found in the fruit is not due to carotene but to a xanthophyll called caricaxantin (Dhekney, 2004).

The papaya plant is also cultivated for papain, an important proteolytic enzyme found in the latex of the unripe fruit. Normally papain production ranges from 245 to 700 kg ha^{-1} (Becker, 1958) and about 250 kg unripe fruit are needed to produce 1 kg of raw papain. If ingested after meals, this enzyme improves digestion and can cure ulcers. Papain is also used in the textile industry to soften wool (Villegas, 1997), in the preparation of concentrated fish protein for animal feed, in the cosmetic industry (toilet soap, toothpaste, and shampoo), for meat tenderizing, and in the brewing industry (Seelig, 1970). In terms of evolution, papain may be associated with protection against herbivorous and fructivorous predators (El Moussaoui et al., 2001; Konno et al., 2004).

Papaya is a diploid plant with nine pairs of chromosomes and has a small genome of 372 Mb (Arumuganathan and Earle, 1991). This species is polygamous with three types of genders: male, female, or hermaphrodite plants. In various regions of the world, hermaphrodite papaya plants are the most grown in commercial plantations because they are more productive. If the plantation consists of only female plants, there is a 6%–10% loss of the productive area. This loss is due to cultivating male plants, because these plants will serve as pollinating agents in the areas were only female plants are cultivated (Ming et al., 2008). Another aspect related to the predominance of plant field cultivation of hermaphrodite plants is the characteristic of the fruit. Hermaphrodite plants produce fruit with a smaller empty internal space that gives the fruit a higher pulp/volume ratio compared to female plants. However, in subtropical regions, female plant cultivation predominates because in these locations, at low temperatures the anthers of the flowers of the hermaphrodite plants can fuse with the carpels and deformed fruit can be produced (Ming et al., 2008).

Sex segregation in papaya plantlets has been highly problematic in commercial plantations. According to Ming et al. (2008), the producers/producing companies using hermaphrodite plant seeds need to plant about five plantlets to a hole to guarantee not more than 3% female plants. These five plantlets only flower after 4–6 months when it is necessary to determine the sex by the roguing procedure to obtain maximum productivity. However, this process is inefficient in terms of time, and requires a high expenditure of water and mineral nutrients. Furthermore, the process of maintaining about three to five plants per hole for 4–6 months results in intense competition for light, water, and mineral nutrients, which can increase costs and inhibit initial papaya plant growth. When seeds are planted from fruit of female plants, producers need to plant four plantlets per hole to guarantee about 6%–10% male flowers. This fact shows the seriousness of the problem when the papaya sex is not determined early. Table 19.1 shows the percentage of female plants expected in the sexing period in relation to the number of plants placed per hole.

Papaya is classified as a plant with C_3 metabolism (Imai et al., 1982; Marler et al., 1994; Campostrini, 1997; Marler and Mickelbart, 1998; Jeyakumar et al., 2007). This is based on the leaf anatomy (no presence of bundle sheath cell), net photosynthetic rate (P_n) in saturating light ($\cong$25 μmol CO_2 m^{-2} s^{-1} in 2000 μmol photons m^{-2} s^{-1}), and on the carbon isotopic discrimination value (−27.5‰) (Buisson and Lee, 1993; Campostrini, 1997; Marler and Mickelbart, 1998; Torres Netto, 2005; Jeyakumar et al., 2007; Lima, 2014; Lima et al., 2015). Thus, environmental and edaphic factors such as light, wind, soil physical and chemical

Table 19.1 Percentage of female flowers expected in the sexing period in relation to the number of plants placed per hole at planting

% Female flowers expected → number of plant/hole

Plants/hole	Hybrids (%)	Varieties - selfing (%)
1	50.00	33.00
2	25.00	10.89
3	12.50	3.59
4	6.25	1.19
5	3.13	0.39

characteristics, air temperature and relative humidity (air vapor pressure deficit [VPD_{air}]), mineral nutrients, and soil water content can intensely affect the photosynthetic process and growth of this species (El-Sharkawy et al., 1985; Aiyelaagbe et al., 1986; Grau and Halloy, 1997; Srinivas, 1996; Clemente and Marler, 1996, 2001; Marler et al., 1994; Weng and Lai, 2005; Marler and Clemente, 2006; Mahouachi et al., 2006; Campostrini et al., 2010; Carr, 2014; Vincent et al., 2015; Lima et al., 2015; Thani et al., 2016). Thus, understanding the effects of environmental factors and their interactions with the physiological processes in papaya is extremely important for the economic sustainability of crop production in nurseries, protected culture, and field cultivation. By basing papaya crop management strategies on the results of scientific studies, particularly those assessing the impact of environmental stresses such as water deficit stress on physiology of this crop, carbon assimilation can be optimized and fruit quality and yield can be increased.

2. POTENTIAL CONSEQUENCES OF WATER AND OTHER ENVIRONMENTAL LIMITATIONS RESULTING FROM PREDICTED CLIMATE CHANGE

Improving agricultural sustainability in order to guarantee food security and safety, while preserving natural resources, is a major global priority and challenge for the next century (Chaves and Zarrouk, 2013). Scarcity of renewable resources is increasingly becoming a threat to global food production. For example, it is estimated that up to 50% of agricultural land may be endangered by the year 2050 due to salinization of arable land (Wang et al., 2003; Scotti-Campos et al., 2016). Therefore, expanding the scientific basis for mitigation and adaptation agricultural crops to increasing environmental limitations is a critical concern. Under field conditions, limitation of water availability is often superimposed on other environmental stresses such as salinity, high irradiance, or extreme temperatures, creating multiple stress conditions. In this context, it is important to consider reduced water availability together with other climatic and soil variables. Only this view will lead to a better understanding plant responses to climate change, thus allowing for the development of biological (i.e., genetics and breeding) and management strategies (i.e., site selection, pruning, etc.) to improve crop use efficiency of natural resources, including water (Chaves et al., 2009; Chaves and Zarrouk, 2013). This situation is particularly important in the tropical and semiarid environment where drought events are often exacerbated by high temperatures, resulting in multidimensional stress (DaMatta, 2003) endangering crop development, yield, and sustainability.

Crop growth and yield, as well as crop quality, are expected to be affected by global climate change. Therefore, a high degree of uncertainty remains regarding food production and security under future climate scenarios, which in addition to global warming, will include altered pest and disease incidences, as well as changes in the magnitude and (seasonal and yearly) distribution of rainfall, leading to increased flooding and drought incidences worldwide. This will undoubtedly affect biogeochemical cycles, primary productivity, phenology, growth, and yields of agricultural crops (Easterling et al., 2007; Gregory et al., 2009; DaMatta et al., 2010; IPCC, 2013; Feng et al., 2013).

In C_3 plants, elevated atmospheric CO_2 is expected to increase intrinsic water use efficiency, defined as the carbon gain per unit of water transpired or the ratio between P_n and stomatal conductance to water vapor (g_s). This results in a large increase in P_n with g_s not affected or decreased by elevated CO_2 (Ainsworth and Long, 2005; Kirschbaum, 2011; Ramalho et al., 2013; Rodrigues et al., 2016). To preserve high yields under water limitation, it is critical to select plants that display high water-use efficiency, which can be achieved either by higher transpiration efficiency or by an increased partitioning of biomass to the harvested plant tissues (higher harvest index). Other characteristics should also be considered, such as phenological cycle that enables plants to escape the most severe water shortage periods, in particular regarding the flowering and fruit/seed set and development stages. Altogether this can be achieved through improved water management practices; it can also be achieved by adjustment of crop developmental/phenological stages to the prevalent environmental conditions, and/or breeding programs to obtain plants with the ability to acclimate to environmental stress factors (Chaves and Zarrouk, 2013). Production of fruit crops under conditions of water scarcity has been improved primarily by optimizing management techniques such as deficit irrigation (Johnson and Handley, 2000), and fruit quality can even improve under moderate water deficit conditions if proper crop management is implemented (Chaves et al., 2010).

The increase of atmospheric CO_2 is estimated to promote increases in global surface temperatures from 0.3 to 1.7°C (best scenario) or from 2.6 to 4.8°C (worst scenario, without additional world mitigation efforts) relative values observed in 1986–2005 (IPCC, 2013, 2014). Also, increased surface temperatures are expected to be accompanied by a higher frequency and intensity of heat waves, particularly in the tropics (IPCC, 2014), which will likely exceed the temperature limits of many plant species (O'Sullivan et al., 2016). Therefore, the potential beneficial impact of atmospheric CO_2 enrichment on crop photosynthesis, growth, and production will be greatly offset by other climatic changes such as global warming and lower water availability. High ambient CO_2 can play an important role in mitigating heat stress in some tropical crops such as coffee by altering its heat tolerance (Martins et al., 2014, 2016; Rodrigues et al., 2016). However, if temperatures become supraoptimal, gas diffusion in the mesophyll can be reduced (Lambers et al., 2008), the biochemical reactions of photosynthesis can be impaired (Pastenes and Horton, 1996; Rodrigues et al., 2016), water relations will be altered due to increased evaporative demand, and changes in membrane fluidity and stability as well as hormone balance may occur (Wahid et al., 2007). By increasing the evaporative demand, high temperatures can promote stomatal closure, decreasing P_n due to reduced CO_2 influx into leaves (DaMatta et al., 2010). Under high irradiation conditions, decreased stomatal opening could also increase in leaf temperatures by as much as 20°C, as is the case of *Coffea arabica* L. (Maestri and Barros, 1977). Moreover, plant responses to high temperatures are often linked to the uncontrolled production and presence of reactive oxygen species (Hasanuzzaman et al., 2013). These highly reactive molecules have the potential to promote lipoperoxidation, pigment bleaching, protein oxidation, enzyme inactivation, as well as DNA degradation (Logan, 2005; DaMatta and Ramalho, 2006; Martins et al., 2016). In turn, cellular homeostasis, photosynthesis, and other major physiological processes can be impaired, negatively affecting crop growth and yields (Prasad et al., 2003; Long et al., 2006; Kirschbaum, 2011; Martins et al., 2016).

Carbon assimilation and cell growth are among the first processes to be affected by water deficit (Chaves et al., 2009) and are gradually reduced as drought severity increases in intensity and/or duration, with a consequent reduction in relative plant water content (RWC) and water potential. These effects are related to reduced CO_2 availability at the chloroplast carboxylation sites, linked first to stomatal closure and afterward to metabolic limitations. Secondary impacts, such as effects on oxidative metabolism, also play a role (Chaves et al., 2009). Stomatal closure, which reduces water lost by transpiration, is among the first whole plant physiological responses to water deficit conditions. Stomatal closure contributes to maintaining leaf RWC within a range where the photosynthetic capacity and quantum yield are not significantly impacted, although carbon assimilation may be downregulated. If water is limited for an entire growing period, additional physiological damage will occur involving an array of mesophyll reactions, and permanent damage will be unavoidable (Nunes et al., 1992; Ramalho and Chaves, 1992; Chaves et al., 2003, 2009; Muller et al., 2011; Ramalho et al., 2014). Permanent damage will likely occur if RWC is within the range of 50%–70%, when loss of most protoplasmic water and modifications in source-sink balance occur (Matos et al., 2010; Muller et al., 2011). At this stage, molecular changes such as up- and downregulation of gene expression can occur. Prolonged water deficits can also impair the functioning and/or content of photosynthetic pigments, photosystem (PS) II and PSI activities, as well as reduce RuBisCO activity and carbohydrate content (Lu and Zhang, 1999; Chaves et al., 2003, 2009; Ramalho et al., 2014). These impacts are frequently linked to the triggering of oxidative stress conditions at the chloroplast (and the whole cell), when energy capture largely exceeds its photochemical use. Therefore, drought tolerance is frequently linked to triggering the production of antioxidants (Logan, 2005; Zlatev et al., 2006) and repair mechanisms. Specific physiological reactions to water deficits are strongly species and genotype dependent, considering plant types (herbaceous, shrubs, and trees), ecophysiological history, and other concurrent environmental stressors (salinity, high irradiance, supraoptimal temperatures) (Nunes et al., 1992; Quick et al., 1992; Ramalho et al., 2000, 2014; Chaves et al., 2003, 2009).

In the papaya, damage to photosynthetic carbon assimilation due to environmental stress, including water deficits, can reduce biomass production (Lima et al., 2015) and net carbon assimilation (Clemente and Marler, 1996; Marler and Mickelbart, 1998; Mahouachi et al., 2006; Lima et al., 2015, 2016), which can influence growth, fruit yield, and fruit quality (Salazar, 1978). Therefore, as drought conditions are expected to become more frequent worldwide due to climate change, understanding the effects of water deficits on papaya physiology, especially the photosynthetic processes, will be critical to maintaining productivity of this crop under water-limiting conditions.

3. ENVIRONMENTAL FACTORS

3.1 Water

Global changes in water availability for crop use have occurred mainly due to less rainfall and therefore a decrease in the amount captured for irrigation use. Although papaya is considered to be a relatively drought-tolerant plant (this species is capable of osmotic adjustment by increasing Na, K, and Cl in the tissues to avoid dehydration) (Marler et al., 1994; Mahouachi et al., 2006), soil water shortage can negatively impact photosynthetic processes, growth (Marler and Mickelbart, 1998; Mahouachi et al., 2006; Campostrini et al., 2010; Lima et al., 2015, 2016), and fruit yield and quality (Salazar, 1978). Under drought conditions, g_s is reduced as a result of stomatal closure. Stomatal closure as a result of drought has been related to increased abscisic acid concentrations in the leaf and root (Mahouachi et al., 2007), but other factors may be involved.

In papaya, g_s is generally reduced at soil water tensions of −50 to −60 kPa (Marler and Mickelbart, 1998; Lima, 2014). In 103-day-old papaya plants grown in 30-L pots, an increase in soil water tension from −10 to −50 kPa caused a reduction in g_s from 0.7 to 0.1 mol m^{-2} s^{-1} (an 85% reduction), and photosynthesis decreased from 18 to 6 μmol m^{-2} s^{-1} (a 66% reduction) (Lima, 2014; Lima et al., 2015).

In 6-month-old papaya plants grown in the field, P_n decreased by 50% when soil water tension dropped from −20 to −68 kPa. Although enzymatic activity related to photosynthetic processes was not assessed in that study, the decrease in P_n was primarily caused by stomatal closure rather than biochemical factors because the photochemical efficiency was not affected by soil water shortage (Marler and Mickelbart, 1998). In that study, soil water shortage (soil water tension of −68 kPa) caused about a 50% reduction in apparent quantum yield (initial slope of the photosynthetic light response curve) (the molar amount of O_2 evolved or CO_2 absorbed per mol of photons absorbed by the photosynthetic apparatus) (Φ_i) a 50% increase in the light compensation point, and about a 30% reduction in the light saturation point. Marler and Mickelbart (1998) observed that water shortage caused a greater reduction in photosynthetic carbon assimilation on cloudless days than on cloudy days, which was primarily due to stomatal factors. Similar results were obtained by Reis and Campostrini (2008).

In terms of soil water quantity, Srinivas (1996) reported that in India, applying water above 60%−100% of the Class A pan evaporation to papaya plants grown in the field was necessary for obtaining sufficient plant height, trunk diameter, number of fruits per plant, mean fruit weight, and total fruit production (t ha^{-1}). The author also reported that applying subsurface water with nozzles placed at a depth of 25 cm caused about a 10% increase in total fruit production (t ha^{-1}). Thus, applying subsurface water may be promising for papaya production. Indeed, placing water as close as possible to the root decreases evaporation and can significantly improve the crop's water-use efficiency.

Mahouachi et al. (2006, 2007) used time domain reflectometry probes in the soil and found that 20% soil moisture was optimum for papaya growth and yield. When the percentage of soil water declined to 10% at about 35 days after suspending water application to papaya plants in 30-L pots, P_n reached values of around 2 $\mu mol\ m^{-2}\ s^{-1}$, while optimum P_n values at 20% soil water were around 10 $\mu mol\ m^{-2}\ s^{-1}$. This 10% reduction in soil water content resulted in decreased g_s (which was associated with a concomitant increase in leaf and root abscisic acid concentrations), transpiration, central leaflet length, leaf dry weight, and number of leaves. A reduction in trunk diameter, plant height, and dry matter gain per day occurred when the soil water content decreased to about 12%. Leaf water potential also started to decrease after the soil water content reached 12%, whereas P_n started to decline at 15% soil moisture.

The quantity of water in the air, assessed by the air vapor pressure deficit (VPD_{air}), may significantly alter stomatal opening and indirectly reduce photosynthetic carbon assimilation (El-Sharkawy et al., 1985; Marler and Mickelbart, 1998; Reis and Campostrini, 2008; Campostrini et al., 2010). El Sharkawy et al. (1985) reported that when the VPD_{air} increased from 1−1.5 kPa to 3.5−4.5 kPa, g_s was reduced by 48% and P_n decreased by 46%. A VPD_{air} above 1 kPa is critical for papaya because above this VPD_{air} threshold stomata close, thereby limiting photosynthetic carbon assimilation.

Thus, the higher the leaf temperature and the drier the air outside the leaf, the greater the $VPD_{leaf\text{-}air}$. Hence a critical value of the $VPD_{leaf\text{-}air}$ for papaya can be considered to be about 2 kPa. At values above 2 kPa the stomatal closing process begins (reductions in g_s) with significant reductions in P_n. If a leaf receives high solar radiation, the excess radiation can cause leaf heating and thus raise the $VPD_{air\text{-}leaf}$, which may reduce g_s and P_n. This explains the sharp midday depression of P_n of papaya frequently observed on hot, cloudless days, which even occurs in irrigated plants with an adequate soil water quantity/tension.

Management strategies can be adopted to reduce leaf heating (reduce $VPD_{leaf\text{-}air}$) and reduce stomatal closure caused by reduced $VPD_{leaf\text{-}air}$. Reis and Campostrini (2011) showed that applying water (microjet irrigation) to the canopy of 7-month-old papaya plants in the field at a time of day when air temperatures are highest (1200−2000 HR.) reduced leaf temperature and $VPD_{leaf\text{-}air}$. Thus, the stomata remained open at midday and P_n was close to 21 μmol m^{-2} s^{-1}, which is about optimum for papaya. In plants that did not receive water by microjet irrigation, leaf temperature increased together with the $VPD_{leaf\text{-}air}$, and g_s, transpiration, and photosynthetic carbon assimilation decreased significantly.

Regarding water application to commercial Formosa papaya plantations, in protected cultivation on the Canary Islands with drip irrigation, in a 24-month cycle, at a spacing of 1600 plants ha^{-1} and an average crop yield of 60 kg $plant^{-1}$, water uptake averaged 10,500 m^3 $year^{-1}$ (Eudaldo Perez Hernanez, personal communication). From these values, it can be estimated that on average the "water footprint" for the papaya in this region of Spain is 100 L H_2O kg^{-1} of papaya produced (Canary Agrarian Research Institute, personal communication). However, this uptake can vary from 75 to 125 L H_2O kg^{-1} of papaya produced. On average, in the Canary Islands in the summer, a Formosa papaya plant takes up about 18 L water day^{-1}.

Under field conditions in the North of Espírito Santo State, Brazil, with 1852 Golden papaya plants ha^{-1}, after 7 years of data collection, with a yield of 27 kg fruit per plant under drip irrigation added to the rainfall, the value of water uptake to produce 1 kg papaya (water footprint) was 650 L kg^{-1}. Rainfall only (without irrigation) yielded about 1 kg of fruit using 250 L of water (Geraldo Ferreguetti, personal communication). A papaya plant takes up on average 10 L water per plant per day. In the same region and in the first year of planting with little rain, 1.5 m × 3.6 m spacing (1852 plants ha^{-1}), and drip irrigation, the water uptake including rainfall to produce 1 kg papaya was 220 L (Lima et al., 2015). When water uptake was quantified taking into account only rainfall, the value decreased to 80 L kg fruit $^{-1}$ (Lima et al., 2015).

On average, in the north of Espírito Santo state, Brazil, a Golden papaya plant grown in the field and including rainfall takes up about 8.3 L water per day. For plants that were not irrigated and only received rainwater, this value was 2.5 L water per plant per day (Lima et al., 2015). The plants that were not irrigated in the study period (for 18 months) had very low production in the second harvest after about 2 months without rain (Lima et al., 2015).

Ferraz (2008) and Ferraz et al. (2015, 2016) used 3.4 m^3 Mylar balloons connected to an infrared gas analyzer and ventilators in the field (Fig. 19.3) to measure the gas exchange of whole papaya plants. These measurements may be very important because they quantify photosynthetic carbon assimilation and transpiration directly at the fruit production site and

Figure 19.3 Structure used to measure gas exchange of whole plants grown in the field. Golden papaya plants grown in the Caliman Agrícola SA company.

therefore may be of great use to future projects related to carbon credit for papaya-producing companies and farms for projects such as irrigation management.

In studies by Ferraz (2008) and Ferraz et al. (2015, 2016) in commercial plantations in the municipality of Linhares, Espirito Santo, Brazil in summer, Golden papaya cultivar (4 m^2 leaf area per plant) on cloudless days (maximum value 2400 $\mu mol\ m^{-2}\ s^{-1}$, 38°C max. temperature, 4 kPa maximum VPD_{air}, with a drip irrigation system applying about 16 L water per plant per day) assimilated 50.7 g CO_2 per day and transpired about 10 L water per day (Fig. 19.4). In the same study, but in the winter, each Golden

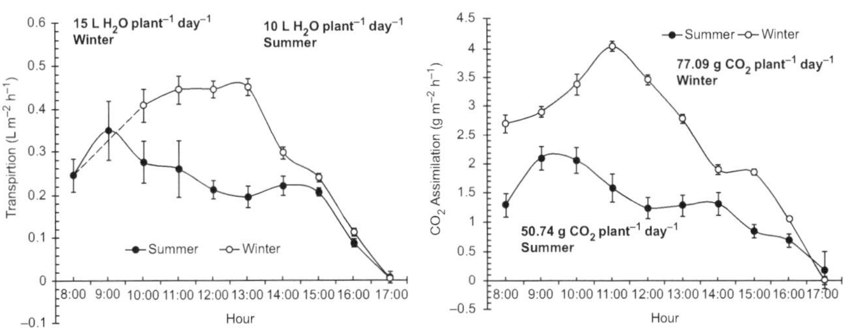

Figure 19.4 Photosynthesis and transpiration of whole Golden papaya plants grown in the field in Linhares, Espirito Santo, Brazil.

papaya plant with 3.5 m^2 total leaf area on cloudless days (max. 1400 μmol m^{-2} s^{-1}), 33°C max. temperature, 3.5 kPa maximum VPD_{air}, with a drip irrigation system applying about 10 L water per plant per day assimilated 77 g CO_2 per day and transpired 15 L water per plant per day (Fig. 19.4).

From the data obtained by Ferraz (2008) and Ferraz et al. (2015), the transpiration ratio (TR) can be calculated for papaya plants grown in Linhares, Brazil. This ratio is the quantity of water transpired divided by the quantity of CO_2 assimilated by the plant. In this case, the ratios were 10 L/50.7 = 0.197 (197 g H_2O g^{-1} CO_2) and 15 L/77 = 0.194 (194 g H_2O g^{-1} CO_2), for the summer and winter, respectively. The TR values were lower than those reported in the literature for C_3 plants (TR = 500). From the results obtained by Ferraz (2008) and Ferraz et al. (2015), it can be concluded that in commercial papaya plantations in the north of Espírito Santo State, Brazil, the transpiration ratio was 2.5 and 4.2 L H_2O m^{-2} leaf area day^{-1} $plant^{-1}$, in the summer and winter, respectively. These results show that in both seasons papaya grown in the region of Linhares, Brazil, have significant water savings.

The results of whole-plant gas exchange measurements are important because they show that although in the summer atmospheric conditions cause a high transpiration demand, the high photosynthetic photon flow, associated with high air and leaf temperatures, caused more stomatal closure that reduced photosynthetic carbon assimilation and transpiration. In summer, under conditions at Linhares, Brazil, papaya exhibited stomatal adjustment to reduce water loss and maintain leaf water status. These results indicate that a high water demand by the surrounding atmosphere is not always reflected in high water loss by papaya plants. As papaya can close stomata when VPD is high, this species can control water loss and thus avoid intense dehydration under field conditions. Based on the results of whole-plant transpiration measurements (Fig. 19.3), the best time to apply fertigation to papaya plants in summer is in the morning, because at this time (until 900 HR) nutrients applied to the soil quickly reach the leaves. In winter, fertigation can be applied until 1300 HR.

In papaya, the negative effects of water shortage on the photosynthetic processes depend on the time of day (at midday the negative effect is very accentuated) (Marler and Mickelbart, 1998), the number of clouds in the sky (under cloudless sky the negative effect is stronger) (Marler and Mickelbart, 1998; Reis and Campostrini, 2008), and the volume of soil exploited by the root system (Marler and Mickelbart, 1998). In plants with a small root system, the effect of water shortage on photosynthetic processes is extremely marked compared with papaya plants where the volume of soil exploited by the roots is much greater (Marler and Mickelbart, 1998; Lima et al., 2015).

In the current scenario of global climatic change, the search for management strategies that can increase the effective use of water is of fundamental importance for papaya production. Thus, nowadays, applying the minimum amount of water to papaya plants so that photosynthetic and other processes are minimally affected is the strategy that should be adopted. In this context, Lima et al. (2015) showed that in field conditions, comparing the use of 100% ET_0 applied on the two sides of the roots, the quantity of water applied can be reduced by up to 30%, when using an irrigation technique known as partial root zone drying (PRD) or water reduction on both sides of the root (regulated deficit irrigation, RDI), without affecting yield and fruit quality. These same authors reported that under the conditions where the root system may be confined, for example, in papaya planted in pots or in a field with shallow soil, the PRD and RDI irrigation techniques may inhibit leaf gas exchange and growth characteristics. This damage occurs more on sunny days with high air temperature.

3.2 Temperature

Because papaya cultivation is advancing into areas such as southern Spain, Japan, Israel, China, Turkey, and Australia (Galán Saúco and Rodríguez Pastor, 2007), and semiarid regions of Brazil in the states of Ceará and Rio Grande do Norte, it is important to know the effects of infra- and superoptimum air temperature on the photosynthetic metabolism and other physiological processes, such as mineral nutrition, growth, and yield. According to Knight (1980), the adequate air temperature range for papaya is 21–33°C, while Lassoudiere (1968) reported 22–26°C. Air temperature negatively affects the photosynthetic processes in papaya through the stomatal and nonstomatal effects. The effect of air temperature on papaya stomata is shown by the action of temperature on VPD_{air} and VPD_{leaf}. High temperatures cause an increase in VPD, which reduces g_s, thus indirectly inhibiting photosynthetic carbon assimilation. For papaya grown in the field, when air temperature increased from 20 to 40°C, the $VPD_{leaf\text{-}air}$ increased from 2 to 6 kPa, causing a reduction in P_n from 20 to 5 μmol m^{-2} s^{-1} (Campostrini et al., 2010). Allan and Jager (1978) reported that P_n increased when air temperature increased from 16 to 30°C. When air temperature reached 41°C, P_n decreased by half compared to the values observed at 30°C.

In field-grown papaya plants, high air temperature in association with high solar radiation caused an increase in leaf temperature (with maximum values of 42°C), and transpiration decreased throughout the plant (Ferraz et al., 2015). In the winter, at leaf temperatures of 36°C, maximum transpiration was about 0.44 L m^{-2} leaf h^{-1}. In the summer, when the leaf temperature was close to 42°C, transpiration decreased to 0.20 L m^{-2} leaf h^{-1} (Ferraz et al., 2015).

The effect of supraoptimal temperature on photochemical processes (nonstomatal effects), and indirectly on the photosynthetic process, can be assessed by a quick, precise, and nondestructive method known as chlorophyll fluorescence emission. Using this technique, it was verified that the photochemical machinery of papaya can tolerate a temperature of up to 44°C (Weng and Lai, 2005). Because supraoptimal temperature first affects the stomata, temperatures above 44°C result in greater fluidity of membrane lipids that promotes a disorganization of the photosystems that are inserted in the thylakoid lipid matrix (Li et al., 2009). This disorganization affects the passage of the antenna system energy to the reaction centers, both located in PSII, and thus the (initial fluorescence) F_0 increases.

A strategy to prevent the negative effect of supraoptimal temperature on g_s is to raise the relative humidity around the papaya canopy. In higher relative humidity, for example, 80%–90%, the effects of this supraoptimal temperature on stomatal closure are much less pronounced. Thus, on days with high temperatures that occur in the summer, an important strategy for papaya plants in protected cultivation is to apply water to the canopy via microsprinklers to cool the leaves and reduce the $VPD_{leaf-air}$. This type of stomatal effect related to supraoptimal temperature is very common in the summer in plantations in southern Spain and the Canary Islands. Obviously, under conditions of more leaf wetting inside protected cultivation, special attention should be paid to fungal diseases.

In some subtropical regions where papaya is cultivated, a common problem in winter is low air temperature that causes significant negative effects on photosynthetic processes, growth, and productivity. According to Nakasone and Paull (1998), air temperatures below c.12–14°C strongly limit fruit ripening and consequently negatively affect plant production.

The effects of low temperature on papaya growth can be indirect via decreased photosynthetic carbon assimilation (Grau and Halloy, 1997).

When the air temperature changed from 25/15°C (day/night) to 15/5°C (day/night) there was a significant reduction in g_s (the stomata closed quickly), and there was complete closing after 5 days of applying the treatment. The nonstomata effects associated with infraoptimal temperatures, obtained by the curve of P_n versus internal CO_2 concentration (the concentration of CO_2 in the substomatal space waiting to be fixed), were also observed 4 days after applying the treatments. This showed that temperatures of 25/15°C (day/night) to 15/5°C (day/night) lowered P_n through the stomatal and nonstomatal effects (Grau and Halloy, 1997).

Regarding the action of temperature on the reproductive processes of papaya, at temperatures above 35°C, hermaphrodite flowers tend to become male, and the female parts of these hermaphrodite flowers become nonfunctional (Chan, 2009). Low temperatures (<20°C) caused a decrease in the number of stamens in the hermaphrodite flowers, a phenomenon known as carpeloidy (Chan, 2009). In hermaphrodite cultivars such as "Solo," temperatures below 17°C can cause up to 100% carpeloid flowers. At temperatures above 35°C, the male and female parts of hermaphrodite flowers tend to develop complications in development and to be nonfunctional (Chan, 2009). O'Hare (1993) reported that 15°C is the minimum soil temperature limit for papaya plant growth.

4. ECONOMIC IMPACT OF WATER SHORTAGE ON PAPAYA

Water scarcity has a tremendous potential to be harmful for papaya producers given the high water requirements of this crop (Goenaga et al., 2004; Posse et al., 2008; Santana et al., 2008). Depending on its duration and intensity, water scarcity might lead to situations of water deficit, which in turn affects key physiological processes such as stomatal conductance and photosynthesis (Clemente and Marler, 1996; Campostrini and Glenn, 2007). Lack of vegetative growth, flower sterility and malformation, fruit set failure, fruit growth and sink strength, and fruit quality will be affected by water scarcity (Lopes de Siqueira and Botrel, 1986; Ribeiro da Silva et al., 2013). If drought persists for a long period, even the survival of papaya plants can be compromised.

Papaya is a giant herb that exhibits monopodial continuous growth (Malo and Campbell, 1986) when environmental conditions (temperature, radiation, water availability, etc.) are not limiting. After papaya plants go from the juvenile to the adult stage, they form new flowers (initially one, later more) in the axil of each new node (Carr, 2014). Consequently, if apical growth is arrested by severe water stress, new nodes are not formed, growth is stunted, and the production of new flowers almost ceases, strongly reducing yield (Aiyelaagbe et al., 1986; Almeida et al., 2003). Young plants, 7–11 weeks after transplanting, are especially sensitive to severe water deficits (Ribeiro da Silva et al., 2013). Even slight water deficits can negatively affect papaya fruit set and quality.

Lack of fruit set is common in fruit crops because of water stress. In papaya, fruiting failure is considered to occur when a portion of the trunk lacks developing fruit. In many papaya cultivation areas, this failure is observed in summer and winter. Sterile, functionally male flowers are often initiated during summer, although effects of transplanting dates and plant age have also been noted (Salinas personal communication). The number of male flowers cannot completely explain the lack of fruit set in summer and flower abscission (Aiyelaagbe et al., 1986); reduced flower fertilization probably plays a role too. Since self-fertilization is the primary mode of reproduction in hermaphrodite papaya genotypes, lack of fruit set is thought to be due to limited stigma receptivity under drought. When pollen tube growth and/or ovule longevity are negatively affected, seed formation is reduced, which in turn limits sink strength and hence fruit size (length and diameter, but also weight). Most common papaya cultivars produce large fruit, and production of smaller fruit limits profitability. However, if seed number is reduced due to water deficit, very small, noncommercial fruit are produced. Malformed, carpeloid flowers are more commonly developed in winter at temperatures below 15°C. A link to water scarcity has not been established for this feature of papaya.

Fruit size could also be reduced by source limitation. In papaya, growing fruit depend largely for their growth on the size and activity of subjacent leaves. For example, a reduction of photosynthetic tissue by more than 50%, achieved by defoliation, may reduce fruiting in papaya (Zhou et al., 2000), although stomatal closure in response to water stress is probably more important in limiting photosynthesis. When water is limited, smaller leaves (limbs) are formed (Olubode et al., 2016), thus reducing the photosynthetic surface area, which can negatively impact crop yield.

In addition to fruit size, sweetness can also be negatively affected by a period of drought due to the negative effects on CO_2 fixation and photoassimilate production and transport. Under such conditions, papaya commercialization can be negatively affected given the critical importance that fruit sugar content has on sales. Certainly, moderate water stress results in the production of sweeter fruit in many crops due to the sugars being more concentrated in smaller fruits, but this situation has not been proven in papaya. In fact, Goenaga et al. (2004) did not find an influence of the amount of irrigation water (testing Class A pan factors from 0.25 to 1.25) on papaya sweetness.

In summary, despite the fact that papaya is fairly tolerant to drought, the implementation of irrigation programs is recommended where the amount and/or distribution of rain does not cover papaya water requirements estimated to be between 1.1 and 1.25 Class A pan factor (Goenaga et al., 2004; Santana et al., 2008). If these requirements are not fulfilled, fruit number, size, and quality will be reduced, thereby negatively affecting farmers' profits. Despite previous comments about the negative effect of water scarcity on papaya, this crop is also extremely sensitive to waterlogging. Submergence of 75% or 100% of the root system of papaya for a little as 2 days resulted in a significant reduction of photosynthesis, stomatal conductance and leaf chlorophyll content (Thani et al., 2016). If anoxia conditions persist, plant loss occurs (Khondaker and Ozawa, 2007), especially in young papaya plants with a limited root system.

5. CONCLUSIONS

The environmental factors discussed in this chapter can affect photosynthetic processes negatively and reduce papaya yield. In scenarios of global climatic change, these factors can have even greater positive or negative impacts on the physiology of this important fruit species. Thus, whether grown in the field or in protected structures, understanding the action of environmental factors on photosynthetic carbon assimilation is extremely important for commercial papaya production. With this knowledge, it will be possible to develop management strategies to most efficiently utilize environmental variables for optimum crop growth and yield and prevent damaging effects of these on papaya physiology. Information in this chapter may also help in genetic and breeding programs for papaya and with agricultural site selection for future *Carica papaya* cultivation.

ACKNOWLEDGMENTS

The authors thank Dr. David Michael Glenn (Appalachian Fruit Research Station, Kearneysville, WV, USDA, USA) and Prof. Elias Fernandes de Souza for their collaboration in projects addressed in this review. We also thank the Coordination for University Personnel Training (Coordenação de Aperfeiçoamento de Pessoal de Nível Superior (CAPES), the National Council for Scientific and Technological Development (CNPq), the Foundation for Research Support of Rio de Janeiro state (Fundação de Amparo à Pesquisa do Estado do Rio de Janeiro [FAPERJ]), and the Study and Projects Financier (Financiadora de Estudos e Projetos [FINEP]) for financial support of the research.

REFERENCES

Almeida, F.T., Bernardo, S., Sousa, E.F., Marin, S.L.D., 2003. Growth and yield of pawpaw under irrigation. Scientia Agricola 60, 419–424.

Ainsworth, E.A., Long, S.P., 2005. What have we learned from 15 years of free-air CO_2 enrichment (FACE)? A meta-analytic review of the responses of photosynthesis, canopy properties and plant production to rising CO_2. The New Phytologist 165, 351–372.

Aiyelaagbe, I.O.Q., Falusi, M.O.A., Babalola, O., 1986. Growth, development and yield of pawpaw (*Carica papaya* L.). 'Homestead selection' in response to soil moisture stress. Plant and Soil 93, 427–435.

Allan, P., Jager, J., 1978. Net photosynthesis in macadamia and papaw and the possible alleviation of heat stress. Crop Protection 7, 125–128.

Arumuganathan, K., Earle, E.D., 1991. Nuclear DNA content of some important plant species. Plant Molecular Biology Reporter 9, 208–218.

Australian Government, 2008. The Biology of *Carica papaya* L. (Papaya, Papaw, Paw Paw). Department of Health and Ageing. Office of the Gene Technology Regulator, p. 55. http://www.ogtr.gov.au/internet/ogtr/publishing.nsf/Content/5DCF28AD2F3779C4CA257D4E001819B9/$File/biology-papaya2008.pdf.

Becker, H.S., 1958. Problems of inference and proof in participant observation. American Sociological Review 23, 652–660.

Buisson, D., Lee, D.W., 1993. The development responses of papaya leaves to simulated canopy shade. American Journal of Botany 80 (8), 947–952.

Campostrini, E., Glenn, D.M., 2007. Ecophysiology of papaya: a review. Brazilian Journal of Plant Physiology 19 (4), 413–424.

Campostrini, E., Glenn, D.M., Yamanishi, O.K., 2010. Papaya: ecophysiology of growth and production. In: DaMatta, F. (Ed.), Ecophysiology of Tropical Tree Crops. Nova Science Publishers, New York, pp. 287–307.

Campostrini, E., 1997. Comportamento de quarto genótipos de mamoeiro (*Carica papaya* L.) sob restrição mecânica ao crescimento do sistema radicular (Ph.D. thesis). UENF, Campos dos Goytacazes, RJ, 166 p.

Carr, M.K.V., 2014. The water relations and irrigation requirements of papaya (*Carica papaya* L.): a review. Experimental Agriculture 50 (2), 270–283.

Chan, Y.K., 2009. Breeding papaya (*Carica papaya* L.). In: Jain, S.M., Priyadarshan, P.M. (Eds.), Breeding Plantation Tree Crops: Tropical Species. Springer Science, pp. 121–159.

Chaves, M.M., Flexas, J., Pinheiro, C., 2009. Photosynthesis under drought and salt stress: regulation mechanisms from whole plant to cell. Annals of Botany 103, 551–560.

Chaves, M.M., Maroco, J.P., Pereira, J.S., 2003. Understanding plant responses to drought – from genes to the whole plant. Functional Plant Biology 30, 239–264.

Chaves, M.M., Zarrouk, O., 2013. Crop responses to available soil water. In: Encyclopedia of Sustainability Science and Technology. Springer-Verlag, Berlin, Heidelberg, New York, pp. 2638–2661.

Chaves, M.M., Zarrouk, O., Francisco, R., Costa, J.M., Santos, T., Regalado, A.P., Rodrigues, M.L., Lopes, C.M., 2010. Grapevine under deficit irrigation: hints from physiological and molecular data. Annals of Botany 105, 661–676.

Clemente, H.S., Marler, T.E., 1996. Drought stress influences gas-exchange responses of papaya leaves to rapid changes in irradiance. Journal of the American Society for Horticultural Science 121 (2), 292–295.

Clemente, H.S., Marler, T.E., 2001. Trade winds reduce growth and influence gas exchange patterns in papaya seedlings. Annals of Botany 88, 379–385.

Crane, J.H., 2005. Papaya Growing in the Florida Home Landscape. University of Florida, IFAS Extension.

DaMatta, F.M., 2003. Drought as a multidimensional stress affecting photosynthesis in tropical tree crops. In: Hemantaranjan, A. (Ed.), Advances in Plant Physiology, vol. 5. Scientific Publishers, Jodhpur, pp. 227–265.

DaMatta, F.M., Grandis, A., Arenque, B.C., Buckeridge, M.S., 2010. Impacts of climate changes on crop physiology and food quality. Food Research International 43 (7), 1814–1823.

DaMatta, F.M., Ramalho, J.C., 2006. Impacts of drought and temperature stress on coffee physiology and production: a review. Theoretical and Experimental Plant Physiology (Ex-Brazilian Journal of Plant Physiology) 18 (1), 55–81.

Dhekney, S., 2004. Molecular Investigations, Cryopreservation and Genetic Transformation Studies in Papaya (*Carica Papaya*) for Cold Hardiness (Ph.D. thesis). University of Florida, USA.

El Moussaoui, A., Nijs, M., Paul, C., Wintjens, R., Vincentelli, J., Azarkan, M., Looze, Y., 2001. Revisiting the enzymes stored in the laticifers of *Carica papaya* in the context of their possible participation in the plant defense mechanism. Cellular and Molecular Life Sciences 58, 556–570.

El-Sharkawy, M.A., Cock, M.J.H., Hernandez, A.D.P., 1985. Stomatal response to air humidity and its relation to stomatal density in a wide range of warm climate species. Photosynthesis Research 7, 137–149.

Easterling, W.E., Aggarwal, P.K., Batima, P., Brander, L.M., Erda, L., Howden, S.M., Kirilenko, A., Morton, J., Soussana, J.F., Schmidhuber, J., Tubiello, F.N., 2007. Food, fibre and forest products. In: Parry, M.L., Canziani, O.F., Palutikof, J.P., van der Linden, P.J., Hanson, C.E. (Eds.), Climate Change 2007: Impacts, Adaptation and Vulnerability. Contribution of Working Group II to the Fourth Assessment Report of the Intergovernmental Panel on Climate Change. Cambridge University Press, Cambridge, pp. 273–313.

Feng, X., Porporato, A., Rodriguez-Iturbe, I., 2013. Changes in rainfall seasonality in the tropics. Nature Climate Change 3, 811–815.

Ferraz, T.M., 2008. Trocas gasosas em mamoeiro Golden: relações entre plantas inteiras e folhas individuais (Masters dissertation). UENF, Campos dos Goytacazes, RJ, 69 p.

Ferraz, T.M., Torres Netto, A., Reis, F.O., Peçanha, A.L., de Sousa, E.F., Machado Filho, J.A., Rodrigues, W.P., Glenn, D.M., Campostrini, E., 2015. Relationships between sap-flow measurements, whole canopy transpiration and reference evapotranspiration in field-grown papaya (*Carica papaya* L.). Theoretical and Experimental Plant Physiology 27, 251–262.

Ferraz, T.M., Rodrigues, W.P., Torres Netto, A., Reis, F.O., Peçanha, A.L., Figueiredo, F.A.M.M.A., Sousa, E.F., Glenn, D.M., Campostrini, E., 2016. Comparison between single-leaf and whole-canopy gas exchange measurements in papaya (*Carica papaya* L.) plants. Scientia Horticulturae (Amsterdam) 209, 73–78.

Galán Saúco, V., Rodríguez Pastor, M.A.C., 2007. Greenhouse cultivation of papaya. Acta Horticulturae 740, 191–195.

Goenaga, R., Rivera, E., Almodóvar, C., 2004. Yield of papaya irrigated with fractions of Class A pan evaporation in a semiarid environment. The Journal of Agriculture of the University of Puerto Rico 88 (1–2), 1–10.

Grau, A., Halloy, S., 1997. Effect of chilling on CO_2 gas-exchange in *Carica papaya* L. And *Carica quercifolia* (A. St. Hil.) solms. Journal of Plant Physiology 150 (4), 475–480.

Gregory, P.J., Johnson, S.N., Newton, A.C., Ingram, J.S.I., 2009. Integrating pests and pathogens into the climate change/food security debate. Journal of Experimental Botany 2827–2838.

Hasanuzzaman, M., Nahar, K., Fujita, M., 2013. Extreme temperature responses, oxidative stress and antioxidant defense in plants. In: Kourosh Vahdati, D. (Ed.), Abiotic Stress – Plant Responses and Applications in Agriculture, pp. 169–205. InTech.

Imai, K., Ogura, F., Murata, Y., 1982. Photosynthesis and respiration of papaya (*Carica papaya* L.) leaves. Acta Oecologica/Oecologia Plantarum 4, 399–407.

IPCC, 2013. Climate change 2013: the physical science basis. Summary for policymakers, technical summary and frequent asked questions. In: Stocker, T.F., Qin, D., Plattner, G.K., Tignor, M.M.B., Allen, S.K., Boschung, J., Nauels, A., Xia, Y., Bex, V., Midgley, P.M. (Eds.), Part of the Working Group I Contribution to the Fifth Assessment Report of the Intergovernmental Panel on Climate Change. Intergovernmental Panel on Climate Change, 203 p.

IPCC, 2014. Climate change 2014: mitigation of climate change. In: Edenhofer, O., Pichs-Madruga, R., Sokona, Y., Farahani, E., Kadner, S., Seyboth, K., Adler, A., Baum, I., Brunner, S., Eickemeier, P., Kriemann, B., Savolainen, J., Schlömer, S., von Stechow, C., Zwickel, T., Minx, J.C. (Eds.), Contribution of Working Group III to the Fifth Assessment Report of the Intergovernmental Panel on Climate Change. Cambridge University Press, Cambridge, UK and New York, USA, 1435 p.

Jeyakumar, P., Kavino, M., Kumar, N., 2007. Physiological performance of papaya cultivars under abiotic stress conditions. Acta Horticulturae 740, 209–214.

Johnson, R.S., Handley, D.F., 2000. Using water stress to control vegetative growth and productivity of temperate fruit trees. HortScience 35, 1048–1050.

Khondaker, N.A., Ozawa, K., 2007. Papaya plant growth as affected by soil air oxygen deficiency. Acta Horticulturae 740, 225–232.

Kirschbaum, M.U.F., 2011. Does enhanced photosynthesis enhance growth? Lessons learned from CO_2 enrichment studies. Plant Physiology 155, 117–124.

Knight Jr., R.J., 1980. Origin and world importance of tropical and subtropical fruit crops. In: Nagy, S., Shaw, P.W. (Eds.), Tropical and subtropical fruits. Composition, properties and uses. AVI Publishing, Westport, Connecticut, 120 p.

Konno, K., Hirayama, C., Nakamura, M., Tateishi, K., Tamura, Y., Hattori, M., Kohno, K., 2004. Papain protects papaya trees from herbivorous insects: role of cysteine proteases in latex. The Plant Journal 37, 370–378.

Lambers, H., Chapin III, F.S., Pons, J.L., 2008. Plant Physiological Ecology, second ed. Springer-Verlag, New York. 604 p.

Lassoudiere, A., 1968. Le papayer (Deuxieme partie). Fruits 23, 585–596.

Li, P., Cheng, L., Gao, H., Jiang, C., Peng, T., 2009. Heterogeneous behavior of PSII in soybean (*Glycine max*) leaves with identical PSII photochemistry efficiency under different high temperature treatments. Journal of Plant Physiology 166, 1607–1615.

Lima, R.S.N., 2014. Partial rootzone drying (PRD) and regulated deficit irrigation (RDI) in *Carica papaya* L.: photosynthetic capacity, growth and water use efficiency (PhD thesis). UENF, Campos dos Goytacazes, RJ, 130 p.

Lima, R.S.N., Figueiredo, F.A.M.M.A., Martins, A.O., Deus, B.C.S., Ferraz, T.M., Gomes, M.M.A., Sousa, E.F., Glenn, D.M., Campostrini, E., 2015. Partial rootzone drying (PRD) and regulated deficit irrigation (RDI) effects on stomatal conductance, growth, photosynthetic capacity, and water-use efficiency of papaya. Scientia Horticulturae (Amsterdam) 183, 13–22.

Lima, R.S.N., García-Tejeroc, I., Lopes, T.S., Costa, J.M., Vaz, M., Durán-Zuazoc, V.H., Chaves, M., Glenn, D.M., Campostrini, E., 2016. Linking thermal imaging to physiological indicators in *Carica papaya* L. under different watering regimes. Agricultural Water Management 164, 148–157.

Logan, B.A., 2005. Reactive oxygen species and photosynthesis. In: Smirnoff, N. (Ed.), Antioxidants and Reactive Oxygen in Plants. Blackwell Publishing, Oxford, UK, pp. 250–267.

Long, S.P., Zhu, X., Naidu, S.L., Ort, D.R., 2006. Can improvement in photosynthesis increase crop yields? Plant Cell and Environment 29, 315–330.

Lopes de Siqueira, D., Botrel, N., 1986. Clima e solo para a cultura do mamoeiro. Informe Agropecuário 12 (134), 8–10.

Lu, C., Zhang, J., 1999. Effects of water stress on photosystem II photochemistry and its thermostability in wheat plants. Journal of Experimental Botany 50, 1199–1206.

Maestri, M., Barros, R.S., 1977. Coffee. In: Alvim, P.T., Kozlowski, T.T. (Eds.), Ecophysiology of Tropical Crops. Academic Press, London, pp. 249–278.

Mahouachi, J., Socorro, A.R., Talon, M., 2006. Responses of papaya seedlings (*Carica papaya* L.) to water stress and re-hydration: growth, photosynthesis and mineral nutrient imbalance. Plant and Soil 281, 137–146.

Mahouachi, J., Arbona, V., Gómez-Cadenas, A.G., 2007. Hormonal changes in papaya seedlings subjected to progressive water stress and re-watering. Plant Growth Regulation 53, 43–51.

Malo, S.E., Campbell, C.W., 1986. The Papaya. University of Florida, Cooperative Extension Service Fruits Crops Fact Sheet FC-11, Gainesville.

Marler, T.E., Clemente, H.S., 2006. Papaya seedling growth response to wind and water deficit is additive. HortScience 41, 96–98.

Marler, T.E., Mickelbart, M.V., 1998. Drought, leaf gas exchange, and chlorophyll fluorescence of field grown papaya. Journal of the American Society for Horticultural Science 23 (4), 714–718.

Marler, T.E., George, A.P., Nissen, R.J., Andersen, P.C., 1994. Miscellaneous tropical fruits. In: Schaffer, B., Andersen, P.C. (Eds.), Handbook of Environmental Physiology of Fruits Crops. V.II, Subtropical and Tropical Crops. CRC Press, Florida, pp. 199–224.

Martins, L.D., Tomaz, M.A., Lidon, F.C., DaMatta, F.M., Ramalho, J.C., 2014. Combined effects of elevated [CO_2] and high temperature on leaf mineral balance in *Coffea* spp. plants. Climatic Change 126 (3–4), 365–379.

Martins, M.Q., Rodrigues, W.P., Fortunato, A.S., Leitão, A.E., Rodrigues, A.P., Pais, I.P., Martins, L.D., Silva, M.J., Reboredo, F.H., Partelli, F.L., Campostrini, E., Tomaz, M.A., ScottiCampos, P., Ribeiro-Barros, A.I., Lidon, F.C., DaMatta, F.M., Ramalho, J.C., 2016. Protective response mechanisms to heat stress in interaction with high [CO_2] conditions in *Coffea* spp. Frontiers in Plant Science 7, 947.

Matos, M.C., Campos, P.S., Passarinho, J.A., Semedo, J.M., Marques, N.M., Ramalho, J.C., Ricardo, C.P., 2010. Drought effect on photosynthetic activity, osmolyte accumulation and membrane integrity of two *Cicer arietinum* genotypes. Photosynthetica 48, 303–312.

Ming, R., Hou, S., Feng, Y., Yu, Q., Dionne-Laporte, A., Saw, J.H., Senin, P., Wang, W., Ly, B.V., Lewis, K.L.T., Salzberg, S.L., Feng, L., Psii, M.R., Skelton, R.L., Murray, J.E., Chen, C., Qian, W., Shen, J., Du, P., Eustice, M., Tong, E., Tang, H., Lyons, E., Paull, R.E., Michael, T.P., Wall, K., Rice, D.W., Albert, H., Wang, M.L., Zhu, Y.J., Schatz, M., Nagarajan, N., Acob, R.A., Guan, P., Blas, A., Wai, C.M., Ackerman, C.M., Ren, Y., Liu, C., Wang, J., Wang, J., Na, J.K., Shakirov, E.V., Haas, B., Thimmapuram, J., Nelson, D., Wang, X., Bowers, J.E., Gschwend, A.R., Delcher, A.L., Singh, R., Suzuki, J.Y., Tripathi, S., Neupane, K., Wei, H., Irikura, B.,

Paidi, M., Jiang, N., Zhang, W., Presting, G., Windsor, A., Navajas-Perez, R., Torre, M.J., Feltus, F.A., Porter, B., Li, Y., Burroughs, A.M., Luo, M.C., Liu, L., Christopher, D.A., Mount, S.M., Moore, P.H., Sugimura, T., Jiang, J., Schuler, M.A., Friedman, V., Mitchell-Olds, T., Shippen, D.E., De Pamphilis, C.W., Palmer, J.D., Freeling, M., Paterson, A.H., Gonsalves, D., Wang, L., Alam, M., 2008. The draft genome of the transgenic tropical fruit tree papaya (*Carica papaya* Linnaeus). Nature 452, 991–997.

Ming, R., Yu, Q., Moore, P.H., 2007. Sex determination in papaya. Seminars in Cell and Developmental Biology 18, 401–408.

Muller, B., Pantin, F., Genard, M., Turc, O., Freixes, S., Piques, M., Gibon, Y., 2011. Water deficits uncouple growth from photosynthesis, increase C content, and modify the relationships between C and growth in sink organs. Journal of Experimental Botany 62, 1715–1729.

Nakasone, H.Y., Paull, R.E., 1998. Tropical Fruits. CAB International, Wallingford, England.

Nunes, M.A., Rijo, P., Ramalho, J.C., 1992. Seasonal changes in the photosynthetic properties of *Ceratonia siliqua* L. (carob tree) leaves under natural conditions. Physiologia Plantarum 86, 381–387.

O'Hare, P., 1993. Growing Pawpaws in South Queensland. Queensland Department of Primary Industries, Brisbane, Australia.

Olubode, O.O., Odeyemi, O.M., Aiyelaagbe, I.O.O., 2016. Influence of environmental factors and production practices on the growth and productivity of pawpaw (*Carica papaya* L.) in south western Nigeria – a review. Fruits 71 (6), 341–361.

O'Sullivan, O.S., Heskel, M.A., Reich, P.B., Tjoelker, M.G., Weerasinghe, L.K., Penillard, A., Zhu, L., Egerton, J.J.G., Bloomfield, K.J., Creek, D., Bahar, N.H.A., Griffin, K.L., Hurry, V., Meir, P., Turnbull, M.H., Atkin, O.K., 2016. Thermal limits of leaf metabolism across biomes. Global Change Biology 23, 209–223.

Pastenes, C., Horton, H., 1996. Effect of high temperature on photosynthesis in beans (II. CO_2 assimilation and metabolite contents). Plant Physiology 112 (3), 1253–1260.

Posse, R.P., Bernardo, S., Sousa, E.F., de Gottardo, R.D., 2008. Evapotranspiração e coeficiente da cultura do mamoeiro. Engenharia Agrícola 28 (4), 681–690.

Prasad, P.V.V., Boote, K.J., Allen Jr., L.H., Thomas, J.M.G., 2003. Super-optimal temperatures are detrimental to peanut (*Arachis hypogaea* L.) reproductive processes and yield at both ambient and elevated carbon dioxide. Global Change Biology 9 (12), 1775–1787.

Quick, W.P., Chaves, M.M., Wendler, R., David, M., Rodrigues, M.L., Passarinho, J.A., Pereira, J.S., Adcock, M.D., Leegood, R.C., Stitt, M., 1992. The effect of water stress on photosynthetic carbon metabolism in four species grown under field conditions. Plant, Cell and Environment 15, 25–35.

Ramalho, J.C., Chaves, M.M., 1992. Drought effects on plant water relations and carbon gain in two lines of *Lupinus albus* L. European Journal of Agronomy 1 (4), 271–280.

Ramalho, J.C., Lauriano, J.A., Nunes, M.A., 2000. Changes in photosynthetic performance of *Ceratonia siliqua* in summer. Photosynthetica 38 (3), 393–396.

Ramalho, J.C., Rodrigues, A.P., Semedo, J.N., Pais, I.P., Martins, L.D., Simões-Costa, M.C., Leitão, A.E., Fortunato, A.S., Batista-Santos, P., Palos, I.M., Tomaz, M.A., Scotti-Campos, P., Lidon, F.C., DaMatta, F.M., 2013. Sustained photosynthetic performance of *Coffea spp.* Under long-term enhanced [CO_2]. PLoS One 8 (12), e82712.

Ramalho, J.C., Zlatev, Z.S., Leitão, A.E., Pais, I.P., Fortunato, A., Lidon, F.C., 2014. Moderate water stress causes differential stomatal and non-stomatal changes in the photosynthetic functioning of *Phaseolus vulgaris* L. genotypes. Plant Biology 16, 133–146.

Reis, F.O., Campostrini, E., 2008. Trocas gasosas e eficiência fotoquímica potencial em mamoeiro do grupo 'Formosa' cultivados em condições de campo. Bragantia 67 (4), 815–822.

Reis, F.O., Campostrini, E., 2011. Microaspersão de água sobre a copa: um estudo relacionado às trocas gasosas e à eficiência fotoquímica em plantas de mamoeiro. Revista Brasileira de Agrociência 17284–17295.

Ribeiro da Silva, J.B., Ferreira, E., Mara, P., da Costa, E.L., Carvalho, G.C., 2013. Irrigação. In: Castricini, A., Rodrigues, M.G.V. (Eds.), Cultivo Do Mamoeiro. Informe Agropecuario EPAMIG, vol. 34, pp. 48–57, 275.

Rodrigues, W.P., Martins, M.Q., Fortunato, A.S., Rodrigues, A.P., Semedo, J.N., Simões-Costa, M.C., Pais, I.P., Leitão, A.E., Colwell, F., Goulao, L., Máguas, C., Maia, R., Partelli, F.L., Campostrini, E., Scotti-Campos, P., Ribeiro-Barros, A.I., Lindon, F.C., DaMatta, F.M., Ramalho, J.C., 2016. Long-term elevated air [CO_2] strengthens photosynthetic functioning and mitigates the impact of supra-optimal temperatures in tropical *Coffea arabica* and *C. canephora* species. Global Change Biology 22 (1), 415–431.

Salazar, R., 1978. Determination of photosynthesis in commercial varieties of papaw (*Carica papaya* L.) and its possible relationship with the production and quality of the fruits. Revista de la Facultad de Ciencias Agrarias 13, 291–295.

Santana, J.L., Sousa, M.J., Rodriguez, M.C., Suarez, C.L., 2008. Response of the papaya (*Carica papaya* L.) plant to deficit irrigation in the Canary Islands. Acta Horticulturae 792, 559–565.

Scotti-Campos, P., Duro, N., da Costa, M., Pais, I.P., Rodrigues, A.P., Batista-Santo, P., Semedo, J.N., Leitão, A.E., Lidon, F.C., Pawlowski, K., Ramalho, J.C., Ribeiro-Barros, A.I., 2016. Antioxidative ability and membrane integrity in salt-induced responses of *Casuarina glauca* Sieber ex Spreng. in symbiosis with N_2-fixing Frankia Thr or supplemented with mineral nitrogen. Journal of Plant Physiology 196, 60–69.

Seelig, R.A., 1970. Papaya-Fruit and Vegetable Facts and Pointers. United Fresh Fruit and Vegetable Association, Washington, DC.

Srinivas, K., 1996. Plant water relations, yield, and water use of papaya (*Carica papaya* L.) at different evaporation-replenishment rates under drip irrigation. Tropical Agriculture 73, 264–269.

Thani, Q.A., Schaffer, B., Liub, G., Vargas, A.I., Crane, J.H., 2016. Chemical oxygen fertilization reduces stress and increases recovery and survival of flooded papaya (*Carica papaya* L.) plants. Scientia Horticulturae (Amsterdam) 202, 173–183.

Torres Netto, A., 2005. Atributos fisiológicos e relações hídricas em genótipos de mamoeiro (*Carica papaya* L.) na fase juvenil (Ph.D. thesis). UENF, Campos dos Goytacazes, RJ, 116 p.

Villegas, V.N., 1997. *Carica papaya* L. In: Verheij, E.W.M., Coronel, R.E. (Eds.), Plant Resources of South-East Asia, Edible Fruits and Nuts, vol. 2. Wageningen University, The Netherlands.

Vincent, C., Rowland, D.L., Schaffer, B., 2015. The potential for primed acclimation in papaya (*Carica papaya* L.): determination of critical water deficit thresholds and physiological response variables. Scientia Horticulturae (Amsterdam) 194, 344–352.

Wahid, A., Gelani, S., Ashraf, M., Foolad, M.R., 2007. Heat tolerance in plants: an overview. Environmental and Experimental Botany 61 (3), 199–223.

Wang, W., Vinocur, B., Altman, A., 2003. Plant responses to drought, salinity and extreme temperatures: towards genetic engineering for stress tolerance. Planta 218, 1–14.

Weng, J., Lai, M., 2005. Estimating heat tolerance among plant species by two chlorophyll fluorescence parameters. Photosynthetica 43 (3), 439–444.

Zhou, L., Christopher, D.A., Paull, R.E., 2000. Defoliation and fruit removal effects on papaya fruit production, sugar accumulation, and sucrose metabolism. Journal of the American Society for Horticultural Science 125 (5), 644–652.

Zlatev, Z.S., Lidon, F.C., Ramalho, J.C., Yordanov, I., 2006. Comparison of resistance to drought of three bean cultivars. Biologia Plantarum 50, 389–394.

CHAPTER 20

Fleshy Fruit Epidermis is a Protective Barrier Under Water Stress

Olfa Zarrouk[1], Carla Pinheiro[1,2], Chandra S. Misra[1], Victoria Fernández[3], Maria M. Chaves[1]
[1]Universidade NOVA de Lisboa, Oeiras, Portugal; [2]Universidade NOVA de Lisboa, Caparica, Portugal; [3]Technical University of Madrid, Madrid, Spain

Contents

1. INTRODUCTION

Fruit growth and ripening are complex developmental processes involving multiple metabolic adjustments. In all growth phases, the epidermis, which is formed by a single layer of cells covered by the cuticle, deeply influences the final quality and attractiveness of the fruit since it acts as a barrier between internal and external environments (Domínguez et al., 2011a; Martin and Rose, 2014). Fruits are exposed to a variety of unfavorable mechanical, climatic, and biological factors that may affect fruit integrity during growth, harvest, transport, and storage (Jenks et al., 1994; Markstädter et al., 2000). In order to retain its functions, the structural integrity of cuticle

Water Scarcity and Sustainable Agriculture in Semiarid Environment
ISBN 978-0-12-813164-0
https://doi.org/10.1016/B978-0-12-813164-0.00020-X

through fruit development and expansion has to remain firm (Edelmann et al., 2005).

The plant cuticle is a physical structure that serves numerous important functions for plant growth and survival (Kerstiens, 1996). It is a hydrophobic membrane covering the aerial surfaces of plant organs and is inarguably one of the most significant innovations in the history of plant evolution. In order to adapt to the new terrestrial environment around 450 million years ago, plants developed the ability to retain water in otherwise highly desiccating external environment (Leliaert et al., 2011; Waters, 2003; Sørensen et al., 2011; Budke et al., 2012).

The classical understanding of the cuticle as a bilayered, independent area of the epidermal cell wall has been recently challenged, highlighting the presence and importance of polysaccharides in the cuticle structure (Fernández et al., 2016; Segado et al., 2016). As major chemical constituents, the cuticle contains waxes either embedded (intracuticular) or deposited onto (epicuticular) the cutin and polysaccharide matrix (Jetter et al., 2006; Fernández et al., 2016). Minor amounts of flavonoids and other phenols have also been identified in the cuticle of several fruit species, such as peach (Fernández et al., 2011) or tomato (España et al., 2014).

In Fig. 20.1, the structural diversity of fruit surfaces is shown with nectarine, plum, apple, and peach fruit as examples. In general, surfaces may be heterogeneous due to the presence of stomata (e.g., plum and nectarine; Fig. 20.1A and B) or trichomes (e.g., peach, Fig. 20.1D. Such surface is covered by a continuous cuticle also covering the trichomes; Fernández et al., 2011) or lenticels (apple; see detail of a lenticel surrounded by the cuticle in Fig. 20.1C).

The main functions attributed to the fruit cuticle are to minimize the water loss and to serve as a protective barrier against physical, chemical, and biological attack, as well as provide the mechanical framework to support the integrity of the fruit. Additionally, some studies in grapes suggest that cuticular density correlates with susceptibility of the berries to *Botrytis* spp. at véraison and that cluster microclimate may affect the development of cuticle and its protective role under biotic and abiotic stresses (Yeats and Rose, 2013). In this context, despite the greater relative thickness of the fruit versus leaf cuticle, the rates of water loss of fruit may be higher than in leaves. This may have a major significance for fruit development and physiology (Schreiber and Riederer, 1996) since the cuticle biophysical

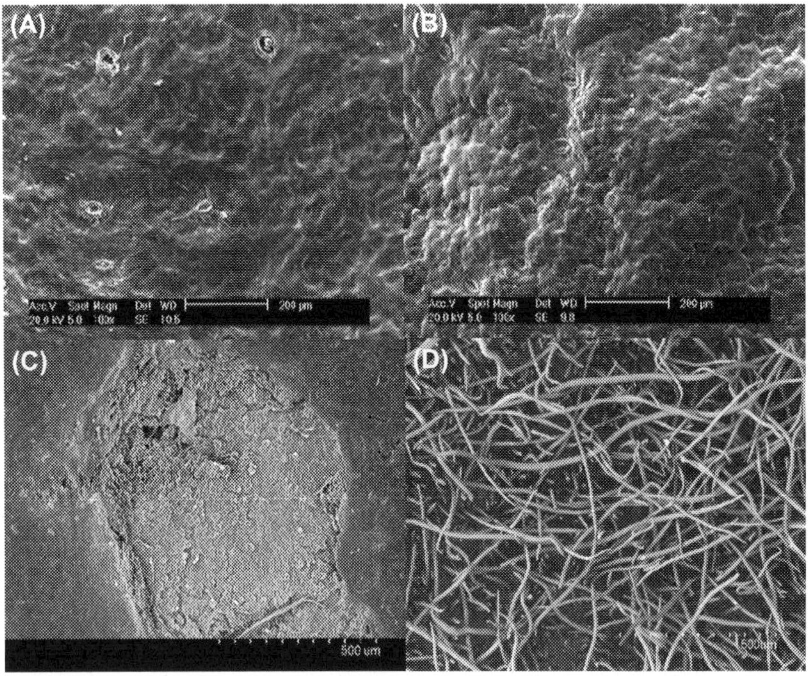

Figure 20.1 Scanning electron microscope (SEM) micrographs showing the structural diversity of fruit surfaces. (A and B) Presence of stomata on plum (*Prunus salicina* L.) and nectarine (*Prunus persica* L.) surfaces; (C) presence of lenticels on apple (*Malus domestica* Borkh.) surface; (D) presence of trichomes on peach (*Prunus persica* L.) surface.

properties also depend on environmental conditions, such as temperature and relative humidity (Edelmann et al., 2005; Matas et al., 2005). This effect can be illustrated by the cuticular permeability differences measured for sunlit and shaded grapefruit and mango fruits, which were associated with differences in epicuticular wax morphology and composition (Léchaudel et al., 2013). Also, Rosenquist and Morrison (1989) reported that grape berries from shaded clusters had less cuticle than berries from sun-exposed clusters. Strength and rigidity of cuticle inversely varies with the increase of temperature (Domínguez et al., 2011a,b). Temperature and chiefly relative humidity have been shown to modify the mechanical properties of the cuticle, leading to a plasticizing effect (Domínguez et al., 2011a,b). Cuticle plasticizing may be important for the adjustments to external stimuli (Domínguez et al., 2011b).

Water scarcity is considered the most limiting factor in Mediterranean crop production (Chaves and Zarrouk, 2012). Water stress appears to play a role in cuticular wax content and biosynthesis (Kosma et al., 2009; Provost et al., 2013). To survive under such conditions fruits have evolved protective mechanisms through integrated molecular and cellular responses. Among the physical adaptations, cuticular wax provides an essential barrier to protect plants from drought stress (Samuels et al., 2008; Lee and Suh, 2015). In grape berries, the preharvest weight loss observed in some grape varieties, accentuated by water deficit, was attributed to cuticle changes (Rogiers et al., 2004).

In spite of the importance of the cuticle under the accepted scenario of global environmental shift, little is known about its function in fruits. The majority of studies carried out on the role of cuticle in plants have been carried out with vegetative tissues, and most of the cuticle biosynthesis investigations focused on the model plant Arabidopsis. Nevertheless, in recent years there has been a growing interest in the biological role of cuticle in fruits, namely tomato (e.g., Segado et al., 2016) and, to a small extent, cherry (Alkio et al., 2012; Isaacson et al., 2009; Knoche and Peschel, 2006; Leide et al., 2007; Shi et al., 2011; Yeats and Rose, 2013; Belge et al., 2014b); peach (Belge et al., 2014a; Fernández et al., 2011); plum (Knoche and Peschel, 2007); and apple (Knoche and Grimm, 2008; Knoche et al., 2011; Legay et al., 2015) fruits. In the case of tomato, it is a climacteric fruit where ethylene phytohormone plays a pivotal role in the ripening and development in contrast to nonclimacteric fruits, such as cherry and grape berries where fruit ripening is reported to be significantly controlled by abscisic acid (ABA) (Wheeler et al., 2009; Zarrouk et al., 2016). This is particularly relevant since recent findings suggest the possible role of ABA in the biosynthesis of epicuticular waxes and cuticular water permeability, as well as in the development and composition of fruits waxy layer, thereby enhancing the resistance to pathogens and pests (Curver et al., 2010; Romero et al., 2012).

To address some of these questions, this chapter aimed to revise the available literature on fruit cuticle physiology, composition, and properties as affected by abiotic stress factors with emphasis on water stress. We also focus on recent molecular findings concerning cuticle biosynthesis and function and highlight some of the major challenges to be elucidated in future studies.

2. WATER PERMEABILITY OF FRUIT CUTICLE AND ITS RESPONSE TO WATER STRESS

2.1 Cuticle as a Fruit Barrier to Water Loss

When vascular plants inhabited dry land, they developed special mechanisms to protect against desiccation as provided by the cuticle (Burghardt and Riederer, 2006; Riederer and Schreiber, 2001). The main physiological attribute of this layer basically lies in adequately controlling water loss. Schreiber and Riederer (1996) have reported the cuticular permeability of leaves and fruits of several plant species from different habitats. In general, fruit cuticles had the highest rate of cuticular transpiration as compared to leaves. This finding points out to cuticle as transport barrier and for the regulation of fruit growth by cuticular transpiration. They were also able to demonstrate that more than cuticle thickness, the waxes seem to act as the effective transport barrier, which was later supported by several studies (Riederer and Schreiber, 2001; Schreiber and Riederer, 1996). The question that arises is: "What is the physiological significance of cuticle when its serves as somewhat less than a perfect barrier to water loss?" This has been the inspiration for the investigation of the physiology of cuticle and its role in stress mechanism especially in fruits under environmental constraints.

2.2 Cuticular Permeance

Unlike tomato or pepper, the surface of many fruits can be heterogeneous and in addition to the cuticle may contain stomata, trichomes of lenticels as exemplified in Fig. 20.1. Subsequently, fruit water loss or absorption of solutes may occur via the cuticle, cuticular irregularities, stomata, trichomes, and other structures such as lenticels (apple) (Fernández and Eichert, 2009; Fernández et al., 2013). While lenticels appear to have a major role on permeability (e.g., Maguire et al., 2000), the role of stomata and trichomes on water permeability has received limited attention and should be clarified in future research.

So far, studies on cuticle permeance have been mostly restricted to fully developed leaves and fruits of dicotyledonous plants (Kerstiens, 1996, 2006; Riederer and Schreiber, 2001). The possible relationship between cuticular waxes and permeance and the potential role of cutin as transport barrier have been assessed either by the comparison of intact surfaces with those whose waxes had been mechanically removed or altered (Jetter et al., 2000) or through analysis of waxes that are solvent extracted and reconstituted (Kirsch et al., 1997; Šantrůček et al., 2004; Schreiber and Schonherr, 1993).

Tomato fruit cuticle represents a good system for cuticular permeability studies, and a few available tomato mutants have already been successfully used in such studies (Vogg et al., 2004; Leide et al., 2007). The cuticular permeance study conducted by Leide et al. (2007) on the role of cuticular waxes in the transpirational water loss of "MicroTom" and its *lecer6* mutant tomato fruit (*Lycopersicon esculentum*) over the different developmental stages showed that even though both the wild type and mutant showed similar patterns of wax accumulation, they exhibited different water permeances. Therefore, cuticular wax quality rather than quantity predominately affects the water barrier properties, as previously reported by Riederer and Schreiber (2001). Vogg et al. (2004) also followed the developmental stage of dwarf tomato "MicroTom" and lecer6 mutant in order to elucidate the quantitative and qualitative modification of cuticular wax composition during fruit maturation. They established that water permeability of cuticle in tomato fruits was largely dependent on the aliphatic component of intracuticular wax layer, which was modified by the variable amount of triterpenoids during the various maturation stages. In this context, Knoche et al. (2000) estimated that the epicuticular contribution to the wax resistance can be inferred to be approximately 50%, probably due to a high concentration of triterpenoids in the intracuticular wax of the fruit and low amounts of fatty acid derivatives. Tsubaki and Azuma (2013) showed that triterpenoids assembled as nanoparticles are able to fill the gaps in cutin matrix, thus providing a strengthening of the cuticular membrane, which could explain the importance of these components in cuticle permeability, as reported for tomato fruit (España et al., 2014). Recent studies in grape berries, under different deficit irrigation regimes and from different vine cluster positions, provided evidence of the importance of both quantity and quality of waxes on the berry cuticle permeance along the ripening process (Zarrouk et al. unpublished results). In addition, in cross-sections of berries analyzed by transmission electron microscopy, an increase of the pectin layer (i.e., the middle lamella) was observed in the stressed berries, suggesting their role in berry water loss control under abiotic stress conditions.

From the different studies, it thereby appears that water permeability does not seem to correlate with the amount of cutin but rather with changes in cuticular wax composition.

3. FRUIT EPIDERMIS ULTRASTRUCTURE IN RESPONSE TO WATER STRESS

Cuticle architectural organization can be discerned using a number of microscopic techniques. Scanning electron microscopy (SEM) can reveal the elaborate and diverse morphologies of epicuticular wax crystals (Jeffree, 2006) whereas transmission electron microscopy (TEM) is helpful in examining cuticular ultrastructure. Cuticles vary considerably in their architectural design and, depending on the species and varieties, differ dramatically in their thickness from nanometer to the micrometer scale (Jeffree, 2006). Light microscopy can also be used on micrometer scale to elucidate the fine structure of cuticle and epidermal cell wall as reported by Yeats and Rose (2013). Histochemical staining coupled with confocal microscopy can further help to resolve the three-dimensional architecture of cuticle (Buda et al., 2009).

Casado and Heredia (2001) studied the ultrastructure of the cuticle in grape berry during development using electron microscopy techniques, observing that the cuticle developed over the entire grape berry with an epicuticular wax layer of 0.5 μm during the period of rapid expansion. However, the final stage of berry growth was characterized by the occurrence of a homogenous cuticle with a thickness of 3 μm (Casado and Heredia, 2001). The existing fruit cuticle studies showed that cuticle thickness is highly variable, but to date not much information could be obtained regarding the patterns of cuticular deposition and what are factors that regulate cuticular architecture. Therefore, future investigations should improve our understanding of the structure-function relationship of the cuticle of fruits as affected, for example, by the state of development or stress factors.

Light microscopy studies indicated that peach fruit from deficit and optimum irrigation had a continuous and much thicker cuticle than fruit from trees subjected to an excess irrigation treatment. These differences in exodermis structure may explain the high percentage of water loss from fruit belonging to the excess irrigation treatment compared to the deficit and optimum irrigation treatments (Crisosto et al., 1994). Additionally, water-deficit treatment was shown to limit cuticular cracks in nectarine fruits (Gibert et al., 2007). In this sense, Saladié et al. (2007) pointed out that other ripening-related physiological processes, such as cell turgor and morphology, should play critical roles in fruit softening. They proposed a model in which the cuticle affects the softening of intact fruits both directly, by providing a physical support, and indirectly, by regulating the water status. They also highlighted the mechanistic distinction between a reduction in firmness, or resistance to compression, of intact fruit and ripening-related textural changes in the pericarp tissue. In recent investigations, TEM analysis in grape berries showed that water stress increased the epicuticular waxes as well as pectin deposition along the berry development, which highlighted the role of these cuticular structures in regulating the fruit water loss under stress. In addition, SEM analysis showed a higher wax density at the surface berries developing under water stress, at all ripening stages (Zarrouk et al., unpublished results).

4. FRUIT CUTICLE CHEMICAL CONSTITUENTS AND ITS MODULATION UNDER WATER STRESS

4.1 Constituents and Structure of Cuticle

Plant cuticles are complex structures and are composed of a macromolecular scaffold of cutin and a variety of organic solvent-soluble lipids that are generally named waxes (Yeats and Rose, 2013), together with polysaccharides (Fernández et al., 2016; Segado et al., 2016). The plant cuticle has been traditionally considered to be an independent structure, distinct from the polysaccharide cell wall underneath (Yeats and Rose, 2013), but the two structures are physically associated and have some overlapping functions. Therefore, it would not be wrong to consider the cuticle as a specialized lipid-modification of cell wall, just as lignification is a common modification of secondary cell wall in plants (Yeats and Rose, 2013; Fernández et al., 2016; Niklas et al., 2017).

Waxes are either deposited within the cuticle, which are called intracuticular wax, or accumulate on its surface as epicuticular wax crystals, or films. The epicuticular waxes confer distinct macroscopic surface properties to the plant surface and are responsible for the glossy appearance of many leaves and fruits, while epicuticular wax crystals account for the dull, glaucous appearance (Yeats and Rose, 2013) and serve as the interface of the plant with the external environment. These epicuticular waxes are highly apolar in nature, may provide a major degree of roughness (Fernandez and Khayet, 2015), and are efficient barriers to transpirational loss (Bernard and Joubès, 2013).

The structure and composition of cuticle is highly complex and can vary widely among plant species and within plant species in different organ and developmental stages. This is well illustrated in the typical range of thickness (1–10 μm) and quantity (100–1000 μg cm^{-2}) of deposited cuticle (Riederer and Muller, 2006).

4.2 Composition of Cuticle Components: Waxes and Cutin

As noted earlier, the cuticle contains waxes and cutin as major constituents, apart from polysaccharides (Guzmán et al., 2014; Segado et al., 2016).

The composition of waxes may differ substantially with species, ontogeny, and environmental conditions (Jenks and Ashworth, 1999). In most cases, the majority of compounds constituting the cuticular wax are derived from very-long-chain fatty acids (VLCFAs; C_{20}–C_{34}), including alkanes, aldehydes, primary and secondary alcohols, ketones, and esters. Several lipophilic secondary metabolites such as pentacyclic triterpenoids, flavonoids, and tocopherols have also been found to be substantial components (Jetter et al., 2006). Knowledge about the molecular biology underlying VLCFA-derived wax biosynthesis has made tremendous progress, and to this end *Arabidopsis* has provided an excellent experimental model (reviewed in Bernard and Joubès, 2013).

Extensive studies were done as early as the 1970s and 1980s (Kolattukudy, 2001) to identify the chemical composition of plant cutin. However, only in the 2000s was the composition of *Arabidopsis* cutin determined (Bonaventure et al., 2004; Franke et al., 2005), and even more recently for fruits such as tomato fruits. Cutin is composed of interesterified hydroxy fatty acids with potentially minor amounts of glycerol, phenylpropanoids, and dicarboxylic acids (Kolattukudy, 2001). Chemical breakdown of ester bonds, using processes such as saponification, readily releases these monomeric constituents, although in some species an additional lipidic polymer referred to as cutan remains recalcitrant to such treatments. Cutan is rich in ether and C–C bond, but its structure remains unknown and restricted to relatively few species (Gupta et al., 2006). The hydroxyl fatty acids of cutin are typically comprised of ω-hydroxy fatty acids, usually with one or two additional midchain hydroxyl groups or an epoxy group.

Detailed analyses of the composition and biosynthesis of fruit cutin and cuticular waxes are summarized in recent reviews (e.g., Lara et al., 2015; Lee and Suh, 2015). However, most of the cuticular biosynthesis and composition studies were conducted with vegetative plant tissues, and little information is available in relation to both fleshy and nonfleshy fruits. The significant differences observed over the last 40 years in the study of cuticle and its components of different organs of the same plant, both cutin (Espelie et al., 1979; Järvinen et al., 2010; Marga et al., 2001) and waxes

(Baker et al., 1975; Radler, 1965, 1970), provided sufficient evidence for the assumption that fruit cuticle composition cannot be inferred from that of vegetative organs.

Tomato fruit emerges as an interesting model for fleshy fruit cuticles. It provided valuable cuticle-related information as fruits are astomatous and comparatively thick, enabling the isolation of considerable amount of "unperforated" material for biochemical, physical, or structural characterization (Martin and Rose, 2014). Albeit less intensively, other fruit cuticles corresponding to the crops of economic relevance, namely peach and cherry, have also been investigated and have revealed significant differences across genotypes. The properties of the different cuticles and the understanding of the functions and impact of this outer layer on quality, storage potential, and shelf life of produce will require its characterization as a first step (Lara et al., 2015). The predominant cuticle components of some of the most common fruits are summarized in Table 20.1.

Tomato fruit cuticle has well-known composition, structure, and properties and is therefore a good model system of fleshy climacteric fruit. However, direct extrapolation of results to cuticles of the other fruits could be wrong (Heredia-Guerrero et al., 2014). As an example, the cuticle of grape berry (a nonclimacteric fruit) is mostly comprised of waxes (up to 40%) (Casado and Heredia, 1999) and a small amount of cutin. This is contrary to the tomato fruit whose cuticle comprises more than 80% of cutin and up to 5% of cuticular waxes (Ramírez et al., 1992).

Waxes from the grape berry cuticle mostly contain aldehydes, where both straight chain, even chain length type predominates. As early as 1965, it was reported in two separate studies that the surface waxes of ripe sultana grape variety consists largely of the oleanolic acid (a triterpenoid), with relatively lesser amounts of n-alcohols (C_{24}, C_{26}, and C_{28}), n-aldehydes (largely C_{28} and C_{30}), esters, free acids (mostly C_{24}, C_{26}, and C_{28}) (Radler, 1965) and n-hydrocarbons (predominantly C_{25}, C_{27}, C_{29}, and C_{31}) (Radler, 1965; Radler and Horn, 1965). Around 30% of oleanolic acid was also found in the cuticular waxes of "Palomino fino" grapes (Casado and Heredia, 1999). However, the authors did not find any aldehydes and the *n*-alkanes represented less than 5% of the total mass (Casado and Heredia, 1999). Oleanolic acid was also found to be dominant component in the cuticular waxes of "Pinot noir" grapes along all phenological stages of berry development (Commenil et al., 1997). Oleanolic acid was also reported to be present in all eight different cultivars investigated in the recent study by Pensec et al. (2014) in which authors reported triterpenoid content to be as high as around 42%–80% of the total cuticular waxes depending

Table 20.1 Predominant cuticular wax and cutin monomers in different fruit species

Fruit	Ripening type	Predominant wax components	Predominant cutin monomers	References
Tomato (*Solanum lycopersicum*)	Climacteric	n-hentriacontane (C_{31}) triterpenoid alcohols (α-, β-, and δ-amyrin)	C_{16} monomers (10,16-dihydroxy $C_{16:0}$)	Isaacson et al. (2009), Leide et al. (2007), Petit et al. (2014), Shi et al. (2013) and Yeats et al. (2012a)
Plum (*Prunus domestica*)	Climacteric	n-nonacosane (C_{29}) triterpenoid acids (oleanolic acid)	Not reported	Ismail et al. (1977)
Peach (*Prunus persica*)	Climacteric	n-tricosane (C_{23}), n-pentacosane (C_{25}) n-heptacosane (C_{27}) triterpenoid acids (ursolic acid, oleanolic acid)	C_{18} monomers (ω-hydroxy $C_{18:1}$)	Belge et al. (2014a) and Fernández et al. (2011)
Sweet cherry (*Prunus avium*)	Nonclimacteric	n-nonacosane (C_{29}) triterpenoid acids (ursolic acid)	Cultivar-dependent (dihydroxy $C_{16:0}$ or dihydroxy $C_{18:x}$ reported mainly)	Belge et al. (2014b) and Peschel et al. (2007)
Strawberry (*Fragaria × ananassa*)	Nonclimacteric	Not reported	C16 monomers (9(10), 16- dihydroxy C16:0)	Järvinen et al. (2010)
Olive (*Olea europaea*)	Nonclimacteric	n-heptacosane (C_{27}), triterpenoid acids (oleanolic acid)	Not reported	Bianchi et al. (1992)
Grape (*Vitis vinifera*)	Nonclimacteric	n-pentacosane (C_{25}) triterpenoid acids (oleanolic acid), n-heptacosane (C_{27}) n-nonacosane (C_{29})	Not reported	Casado and Heredia (1999), Commenil et al. (1997), Pensec et al. (2014), Radler (1965, 1970) and Radler and Horn (1965)

on the genotype. In addition, only minor quantitative differences were found in the cuticular wax fraction of the fruits of the European wild vine (*Vitis vinifera* ssp. *sylvestris*) and the cultivated vine (*V. vinifera* ssp. *vinifera*) suggesting their origin early in the evolution (Radler, 1970). However, despite the wax composition of the grape berry cuticle that has been shown in several studies, there is still no data on grape berry cutin composition.

In olive fruits, epicuticular waxes are composed of benzyl esters from C_{22} to C_{30}, esters (C_{31} to C_{44}), diterpene esters such as geranylgeranyl and phytyl esters from C_{16} to C_{22}, fatty acids from C_{16} to very-long-chain fatty acids such as C_{32}, hydroxylated and dicarboxylic fatty acids, and dioic fatty acids from C_{16} to C_{28}, monoglycerols, diglycerols and triacylglycerols, triterpenic acids such as oleanolic, maslinic and ursolic acids and hydrocarbons. The olive cutin polymer matrix consists mainly of esterified hydroxy, polyhydroxy, epoxy, and dioic fatty acids (Vichi et al., 2015).

In sweet cherry fruit (*Prunus avium*), triterpenes, alkanes, and alcohols accounted for 75.6%, 19.1%, and 1.2% of total fruit wax, respectively. The most abundant constituents were the triterpenes ursolic (60.0%) and oleanolic acid (7.5%), the alkanes nonacosane (13.0%) and heptacosane (3.0%), and the secondary alcohol nonacosan-10-ol (1.1%). When cherry fruit developed, triterpenes per unit area decreased, but alkanes and alcohols remained constant. The cutin fraction consisted of mostly C16 (69.5%) and, to a lower extent, C18 monomers (19.4%) comprising alkanoic, ω-hydroxyacids, α,ω-dicarboxylic, and midchain hydroxylated acids. The most abundant constituents were 9(10),16-dihydroxy-hexadecanoic acid (53.6%) and 9,10,18-trihydroxy-octadecanoic acid (7.8%) (Peschel et al., 2007).

The chemical composition of fruit cuticle components shows significant variability among fruit tree species, varieties, and along fruit phenology. This variability in part results from environmental factors. However, little is known about how compositional differences in wax and cutin relate to fruit cuticle response to water stress. Such studies are increasingly needed as a consequence of increasingly limited water supplies worldwide, growing demands to enhance agricultural production, and the limitations imposed by climate change.

5. GENETIC REGULATION OF FRUIT CUTICLE ASSEMBLY AND ITS ADAPTATION TO WATER STRESS

Most cuticle biosynthesis studies have been carried out with the model plant *Arabidopsis* (Pollard et al., 2008; Samuels et al., 2008; Schreiber, 2010).

Using the information available from *Arabidopsis*, and through homology search, it was possible to identify in a large number of fleshy fruit species homologous genes with identical predicted function (Albert et al., 2013; Alkio et al., 2012; Mintz-Oron et al., 2008). This is possible due to transcriptomics approaches and is dependent on the bioinformatics tools available and on the presence of a given protein in *Arabidopsis* cuticle, which is relatively small and thin. Furthermore, homology search is very useful for the identification of possible mechanistic conservation but does not lead to the discovery of the fruit-specific molecular pathways and processes.

The cuticle of fleshy fruits is generally thicker as compared to *Arabidopsis* leaf cuticle and therefore is a more suitable model for identifying new genes involved in formation and restructuring of other fruit cuticles. The recent strides made in the field of genetics and advancement of molecular techniques had made tomato fruit a suitable model, which made it possible to characterize the cuticle-related genes. Several transgenic tomato lines have been produced to explore and confirm the cuticle-related gene function and are well documented in the available literature (Girard et al., 2012; Shi et al., 2013; Vogg et al., 2004; Yeats et al., 2012b). Although key ABC transporters for cutin precursors export across the plasma membrane have been identified (Bird et al., 2007; Chen et al., 2011; Fabre et al., 2016; Pighin et al., 2004), the molecular mechanism underlying the intracellular and extracellular transport of wax and cutin precursors remains poorly understood. To date, and despite tremendous recent efforts, several key areas of cuticle biogenesis and assembly remain unclear.

The precursors of the cuticular wax biosynthesis are very long chains of fatty acids (VLCFAs), which are generated by the de novo fatty acid biosynthesis in the plastid, whereas fatty acid elongation occurs in the endoplasmic reticulum (ER) of the epidermal cells. The C_{16} or C_{18} acyl-CoAs serve as precursors for the biosynthesis of VLCFAs ($C_{20}-C_{34}$) by the addition of C_2 units derived from malonyl-CoA. This formation of VLCFAs is usually carried out by the enzymes of fatty acid elongase (FAE) complex in the ER (Kunst and Samuels, 2009; Li-Beisson et al., 2013). The FAE complex consists of β-ketoacyl-CoA synthase (KCS), β-ketoacyl-CoA reductase (KCR), 3-hydroxyacyl-CoA dehydratase (HCD), and trans-2,3-enoyl-CoA reductase (ECR), which catalyze the sequential reactions: condensation, reduction, dehydration, and a second reduction, respectively (Kunst and Samuels, 2009; Li-Beisson et al., 2013). The aliphatic components of the wax are derived from VLCFAs with the chain length of 24−34 carbons (Samuels et al., 2008), and biosynthesis requires a unique set of condensing enzymes: KCS1, CER6/KCS6/CUT1, and KCS9 (Kim et al., 2013; Todd et al., 1999). Other classes of VLCFAs synthesis genes are also described, namely LACS (Lü et al., 2009; Weng et al., 2010), which are involved in the synthesis of VLCFAs required for pollen coat formation (Jessen et al., 2011). There are other cuticle-related genes that affect the cuticle biosynthesis and transport, which are summarized in Table 20.2.

Wax formation involves various steps from its synthesis in plastids to its elongation in ER from where it is transported to the plasma membrane (PM). Beyond plasma membrane, the hydrophobic wax components have to cross the hydrophilic cell wall, this function being probably facilitated by the cell wall—associated lipid transfer proteins (Bernard and Joubès, 2013; Kunst and Samuels, 2009). Two glycosyl-phosphatidylinositol (GPI)-anchored lipid transfer proteins (LTPGs), LTPG1 and LTPG2, were shown to be required for wax export from the PM to the cell surface (DeBono et al., 2009; Kim et al., 2012), but the mechanism underlying their roles in wax export is still not clear.

Several lipid-transfer proteins have been reported to modulate plant responses to environmental stress. Recently reports showed that wax carrier gene such as *LTP3*, positively regulated by *MYB96* transcription factor (modulator of wax biosynthesis), is significantly induced by water stress and ABA application (Guo et al., 2013; Seo et al., 2011). LTPs are the most important lipid transfer protein for the extrusion of wax beyond PM and study of its relative expression under water deficit might unveil the adaptive response of grape berries under stress and how cuticular wax load is modulated.

Cutin is built via 2-monoacylglycerol units polymerization, and the process had eluded many scientists to date. A recent study of tomato cutin deficient (CD1) mutant, which has limited amount of cutin and accumulates 2-monoacylglycerol at the fruit surface, led to the identification of the member of glycine—aspartic acid—serine—leucine (GDSL) motif lipase/hydrolase superfamily that acts as the cutin synthase (Beisson and Ohlrogge, 2012; Girard et al., 2012). This *CD1* gene is expressed in the epidermal cells, and the encoded GDSL protein is localized extracellularly within the cuticle, which allows cutin polymerization (Yeats et al., 2012b).

Plant cuticle biosynthesis, transport, and deposition is a complex and highly regulated network of genetic interplay and is influenced by environmental stress, but details of this regulation are not well understood (Bernard and Joubès, 2013; Nawrath et al., 2013). Apart from the network of transcription factors that regulate cuticle biosynthesis, namely

Table 20.2 Genes associated with cuticle biosynthesis

Gene	Protein family name	Affected cuticular components	References
KCS1	β-Ketoacyl-coenzyme A synthase	Cuticular wax biosynthesis	Todd et al. (1999)
KCS6/CER6/CUT1	β-Ketoacyl-coenzyme A synthase	Cuticular wax biosynthesis	Hooker et al. (2002) and Millar et al. (1999)
FATB	Fatty acyl-ACP thioesterase B	Cuticular wax biosynthesis	Bonaventure et al. (2003)
LACS1/CER8	Long-chain acyl-CoA synthetase	Cuticular wax biosynthesis	Lü et al. (2009) and Weng et al. (2010)
LACS2	Long-chain acyl-CoA synthetase	Cuticular wax biosynthesis	Lü et al. (2009) and Weng et al. (2010)
LCAS4	Long-chain acyl-CoA synthetase	Cuticular wax biosynthesis	Jessen et al. (2011)
KCS2/DAISY	β-Ketoacyl-coenzyme A synthase	Cuticular wax biosynthesis	Franke et al. (2009) and Lee et al. (2009)
KCS9	β-Ketoacyl-coenzyme A synthase	Cuticular wax biosynthesis	Kim et al. (2013)
KCR1	β-Ketoacyl-coenzyme A reductase	Cuticular wax biosynthesis	Beaudoin et al. (2009)
CER2	BAHD acyltransferase	Cuticular wax biosynthesis	Haslam et al. (2012, 2015) and Xia et al. (1996)
CER1	Alkane-forming pathway	Cuticular wax biosynthesis	Aarts et al. (1995) and Bourdenx et al. (2011)

Continued

Table 20.2 Genes associated with cuticle biosynthesis—cont'd

Gene	Protein family name	Affected cuticular components	References
CER3/WAX2	VLC-Acyl-CoA reductase putative	Cuticular wax biosynthesis	Chen et al. (2003)
ABCG12/CER5	ABC transporter	Cuticular wax transport	Pighin et al. (2004)
LTPG1/LTPG2/ LTPG3	GPI-anchored lipid transfer protein (LTPG)	Cuticular wax transport	DeBono et al. (2009) and Kim et al. (2012)
LTP3	Lipid transfer protein	Cuticular wax transport	Guo et al. (2013) and Seo et al. (2011)
CER9	Putative E3 ubiquitin ligase	Posttranslational regulation of wax biosynthesis	Lü et al. (2012) and Zhao et al. (2014)
GDSL/CD1	Cutin synthase/ hydroxyacylglycerol transesterase	Cutin and wax biosynthesis and epidermal patterning	Beisson and Ohlrogge (2012), Buxdorf et al. (2014), Shi et al. (2013) and Yeats et al. (2012b)
WSD1	Wax synthase	Wax biosynthesis	Li et al. (2008)

WIN1/SHN1, *MYB106*, and *MYB16* (Aharoni et al., 2004; Broun et al., 2004; Oshima et al., 2013), an additional regulatory mechanism that does not involve direct transcriptional activation or repression by promoter binding has recently been discovered. One such example are the studies from *CER9*, which is the first described gene related to cuticle biosynthesis whose loss of function improves plant response to water deficit (Lü et al., 2012). The exact role and mechanism of *CER9*, which caused these changes, in still not clear, but recent findings by Zhao et al. (2014) reinstated the role of *CER9* in cuticle biosynthesis and its role in water stress tolerance.

Waxes of several fruits have shown the presence of secondary metabolites, namely oleanolic acid as one of the major components (Casado and Heredia, 1999). Fukushima et al. (2011) have recently identified two CYP716A12 genes—*CYP716A15* and *CYP716A17*—and showed their ability to promote terpenoids production, such as oleanolic acid, in grapes. Both transcripts were upregulated in grape berries subjected to water and heat stresses (Zarrouk et al., unpublished results) suggesting the role of triterpenoids in cuticle response to environmental stresses. *CYP716A* subfamily members are multifunctional oxidases involved in triterpenoid biosynthesis (Fukushima et al., 2011). They are Cytochrome P450 monooxygenases/hydroxylases (P450s), which are involved in secondary metabolism and catalyze the wide variety of monooxygenation/hydroxylation reactions (Mizutani and Sato, 2011; Shibuya et al., 2006). As grape berry skin, particularly its cuticle, is rich in terpenoids (oleanolic acid), it can be postulated that the transcription of P450 genes and the activity of the coded proteins are needed for its production (Seki et al., 2011; Yonekura-Sakakibara and Saito, 2009).

The list of genes involved in cuticle biosynthesis has been an extensive topic of research in both model and nonmodel plants. With more and more genome-based techniques and genetic resources available, it will be feasible to study the existing list of potential candidate genes that are associated with cuticle biosynthesis, transport, and deposition, and shed more light on the dynamics of cuticle biology in the near future (Martin and Rose, 2014). Understanding cuticle biosynthetic pathways will bring us closer to selectively modifying cuticle properties and enhancing agricultural productivity in fruits such as grapevine. Nevertheless, an in-depth understanding of the complexity of cuticle function at the molecular level is still required, as well as the understanding of its adaptive response to environmental stress (Yeats et al., 2012a; Yeats and Rose, 2013). These findings might reveal the cuticle complexities that underscore the fact that the cuticle is much more than just a barrier to water loss (Yeats and Rose, 2013).

6. CONCLUSIONS

The fruit cuticle is receiving growing attention because of its significant role in fruit quality and storage, the fruit skin being also a source of biologically active compounds (e.g., triterpenoids). Nevertheless, little is known about its structure, biosynthesis, and composition. Even less is known about the effects of water stress on fruit cuticle. From the existing literature in vegetative organs, the key role of cuticular wax on the protection of plants during periodic drying and water stress is clearly ascertained. However, further work is needed to fully grasp the cell-signaling networks involved in these processes and the impact of stress-induced alterations, namely in the fruit cuticle and its quality.

Fleshy fruit surfaces are heterogeneous, and in addition to the cuticle may contain stomata, trichomes of lenticels. Subsequently, a direct extrapolation of results generated in the model fruit tomato to cuticles of the other fruits could be wrong. Additionally, the potential differences between climacteric and nonclimacteric fruits are also not fully understood. Specifically, understanding how, in nonclimacteric fruits, the developmental signaling interacts with the activities of ABA in cuticular wax metabolic regulation will shed light for future research.

On the other hand, much of the knowledge regarding cuticular wax-related genes responding to water stress has been generated using vegetative plant organs, and information on the detailed function and regulation of these genes, namely on the key genes of wax metabolism, is scarce. Moreover, functional proteins related to cuticular wax transport and other proteins regulating the expression of genes related to cuticular wax synthesis under drought conditions remain to be revealed. To date, the molecular mechanism underlying the intracellular and extracellular transport of wax and cutin precursors remains poorly understood, and despite tremendous recent efforts, several key areas of cuticle biogenesis and assembly remain unclear.

A comprehensive understanding of the physiological and biochemical impacts of limited water availability on economically significant fruit production traits (e.g., yield and quality) is prerequisite to the advanced breeding and other strategies necessary for enhancing fruit crop tolerance to limited water availability.

ACKNOWLEDGMENTS

This work was supported by the European Community's Seventh Framework Program (FP7/2007–2013) under the grant agreement n° FP7-311775, Project INNOVINE and was integrated into the COST (European Cooperation in Science and Technology) Action FA1106 "Quality fruit" and COST Action FA1306 "The quest for tolerant varieties: phenotyping at plant and cellular level." OZ is supported by postdoctoral fellowships from FCT (SFRH/BPD/111693/2015). This work was supported by FCT, through R&D Unit, UID/Multi/04551/2013 (GreenIT).

REFERENCES

Aarts, M.G., Keijzer, C.J., Stiekema, W.J., Pereira, A., 1995. Molecular characterization of the CER1 gene of *Arabidopsis* involved in epicuticular wax biosynthesis and pollen fertility. Plant Cell 7, 2115–2127.

Aharoni, A., Dixit, S., Jetter, R., Thoenes, E., van Arkel, G., Pereira, A., 2004. The SHINE clade of AP2 domain transcription factors activates wax biosynthesis, alters cuticle properties, and confers drought tolerance when overexpressed in Arabidopsis. The Plant Cell 16, 2463–2480.

Albert, Z., Ivanics, B., Molnár, A., Miskó, A., Tóth, M., Papp, I., 2013. Candidate genes of cuticle formation show characteristic expression in the fruit skin of apple. Plant Growth Regulation 70, 71–78.

Alkio, M., Jonas, U., Sprink, T., Van Nocker, S., Knoche, M., 2012. Identification of putative candidate genes involved in cuticle formation in *Prunus avium* (sweet cherry) fruit. Annals of Botany 110, 101–112.

Baker, E.A., Procopiou, J., Hunt, G.M., 1975. The cuticles of citrus species. Composition of leaf and fruit waxes. Journal of the Science of Food and Agriculture 26, 1093–1101.

Beaudoin, F., Wu, X., Li, F., Haslam, R.P., Markham, J.E., Zheng, H., Napier, J.A., Kunst, L., 2009. Functional characterization of the *Arabidopsis* β-ketoacyl-coenzyme A reductase candidates of the fatty acid elongase. Plant Physiology 150, 1174–1191.

Beisson, F., Ohlrogge, J., 2012. Plants: knitting a polyester skin. Nature Chemical Biology 8, 603–604.

Belge, B., Llovera, M., Comabella, E., Graell, J., Lara, I., 2014a. Fruit cuticle composition of a melting and a nonmelting peach cultivar. Journal of Agricultural and Food Chemistry 62 (15), 3488–3495.

Belge, B., Llovera, M., Comabella, E., Gatius, F., Guillén, P., Graell, J., Lara, I., 2014b. Characterization of cuticle composition after cold storage of "Celeste" and "Somerset" sweet cherry fruit. Journal of Agriculture and Food Chemistry 62, 8722–8729.

Bernard, A., Joubès, J., 2013. Arabidopsis cuticular waxes: advances in synthesis, export and regulation. Progress in Lipid Research 52, 110–129.

Bianchi, G., Murelli, C., Vlahov, C., 1992. Surface waxes from olive fruits. Phytochemistry 31, 3503–3506.

Bird, D., Beisson, F., Brigham, A., Shin, J., Greer, S., Jetter, R., Kunst, L., Wu, X., Yephremov, A., Samuels, L., 2007. Characterization of *Arabidopsis* ABCG11/WBC11, an ATP binding cassette (ABC) transporter that is required for cuticular lipid secretion. Plant Journal 52, 485–498.

Bonaventure, G., Salas, J.J., Pollard, M.R., Ohlrogge, J.B., 2003. Disruption of the FATB gene in *Arabidopsis* demonstrates an essential role of saturated fatty acids in plant growth. Plant Cell 15, 1020–1033.

Bonaventure, G., Beisson, F., Ohlrogge, J., Pollard, M., 2004. Analysis of the aliphatic monomer composition of polyesters associated with *Arabidopsis* epidermis: occurrence of octadeca-cis-6, cis-9-diene-1,18-dioate as the major component. Plant Journal 40, 920–930.

Bourdenx, B., Bernard, A., Domergue, F., Pascal, S., Léger, A., Roby, D., Pervent, M., Vile, D., Haslam, R.P., Napier, J.A., Lessire, R., Joubès, J., 2011. Overexpression of *Arabidopsis* ECERIFERUM1 promotes wax very-long-chain alkane biosynthesis and influences plant response to biotic and abiotic stresses. Plant Physiology 156, 29–45.

Broun, P., Poindexter, P., Osborne, E., Jiang, C.Z., Riechmann, J.L., 2004. WIN1, a transcriptional activator of epidermal wax accumulation in Arabidopsis. Proceedings of the National Academy of Sciences of the United States of America 101, 4706–4711.

Buda, G., Isaacson, T., Matas, A.J., Paolillo, D.J., Rose, J.K.C., 2009. Three dimensional imaging of plant cuticle architecture using confocal scanning laser microscopy. Plant Journal 60, 378–385.

Budke, J.M., Goffinet, B., Jones, C.S., 2012. The cuticle on the gametophyte calyptra matures before the sporophyte cuticle in the moss *Funaria hygrometrica* (*Funariaceae*). American Journal of Botany 99, 14–22.

Burghardt, M., Riederer, M., 2006. Cuticular transpiration in biology of the plant cuticle. In: Riederer, M., Müller, C. (Eds.). Blackwell Publishing, Oxford, pp. 292–311.

Buxdorf, K., Rubinsky, G., Barda, O., Burdman, S., Aharoni, A., Levy, M., 2014. The transcription factor SlSHINE3 modulates defense responses in tomato plants. Plant Molecular Biology 84, 37–47.

Casado, C.G., Heredia, A., 1999. Structure and dynamics of reconstituted cuticular waxes of grape berry cuticle (*Vitis vinifera* L.). Journal of Experimental Botany 50, 175–182.

Casado, C.G., Heredia, A., 2001. Ultrastructure of the cuticle during growth of the grape berry (*Vitis vinifera* L.). Physiologia Plantarum 111, 220–224.

Chaves, M.M., Zarrouk, O., 2012. Crop responses to available soil water. In: Meyers, R., SpringerReference (Eds.), Encyclopedia of Sustainability Science and Technology. Springer-Verlag Berlin Heidelberg, pp. 2638–2661. https://doi.org/10.1007/978-1-4419-0851-3_194. www.springerreference.com.

Chen, X., Goodwin, S.M., Boroff, V.L., Liu, X., Jenks, M.A., 2003. Cloning and characterization of the *WAX2* gene of *Arabidopsis* involved in cuticle membrane and wax production. Plant Cell 15, 1170–1185.

Chen, G., Komatsuda, T., Ma, J.F., Nawrath, C., Pourkheirandish, M., Tagiri, A., Hu, Y.G., Sameri, M., Li, X., Zhao, X., Liu, Y., Li, C., Ma, X., Wang, A., Nair, S., Wang, N., Miyao, A., Sakuma, S., Yamaji, N., Zheng, X., Nevo, E., 2011. An ATP-binding cassette subfamily G full transporter is essential for the retention of leaf water in both wild barley and rice. Proceedings of the National Academy of Sciences of the United States of America 108, 12354–12359.

Commenil, P., Brunet, L., Audran, J.C., 1997. The development of the grape berry cuticle in relation to susceptibility to bunch rot disease. Journal of Experimental Botany 48, 1599–1607.

Crisosto, C.H., Johnson, R.S., Luza, J.G., Crisosto, G.M., 1994. Irrigation regimes affect fruit soluble solids concentration and rate of water loss of 'O'Henry' peaches. HortScience 29, 1169–1171.

Curver, K., Seifi, H., Mouille, G., de Rycke, R., Asselbergh, B., Van Hecke, A., Vanderschaeghe, D., Höfte, H., Callewaert, N., Van Breusegem, F., Höfte, M., 2010. Abscissic acid deficiency causes changes in cuticle permeability and pectin composition that influence tomato resistance to *Botrytis cinerea*. Plant Physiology 154, 847–860.

DeBono, A., Yeats, T.H., Rose, J.K., Bird, D., Jetter, R., Kunst, L., Samuel, L., 2009. Arabidopsis LTPG is a glycosyl phosphatidyl inositol anchored lipid transfer protein required for export of lipids to the plant surface. The Plant Cell 21, 1230–1238.

Domínguez, E., Cuartero, J., Heredia, A., 2011a. An overview of plant cuticle biomechanics. Plant Science 181, 77–84.

Domínguez, E., Heredia-Guerrero, J.A., Heredia, A., 2011b. The biophysical design of plant cuticles: an overview. New Phytologist 189, 938–949.

Edelmann, H.G., Neinhuis, C., Bargel, H., 2005. Influence of hydration and temperature on the rheological properties of plant cuticles and their impact on plant organ integrity. Journal of Plant Growth and Regulation 24, 116–126.

España, L., Heredia-Guerrero, J.A., Segado, P., Benítez, J.J., Heredia, A., Domínguez, E., 2014. Biomechanical properties of the tomato (*Solanum lycopersicum*) fruit cuticle during development are modulated by changes in the relative amounts of its components. New Phytologist 202, 790–802.

Espelie, K.E., Dean, B.B., Kolattukudy, P.E., 1979. Composition of lipid-derived polymers from different anatomical regions of several plant species. Plant Physiology 64, 1089–1093.

Fabre, G., Garroum, I., Mazurek, S., Daraspe, J., Mucciolo, A., Sankar, M., Humbel, B.M., Nawrath, C., 2016. The ABCG transporter PEC1/ABCG32 is required for the formation of the developing leaf cuticle in *Arabidopsis*. New Phytologist 209 (1), 192–201.

Fernández, V., Eichert, T., 2009. Uptake of hydrophilic solutes through plant leaves: current state of knowledge and perspectives of foliar fertilization. Critical Reviews in Plant Sciences 28 (1), 36–68.

Fernández, V., Guzmán-Delgado, P., Graça, J., Santos, S., Gil, L., 2016. Cuticle structure in relation to chemical composition: re-assessing the prevailing model. Frontiers in Plant Science 7, 427. http://doi.org/10.3389/fpls.2016.00427.

Fernández, V., Khayet, M., Montero-Prado, P., Heredia-Guerrero, J.A., Liakopoulos, G., Karabourniotis, G., Del Río, V., Domínguez, E., Tacchini, I., Nerín, C., Val, J., Heredia, A., 2011. New insights into the properties of pubescent surfaces: peach fruit as a model. Plant Physiology 156, 2098–2108.

Fernández, V., Sotiropoulos, T., Brown, P.H., 2013. Foliar Fertilisation: Principles and Practices. International Fertiliser Industry Association (IFA), Paris, ISBN 979-10-92366-00-6, 140 p.

Fernandez, V., Khayet, M., 2015. Evaluation of the surface free energy of plant surfaces: towards standardizing the procedure. Frontiers in Plant Science 6, 510. https://doi.org/10.3389/fpls.2015.00510.

Franke, R., Briesen, I., Wojciechowski, T., Faust, A., Yephremov, A., Nawrath, C., Schreiber, L., 2005. Apoplastic polyesters in *Arabidopsis* surface tissues-a typical suberin and a particular cutin. Phytochemistry 66, 2643–2658.

Franke, R., Höfer, R., Briesen, I., Emsermann, M., Efremova, N., Yephremov, A., Schreiber, L., 2009. The *DAISY* gene from *Arabidopsis* encodes a fatty acid elongase condensing enzyme involved in the biosynthesis of aliphatic suberin in roots and the chalaza-micropyle region of seeds. Plant Journal 57, 80–95.

Fukushima, E.O., Seki, H., Ohyama, K., Ono, E., Umemoto, N., Mizutani, M., Saito, K., Muranaka, T., 2011. CYP716A subfamily members are multifunctional oxidases in triterpenoid biosynthesis. Plant and Cell Physiology 52, 2050–2061.

Gibert, C., Chadœuf, J., Vercambre, G., Génard, M., Lescourret, F., 2007. Cuticular cracking on nectarine fruit surface: spatial distribution and development in relation to irrigation and thinning. Journal of the American Society for Horticultural Science 132, 583–591.

Girard, A.L., Mounet, F., Lemaire-Chamley, M., Gaillard, C., Elmorjani, K., Vivancos, J., Runavot, J.L., Quemener, B., Petit, J., Germain, V., Rothan, C., Marion, D., Bakan, B., 2012. Tomato GDSL1 is required for cutin deposition in the fruit cuticle. The Plant Cell 24, 3119–3134.

Guo, L., Yang, H., Zhang, X., Yang, S., 2013. Lipid transfer protein 3 as a target of MYB96 mediates freezing and drought stress in *Arabidopsis*. Journal of Experimental Botany 64, 1755–1767.

Gupta, N.S., Collinson, M.E., Briggs, D.E.G., Evershed, R.P., Pancost, R.D., 2006. Reinvestigation of the occurrence of cutan in plants: implications for the leaf fossil record. Paleobiology 32, 432–449.

Guzmán, P., Fernández, V., Graça, J., Cabral, V., Kayali, N., Khayet, M., Gil, L., 2014. Chemical and structural analysis of *Eucalyptus globulus* and *E. camaldulensis* leaf cuticles: a lipidized cell wall region. Frontiers in Plant Science 5, 481. https://doi.org/10.3389/fpls.2014.00481.

Haslam, T.M., Mañas-Fernández, A., Zhao, L., Kunst, L., 2012. *Arabidopsis* ECERIFERUM2 is a component of the fatty acid elongation machinery required for fatty acid extension to exceptional lengths. Plant Physiology 160, 1164–1174.

Haslam, T.M., Haslam, R., Thoraval, D., Pascal, S., Delude, C., Domergue, F., Fernández, A.M., Beaudoin, F., Napier, J.A., Kunst, L., Joubès, J., 2015. ECERIFERUM2-LIKE proteins have unique biochemical and physiological functions in very-long-chain fatty acid elongation. Plant Physiology 167, 682–692.

Heredia-Guerrero, J.A., Benítez, J.J., Domínguez, E., Bayer, I.S., Cingolani, R., Athanassiou, A., Heredia, A., 2014. Infrared and Raman spectroscopic features of plant cuticles: a review. Frontiers in Plant Science 5, 305. https://doi.org/10.3389/fpls.2014.00305.

Hooker, T.S., Millar, A.A., Kunst, L., 2002. Significance of the expression of the *CER6* condensing enzyme for cuticular wax production in *Arabidopsis*. Plant Physiology 129, 1568–1580.

Isaacson, T., Kosma, D.K., Matas, A.J., Buda, G.J., He, Y., Yu, B., Pravitasari, A., Batteas, J.D., Stark, R.E., Jenks, M.A., Rose, J.K., 2009. Cutin deficiency in the tomato fruit cuticle consistently affects resistance to microbial infection and biomechanical properties, but not transpirational water loss. Plant Journal 60, 363–377.

Ismail, H.M., Brown, G.A., Tucknott, O.G., Holloway, P.J., Williams, A.A., 1977. Nonanal in epicuticular wax of golden egg plums (*Prunus domestica*). Phytochemistry 16, 769–770.

Järvinen, R., Kaimainen, M., Kallio, H., 2010. Cutin composition of selected northern berries and seeds. Food Chemistry 122, 137–144.

Jeffree, C.E., 2006. The fine structure of the plant cuticle. In: Riederer, M., Müller, C. (Eds.), Biology of the Plant Cuticle, Annual Plant Reviews Series, vol. 23. Blackwell Publishing Ltd., Oxford, pp. 11–125.

Jenks, M.A., Rich, P.J., Ashworth, E.N., 1994. Involvement of cork cells in the secretion of epicuticular wax filaments on *Sorghum bicolor* (L.) Moench. International Journal of Plant Science 155, 506–518.

Jenks, M.A., Ashworth, E.N., 1999. Plant epicuticular waxes: function, production and genetics. Horticultural Reviews 23, 1–68.

Jessen, D., Olbrich, A., Knüfer, J., Krüger, A., Hoppert, M., Polle, A., et al., 2011. Combined activity of LACS1 and LACS4 is required for proper pollen coat formation in *Arabidopsis*. Plant Journal 68, 715–726.

Jetter, R., Kunst, L., Samuels, A.L., 2006. Composition of plant cuticular waxes. In: Riederer, M., Müller, C. (Eds.), Biology of the Plant Cuticle, Annual Plant Reviews Series, vol. 23. Blackwell Publishing Ltd., Oxford, pp. 145–181.

Jetter, R., Schäffer, S., Riederer, M., 2000. Leaf cuticular waxes are arranged in chemically and mechanically distinct layers: evidence from *Prunus laurocerasus* L. Plant Cell and Environment 23, 619–628.

Kerstiens, G., 1996. Cuticular water permeability and its physiological significance. Journal of Experimental Botany 47, 1813–1832.

Kerstiens, G., 2006. Water transport in plant cuticles: an update. Journal of Experimental Botany 57, 2493–2499.

Kim, H., Lee, S.B., Kim, H.J., Min, M.K., Hwang, I., Suh, M.C., 2012. Characterization of glycosylphosphatidylinositol-anchored lipid transfer protein 2 (LTPG2) and overlapping function between LTPG/LTPG1 and LTPG2 in cuticular wax export or accumulation in *Arabidopsis thaliana*. Plant and Cell Physiology 53, 1391–1403.

Kim, J., Jung, J.H., Lee, S.B., Go, Y.S., Kim, H.J., Cahoon, R., Markham, J.E., Cahoon, E.B., Suh, M.C., 2013. Arabidopsis 3-ketoacylcoenzyme A synthase9 is involved in the synthesis of tetracosanoic acids as precursors of cuticular waxes, suberins, sphingolipids and phospholipids. Plant Physiology 162, 567–580.

Kirsch, T., Kaffarnik, F., Riederer, M., Schreiber, L., 1997. Cuticular permeability of the three tree species *Prunus laurocerasus* L., *Ginko biloba* L. and *Juglans regia* L.: comparative investigation of the transport properties of intact leaves, isolated cuticles and reconstituted cuticular waxes. Journal of Experimental Botany 48, 1035–1045.

Knoche, M., Peschel, S., Hinz, M., Bukovac, M.J., 2000. Studies on water transport through the sweet cherry fruit surface: characterizing conductance of the cuticular membrane using pericarp segments. Planta 212, 127–135.

Knoche, M., Peschel, S., 2006. Water on the surface aggravates microscopic cracking of the sweet cherry fruit cuticle. Journal American Society Horticultural Science 131, 192–200.

Knoche, M., Peschel, S., 2007. Deposition and strain of the cuticle of developing European plum fruit. Journal American Society Horticultural Science 132, 597–602.

Knoche, M., Grimm, E., 2008. Surface moisture induces microcracks in the cuticle of 'Golden Delicious' apple. HortScience 43, 1929–1931.

Knoche, M., Khanal, B.P., Stopar, M., 2011. Russeting and microcracking of 'Golden Delicious'apple fruit concomitantly decline due to gibberellin A4+7 application. Journal American Society Horticultural Science 136, 159–164.

Kolattukudy, P.E., 2001. Polyesters in higher plants. Advances in Biochemical Engineering/Biotechnology 71, 1–49.

Kosma, D.K., Bourdenx, B., Bernard, A., Parsons, E.P., Lü, S., Joubès, J., Jenks, M.A., 2009. The impact of water deficiency on leaf cuticle lipids of *Arabidopsis*. Plant Physiology 151, 1918–1929.

Kunst, L., Samuels, L., 2009. Plant cuticles shine: advances in wax biosynthesis and export. Current Opinion in Plant Biology 12, 721–727.

Lara, I., Belge, B., Goulao, L.F., 2015. A focus on the biosynthesis and composition of cuticle in fruits. Journal of Agriculture and Food Chemistry 63, 4005–4019.

Léchaudel, M., Lopez-Lauri, F., Vidal, V., Sallanon, H., Joas, J., 2013. Response of the physiological parameters of mango fruit (transpiration, water relations and antioxidant system) to its light and temperature environment. Journal of Plant Physiology 170, 567–576.

Lee, S.B., Jung, S.J., Go, Y.S., Kim, H.U., Kim, J.K., Cho, H.J., Park, O.K., Suh, M.C., 2009. Two *Arabidopsis* 3-ketoacyl CoA synthase genes, KCS20 and KCS2/DAISY, are functionally redundant in cuticular wax and root suberin biosynthesis, but differentially controlled by osmotic stress. Plant Journal 60, 462–475.

Lee, S.B., Suh, M.C., 2015. Advances in the understanding of cuticular waxes in *Arabidopsis thaliana* and crop species. Plant Cell Reports 34, 557–572.

Leide, J., Hildebrandt, U., Reussing, K., Riederer, M., Vogg, G., 2007. The developmental pattern of tomato fruit wax accumulation and its impact on cuticular transpiration barrier properties: effects of a deficiency in a β-ketoacyl-coenzyme A synthase (LeCER6). Plant Physiology 144, 1667–1679.

Legay, S., Guerriero, G., Deleruelle, A., Lateur, M., Evers, D., André, C.M., Hausman, J.F., 2015. Apple russeting as seen through the RNA-seq lens: strong alterations in the exocarp cell wall. Plant Molecular Biology 88, 21–40. https://doi.org/10.1007/s11103-015-0303-4.

Leliaert, F., Verbruggen, H., Zechman, F.W., 2011. Into the deep: new discoveries at the base of the green plant phylogeny. Bioessays 33, 683–692.

Li, F., Wu, X., Lam, P., Bird, D., Zheng, H., Samuels, L., Jetter, R., Kunst, L., 2008. Identification of the wax ester synthase/acyl-coenzyme A: diacylglycerol acyltransferase WSD1 required for stem wax ester biosynthesis in *Arabidopsis*. Plant Physiology 148, 97–107.

Li-Beisson, Y., Shorrosh, B., Beisson, F., Andersson, M.X., Arondel, V., Bates, P.D., et al., 2013. Acyl-lipid metabolism. In: Arabidopsis Book, vol. 11, p. e0161.

Lü, S., Song, T., Kosma, D.K., Parsons, E.P., Rowland, O., Jenks, M.A., 2009. *Arabidopsis* CER8 encodes long-chain acyl CoA synthetase 1 (LACS1) that has overlapping functions with LACS2 in plant wax and cutin synthesis. Plant Journal 59, 553–564.

Lü, S., Zhao, H., Des Marais, D.L., Parsons, E.P., Wen, X., Xu, X., Bangarusamy, D.K., Wang, G., Rowland, O., Juenger, T., Bressan, R.A., Jenks, M.A., 2012. Arabidopsis ECERIFERUM9 involvement in cuticle formation and maintenance of plant water status. Plant Physiology 159, 930–944.

Maguire, K.M., Banks, N.H., Lang, A., Gordon, I.L., 2000. Harvest date, cultivar, orchard and tree effects on water vapour permeance in apples. Journal of American Society for Horticultural Science 125, 100–104.

Marga, F., Pesacreta, T.C., Hasenstein, K.H., 2001. Biochemical analysis of elastic and rigid cuticles of *Cirsium horridulum*. Planta 213, 841–848.

Markstädter, C., Federle, W., Jetter, R., Riederer, M., Hölldobler, B., 2000. Chemical composition of the slippery epicuticular wax blooms on *Macaranga* (Euphorbiaceae) ant-plants. Chemoecology 10, 33–40.

Martin, L.B.B., Rose, J.K.C., 2014. There's more than one way to skin a fruit: formation and functions of fruit cuticles. Journal of Experimental Botany 65, 4639–4651.

Matas, A.J., López-Casado, G., Cuartero, J., Heredia, A., 2005. Relative humidity and temperature modify the mechanical properties of isolated tomato fruit cuticles. American Journal of Botany 92, 462–468.

Millar, A.A., Clemens, S., Zachgo, S., Giblin, E.M., Taylor, D.C., Kunst, L., 1999. *CUT1*, an *Arabidopsis* gene required for cuticular wax biosynthesis and pollen fertility, encodes a very-long-chain fatty acid condensing enzyme. Plant Cell 11, 825–838.

Mintz-Oron, S., Mandel, T., Rogachev, I., Feldberg, L., Lotan, O., Yativ, M., Wang, Z., Jetter, R., Venger, I., Adato, A., Aharoni, A., 2008. Gene expression and metabolism in tomato fruit surface tissues. Plant Physiology 147, 823–851.

Mizutani, M., Sato, F., 2011. Unusual P450 reactions in plant secondary metabolism. Archives of Biochemistry and Biophysics 507, 194–203.

Nawrath, C., Schreiber, L., Franke, R.B., Geldner, N., 2013. Apoplastic diffusion barriers in *Arabidopsis*. The Arabidopsis book/American Society of Plant Biologists 11, e0167.

Niklas, K.J., Cobb, E.D., Matas, A.J., 2017. The evolution of hydrophobic cell wall biopolymers: from algae to angiosperms. Journal of Experimental Botany 68, 5369–5387.

Oshima, Y., Shikata, M., Koyama, T., Ohtsubo, N., Mitsuda, N., Ohme-Takagi, M., 2013. MIXTA-like transcription factors and WAX INDUCER1/SHINE1 coordinately regulate cuticle development in *Arabidopsis* and *Torenia fournieri*. The Plant Cell 25, 1609–1624.

Pensec, F., Pęczkowski, C., Grabarczyk, M., Wozniak, A., Benard-Gellon, M., Bertsch, C., Chong, J., Szakiel, A., 2014. Changes in the triterpenoid content of cuticular waxes during fruit ripening of eight grape (*Vitis vinifera*) cultivars grown in the upper Rhine valley. Journal of Agriculture and Food Chemistry 62, 7998–8007.

Peschel, S., Franke, R., Schreiber, L., Knoche, M., 2007. Composition of the cuticle of developing sweet cherry fruit. Phytochemistry 68 (7), 1017–1025.

Petit, J., Bres, C., Just, D., Garcia, V., Mauxion, J.P., Marion, D., Bakan, B., Joubès, J., Domergue, F., Rothan, C., 2014. Analyses of tomato fruit brightness mutants uncover both cutin-deficient and cutin-abundant mutants and a new hypomorphic allele of GDSL lipase. Plant Physiology 164, 888–906.

Pighin, J.A., Zheng, H., Balakshin, L.J., Goodman, I.P., Western, T.L., Jetter, R., Kunst, L., Samuels, A.L., 2004. Plant cuticular lipid export requires an ABC transporter. Science 306, 702–704.

Pollard, M., Beisson, F., Li, Y.H., Ohlrogge, J.B., 2008. Building lipid barriers: biosynthesis of cutin and suberin. Trends in Plant Science 13, 236–246.

Provost, G., Domergue, F., Lalanne, C., Campos, P.R., Grosbois, A., Bert, D., Meredieu, C., Danjon, F., Plomion, C., Gion, J.M., 2013. Soil water stress affects both cuticular wax content and cuticle-related gene expression in young saplings of maritime pine (*Pinus pinaster* Ait). BMC Plant Biology 13, 95.

Radler, F., Horn, D.H.S., 1965. The composition of grape cuticle wax. Australian Journal of Chemistry 18, 1059–1069.

Radler, F., 1965. The surface waxes of the sultana vine (*Vitis vinifera* cv. Thompson Seedless). Australian Journal Biological Science 18, 1045–1056.

Radler, F., 1970. Investigations on the cuticular wax of *Vitis vinifera* ssp. *sylvestris* and *Vitis vinifera* ssp. *vinifera*. Angewandt Botanik 44, 187–195.

Ramírez, F.J., Luque, P., Heredia, A., Bukovac, M.J., 1992. Fourier transform IR of enzymatically isolated tomato fruit cuticular membrane. Biopolymers 32, 1425–1429.

Riederer, M., Muller, C., 2006. Annual Plant Reviews, Biology of the Plant Cuticle. John Wiley & Sons, Oxford. Blackwell Publishing.

Riederer, M., Schreiber, L., 2001. Protecting against water loss: analysis of the barrier properties of plant cuticles. Journal of Experimental Botany 52, 2023–2032.

Rogiers, S.Y., Hatfield, J.M., Jaudzems, V.G., White, R.G., Keller, M., 2004. Grape berry cv. Shiraz epicuticular wax and transpiration during ripening and preharvest weight loss. Amerian Journal of Enology and Viticulture 55, 121–127.

Romero, P., Rodrigo, M.J., Alférez, F., Ballester, A.R., González-Candelas, L., Zacarías, L., Lafuente, M.T., 2012. Unravelling molecular responses to moderate dehydration in harvested fruit of sweet orange (*Citrus sinensis* L. Osbeck) using a fruit-specific ABA-deficient mutant. Journal of Experimental Botany 63, 2753–2767.

Rosenquist, J.K., Morrison, J.C., 1989. Some factors affecting the cuticle and wax accumulation on grape berries. American Journal of Enology and Viticulture 40, 241–244.

Saladié, M., Matas, A.J., Isaacson, T., Jenks, M.A., Goodwin, S.M., Niklas, K.J., Xiaolin, R., Labavitch, J.M., Shackel, K.A., Fernie, A.R., Lytovchenko, A., O'Neill, M.A., Watkins, C.B., Rose, J.K., 2007. A reevaluation of the key factors that influence tomato fruit softening and integrity. Plant Physiology 144, 1012–1028.

Samuels, L., Kunst, L., Jetter, R., 2008. Sealing plant surfaces: cuticular wax formation by epidermal cells. Annual Review of Plant Biology 59, 683–707.

Šantrůček, J., Šimáňová, E., Karbulková, J., Šimková, M., Schreiber, L., 2004. A new technique for measurement of water permeability of stomatous cuticular membranes isolated from *Hedera helix* leaves. Journal of Experimental Botany 55, 1411–1422.

Schreiber, L., 2010. Transport barriers made of cutin, suberin and associated waxes. Trends in Plant Science 15, 546–553.

Schreiber, L., Riederer, M., 1996. Ecophysiology of cuticular transpiration: comparative investigation of cuticular water permeability of plant species from different habitats. Oecologia 107, 426–432.

Schreiber, L., Schonherr, J., 1993. Mobilities of organic compounds in reconstituted cuticular wax of barley leaves: determination of diffusion coefficients. Pesticide Science 38, 353–361.

Segado, P., Domínguez, E., Heredia, A., 2016. Ultrastructure of the epidermal cell wall and cuticle of tomato fruit (*Solanum lycopersicum* L.) during development. Plant Physiology 170, 935–946.

Seki, H., Sawai, S., Ohyama, K., Mizutani, M., Ohnishi, T., Sudo, H., Fukushima, E.O., Akashi, T., Aoki, T., Saito, K., Muranaka, T., 2011. Triterpene functional genomics in licorice for identification of CYP72A154 involved in the biosynthesis of glycyrrhizin. The Plant Cell 23, 4112–4123.

Seo, P.J., Lee, S.B., Suh, M.C., Park, M.J., Go, Y.S., Park, C.M., 2011. The MYB96 transcription factor regulates cuticular wax biosynthesis under drought conditions in *Arabidopsis*. The Plant Cell 23, 1138–1152.

Shi, J.X., Adato, A., Alkan, N., He, Y., Lashbrooke, J., Matas, A.J., Meir, S., Malitsky, S., Isaacson, T., Prusky, D., Leshkowitz, D., Schreiber, L., Granell, A.R., Widemann, E., Grausem, B., Pinot, F., Rose, J.K., Rogachev, I., Rothan, C., Aharoni, A., 2013. The tomato SlSHINE3 transcription factor regulates fruit cuticle formation and epidermal patterning. New Phytologist 197, 468–480.

Shi, J.X., Malitsky, S., Oliveira, S.D., Branigan, C., Franke, R., Schreiber, L., Aharoni, A., 2011. SHINE transcription factors act redundantly to pattern the archetypal surface of Arabidopsis flower organs. PLoS Genetics 7, e1001388.

Shibuya, M., Hoshino, M., Katsube, Y., Hayashi, H., Kushiro, T., Ebizuka, Y., 2006. Identification of β-amyrin and sophoradiol 24-hydroxylase by expressed sequence tag mining and functional expression assay. FEBS Journal 273, 948–959.

Sørensen, I., Pettolino, F.A., Bacic, A., Ralph, J., Lu, F., O'Neill, M.A., Fei, Z., Rose, J.K., Domozych, D.S., Willats, W.G., 2011. The charophycean green algae provide insights into the early origins of plant cell walls. Plant Journal 68, 201–211.

Todd, J., Post-Beittenmiller, D., Jaworski, J.G., 1999. KCS1 encodes a fatty acid elongase 3-ketoacyl-CoA synthase affecting wax biosynthesis in *Arabidopsis thaliana*. Plant Journal 17, 119–130.

Tsubaki, S., Azuma, J., 2013. Total fractionation of green tea residue by microwave-assisted alkaline pretreatment and enzymatic hydrolysis. Bioresource Technology 131, 485–491.

Vichi, S., Cortés-Francisco, N., Romero, A., Caixach, J., 2015. Direct chemical profiling of olive (*Olea europaea*) fruit epicuticular waxes by direct electrospray-ultrahigh resolution mass spectrometry. Journal Mass Spectrometry 50, 558–566.

Vogg, G., Fischer, S., Leide, J., Emmanuel, E., Jetter, R., Levy, A.A., Riederer, M., 2004. Tomato fruit cuticular waxes and their effects on transpiration barrier properties: functional characterization of a mutant deficient in a very-long-chain fatty acid β-ketoacyl-CoA synthase. Journal of Experimental Botany 55, 1401–1410.

Waters, E.R., 2003. Molecular adaptation and the origin of land plants. Molecular Phylogenetic and Evolution 29, 456–463.

Weng, H., Molina, I., Shockey, J., 2010. Organ fusion and defective cuticle function in a lacs1 lacs2 double mutant of *Arabidopsis*. Planta 231, 1089–1100.

Wheeler, S., Loveys, B., Ford, C., Davies, C., 2009. The relationship between the expression of abscisic acid biosynthesis genes, accumulation of abscisic acid and the promotion of *Vitis vinifera* L. berry ripening by abscisic acid. Australian Journal Grape Wine Research 15, 195–204.

Xia, Y., Nikolau, B.J., Schnable, P.S., 1996. Cloning and characterization of *CER2*, an *Arabidopsis* gene that affects cuticular wax accumulation. Plant Cell 8, 1291–1304.

Yeats, T.H., Rose, J.K., 2013. The formation and function of plant cuticles. Plant Physiology 163, 5–20.

Yeats, T.H., Buda, G.J., Wang, Z., Chehanovsky, N., Moyle, L.C., Jetter, R., Schaffer, A.A., Rose, J.K., 2012a. The fruit cuticles of wild tomato species exhibit architectural and chemical diversity, providing a new model for studying the evolution of cuticle function. Plant Journal 69, 655–666.
Yeats, T.H., Martin, L.B., Viart, H.M., Isaacson, T., He, Y., Zhao, L., et al., 2012b. The identification of cutin synthase: formation of the plant polyester cutin. Nature Chemical Biology 8, 609–611.
Yonekura-Sakakibara, K., Saito, K., 2009. Functional genomics for plant natural product biosynthesis. Natural Product Reports 26, 1466–1487.
Zarrouk, O., Brunetti, C., Egipto, R., Pinheiro, C., Genebra, T., Gori, A., Lopes, C.M., Tattini, M., Chaves, M.M., 2016. Grape ripening is regulated by deficit irrigation/elevated temperatures according to cluster position in the canopy. Frontiers in Plant Science 7, 1640. https://doi.org/10.3389/fpls.2016.01640.
Zhao, H., Zhang, H., Cui, P., Ding, F., Wang, G., Li, R., Jenks, M.A., Lü, S., Xiong, L., 2014. The putative E3 ubiquitin ligase ECERIFERUM9 regulates abscisic acid biosynthesis and response during seed germination and post-germination growth in *Arabidopsis*. Plant Physiology 165, 1255–1268.

FURTHER READING

López-Casado, G., Matas, A.J., Domínguez, E., Cuartero, J., Heredia, A., 2007. Biomechanics of isolated tomato (*Solanum lycopersicum* L.) fruit cuticles: the role of the cutin matrix and polysaccharides. Journal of Experimental Botany 58, 3875–3883.
Luque, P., Bruque, S., Heredia, A., 1995. Water permeability of isolated cuticular membranes – a structural analysis. Archives of Biochemistry and Biophysics 317, 417–422.
Nelson, D.R., 1999. Cytochrome P450 and the individuality of species. Archives of Biochemistry and Biophysics 369, 1–10.
Nelson, D.R., Schuler, M.A., Paquette, S.M., Werck-Reichhart, D., Bak, S., 2004. Comparative genomics of rice and Arabidopsis. Analysis of 727 cytochrome P450 genes and pseudogenes from a monocot and a dicot. Plant Physiology 135, 756–772.
Serra, O., Chatterjee, S., Huang, W., Stark, R.E., 2012. Mini-review: what nuclear magnetic resonance can tell us about protective tissues. Plant Science 195, 120–124.

INDEX

Note: 'Page numbers followed by "f" indicate figures and "t" indicate tables.'

B

C

D

G

P

R

Printed in the United States
By Bookmasters